DYNAMICS OF EXPLOSIONS

Edited by
A. L. Kuhl
R&D Associates
Marina del Rey, California

J. R. Bowen
University of Washington
Seattle, Washington

J.-C. Leyer
Université de Poitiers
Poitiers, France

A. Borisov
USSR Academy of Sciences
Moscow, USSR

Volume 114
PROGRESS IN
ASTRONAUTICS AND AERONAUTICS
Martin Summerfield, Series Editor-in-Chief
Princeton Combustion Research Laboratories, Inc.
Monmouth Junction, New Jersey

Technical papers presented from the Eleventh International Colloquium on Dynamics of Explosions and Reactive Systems, Warsaw, Poland, August 1987, and subsequently revised for this volume.

Published by the American Institute of Aeronautics and Astronautics, Inc., 370 L'Enfant Promenade, S.W., Washington, DC 20024-2518.

American Institute of Aeronautics and Astronautics, Inc.
Washington, D.C.

Library of Congress Cataloging in Publication Data

International Colloquium on Dynamics of Explosions and Reactive
  Systems (11th:1987:Warsaw, Poland)
  Dynamics of explosions/edited by A.L. Kuhl, J.R. Bowen, J.-C.
Leyer, and A. Borisov.

  (Progress in astronautics and aeronautics; v.114)
  "Technical papers presented from the Eleventh International
Colloquium on Dynamics of Explosions and Reactive Systems, Warsaw
Poland, August 1987, and subsequently revised for this volume."
  Companion volume to: Dynamics of reactive systems.
Includes index.
1. Explosions—Congresses. 2. Gasdynamics—Congresses.
I. Kuhl, A.L.　II. Title.　III. Series.
TL507.P75　　vol. 114　　629.1 s —dc19　　[662.'.2]　　88-39582
[QD516]
ISBN 0-930403-47-9

Copyright © 1988 by the American Institute of Aeronautics and
Astronautics, Inc. All rights reserved. Reproduction or translation of any
part of this work beyond that permitted by Sections 107 and 108 of the U.S.
Copyright Law without the permission of the copyright owner is unlawful.
The code following this statement indicates the copyright owner's consent
that copies of articles in this volume may be made for personal or internal
use, on condition that the copier pay the per-copy fee ($2.00) plus the per-
page fee ($0.50) through the Copyright Clearance Center, Inc., 21 Congress
Street, Salem, Mass. 01970. This consent does not extend to other kinds of
copying, for which permission requests should be addressed to the publisher.
Users should employ the following code when reporting copying from this
volume to the Copyright Clearance Center:
　　　　　　　　0-930403-47-9/88 $2.00 + .50

Progress in Astronautics and Aeronautics

Series Editor-in-Chief

Martin Summerfield
Princeton Combustion Research Laboratories, Inc.

Series Editors

A. Richard Seebass
University of Colorado

Allen E. Fuhs
Carmel, California

Assistant Series Editor

Ruth F. Bryans
Ocala, Florida

Norma J. Brennan
Director, Editorial Department
AIAA

Jeanne Godette
Series Managing Editor
AIAA

Table of Contents

Preface xvii

Chapter I.   Gaseous Detonations 1

Numerical Analyses Concerning the Spatial Dynamics of an
Initially Plane Gaseous ZDN Detonation 3
   S. U. Schöffel and F. Ebert, *Universität Kaiserslautern,
   Kaiserlautern, Federal Republic of Germany*

Detonation Parameters for the Hydrogen-Chlorine System 32
   R. Knystautas and J. H. Lee, *McGill University,
   Montreal, Quebec, Canada*

Hydrazine Vapor Detonations 45
   M. D. Pedley, C. V. Bishop, F. J. Benz, C. A. Bennett,
   R. D. McClenagan, and D. L. Fenton, *NASA Johnson Space
   Center, Las Cruces, New Mexico*, R. Knystautas, J. H.
   Lee, O. Peraldi, and G. Dupre, *McGill University,
   Montreal, Quebec, Canada*, and J. E. Shepherd,
   Rensselaer Polytechnic Institute, Troy, New York

Applicability of the Inverse Method to the Determination of
C-J Parameters for Gaseous Mixtures at Elevated Pressures 64
   P. A. Bauer, P. Vidal, N. Manson, and O. Heuzé, *Laboratoire
   d'Energétique et de Détonique, Poitiers, France*

Safe Gap Revisited 77
   H. Phillips, *Health and Safety Executive, Buxton,
   Derbyshire, United Kingdom*

Chapter II.   Detonation Transition and Transmission 97

Concentration and Temperature Nonuniformities
of Combustible Mixtures as Reason for Pressure
Waves Generation 99
   Y. B. Zel'dovich, B. E. Gelfand, S. A. Tsyganov, S. M.
   Frolov, and A. N. Polenov, *USSR Academy of Sciences,
   Moscow, USSR*

Heat Evolution Kinetics in High-Temperature Ignition
of Hydrocarbon/Air or Oxygen Mixtures..124
    A. A. Borisov, V. M. Zamanskii, V. V. Lisyanskii, G. I.
    Skachkov, and K. Y. Troshin, *USSR Academy of Sciences,
    Moscow, USSR*

Fluid Dynamic Effects on the Transition to Detonation
from Turbulent Flame in Unconfined Gas Mixtures......................140
    S. Taki and Y. Ogawa, *Fukui University, Fukui, Japan*

Numerical Simulations of the Development and
Structure of Detonations...155
    E. S. Oran, K. Kailasanath, and R. H. Guirguis, *Naval
    Research Laboratory, Washington, DC*

Transmission of Overdriven Plane Detonations: Critical
Diameter as a Function of Cell Regularity and Size........................170
    D. Desbordes, *Laboratoire d'Energétique et de
    Détonique, Poitiers, France*

Role of an Inhibitor in the Onset of Gas Detonations
in Acetylene Mixtures..186
    M. Vandermeiren and P. J. Van Tiggelen, *Université
    Catholique de Louvain, Louvain-la-Neuve, Belgium*

Experimental and Theoretical Investigation of the Effective Energy
in a Shock Tube...201
    M. Tang and J. Peng, *East China Institute of Technology,
    Nanjing, China*

Chapter III.   Nonideal Detonations and Boundary Effects........209

Nonideal Detonation Waves in Rough Tubes...................................211
    Y. B. Zel'dovich, A. A. Borisov, B. E. Gelfand, S. M.
    Frolov, and A. E. Mailkov, *USSR Academy of Sciences,
    Moscow, USSR*

Influence of Obstacle Spacing on the Propagation
of Quasi-Detonation..232
    L. S. Gu, R. Knystautas, and J. H. Lee, *McGill University,
    Montreal, Quebec, Canada*

**Propagation of Detonation Waves in an Acoustic
Absorbing Walled Tube**..248
  G. Dupré, *Centre National de la Recherche
  Scientifique, Orléans, France*, and O. Peraldi,
  J. H. Lee, and R. Knystautas, *McGill University,
  Montreal, Quebec, Canada*

**Lateral Interaction of Detonating and Detonable
Gaseous Mixtures**..264
  J. C. Liu, C. W. Kauffman, and M. Sichel, *University of
  Michigan, Ann Arbor, Michigan*

**Steady, Plane, Double-Front Detonations in Gaseous
Detonable Mixtures Containing a Suspension of
Aluminum Particles**...284
  B. A. Khasainov, *USSR Academy of Sciences, Moscow,
  USSR*, and B. Veyssière, *Laboratoire d'Energétique et de
  Détonique, ENSMA, Poitiers, France*

Chapter IV.   Condensed-Phase Detonations..........................301

Critical Conditions for Hot Spot Evolution in Porous Explosives..........303
  B. A. Khasainov, A. V. Attetkov, A. A. Borisov,
  B. S. Ermolaev, and V. S. Soloviev, *USSR Academy of
  Sciences, Moscow, USSR*

**Mechanism of Deflagration-to-Detonation Transition in
High-Porosity Explosives**...322
  A. A. Sulimov, B. S. Ermolaev, and V. E. Khrapovski,
  USSR Academy of Sciences, Moscow, USSR

**Effect of Graphite and Diamond Crystal Form and Size on
Carbon-Phase Equilibrium and Detonation
Properties of Explosives**..331
  S. A. Gubin, V. V. Odintsov, and S. S. Sergeev,
  *Moscow Physical Engineering Institute, Moscow,
  USSR*, and V. I. Pepekin, *USSR Academy of
  Sciences, Moscow, USSR*

Two-Phase Steady Detonation Analysis..341
  J. M. Powers, D. S. Stewart, and H. Krier, *University
  of Illinois at Urbana-Champaign, Urbana, Illinois*

Heterogeneous Detonation Along a Wick..........362
B. Plewinsky, W. Wegener, and K.-P. Herrmann, *Bundesanstalt für Materialforschung und -Prüfung, Berlin, Federal Republic of Germany*

Photographically Observed Waves in Detonation of Liquid Nitric Oxide..........372
G. L. Schott and K. M. Chick, *Los Alamos National Laboratory, University of California, Los Alamos, New Mexico*

Chapter V.   Explosions..........387

Overpressures Imposed by a Blast Wave..........389
J. Brossard, P. Bailly, C. Desrosier, and J. Renard, *University of Orléans, Bourges, France*

A Model for Point Explosions with Multistep Kinetics..........401
H. Salem, M. A. Fouad, and M. M. Kamel, *Cairo University, Cairo, Egypt*, and M. A. El Kady, *Al-Azhar University, Cairo, Egypt*

Air-Blast Cumulation in Gaseous Detonating Systems..........419
D. Desbordes, *Laboratoire d' Energétique et de Détonique, Poitiers, France,* and A. L. Kuhl, *R&D Associates, Marina del Rey, California*

Steam Explosions: Major Problems and Current Status..........436
J. H. Lee and D. L. Frost, *McGill University, Montreal, Quebec, Canada*

Dynamics of Explosive Interactions Between Multiple Drops of Tin and Water..........451
D. L. Frost and G. Ciccarelli, *McGill University, Montreal, Quebec, Canada*

Chapter VI.   Vapor-Cloud Explosions and Safety Applications..475

Dispersion of Dense Gaseous Fuels Released into the Atmosphere..........477
O. M. F. Elbahar and M. M. Kamel, *Cairo University, Giza, Egypt*

Experimental Investigations into the Deflagration of Flat, Premixed Hydrocarbon/Air Gas Clouds..........488
  H. Pförtner and H. Schneider, *Fraunhofer-Institut für Chemische Technologie (ICT), Pfinztal-Berghausen, Federal Republic of Germany*

Analysis of a Damage Scenario and Potential Hazards of Liquefied Gaseous Fuel Carriers in Inland Waterways..........499
  O. M. F. Elbahar and M. M. Kamel, *Cairo University, Giza, Egypt*

Influence of Obstacles on the Rate of Pressure Rise in Closed Vessel Explosions..........512
  G. E. Andrews and P. Herath, *University of Leeds, Leeds, England, United Kingdom*

Author Index for Volume 114..........533

List of Series Volumes..........534

Table of Contents for Companion Volume 113: Part I

Preface ....

Chapter I. Ignition Dynamics 1

Ignition Processes in Hydrogen-Oxygen Mixtures and the
Influence of the Uniform Pressure Assumption 3
    U. Maas and J. Warnatz, *University of Heidelberg,
Heidelberg, Federal Republic of Germany*

Relationship Between Ignition Delay and Reaction Zone
Energy Release 19
    M. J. Rabinowitz and M. Y. Frenklach, *The Pennsylvania
State University, University Park, Pennsylvania*

Study of Methane Ignition in Reflected Shock Waves 28
    S. M. Hwang, M. J. Rabinowitz, W. C. Gardiner Jr.,
and D. L. Robinson, *The University of Texas at Austin,
Austin, Texas*

Ignition Processes of Falling Droplets Columns Behind
a Reflected Shock 37
    Y. Mizutani and K. Nakabe, *Osaka University, Osaka, Japan*,
M. Yoshida and H. Nogiwa, *Hitachi, Ltd., Tokyo, Japan*,
and H. Jinrong, *Huazhong University of Science and
Technology, Wuhan, People's Republic of China*

Low-Temperature Ignition of Acetaldehyde Oxygen Mixtures
Initiated by Organic Peroxides Adsorbed on a Reaction
Vessel Surface 58
    A. B. Nalbandyan, I. A. Vardanyan, A. M. Arustamyan,
E. A. Oganesyan, and A. G. Dorunts, *Armenian Academy
of Sciences, Yerevan, USSR*

Chapter II. Flame Chemistry 65

Systematic Reduction of Flame Kinetics:
Principles and Details 67
    N. Peters, *Institut für Technische Mechanik,
RWTH Aachen, Federal Republic of Germany*

Sensitivity Analysis in Aliphatic Hydrocarbon Combustion 87
    U. Nowak, *Konrad-Zuse-Zentrum für Informationstechnik,
Berlin, Federal Republic of Germany*, and J. Warnatz,
*Universität Heidelberg, Heidelberg, Federal Republic
of Germany*

Study of the Influence of Nitrogen Oxides on the
Chemi-ionization in $C_2H_2/O_2/Ar$ Flames 104
    J. Vandooren, F. Mirapalheta, and P. J. Van Tiggelen,
*Université Catholique de Louvain,
Louvain-la-Neuve, Belgium*

Chapter III. Diffusion Flames in Shear Flows..127

Asymptotic Analysis of the Structure and Extinction
  of Methane-Air Diffusion Flames...129
    C. Treviño, *National Autonomous University of Mexico,
    Mexico*, and F. A. Williams, *Princeton University,
    Princeton, New Jersey*

Structure and Extinction Limits of Some Strained
  Premixed Flames...166
    G. Dixon-Lewis, *University of Leeds, Leeds,
    England, United Kingdom*

Influences of a Tangential Shear Flow and Differential Diffusion on
  Hydrodynamic Flame Stability..184
    S. Kadowaki, *Nagoya Institute of Technology, Nagoya, Japan*

Simulation of Stretched Premixed CH_4-Air and C_3H_8-Air
  Flames with Detailed Chemistry..195
    G. Stahl and J. Warnatz, *Universität Heidelberg,
    Heidelberg, Federal Republic of Germany*, and
    B. Rogg, *University of Cambridge, Cambridge,
    England, United Kingdom*

Stability of a Premixed Laminar V-Shaped Flame...215
    D. Escudié, *Ecole Centrale de Lyon,
    Ecully, France*

Flame Propagation in a Nonuniform Mixture: The Structure
  of Anchored Triple-Flames..240
    J. W. Dold, *University of Bristol, Bristol, England,
    United Kingdom*

Chapter IV. Dynamics of Flames..249

Fluid Mechanical Properties of Flames in Enclosures..251
    D. A. Rotman and M. Z. Pindera, *Lawrence Livermore
    National Laboratory, Livermore, California*, and A. K.
    Oppenheim, *University of California, Berkeley, California*

Dynamic Effects of Flame Baroclinicity...266
    M. Z. Pindera, *Lawrence Livermore National Laboratory,
    Livermore, California*

Flame Propagation Model by Use of Finite-Difference Methods..................................275
    Y. Takano, *Tottori University, Tottori, Japan*

Numerical Study on the Reaction Process in a Plane Shear Layer................................289
    T. Hasegawa and S. Yamaguchi, *Nagoya Institute
    of Technology, Nagoya, Japan*

Flame Front Turbulence Behavior in an Accelerating or
  Decelerating Methane-Air Mixture..310
    T. Tsuruda, K. Komatsu, and T. Hirano, *The University
    of Tokyo, Tokyo, Japan*

Mechanism of Gas Flame Acceleration in the Presence of
  Neutral Particles..325
    P. Goral, R. Klemens, and P. Wolanski, *Warsaw University
    of Technology, Warsaw, Poland*

Possible Acoustic Source in Turbulent Combustion...336
    N. Kidin and V. Librovich, *Institute for Problems in Mechanics, Moscow, USSR,* and M. Macquisten, J. Roberts, and M. Vuillermoz, *South Bank Polytechnic, London, United Kingdom*

Structure of Unsteady and Steady Hydrogen/Air Premixed Flames..349
    S. Fukutani, S. Yamamoto, and Hiroshi Jinno, *Kyoto University, Kyoto, Japan*

Flame Flashback for Low Reynolds Number Flows..367
    G. A. Karim, *The University of Calgary, Calgary, Alberta, Canada,* and R. Lapucha, *Aeronautical Institute, Warsaw, Poland*

Chapter V.   Combustion Diagnostics..385

Diagnostics in Reacting Flows..387
    Y. Levy and Y. M. Timnat, *Technion—Israel Institute of Technology, Haifa, Israel*

Multidirection Speckle Photography of Density Gradients in a Flame..403
    G. N. Blinkov, N. A. Fomin, and R. I. Soloukhin, *Heat and Mass Transfer Institute, Minsk, USSR*

Investigations on the Instantaneous Density Field of Turbulent Premixed Conical Flames..417
    A. Boukhalfa, B. Sarh, M. Debbich, and I. Gökalp, *Centre de la Recherche Scientifique, Orléans, France*

Author Index for Volume 113: Part I..432

List of Series Volumes..433

Table of Contents for Companion Volume 113: Part II

Preface...xvii

Chapter I. Combustion of Dust-Air Mixtures..1

Fundamental Characteristics of Laminar Flames in
Cornstarch Dust-Air Mixture...3
    Y. Pu, *Acadima Simica, Beijing, China*

Flame Characteristics of Pine and Cork Dust Suspensions...26
    J. A. Campos and L. Lemos, *University of Coimbra,*
    Coimbra, Portugal, and A. R. Janeiro Borges,
    University of Lisbon, Lisbon, Portugal

New Experimental Apparatus for Studying the
Propagation of Dust-Air Flames...43
    C. Proust and B. Veyssiere, *ENSMA, Poitiers, France*

Turbulent Dust Combustion in a Jet-Stirred Reactor...62
    C. S. Tai, C. W. Kauffman, M. Sichel, and J. A. Nicholls,
    University of Michigan, Ann Arbor, Michigan

Influence of Turbulence on Flammability Limits
of Dust Clouds..87
    M. A. Nettleton, *University of Queensland, St. Lucia,*
    Brisbane, Queensland, Australia

Effect of Nonuniform Coal Particle Distribution on
Combustion Aerodynamics...102
    G. Gmurczyk and R. Klemens, *Warsaw University of*
    Technology, Warsaw, Poland

Chapter II. Liquid Fuel Combustion..113

Analysis of the Ignition and Flame Propagation Caused by
Vapor Radiation Absorption of a Vaporizing Fuel
at Zero Gravity...115
    B. Amos, H. Kodama, and A. C. Fernandez-Pello,
    University of California, Berkeley, California

Two-Dimensional Modeling of Flame Propagation in
Fuel Stream Arrangements...128
    R. H. Rangel and W. A. Sirignano, *University of California,*
    Irvine, California

Flame Propagation in Liquid-Fuel Droplet Arrays at
Elevated Pressure Under Zero Gravity..151
    S. Okajima, *Hosei University, Tokyo, Japan,* and H. Hara,
    Noritz Corporation, Tokyo, Japan

Analysis of Droplet Combustion at Supercritical Conditions.........................168
    X. Chang and T. Fujiwara, *Nagoya University, Nagoya, Japan*, and A. Umemura, *Yamagata University, Yonezawa, Japan*

Rate of Atomization of Liquid Drops in a Gas Flow
Behind a Shock Wave...182
    A. Wierzba, *Institute of Aviation, Warsaw, Poland*,
    T. Yoshida, *Ichinoseki National College of Technology*,
    and K. Takayama, *Tohoku University, Sendai, Japan*

Soot Concentration Field in Flames of Heavy
Liquid Fuels..191
    S. Slupek and J. A. Koziński, *University of Mining and Metallurgy, Kraków, Poland*

Predicting Soot Concentration in a Kerosene Pool Fire.........................204
    A. Bouhafid, C. Breillat, J. P. Vantelon, and W. L. Grosshandler, *Université de Poitiers, Poitiers, France*

Chapter III. Combustion Engines..223

n-Butane Ignition in a Wide Range of Temperatures............................238
    Y. Ohta, *Nagoya Institute of Technology, Nagoya, Japan*,
    A. K. Hayashi and T. Fujiwara, *Nagoya University, Nagoya, Japan*, and H. Takahashi, *Meijo University, Nagoya, Japan*

Analytical Investigations of Plasma Jet Ignition System for
a More Efficient Combustion of Lean Burning SI Engines......................238
    J. K. Z. Kupe and H. Wilhelmi, *Institut für Industrieofenbau und Wärmetechnik, RWTH Aachen, Federal Republic of Germany*, and W. Adams, *FEV Motorentechnik, Aachen, Federal Republic of Germany*

Influence of Jet Characteristics and Effect of Geometry on
Combustion of Methane-Air Mixture in a Constant-
Volume Chamber..263
    G. Gmurczyk and P. Wolanski, *Warsaw University of Technology, Warsaw, Poland*

Mixing of Unburned Mixture with Flame Frontal Zone:
Another Cause of Engine Knock?..277
    Y. Ohta, *Nagoya Institute of Technology, Nagoya Japan*, and H. Takahashi, *Meijo University, Nagoya, Japan*

Measurement of Burning Characteristics of Hydrocarbon-Air
Mixtures at High Temperature and Pressure Achieved
by a Rapid Compression Machine..290
    T. Kawakami, S. Okajima, and K. Iinuma, *Hosei University, Tokyo, Japan*

Chapter IV. Heterogeneous Combustion and Practical Applications.......301

Gasification and Combustion of White Pine and
Cork Dusts..303
    L. A. Araujo, *National Laboratory of Engineering and Industrial Technology, Coimbra, Portugal*, and J. A. Campos, *University of Coimbra, Coimbra, Portugal*

Combustion Reactivity of Coal Chars..320
    W. Rybak and M. Zembrzuski, *Technical University of Wroclaw, Wroclaw, Poland*

Equilibrium NO as a Function of Combustion Parameters...334
    A. Meggyes, *Budapest Technical University, Budapest, Hungary,* and E. Boschan, *Hungarian Hydrocarbon Institute, Szazhalombatta, Hungary*

Model for Dioxin and Furan Production in Municipal-Waste Incinerators..343
    S. S. Penner, C. P. Li, and D. F. Wiesenhahn, *University of California, San Diego, La Jolla, California*

Modeling and Optimization of an Industrial Glass Furnace..363
    M. G. Carvalho, P. Oliveira, and V. Semião, *Instituto Superior Técnico, Lisbon, Portugal*

Simplified Model of a Surface-Combustion Burner with Radiant Heat Emission..385
    A. C. McIntosh, *University of Leeds, Houldsworth School of Applied Science, Leeds, England, United Kingdom*

Effect of Radiation and Convection on Flame Propagation over Solid Fuel Bed..406
    J. Fangrat and P. Wolanski, *Warsaw University of Technology, Warsaw, Poland*

Author Index for Volume 113: Part II..419

List of Series Volumes..420

Preface

Companion volumes, *Dynamics of Explosions* and *Dynamics of Reactive Systems*, present revised and edited versions of 83 out of the 157 papers given at the Eleventh International Colloquium on the Dynamics of Explosions and Reactive Systems held in Warsaw, Poland, in August 1987.

The colloquia originated in 1966 as a result of the widely held belief among leading researchers that revolutionary advances in the understanding of detonation wave structure warranted a forum for the discussion of important findings in the gasdynamics of flow associated with exothermic processes—the essential feature of detonation waves—and other associated phenomena.

Dynamics of Explosions principally concerns the interrelationship between the rate processes of energy disposition in a compressible medium and the concurrent nonsteady flow as it typically occurs in explosion phenomena. *Dynamics of Reactive Systems* (Volume 113, Parts I and II) spans a broader area, encompassing the processes of coupling the dynamics of fluid occurring in any combustion system. The colloquium, then, in addition to embracing the usual topics of explosions, detonations, shock phenomena, and reactive flow, included papers that deal primarily with the gasdynamic aspect of nonsteady flow in combustion systems, the fluid mechanics aspects of combustion, (with particular emphasis on the effects of turbulence), and diagnostic techniques used to study combustion phenomena.

In this volume, *Dynamics of Explosions*, papers have been arranged into chapters on gaseous detonations, detonation transition and transmission, nonideal detonations and boundary effects, condensed-phase detonations, explosions, and vapor-cloud explosions and safety applications. While the brevity of this preface does not permit the editors to do justice to all papers, we offer the following highlights of some of the especially noteworthy contributions.

In Chapter I, Gaseous Detonations, *Schöffel and Ebert* report on two-dimensional, numerical simulations of gaseous detonations. They show that the classical Zeldovich-Neumann-Doring structure for steady detonations is catastrophically unstable, and that transverse waves arise spontaneously in the flow. Numerical simulations were used to study the influence of channel width and initial pressure on the transverse wave structure; results appear to agree with experimental data. *Knystautas and Lee* have measured the detonation parameters for hydrogen-chlorine mixtures. The measured detonation velocities and pressures were significantly larger than the equilibrium C-J values; hence they may be considered pathological detonations, as predicted by Guenoche. *Pedley et al.* report on detonation parameters measured in hydrazine vapor, a fuel extensively used in aerospace applications. The detonation sensitivity was found to be similar to that for propane-oxygen mixtures. *Bauer et al.* describe an inverse method for determining the C-J detonation parameters at elevated pressures, without resorting to an equation of state.

In Chapter II, Detonation Transition and Transmission, *Zel'dovich and coworkers* point out the importance of Concentration and Temperature Nonuniformities (CTN) in the development of turbulent combustion and its transition to detonation. These nonuniformities may be created by imperfections in practical combustion systems, or may even arise in homogeneous systems due to the probabilistic velocity distribution of the molecules. *Borisov et al.* have measured ignition delays and heat release times for a variety of hydrocarbon mixtures. At the detonation limits, heat release times were commensurate with ignition delays, whereas within the limits they were always greater than the induction period. Models are proposed to simulate the dynamics of the heat evolution. *Taki and Ogawa* used comb-shaped grids to induce the fluid-dynamic acceleration of a turbulent flame and thereby trigger its transition to detonation in an unconfined ethane-oxygen mixture. *Oran and coworkers* describe recent numerical simulations of the evolution of instabilities leading to a cellular detonation structure. *Desbordes* reports on detonation transmission experiments in acetylene-oxygen-argon mixtures. He measured dramatic reductions in the critical diameter for strongly overdriven detonations, and inferred from these results that the critical energy required for the initiation of spherical detonations is not unique and must depend on the amount of overdrive.

Chapter III focuses on the subject of Nonideal Detonations and Boundary Effects. According to classical theories, reaction waves can steadily propagate either at the C-J detonation speed or as a slow deflagration. Experiments performed under less idealized conditions have shown, however, that a spectrum of intermediate wave speeds are possible, depending on the boundary conditions of the problem. *Zel'dovich and coworkers* described one such case. Steady detonation velocities of about 900 m/s were measured in a 7-cm diameter, 15-m long, rough-walled tube filled with a propane-air mixture. Reaction-zone models are proposed to explain this effect. *Lee and coworkers* at McGill University offer other examples of nonideal detonations. Flame propagation experiments were performed in geometrically roughened tubes. Five combustion wave regimes were observed: quenching, weak turbulent deflagration, choking, quasidetonation and ideal C-J detonation. Autoignition effects seemed to control the quasidetonation regime. In other experiments, acoustically absorbing walls were used to damp the transverse wave structure of a cellular detonation, thereby causing the reaction wave to propagate in a deflagration mode. *Khasainov and Veyssiere* use a double-detonation-front model to analyze the two reaction zones observed in detonable gases with suspended aluminum particles, and evaluate the controlling parameters of the problem.

In Chapter IV, Condensed-Phase Detonations, *Khasainov et al.* use a visco-plastic model of pore deformation to evaluate the critical conditions for the evolution of hot spots in porous explosives. *Sulimov et al.* report on the transition form deflagration to low-speed detonation measured in granular nitrocellulose propellants. Transition distances were found to be proportional to particle diameter. *Gubin et al.* show that the formation of graphite crystals causes the anomalous behavior of the detonation-velocity-versus-density curve for TNT. *Plewinsky et al.* report on low velocity (400 to 1480 m/s) detonations in cotton wicks soaked with TMDS. *Schott and*

Chick have measured a detonation velocity of 5.5 km/s in cryogenic nitric oxide.

Chapter V, Explosions, provides new information on the consequences of gaseous detonations and vapor explosions. *Brossard and coworkers* have measured the airblast environment created by the detonation of a small hemispherical bubble of a propane-oxygen mixture. The scaled airblast characteristics (e.g., peak pressures, impulses, and durations) are compared with equivalent TNT curves. These results are especially useful in predicting the damage caused by accidental gaseous explosions. *Kamel and coworkers* present a blast wave analysis of point explosions propagating in propane-air mixtures. *Desbordes and Kuhl* report a novel technique for studying the effects of multiple explosions in small-scale experiments. Strong accumulation effects were measured in shock convergence regions due to nonlinear wave interactions. *Lee, Frost and Ciccarelli* review a new phenomena known as vapor explosions. Such explosions involve no chemical energy release; instead, they are caused by the rapid vaporization of a cold liquid due to heat transfer from a hot liquid or surface. Explosion waves traveling at 5–10 m/s were observed using molten tin droplets in water. The collapse of a vapor bubble on one droplet generated a pressure wave that triggered the explosion of the neighboring droplet.

In Chapter VI, Vapor-Cloud Explosions and Safety Applications, *Elbahar and Kamel* present an analytic model of the dispersion of dense, pancake-shaped clouds created by the spill of liquefied gaseous fuels (LGF). *Pförtner and Schneider* report on the blast waves generated by the combustion of large pancake-shaped LGF clouds. Maximum flame speeds of 6–8 m/s were measured in unperturbed clouds; obstacle-generated turbulence increased the flame speed to about 20 m/s. Peak overpressures were approximately proportional to the square of the flame speed. *Elbahar and Kamel* also present a fault tree analysis of the damage resulting from LFG carrier accidents occurring on inland waterways.

The companion volume, *Dynamics of Reactive Systems: Parts I and II*, includes papers on ignition dynamics, flame chemistry, combustion diagnostics, the combustion of dust-air mixtures, liquid fuel combustion, combustion engines, and heterogeneous combustion and practical applications (Volume 113: Parts I and II in the AIAA *Progress in Astronautics and Aeronautics* series).

Both volumes, we trust, will help satisfy the need first articulated in 1966 and will continue the tradition of augmenting our understanding of the dynamics of explosions and reactive systems begun the following year in Brussels with the first colloquium. Subsequent colloquia have been held on a biennial basis (1969 in Novosibirsk, 1971 in Marseilles, 1973 in La Jolla, 1975 in Bourges, 1977 in Stockholm, 1979 in Göttingen, 1981 in Minsk, 1983 in Poitiers, 1985 in Berkeley, and 1987 in Warsaw). The colloquium has now achieved the status of a prime international meeting on these topics, and attracts contributions from scientists and engineers throughout the world.

The proceedings of the first six colloquia have appeared as part of the journal, *Acta Astronautica*, or its predecessor, *Astronautica Acta*. With the publication of the Seventh Colloquium, the proceedings now appear as part of the AIAA *Progress in Astronautics and Aeronautics* series.

Acknowledgments

The Eleventh Colloquium was held under the auspices of the Institute of Heat Engineering, Warsaw University of Technology and of the Polish Academy of Sciences, on August 3–7, 1987. Arrangements in Warsaw were made by Dr. B. Staniszewski and Dr. P. Wolanski. The publication of the Proceedings has been made possible by a grant from the National Science Foundation (USA).

Preparations for the Twelfth Colloquium are underway. The meeting is scheduled to take place July 1989 at the University of Michigan-Ann Arbor.

A. L. Kuhl
J. R. Bowen
J.-C. Leyer
A. Borisov

April 1988

Chapter I. Gaseous Detonations

Numerical Analyses Concerning the Spatial Dynamics of an Intially Plane Gaseous ZDN Detonation

S. U. Schöffel* and F. Ebert†

Universität Kaiserlautern, Kaiserslautern, Federal Republic of Germany

Abstract

The investigation of detonation dynamics and cellular structure is important to find criteria for the establishment and failure of a detonation. Recent large-scale experiments have shown that scale and geometric effects strongly influence the ability of a fuel/air mixture to propagate or sustain a detonation and the chance that a deflagration will undergo a transition to detonation. The starting point of the used numerical code is a steady, choked flow correspondding to a Chapman-Jouguet detonation. The implementation of realistic reaction kinetics leads by means of numerical integration to the well-known Zeldovich-Döring-von Neumann (ZDN) steady detonation structure. In this paper we will show by methods of numerical analysis that the ZDN structure based on thermochemical equilibrium turns out to be catastropically unstable. In a reference system fixed to the detonation front, the applied numerical scheme yields a spontaneous establishment of the transverse wave structure. The influence of the channel width and the initial pressure on the transverse spacing is studied and compared with experimental outcomes. The eigenvalue character of the cell size and its channel width dependence are demonstrated. From the numerical results a correspondence principle can be derived, which expresses the vanishing influence of the channel width on the characteristic spacing with an increasing number of acoustic modes.

Copyright © 1988 by Government of West Germany. Permission to be published by American Institute of Aeronautics and Astronautics. All rights reserved.

* Research Assistant, Department of Mechanical Engineering
† Professor, Department of Mechanical Engineering

Introduction

The cell size of a self-sustained detonation is, according to Lee (1984), the most important dynamic parameter, because it forms a reference value for other critical detonation parameters, such as the critical tube diameter and the initiation energy. The usual way of determining dynamic detonation parameters consists of applying the empirical confirmed assumption that there exists a constant ratio of the detonation cell size and a characteristic reaction-zone thickness. Shchelkin and Troshin (1965) presented empirical values of a scaling factor to estimate the cell size for any gaseous mixture from a chemical length scale. Reaction-zone lengths can be computed by making use of the idealized, one-dimensional, Zeldovich-Döring-von Neumann (ZDN) model of steady detonation structure.

A system of stiff ordinary differential equations describing detailed chemical reaction kinetics must be solved to determine a ZDN structure for a given mixture. Modern detonation theory begins with the introduction of the ZDN model, which consists of a mechanistic refinement of the classical Chapman-Jouguet (CJ) theory and is based on thermochemical equilibrium. Except for a consideration of chemical microkinetics, no macrokinetics is taken into account. Therefore, for a physical model, the integral steady balance equations of thermofluid dynamics are necessary for determining the plane ZDN structure.

To predict characteristic dynamic detonation parameters, such as the cell size, the macrokinetics of a detonation propagation process should be taken into account. Therefore, a more profound theory should be based on nonequilibrium thermodynamics. The unsteady partial balance equations of thermofluid dynamics, together with species balance equations for the chemical microkinetics, will therefore be the starting point for studying the spatial detonation dynamics (see Table 1).

The space of even the largest vector computers is insufficient for resolving all physicochemical length scales with acceptable accuracy in a numerical code. Thus, the microkinetic model applied here is restricted to global two-step chemical reaction kinetics. The validity of our approach is estimated by comparing the calculated reaction-zone ZDN structure based on global kinetics with the results using detailed chemical kinetics (Shepherd and Westbrook 1986). The spatial dynamics of a detonation process in two-dimensional channel geometry is studied in a reference system that moves with the average detonation velocity established by the classical CJ model. At the time we

NUMERICAL ANALYSES OF ZDN DETONATION

Table 1 Development of a dynamic structure theory starting from the static ZDN structure - scope of paper in hand

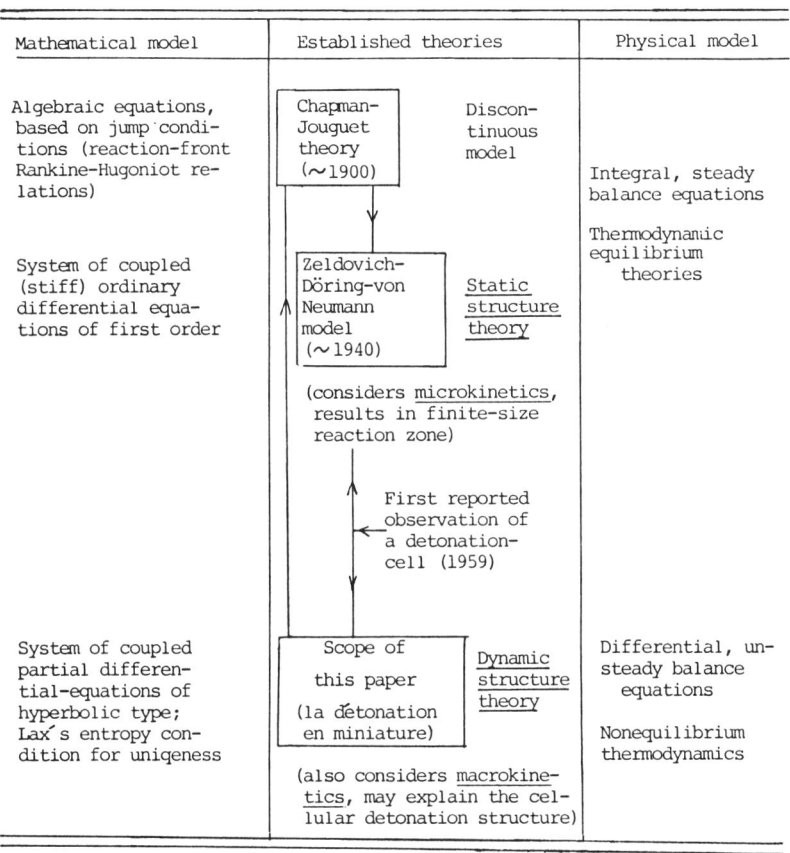

Mathematical model	Established theories		Physical model
Algebraic equations, based on jump conditions (reaction-front Rankine-Hugoniot relations)	Chapman-Jouguet theory (~1900)	Discontinuous model	Integral, steady balance equations
			Thermodynamic equilibrium theories
System of coupled (stiff) ordinary differential equations of first order	Zeldovich-Döring-von Neumann model (~1940)	Static structure theory	
	(considers microkinetics, results in finite-size reaction zone)		
	First reported observation of a detonation-cell (1959)		
System of coupled partial differential-equations of hyperbolic type; Lax's entropy condition for uniqeness	Scope of this paper (la détonation en miniature)	Dynamic structure theory	Differential, unsteady balance equations
			Nonequilibrium thermodynamics
	(also considers macrokinetics, may explain the cellular detonation structure)		

obtained the results presented here, we looked in vain for any technical paper, in which a dynamic detonation process was studied in a so-called Galilei-transformed reference system moving with the average precursor shock velocity.

Review

A spontaneous establishment of the complex transverse cellular wave structure has been observed for a controlled one-dimensional detonation-initiation experiment (Strehlow and Fernandes 1965). The development of the structure is illustrated in a print of a smoke track in Fig.1 of their paper, where finite-amplitude transverse waves appear despite plane initiation. Numerous authors, including K. I. Shchelkin, R. M. Zaidel, J. J. Erpenbeck, G. G. Chernyi et al., and recently Rosales (1987), studied the stability of one-dimensional, laminar detonation waves. An extensive discussion of detonation stability can be found in Chap. 6A of the work of Fickett and Davis (1979). However, all analytical and mostly linear studies of stability menioned there are based on simplified or physically unrealistic models, such as the square-wave detonation model.

Oppenheim and Rosciszewski (1963), in their summary of a numerical study, conclude that "only a thorough understanding of the so-called ´laminar´ wave structure can provide proper basis for the assessment of the effects of turbulence and other time dependent and multidimensional phenomena that may accompany the detonation process." Presently it is generally accepted that the ZDN model constitutes the main feature of the laminar wave structure. The ZDN structure can be readily determined by integration routines for systems of stiff ordinary differential equations. The prediction of the detonation cell size from a calculated reaction-zone length is complicated, since the scaling constant depends on the mixture composition, the selection criterion for a reaction-zone length, and the principal irregularity of the soot imprints. Moreover, simple scaling by a constant of proportionality ignores the complex and nonlinear interaction between thermofluid dynamics and reaction chemistry. As pointed out by Shepherd (1985), a single-parameter approach by matching at a single point (often stoichiometric composition) can only predict cell size within factor 2.

The observed complex multimode structure with the appearance of finite transverse waves suggests that the transverse spacing might not be characterized by a single parameter, which is uniquely related to an idealized reaction-zone length. Consequently, an alternative approach, as shown in Table 1, is based on a dynamic structure theory in which the macrokinetics of a detonation process is also considered. The gasdynamic, hyperbolic Eulerian equations, coupled with a simple chemical reaction mechanism, comprise the underlying mathematical model for the results presented here.

A satisfactory, accurate simulation of explosion and blast-wave phenomena poses serious computational difficulties and places outstanding demands on numerical methods, as seen in the work of Chushkin and Shurshalov (1982) and Book et al. (1982).

Numerical simulation results of the cellular structure of marginal (single-mode) detonations in two-space coordinates were published by Oran et al. (1982), Markov (1981), and Taki and Fujiwara (1982). All of these simulations nicely revealed most of the experimentally confirmed features, such as the finite transverse spacing, triple-shock configurations of shock waves (Mach-stem formation), variation of extents of induction zones, or pulsations of the state variables with average CJ values. In this paper we took advantage of the large memory core of a vector processor to study the experimentally observed multimode detonation structure.

In contrast to the significance of the ZDN model for predicting dynamic detonation parameters, all previous numerical simulations of detonation dynamics avoided the question of hydrodynamic stability of the plane ZDN structure. In the book edited by Zierep and Oertel (1982), quite a few authors underline the (also experimentally) well - known fact that convective instability phenomena in fluids leading to cellular (or coherent) structures are highly dependent on the initial flow conditions. Previous investigations of cellular detonation dynamics are restricted to arbitrary, blast-wave-perturbed CJ shock initial values. Furthermore, unrealistic, large-scale (transverse) perturbations of the order of the combustion channel width are applied to establish the complex wave structure. Theoretical arguments stemming from the work of Rosales and Majda (1983) provide new criteria for the formation of regularly spaced Mach stems. The basic, inhomogeneous ground flow assumed in their study is also an unperturbed ZDN wave state. An essential fact resulting from the theory is that a caustic is needed in analogy to nonlinear geometric optics to establish a transverse wave front behind the perturbed precursor shock. Strehlow (1968), Barthel and Strehlow (1966), and Barthel (1974) already stressed that periodic caustics play an important role in reacting Mach-stem formation. An article by Rosales and Majda (1984) contains the crucial idea that high-frequency nonlinear wave interaction is necessary for a wave bifurcation to Mach stems. The spontaneous breakdown of the smooth precursor shock solution and the catastrophic instability of a ZDN ground flow have never been carefully investigated by numerical analysis. Thus, this paper addresses the challenge of closing this gap.

Fluid Dynamic Phenomena Accompanying a
Self-Sustaining Detonation Propagation Process

According to arguments originating mainly from the works of Shchelkin and Troshin (1965), the instability of a ZDN structure results from the exponential temperature de-

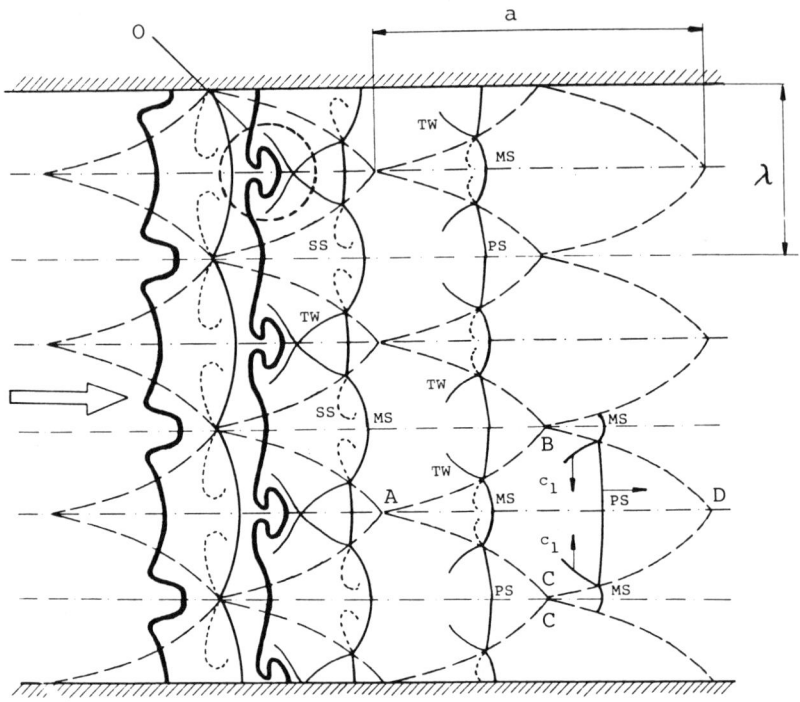

Fig. 1 Schematic sketch of phenomena accompanying a self-sustained two-dimensional detonation propagation process.

pendence of the induction time period. This dependence gives rise to self-excited oscillations of accumulator type according to the classification of Magnus (1976). A necessary condition for sustaining such an oscillation consists of exceeding the choking condition in the CJ plan. Then, an intimate coupling of the reaction zone with the precursor shock wave causes sweeping oscillations. The heat release in the initially plane reaction zone weakens the slightly overcompressed detonation front by rarefaction waves. Consequently, a decrease in post-shock pressure results, which causes acceleration and Rayleigh-Taylor instability of the flame. The flame-folding process resulting from the pressu- decline is illustrated in Fig.1, which shows a schematic sketch of the phenomena accompanying a self-sustained, two-dimensional detonation propagation process.

Flame convolution and wrinkling induced by weak pressure waves has been studied by Ebert and Schöffel (1986). The results nicely compared with Markstein's (1964) experiment of shock-wave flame interaction. As shown by Chu(1955), any accelerating flame gives rise to the establishment of

pressure waves. The flame-induced weak transverse waves
cross each other at a focus, as can be seen in Fig. 1. In
fact, it has been verified experimentally by Sturtevant and
Kulkarny (1976) that focusing weak shock waves generate a
Mach stem near the focus. The mushrooming effect of the accelerated reaction-front and the induction-time increase by
the pressure drop in the reaction-zone leads to rapid quenching of the flame. A half-cell width away reignition appears due to the high-pressure level behind the Mach stem.
The triple-shock track in a channel with the width of $\lambda/2$
is indicated in Fig. 2 by a dashed-dotted line. The slip
surfaces (contact discontinuities) emanating from the triple
points cause the soot tracks, which settle at coal-dust-covered channel walls. As can be seen in Fig. 2, these shear
waves gradually roll up due to Kelvin-Helmholtz instability.
In Fig. 2 the foci of the weak transverse waves at the channel walls can also be discerned.

Evidence of Adequacy of Hyperbolic Mathematical Model

Mathematical models of blast-wave (particularly detonation) phenomena usually neglect molecular transport processes such as viscosity, heat conduction, and species diffusion. This simplified approach can be justified physically,
if the characteristic time scales of nonlinear wave propagation appear to be by far smaller than the ones for the
other transport processes. In fact, this condition is largely fulfilled for the cellular detonation propagation in
rectangular channel geometry.

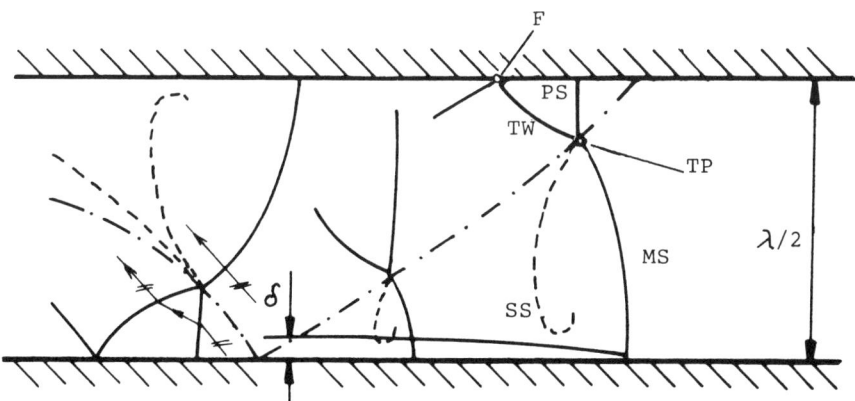

Fig. 2 Self-sustained propagation in a narrow channel illustrating
the focusing of weak transverse waves at the channel wall:
λ, detonation system; δ, boundary-layer thickness;
F, focus; MS, Mach stem; PS, primary shock (incident wave);
RS, reflected shock; SS, slip surface; TP, triple point

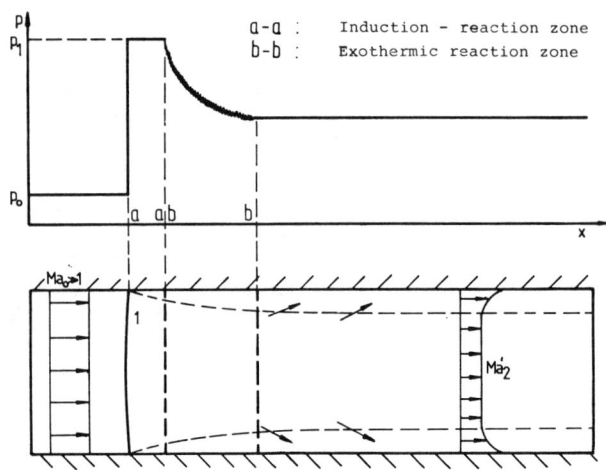

Fig. 3 Principle sketch of the Zeldovich-Döring-von Neumann detonation structure, taking into account boundary-layer effects.

According to observations by Strehlow et al. (1972), the detonation cell size in an 70 % argon/helium-diluted, stoichiometric hydrogen/oxygen mixture is equal to $\lambda \simeq 0.30/p_0$ cm, $\lambda \simeq 0.60/p_0$ cm, respectively, where ¦ is the pressure of the ambient, premixed gas (index 0) in the combustion channel before ignition in units of physical atmospheres. For characteristic time and length scales, respectively, we choose the smallest physicochemical values, which prove to be the induction-reaction time τ_I and induction-reaction thickness l_I.

The CJ detonation Mach number Ma_{CJ} is uniquely determined by the condition that the Rayleigh line RL in Fig. 4 is tangential to the equilibrium Hugoniot curve H_1 labeled by $X = X_{eq}$. Thermochemical equilibrium corresponds to vanishing exothermic reaction velocity w_X according to the next section. Usually detonation velocities $U_{CJ} = Ma_{CJ} c_0$ (where c_0 is sound velocity) are calculated by an equilibrium code (Gordon and McBride 1971), which determines the minimum of the Gibbs free enthalpy in the phase space.

For the induction-reaction length holds the equation:

$$l_I = \int_0^{\tau_I} (U_{CJ} - u)\, dt = (U_{CJ} - u)\, \tau_I \qquad (1),$$

where τ_I obeys simple Arrhenius kinetics (Korobeinikov et al. 1972) or may be obtained by fits to experimental data (Oran et al. 1982). The fluid velocity u in a laboratory reference system is assumed to be constant (indepen-

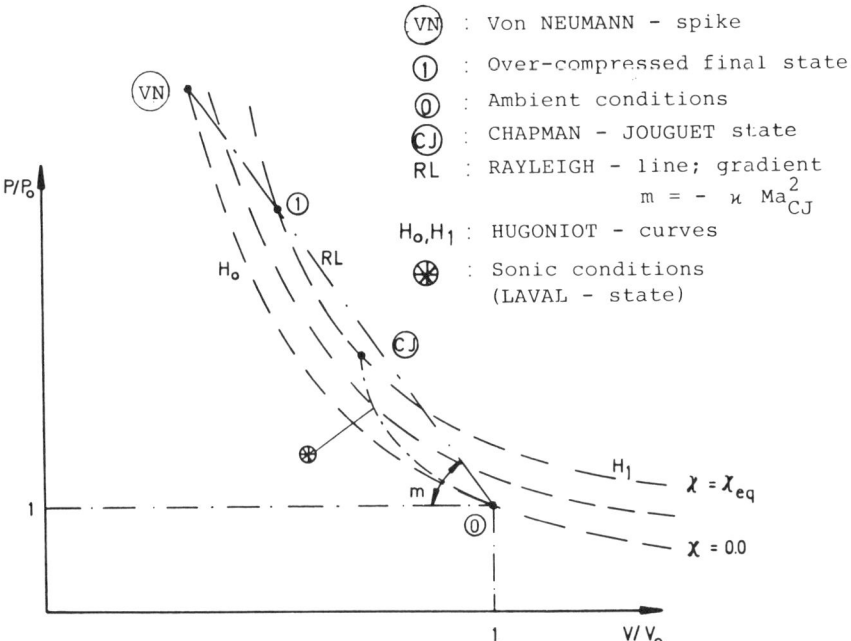

Fig. 4 Graphics for determining thermodynamic states in the steady reaction-zone of a ZDN detonation.

dent of time t) before the end of the induction-reaction process. To obtain l_I the so-called von Neumann state (VN in Fig.4) behind the precursor shock must be determined.

For a polytropic gas with constant specific heat ratio $\varkappa$ the Rankine-Hugoniot relations yield

$$u_{VN}/c_0 = \frac{2}{\varkappa + 1} Ma_{CJ} (1 - Ma_{CJ}^{-2}) \qquad (2)$$

$$T_{VN}/T_0 = 1 + \frac{2(\varkappa-1)}{(\varkappa+1)^2} (\varkappa Ma_{CJ}^2 + 1)(Ma_{CJ}^2 - 1)/Ma_{CJ}^2 \qquad (3)$$

By use of the induction-kinetic data of Korobeinikov et al. (1972), from the cell-size data previously mentioned we get values of $\lambda = 13\ l_I$ and $\lambda = 12\ l_I$, respectively, for argon- and helium-diluted oxyhydrogen.

From an analysis of the soot tracks and also from theoretical considerations by Vasiliev and Nikolaev (1978), it can be concluded that the ration of detonation cell width and cell length a is equal to 0.6. Hence, the period Δt_1, which passes between two successive cell formations, is equal to $\lambda/(0.6\ U_{CJ})$, with $U_{CJ} \approx 2.5$ km/s. For $p_0 = 1$ atm, we get a maximum value of approximately 10^{-5} for the ratio of Δt_1 to the viscous time scale $\Delta t_2 = \lambda^2/\nu$, with the kinematic viscosity $\nu \approx 40 \cdot 10^{-6}$ m^2/s.

On the other hand, the thickness of the boundary layer δ developing behind the Mach stem can easily be evaluated and compared with the channel width $W = \lambda/2$, as seen in Fig. 2. With the value $2\ (\nu \Delta t_I/2)^{1/2}$, we obtain about $3.6\ (\nu/\lambda\ U_{CJ})^{1/2}$ for the ratio δ/W, with a maximum value of about 10^{-2}. A corresponding ratio can be determined for the thermal and diffusion boundary layer by using the Reynolds analogy between molecular momentum, heat, and mass transfer (Schlichting 1982).

The numerical results in two-space coordinate presented here use a mesh size of $\Delta x = \Delta y = 0.2\ l_I$. Therefore, with the applied discretization, any first-order accurate numerical code yields truncation errors exceeding the boundary-layer thickness. However, higher-order accurate methods produce spatial errors below the viscous length scales in smoother regions; thus, it appears reasonable to also include molecular transport effects in the mathematical model.

Nevertheless, our estimation provides encouraging support that the main feature of spatial detonation dynamics, the cellular structure, can be adequately modeled by the hyperbolic balance-equations presented in the next section.

Fundamental Balance Equations

According to the previous estimation, the gasdynamic Eulerian equations are adequate for describing the thermo-fluid dynamic aspect of the spatial detonation dynamics. The Eulerian equations are hyperbolic differential balance equations for mass, momentum, and total energy, namely,

the equation of continuity:

$$\partial \rho / \partial t + \nabla \cdot (\rho \underline{v}) = 0 \qquad (4)$$

the equation of motion:

$$\partial (\rho \underline{v}) / \partial t + \nabla \cdot (\rho \underline{v}^t \underline{v}) + \nabla p = 0 \qquad (5)$$

and the equation of total energy conservation (first law of thermodynamics):

$$\partial (\rho e) / \partial t + \nabla \cdot (\rho e \underline{v} + p \underline{v}) = 0 \qquad (6).$$

Furthermore, we assume polytropic (thermal and caloric ideal) behavior of the gas, which results in an equation of state of the form:

$$e = \varepsilon + \underline{v}^2/2 - \chi q, \quad \text{with} \quad \varepsilon = p/[\rho (\varkappa - 1)] \qquad (7).$$

Additionally, we apply a constant $\varkappa$ across the reaction front. In the equations, ρ denotes the mass density, $\underset{\sim}{v}$ is the fluid velocity, p is pressure, and e is the total energy per mass unit. The total energy is composed of the specific internal energy ε, the specific kinetic energy $\underset{\sim}{v}^2/2$ and the specific heat of reaction q times the parameter χ of global exothermic reaction progress. The ∇ represents the gradient (or Nabla/Del) operator, and $\partial/\partial t$ is the local partial derivative with respect to t. The equation of state
Eq. 7 provides a coupling between the fluid dynamics and chemical reaction kinetics. The global microkinetics applied here is composed of an equation for the (nonexothermic) induction reaction and another for the heat-release reaction. The reaction velocities are defined as Lagrangian (material) derivatives

$$w_\xi = D \xi / Dt \quad (8)$$

and

$$w_\chi = D \chi / Dt \quad (9)$$

of the induction parameter ξ and the reaction progress parameter χ with respect to t.

Using the Reynolds transport theorem

$D/Dt = \partial/\partial t + (\underset{\sim}{v} \cdot \nabla)$ (10), which connects the Lagrangian derivatives D/Dt with Eulerian (local) derivatives $\partial/\partial t$, we obtain from Eqs. (8) and (9) by the equation of continuity, Eq. (4):

$$\partial(\rho \xi)/\partial t + \nabla \cdot (\rho \underset{\sim}{v} \xi) = \rho w_\xi \quad (11)$$

$$\partial(\rho \chi)/\partial t + \nabla \cdot (\rho \underset{\sim}{v} \chi) = \rho w_\chi \quad (12)$$

Equations (11) and (12) are inhomogeneous balance equations for ξ and χ in so-called conservation or divergence form (see Korobeinikov et al. 1972) for comparison. The source terms w_ξ in (11), w_χ in (12), respectively, obey Arrhenius kinetics for the reaction temperature dependence.

The conservation form of a hyperbolic transport equation insures the application of the so-called "shock-capturing" concept. According to this concept, a consistently formulated numerical scheme yields weak convergence to the jump conditions. These conditions hold across gasdynamic discontinuities (particularly shocks) and are known as the Rankine-Hugoniot relations. In case of exothermic chemical reactions the Rankine-Hugoniot equations may be generalized to the reaction-front equations, which constitute the mathematical model of the classical CJ theory.

Numerical Method Applied

The balance Eqs. (1-6) are solved in two-space coordinates by means of a two-step, second-order accurate, explicit predictor-corrector, MacCormack, finite-volume scheme. The inhomogeneous terms in Eqs. (10) and (11) are treated in a manner originating from Dwyer et al. (1974). To prevent spurious oscillations (wiggles and ripples), we employ flux limiters that guarantee monotonous behavior. Harten (1983) introduced the notion of so-called total-variation-diminishing TVD schemes, which combine the smoothing of the high-frequency oscillations with second-order accuracy and high resolution. The critical amounts of artificial viscosity needed to satisfy the TVD property for schemes of Lax-Wendroff type has been investigated by Davis (1984). In case of a linear transport equation the two-step Lax-Wendroff scheme is equivalent to MacCormack's predictor-corrector method.

High-frequency fluctuations are completely damped by monotonous upwind schemes. Moreover, upwind TVD schemes based, for instance, on the MUSCL (monotonous upstream centered conservation law) approach of van Leer (1982) nicely resolve discontinuities in the linearly degenerated case such as contact discontinuities and particularly shear waves. Upwind schemes are superior to symmetric schemes, especially for studying high-Mach-number reacting flow.

The initial ZDN data for Eqs. (1-6) lead to an ill-posed problem in the sense of Hadamard (Garabedian 1964). A high-frequency analysis shows that short wavelength perturbations of the initially inhomogeneous ZDN ground flow have arbitrarily large growth rates. Richtmyer and Morton (1967) point out that, in the case of an ill-posed initial value problem, the applied difference scheme should be unstable to insure consistency. The irregularity introduced by the formation of a singularity (cusp) in the precursor shock front is regularized by imposing a sufficiently high amount of artificial viscosity on the applied symmetric TVD scheme.

For the stable TVD MacCormack scheme applied here the time step Δt is restricted according to a Courant-number of $Co = 0.95 \sqrt{3}/2$, defined as the maximum over the grid of the quantity $(\frac{|u|}{\Delta x} + \frac{|v|}{\Delta y})\Delta t$, where u and v denote the components of the velocity $\underset{\sim}{v}$ in two-dimensional, cartesian geometry. The Courant-Friedrichs-Lewy (CFL) stability condition requires $Co \leq 1$ for any explicit finite-difference method. Linear stability of a numerical method does not preclude that in a basic, steady flow small-scale structures evolve with mesh-size order of magnitude. The nonlinearity of the governing equations allows in the case of an unstable ZDN flow the bifurcation to a self-translating Mach-stem pattern.

The integration of the balance equations is broken down into fractional steps that are combined in an order-preserving Strang-type operator splitting. The complete evolution operator $\mathcal{L}^{2\Delta t}$ is given by:

$$\mathcal{L}^{2\Delta t} = \mathcal{L}_x^{\Delta t}\, \mathcal{L}_y^{\Delta t}\, \mathcal{L}_y^{\Delta t}\, \mathcal{L}_x^{\Delta t} \tag{13}$$

where $\mathcal{L}_x$ advances the solution in the x direction and $\mathcal{L}_y$ in the orthogonal y direction.

Initial and Boundary Conditions

Figure 3 shows a principal sketch of a stationary ZDN detonation structure, also taking boundary-layer effects into account. A numerical computation of the stationary ZDN structure for detailed chemical kinetics was performed first by Westbrook and Urtiew (1982) to determine critical detonation parameters for a wide variety of hydrocarbon mixtures. More recent results concerning the calculation of the ZDN structure have been published by Shepherd (1985).

For the global chemistry applied here the ZDN initial condition is determined by adaptive Romberg integration of the equation of motion of a fluid element in a reference system, which moves with the average CJ detonation velocity U_{CJ}. The equation of motion reads

$$D\hat{x}/Dt = U_{CJ} - u(\chi) \tag{14}$$

where $\hat{x}$ denotes the distance from the precursor shock front. By applying the chain rule of differentiation, it follows that

$$\hat{x}(\chi) = \int_0^{\chi} [U_{CJ} - u(\tilde{\chi})]/w_{\tilde{\chi}}\, d\tilde{\chi} + 1_I \tag{15}$$

with the exothermic reaction velocity w_χ and the initial condition $\hat{x}(0) = 1_I$. Within the delay time τ_I, no chemical energy release takes place, and the reaction-progress parameter is assumed to be zero. For the values $u(\chi)$ or $T(\chi)$ algebraic terms are obtained by determining the intersection of the RL with a gradient $-\varkappa Ma_{CJ}^2$ with the Hugoniot graph for a certain χ according to Fig. 4.

To find the ZDN structure, the inverse function must be determined, which can easily be done numerically by simple Newtonian iteration

$$\chi_{n+1} = \chi_n - [x_n(\chi_n) - x_i]/f(\chi) \tag{16},$$

with $f(\chi_n) = (dx/d\chi)_n$ for a monotonous original $x(\chi)$ and a given discretization x_i ($i = 1,...,N$).

The boundary conditions, which are satisfied by the transport variables at the solid channel walls, are set by use of the concept of symmetrical mirror images. Since we neglect viscous effects, we may apply slip-conditions for the velocity tangential to the wall. Neither momentum, heat, or mass transfer occurs across the channel wall. For the normal velocity an antimetric (reflecting) boundary condition is imposed, whereas the other variables (density, internal energy, and reaction progress parameters) are subject to symmetric (absorbing) conditions. The inflow/outflow boundary conditions depend on the underlying mathematical frame of reference.

In a laboratory reference system the complete CJ state must be implied at the supersonic inflow according to the theory of characteristics. The flow velocity is always supersonic behind the detonation front. The outflow conditions correspond to quiescent, unperturbed fluid, which has not yet been shocked. The main disadvantage of a laboratory reference system for performing numerical analyses are phase errors caused by the relative velocity between the shock wave and the computational mesh. Moreover, the rapidly moving regime of interest requires continuous mesh rezoning to combine spatial accuracy with reasonable computational costs.

A precursor shock-oriented (so-called Galilei-transformed) frame of reference drastically reduces numerical dispersion and the total amount of mesh points needed to obtain accurate results. The inflow conditions correspond to homogeneous, unshocked flow with a fluid velocity U_{CJ}. The flowfield remains completely in a subsonic domain. At the outflow only the equilibrium value of χ is prescribed, and the other transport variables are extrapolated with zeroth, respectively, first order.

Results in a Galilei-Transformed Frame of Reference

The reference values for nondimensionalizing the physical variables and the applied reaction chemistry correspond to the work of Taki and Fujiwara (1982). Thus, we can compare with their results obtained in a laboratory reference system. The high-frequency $2 \Delta y$ fluctuations causing the instability of the plane ZDN structure can be recognized in Figs. 5 and 6.

The self-induced limit cycle oscillations of the thermodynamic pressure can be seen in Figs. 7-11. The onset of instability is also affected by the machine error of the

NUMERICAL ANALYSES OF ZDN DETONATION 17

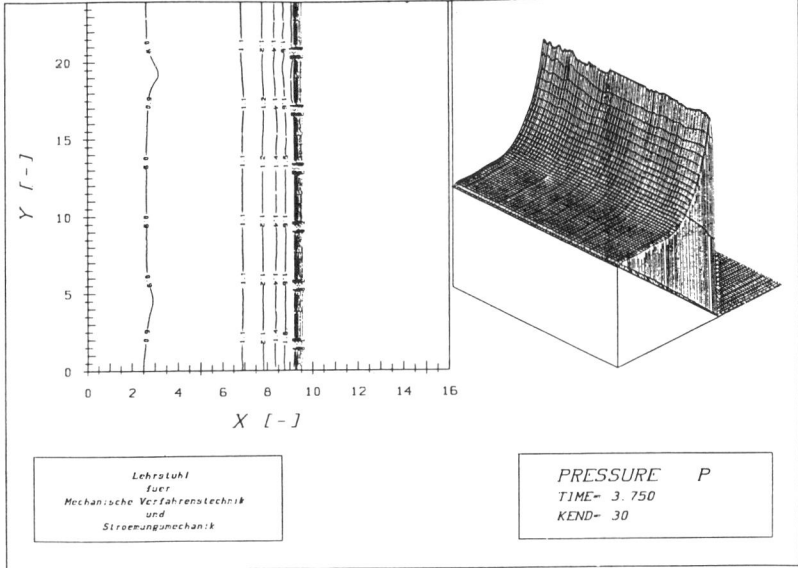

Fig. 5 Transverse high-frequency pressure fluctuations caused by instability of planar ZDN structure for channel width $W/l_I = 24$.

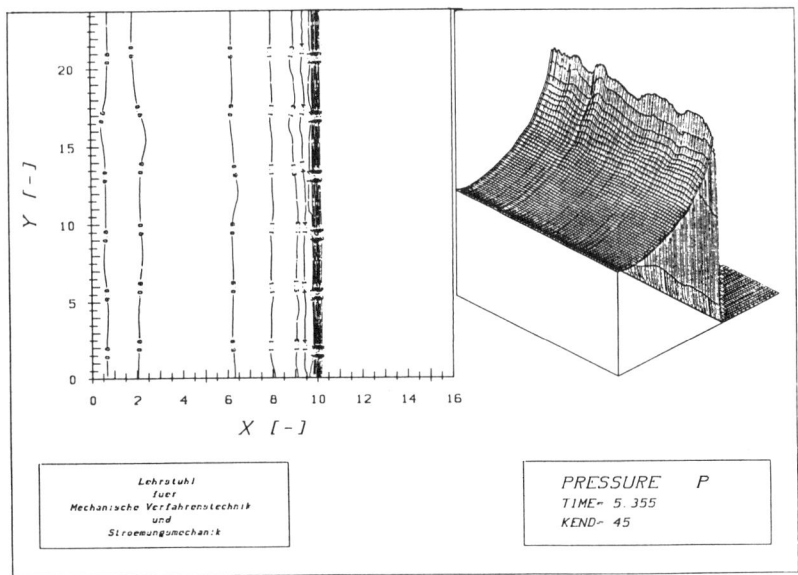

Fig. 6 Formation of long-wave transverse modes due to aliasing from small scale pressure fluctuations according to Fig. 5.

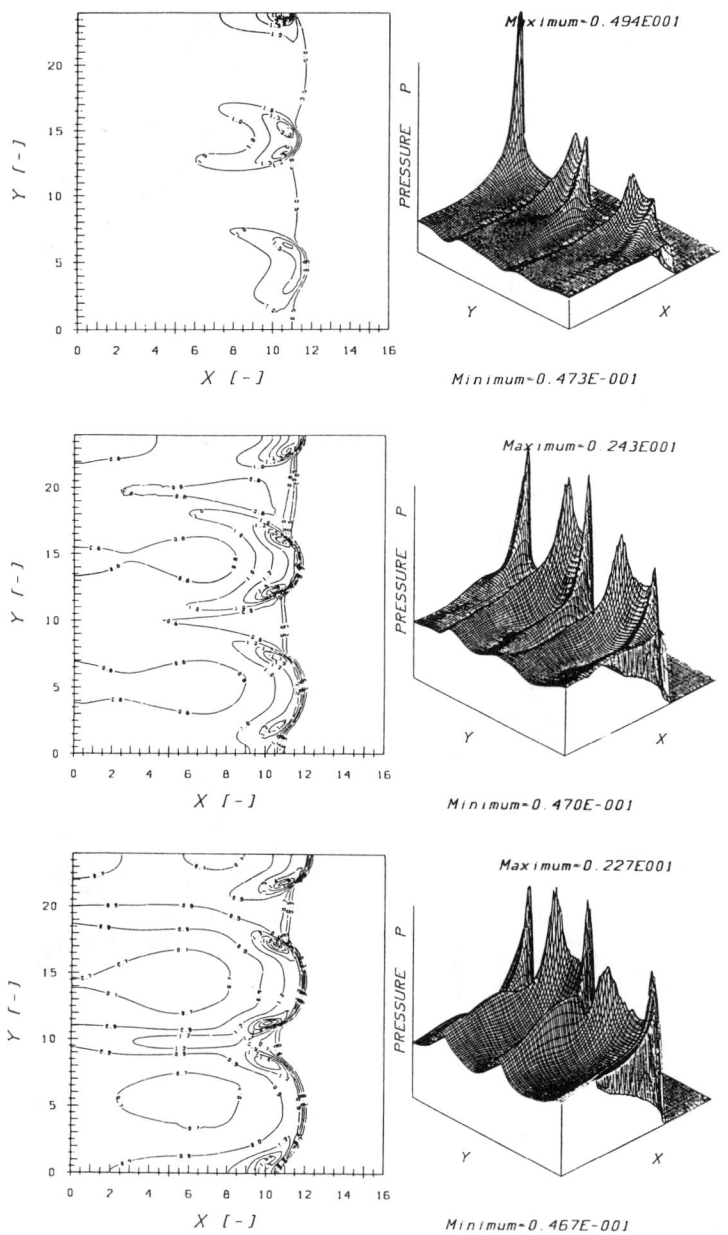

Fig. 7 Pressure distribution for dimensionless times t = 28.913, 29.948, and 29.984 (corresponding to 265, 270, and 275 CFL time steps) with mode number n = 2.5.

NUMERICAL ANALYSES OF ZDN DETONATION

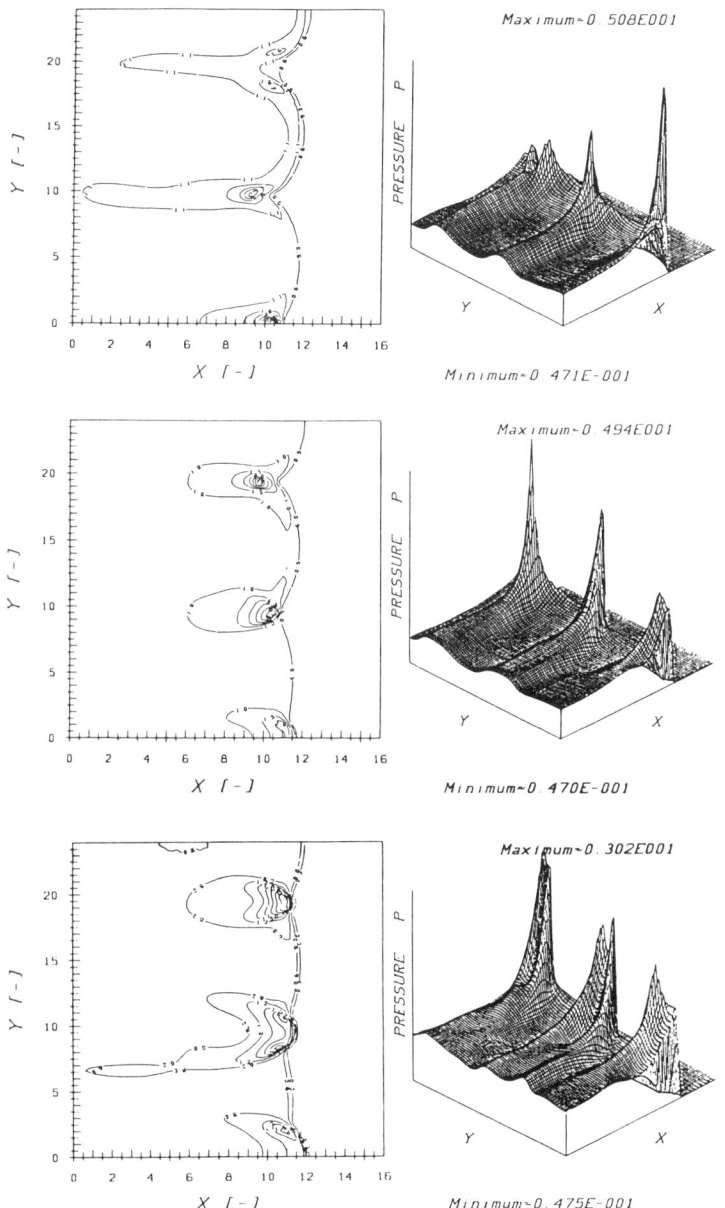

Fig. 8 Pressure distribution for t = 30.516, 31.049, and 31.583 (corresponding to 280, 285, and 290 CFL time steps).

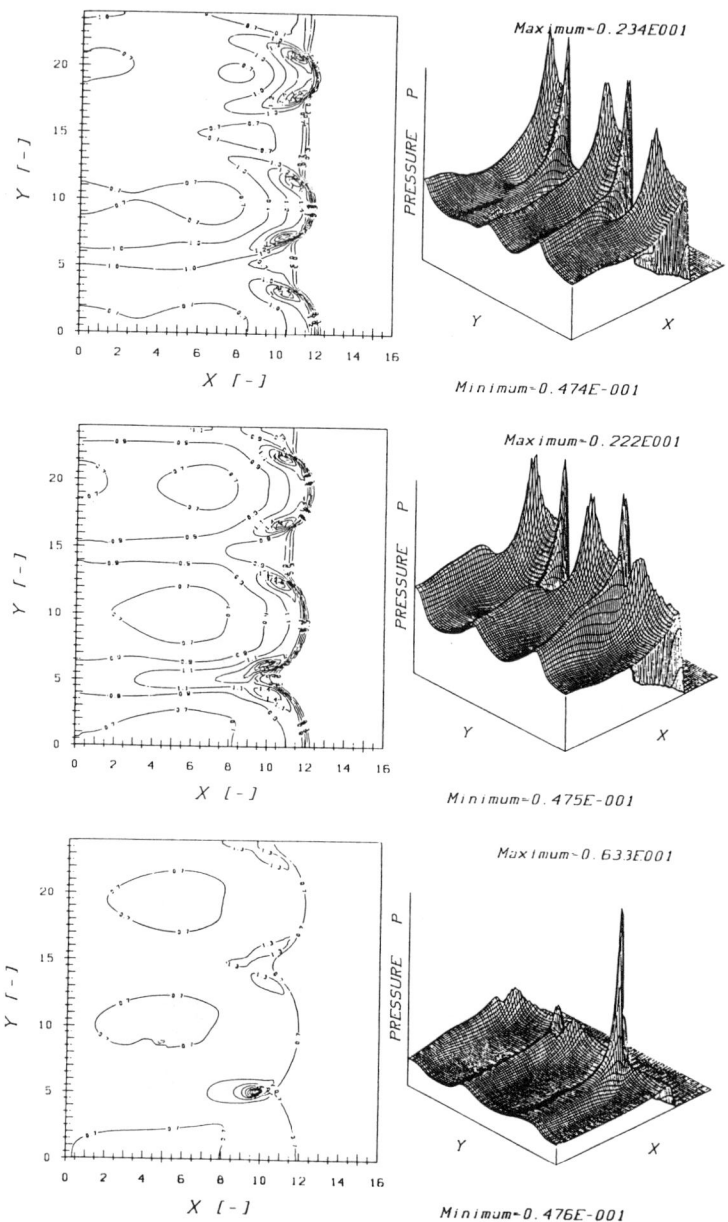

Fig. 9 Pressure distribution for t = 32.115, 32.642, and 33.165 (corresponding to 295, 300 and 305 CFL time steps).

NUMERICAL ANALYSES OF ZDN DETONATION

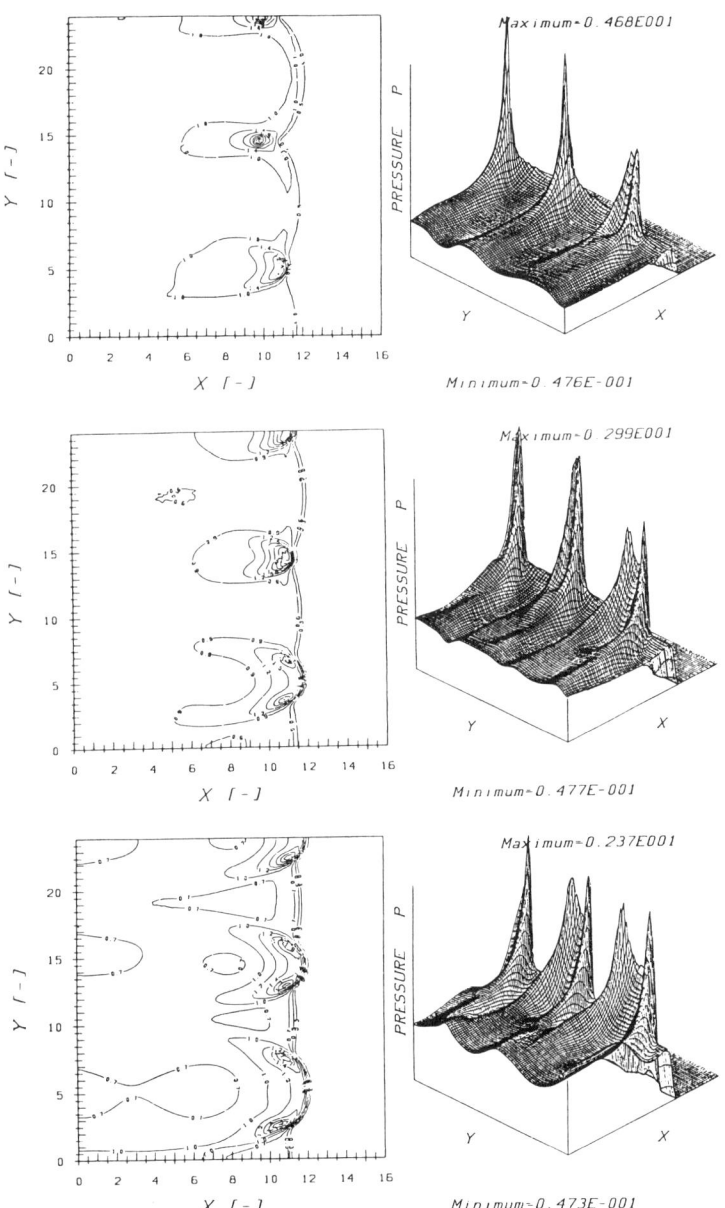

Fig. 10 Pressure distribution for t = 33.687, 34.216, and 34.751 (corresponding to 310, 315, and 320 CFL time steps).

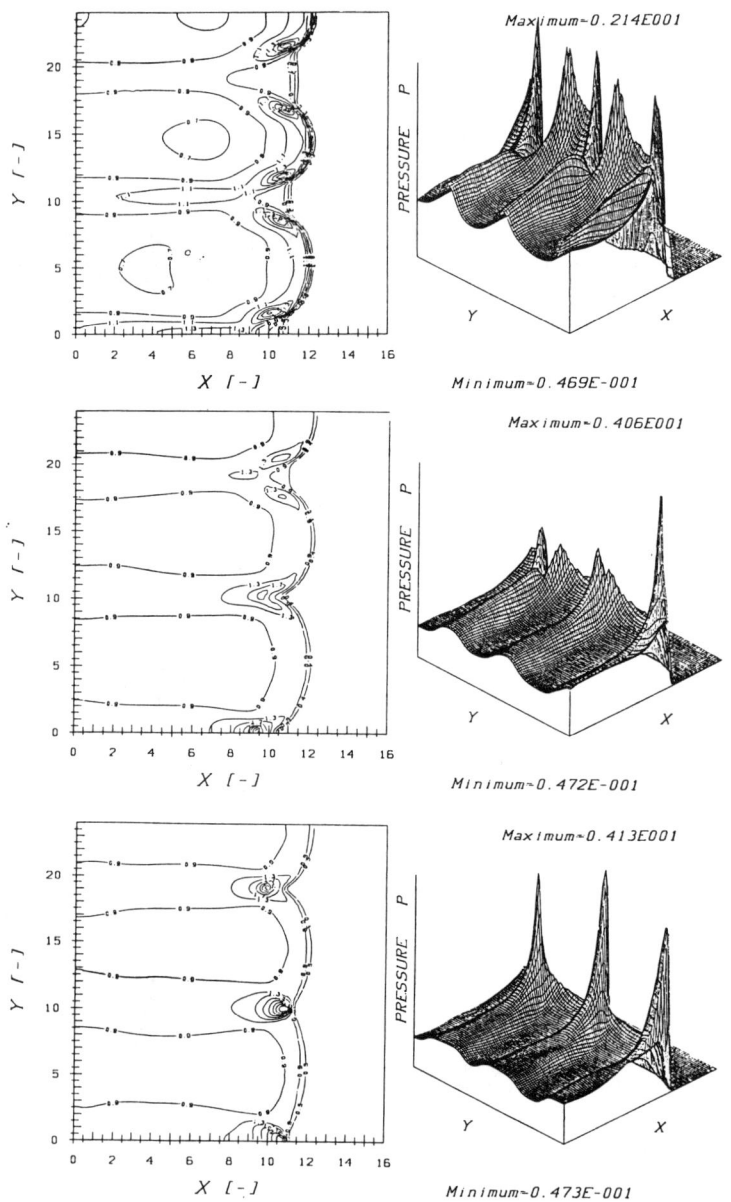

Fig. 11 Pressure distribution for t = 35.285, 35.813, and 36.344 (corresponding to 325, 330 and 335 CFL time steps).

computer that is used. The machine's finite precision arithmetic was increased from single precision (8 decimal digits) to double precision (16 decimal digits) to study the influence of spurious perturbations at the amplitude of the round-off error. The double precision calculation depicts that the high-frequency modes are rather introduced through nonlinearity than through roundoff error perturbations. The ZDN instability and the associated establishment of the transverse wave structure is only slightly influenced by computer arithmetics.

The three-front configurations consisting of Mach-stem, incident shock, and transverse waves (reflected shocks) intersecting at the triple points are well resolved in Figs. 7 - 11. The pressure spikes correspond to the values at the triple points. In the reaction zone, another (weak) pressure maximum may be discerned that is caused by the colliding shear waves emanating from the triple points. The time sequences of the plots in Figs. 7 - 11 range from 265 to 335 steps according to the CFL criterion with a step size of 5 time steps.

Fig. 12 Pressure isolines for channel width W/l_I after Kend = 250 time steps according to CFL criterion (time $t = 53.56$).

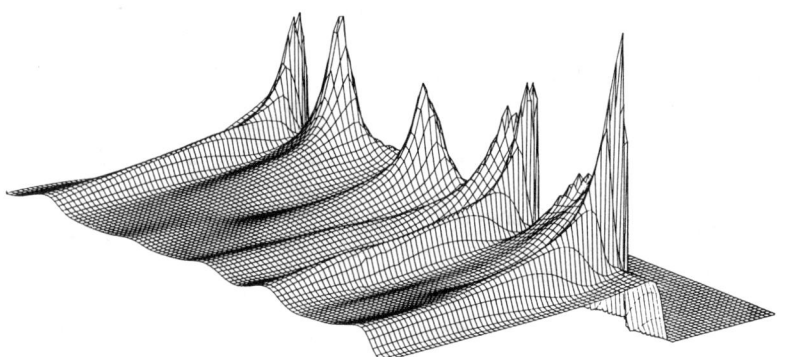

Fig. 13 Pressure relief corresponding to isolines in previous Figure 12.

For the given channel width of 24 l_I, we got a mode number of 2.5, which corresponds to an average of five transverse waves in the reaction zone. Figure 12 shows how the precursor shock wave is convoluted by the triple-shock intersection for a channel width of 34 l_I with a mode number of 4. In Fig. 13 the pertinent three-dimensional pressure distribution is plotted. It can be recognized that the cellular structure appears to be more irregular for the channel width 34 l_I than for the 24 - l_I channel. The reason is the overlapping behavior between the mode numbers 4 and 4.5. Therefore, it can be demonstrated that a unique cell size cannot always be assigned to any channel width W/l_I (see Fig. 14).

For certain values of W/l_I the numerical code yielded a chaotic character of the transverse wave structure. Such a mode gap appeared, for instance, at a channel width of 26 l_I. In this case the characteristic modes due to chemical energy release do not obey the standing wave condition valid for the eigenvalues of the linear acoustic wave equation.

Comparison with Experiment

From the mixture compositions used in the laboratory experiments of Strehlow et al. (1972), molecular weights, specific heat releases, and induction-reaction lengths were determined by means of the Rankine-Hugoniot relations. The results did not show the discrepancy of factor two between experiment and numerical simulation mentioned by Taki and Fujiwara (1978) in a paper concerning cellular detonation dynamics preceding their 1982 paper. Figure 4 illustrates diagrammatically that the experiments of Strehlow et al.

NUMERICAL ANALYSES OF ZDN DETONATION

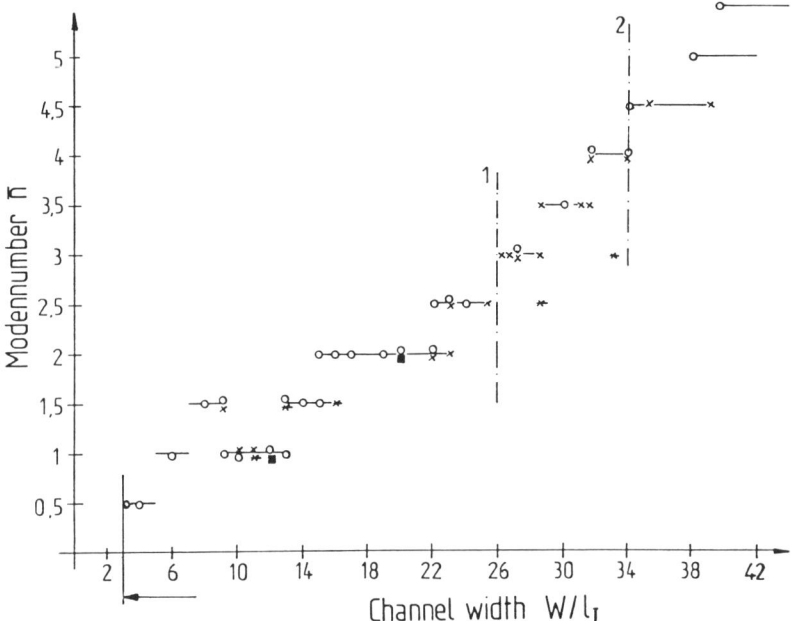

Fig. 14 Comparison of the author's analysis of experiments performed by Strehlow et al. (1972) with numerical computation results:

x ,   experiment of Strehlow et al. (1972) for argon-diluted oxyhydrogen analyzed by Schöffel for the assumed global reaction kinetics;
*,   experiment of Strehlow et al. (1972) for helium-diluted mixture;
o,   results of calculations performed by Schöffel;
■,   computational results obtained by Taki and Fujiwara (1982).

and our numerical simulation results compare nicely indeed. Strehlow et al. (1972) plotted the observed mode number vs. the initial pressure for a $2 H_2 + O_2 + 3 Ar$ mixture. Our Fig. 15 shows a drawing of Fig. 9 of their paper with exchanged coordinate axes.

The overlapping behavior between mode numbers 4 and 4.5 for transverse wave numbers 8 and 9, respectively, resulting from our numerical simulation, nicely corresponds to the experimental outcomes (see Fig. 15 for comparison). Moreover, the mode gap with concomitant chaotic transverse wave structure obtained between mode numbers 2.5 and 3 for wave numbers 5 and 6, respectively, can also be discerned in Fig. 15.

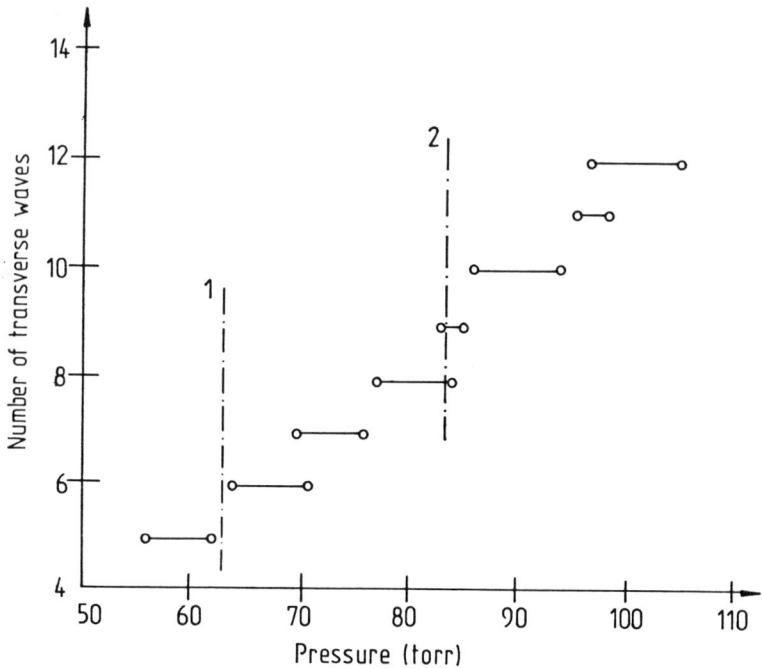

Fig. 15 Number of transverse waves as a function of initial pressure according to a graph in Strehlow et al.'s (1972) experimental work.

Summary

This paper attempts to provide deeper theoretical understanding of the mechanisms leading to the establishment of the dynamic cellular detonation structure. Numerical methods are applied to study the experimentally observed breakdown of smooth shock fronts and the spontaneous development of the transverse wave structure. The results demonstrate the role of caustics for the formation of a self-translating Mach-stem pattern with characteristic spacing. The formation of a triple-shock configuration appears to be caused by focusing weak shock waves in the exothermic reaction zone of a ZDN ground flow. Moreover, our outcomes reveal that the transverse wave structure develops from high-frequency fluctuations caused by the inherent, fluid dynamic instability of the ZDN structure.

The first author is grateful for an interesting discussion with Dr. Phil Colella at the University of California/Berkeley in March 1987. In a most recent paper, Bell, Colella, et al. (1987) also studied numerically the formation of Mach triple points in shock-fixed (Galilei-transformed) frames of reference by a two-dimensional second-order upwind scheme originally conceived for solving the nonreactive, gasdynamic Eulerian equations. Bell, Colella, et al. (1987) only investigated how the transverse wave spacing is controlled by the constant of induction-reaction kinetics and the overall chemical energy released. Beyond their approach, the results presented in our paper demonstrate how the spacing is influenced by the channel width, the mode-transition processes, and the principal irregularity of the transverse structure. In a related publication Schöffel and Ebert (1987) discuss the transition process from a marginal ZDN detonation to quenching conditions and vice versa in terms of the deflagration-to-detonation transition problem. Bell, Colella, et al. (1987) triggered the Mach-stem formation by an artificial perturbation, whereas in our calculation the breakdown of the smooth precursor shock front arises spontaneously. A consequence of the apparently coherent behavior of the high-frequency thermodynamic fluctuations are local explosions at the channel walls. Only for certain channel widths can the blast waves originating from the walls catch up with the inhomogeneous ZDN ground flow to establish a fully developed highly regular cellular structure. For other channel widths the applied numerical scheme either yielded a chaotic or an ambiguous transverse wave structure. In the latter case the cell width continually changes between neighboring, energetically equivalent modes.

Therefore, the irregularity observed for most gaseous mixtures (e.g., hydrogen/air) can also be recognized under certain circumstances for mixtures (e.g., argon-diluted oxyhydrogen) that reveal an utmost high degree of regularity. For channels wider than an inferior, critical value, the main precursor shock front remains at its initial place, underlining the importance of the classical, steady CJ theory applied for setting boundary conditions at the inflow/outflow borders of the computational region. The recurrence to the CJ model, indicated by an arrow in Table 1, is a salient feature of the spatial detonation dynamics.

The following conclusion can be drawn from laboratory experiments by other authors and from the numerical results presented in this paper:

1) Two-dimensional detonation propagation is, on the average, a steady, one-dimensional process sufficiently described by the classical CJ theory.

2) Irregular behavior is a property inherent in all gaseous mixtures, if adequate system parameters are chosen.

3) A certain internal resonance between chemical and acoustic modes must prevail to establish a high degree of regularity (see the discussion of the Rayleigh criterion for thermoacoustical systems by Chu (1956).

4) Integral numbers of transverse waves are preferred instead of odd ones.

5) In wide enough channels the cell size of highly regular mixtures proves to be constant (correspondence principle).

The numerical calculations were performed with a Siemens BS 2000 scalar computer system, and a Fujitsu VP 100 vector processor was used for larger arrays.

An experimental work of Manzhalei and Subbotin (1976) indicates that no transverse wave structure is observed if the von Neumann state (VN in Fig. 4) exceeds a critical compression value. Further investigations are planned to study the stability of a ZDN ground flow in the dependence of the degree of overdrive.

Acknowledgments

S. Schöffel thanks Prof. Joe Shepherd (now at Rensselaer Polytechnic Institute, RPI, Troy, NY) for fruitful discussions and the friendly hospitality during the author's stay at RPI at the end of March, 1987. The Deutsche Forschungsgemeinschaft is acknowledged for sponsoring our work in the Priority Research Program Finite Approximations in Fluid Mechanics since June, 1984.

References

Barthel, H. O. and Strehlow, R. A. (1966) Wave propagation in one-dimensional reactive flows, Phys. Fluids, 9, 1896-1907.

Barthel, H. O. (1974) Predicted spacings in hydrogen-oxygen-argon detonations, Phys. Fluids, 17, 1547-1553.

Bell, J. B., Colella, P., Trangenstein, J. A., and Welcome, M. (1987) Adaptive methods for high Mach number reacting flow, AIAA 8th Computational Fluid Dynamics Conference, Honolulu, HI, June 9-11, AIAA Paper CP 1176-87

Book, D. L., Boris, J. P., Fry, M. A., Guirguis, R. H., and Kuhl, A. L. (1982) Adaptation of flux-corrected transport algorithms for modeling blast waves, 8th International Conference on Numerical Methods in Fluid Mechanics, Vol. 170, edited by E. Krause, Aachen, FRG, Lecture Notes in Physics

Chu, B.-T. (1955) Pressure waves generated by addition of heat in a gaseous medium, NACA TN-3411.

Chu, B.-T. (1956) Stability of systems containing a heat source - the Rayleigh criterion, NACA Research Memorandum 56D27.

Chushkin, P. I. and Shurshalov, L. V. (1982) Numerical computations of explosions in gases, 8th International Conference on Numerical Methods in Fluid Mechanics, Vol. 170, edited by E. Krause, Aachen, FRG, Lecture Notes in Physics.

Dwyer, H., Allen, R., Ward, M., Karnopp, D., and Margolis, D. (1974) Shock capturing finite difference methods for unsteady gas transfer, AIAA Paper 74-521.

Ebert, F. and Schöffel, S.U. (1986) Calculation of the flowfield caused by shock-wave and deflagration interaction, DFG-priority research program, Results 1984-85, Notes on Numerical Fluid Mechanics, Vol. 14, 56-70, edited by: E. H. Hirschel, Vieweg, Braunschweig/Wiesbaden, FRG, see also: Combustion theory: a report on Euromech 203, J. Fluid Mech., 409-414.

Fickett, W. and Davis, W.C. (1979) Detonation, Univ. of California Press, Berkeley, CA.

Garabedian, P.R. (1964) Partial Differential Equations, Wiley New York, p. 109.

Gordon, S. and McBride, B.J. (1971) Computer program for the calculation of complex chemical equilibrium compositions, rocket performance, incident and reflected shocks and Chapman-Jouguet detonations, NASA SP-273.

Harten, A. (1983) High resolution schemes for hyperbolic conservation laws, J. Comput. Phys., 49, 357-393.

Korobeinikov, V. P., Levin, V. A., Markov, V. V., and Chernyi, G. G. (1972) Propagation of blast wave in a combustible gas, Astronaut. Acta, 17, 529-537.

Lee, J. H. (1984) Dynamic parameters of gaseous detonations, Annu. Rev. Fluid Mech., 16, 311-336

Magnus, K. (1976) Schwingungen, Teubner, Stuttgart, FRG

Manzhalei, V. I. and Subbotin, V. A. (1976) Combust. Explos. Shock Waves USSR, 12, No. 6, 819.

Markov, V. V. (1981) Dokl. Akad. Nauk SSSR 258, 314-317.

Markstein, G. H. (1964) Non-Steady Flame Propagation, AGARDograph 75.

Oppenheim, A. K. and Roscziszewski, J. (1963) Determination of the detonation wave structure, 9th Symposium (International) on Combustion, The Combustion Institute, Pittsburgh, PA, pp. 424-441.

Oran, E. S., Boris, J. P., Young, T., Flanigan, M., Burks, T., and Picone, M. (1982) Numerical simulations of detonations in hydrogen-air and methane-air mixtures, 18th Symposium (International) on Combustion, The Combustion Institute, Pittsburgh, PA., pp. 1641-1649.

Richtmyer, R.D. and Morton, K. W. (1983) Difference methods for Initial-Value Problems, Interscience, London, U.K.

Rosales, R. R. (1987) Catastrophic instabilities in square wave models of detonation, SIAM-conference on Numerics of Combustion, San Francisco, March 9-11

Rosales, R. R. and Majda, A. (1983) A theory for spontaneous machstem formation in reacting shock fronts. I. The basic perturbation analysis, SIAM J. Appl. Math., vol.43, No.6, pp. 1310 - 1334

Rosales, R. R. and Majda, A. (1984) II. Steady-wave bifurcations and the evidence for breakdown, Studies in Applied Mathematics, pp. 117-148

Schlichting, H. (1982) Grenzschichttheorie, Braun, Karlsruhe, FRG.

Schöffel, St. and Ebert, F. (1988) A Numerical Investigation of the Reestablishment of a Quenched Gaseous Detonation in a Galilei-Transformed System, Proceedings of the 16th International Symposium on Shock Tubes and Waves, Aachen, FRG, July 26-31, 1987, edited by H. Grönig, VCH Verlagsgesellschaft, Weinheim, FRG, pp. 779-786.

Shchelkin, K. I. and Troshin, Ya. K. (1965) Gasdynamics of Combustion Mono, Baltimore, MD.

Shepherd, J. E. and Westbrook, C. K. (1986) Detailed chemical kinetic models, Progress in Astronautics and Aeronautics: Fuel-Air Detonations, edited by J. H. Lee and R. Knystautas, AIAA, New York

Strehlow, R. A. (1968) Fundamentals of Combustion, International Textbook, Scranton, PA, pp. 287-352.

Strehlow, R. A., Adamczyk, A. A., and Stiles, R. J. (1972) Transient studies of detonation waves, Astron. Acta, 17, 509 - 527.

Strehlow, R. A. and Fernandes, F. D. (1965) Transverse waves in detonations, Combust. Flame, 9, 109-119.

Sturtevant, B. and Kulkarny, V. A. (1976) The focusing of weak shock waves, J. Fluid Mech., 73, 651-671.

Taki, S. and Fujiwara, T. (1982) Numerical simulations of triple shock behavior of gaseous detonation, 18th. Symposium (International) on Combustion, The Combustion Institute, Pittsburgh, PA, pp. 1671-1681.

Van Leer, B. (1982) Flux-Vector Splitting for the Euler Equations, Lect. Notes Phys., 17, 507-512.

Vasiliev, A. A. and Nikolaev, Yu. (1978) Closed theoretical model of a detonation cell, Astronaut. Acta, 5, 983-996.

Westbrook, C. K. and Urtiew, P. A. (1982) Chemical kinetic prediction of critical parameters in gaseous detonations, 19th Symposium (International) on Combustion, The Combustion Institute, Pittsburgh, PA, pp. 615-623.

Zierep, J. and Oertel, H.jr. (1982) Convective Transport and Instability Phenomena, Braun, Karlsruhe, FRG.

Detonation Parameters for the Hydrogen-Chlorine System

R. Knystautas* and J. H. Lee†
McGill University, Montreal, Quebec, Canada

Abstract

Detonation velocity, pressure, cell size, and critical tube diameter were measured simultaneously in four hydrogen-chlorine mixtures. The experiment consisted of a 50 mm diameter tube, 1.5 m long connected to a cylindrical chamber, 150 mm in diameter and 30 cm long. The mixture compositions studied covered the range of 36-60% H_2-Cl_2 and initial pressures p_o = 10-180 Torr. The measured detonation velocities and pressures invariably are above the calculated equilibrium Chapman-Jouguet levels and thereby appear to be in agreement with the predictions of Guenoche et al., who concluded that H_2-Cl_2 detonations fall in the realm of pathological detonations. Detonation cell size results exhibit the characteristic U-shaped behavior as a function of mixture composition similar to other detonable systems. The results also show the usual inverse dependence of cell size on initial pressure. Based on cell size measurements, the relative detonation sensitivity of the stoichiometric H_2-Cl_2 mixture is comparable to that for stoichiometric ethylene-oxygen. Critical tube diameter measurements indicate that the d_c = 13λ law holds for the hydrogen-chlorine system. Predictions of critical energy for direct initiation as deduced from cell size data are in good agreement with the available experimental measurements.

Introduction

Several recent experimental studies (Lee et al. 1972; Akyurtlu 1975) have demonstrated that hydrogen-chlorine

---

Copyright © 1988 by the American Institute of Aeronautics and Astronautics, Inc. All rights reserved.
*Professor, Department of Mechanical Engineering.

mixtures are readily detonable and that the detonations possess the same characteristic cellular structure common to other systems in which oxygen is the oxidizing agent. However, these studies, as well as recent theoretical work [Guenoche et al. (1981)] bring up a number of anomalies that appear to exist in the hydrogen-chlorine system. Guenoche et al. show that the heat release function in the reaction zone of a H_2-Cl_2 detonation has an overshoot and reaches a maximum before chemical equilibrium is achieved. The consequence of this is that the detonation states are higher than the equilibrium values calculated using the conventional Chapman-Jouguet (C-J) model in which the sonic conditions and chemical equilibrium are achieved simultaneously. The predicted excess values should be discernible experimentally (~6% larger than C-J for the detonation Mach number and velocity; ~12% larger for the detonation pressure).

The experiments of Akyurtlu (1975) indicate higher than C-J propagation velocities. However, his observed detonability limits appear to be at variance with an earlier study by Lee et al. (1972). Akyurtlu observes that the most sensitive mixtures are for lean mixtures and that the sensitivity monotonically decreases as one approaches stoichiometry. He found the detonability limit for stoichiometric (equimolar) H_2-Cl_2 to be at about $p_o = 120$ Torr, whereas Lee et al. report detonations in identical mixtures to as low as $p_o = 30$ Torr.

One of the important developments in recent years has been the establishment of empirical correlations between the cell size λ and the other dynamic detonation parameters. Perhaps the most basic of such correlations has been the one relating the cell size λ to the critical tube diameter d_c (i.e., $d_c = 13\lambda$). Almost all of the existing studies have been made on systems with O_2 as the oxidizer. It would therefore be of interest to explore systems other than those involving hydrocarbon oxidation to see if the $d_c = 13\lambda$ law is valid.

The present paper reports the results of a detailed study of detonation in hydrogen-chlorine mixtures in which the detonation velocity, pressure, cell size, and critical tube diameter measurements are used to check directly the $d_c = 13\lambda$ correlation for H_2-Cl_2. The cell size data are further used to estimate the critical initiation energy E_c via the surface energy model. Such estimates for E_c are then compared to the limited but indicative initiation energy results that can be extracted from our earlier experimental study of H_2-Cl_2 detonations [Lee et al. (1972)].

Experimental Details

The apparatus used in the present study consisted of a stainless steel tube and cylindrical chamber. The tube dimensions were 50 mm i.d. and 1.5 m long; the detonation chamber to which the tube was attached had the dimensions of 15 cm i.d. by 30 cm long. One end of the detonation tube was equipped with an initiation plug (electric spark gap), while the other end was inserted into the end flange of the detonation chamber such that the wave traveling along the detonation tube would transmit uninhibited directly into the detonation chamber at its centerline axis.

Stoichiometric and off-stoichiometric mixtures of H_2-Cl_2 were prepared by premixing the constituents in a 12 liter pressure vessel. The vessel was first evacuated to better than 0.1 mbar, then the hydrogen and chlorine were introduced individually from their respective bottles. The composition of the mixture was controlled by monitoring the partial pressures of the constituents via Bourdon gages. The total pressure of the mixture in the tank was 2 atm absolute and the constituents were allowed to mix by diffusion with some external mechanical agitation.

In any given experiment, the detonation tube and chamber were first evacuated using a vacuum pump preceded by a liquid nitrogen trap to condense out the residual Cl_2 and HCl vapors. The trap was disconnected and vented outdoors at the end of a day's experiments. The apparatus was then filled with the premixed mixture to the desired initial pressure, which ranged 10-180 Torr throughout the series of experiments. The mixture compositional range extended from 36%H_2/64%Cl_2 at the lean end to 60%H_2/40%Cl_2 at the rich end. Initiation of detonation in the detonation tube was achieved by the discharge of a condenser bank (0.6 μF, 16-18 kV) through the ignition spark gap. The energy stored in the condenser bank was in the range 80-100 J. Detonation propagation along the detonation tube and its transmission into the detonation chamber was monitored by periodically spaced pressure transducers (PCB113A24, 5 mV/psi nominal). The pressure signals were recorded on a dual-beam oscilloscope. Detonation cell size was determined from smoked foils placed at the inner periphery of the detonation tube just before the exit into the detonation chamber. The smoked foils were prepared from thin Mylar sheets (~0.4 mm thick) smoked with a kerosene flame. To determine the conditions for critical transmission from the detonation tube into the chamber, the critical initial pressure for each mixture composition was

DETONATION PARAMETERS

established from time-of-arrival measurements using pressure transducers in the tube and the chamber.

Results and Discussion

The results for the detonation velocity in the four mixtures studies are displayed in Figs. 1-4. Where available, the results of Akyurtlu (1975) are also plotted. Although there are fluctuations in evidence, the general trend of the presents results indicates that the experimental values are slightly superior to the C-J values calculated. This supports the theoretical work of Guenoche et al. (1981) in which they investigated the influence of the heat release function on the detonation states. Kinetic calculations indicate that, for certain chemical systems, the heat release function does not increase monotonically from zero behind the shock front to its final value at the C-J plane. An overshoot in the heat release function can occur, resulting in thermal choking in the

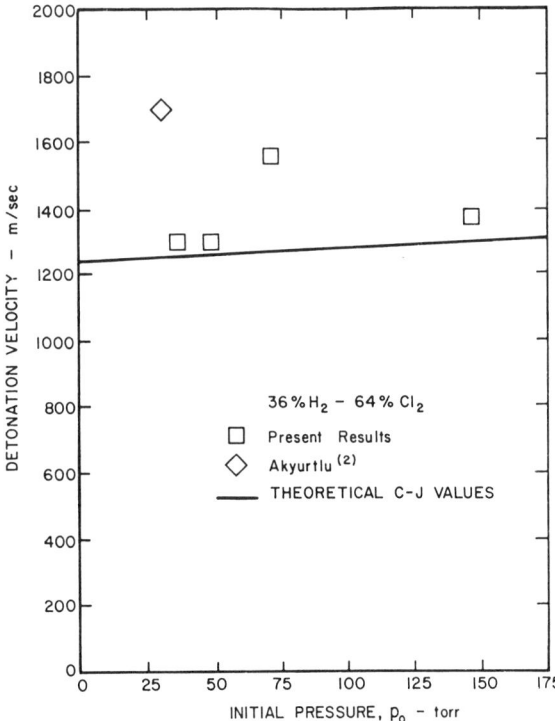

Fig. 1 Variation of detonation velocity with initial pressure for the 36% H_2-Cl_2 mixture.

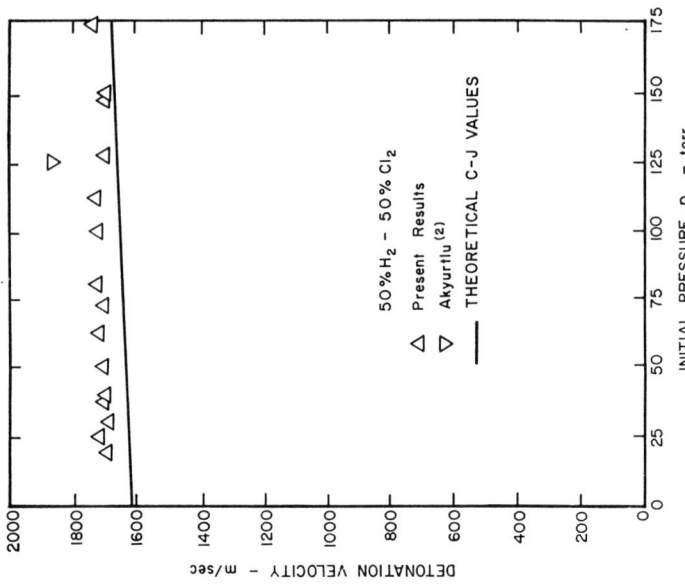

Fig. 3 Variation of detonation velocity with initial pressure for the 50% H_2-Cl_2 mixture.

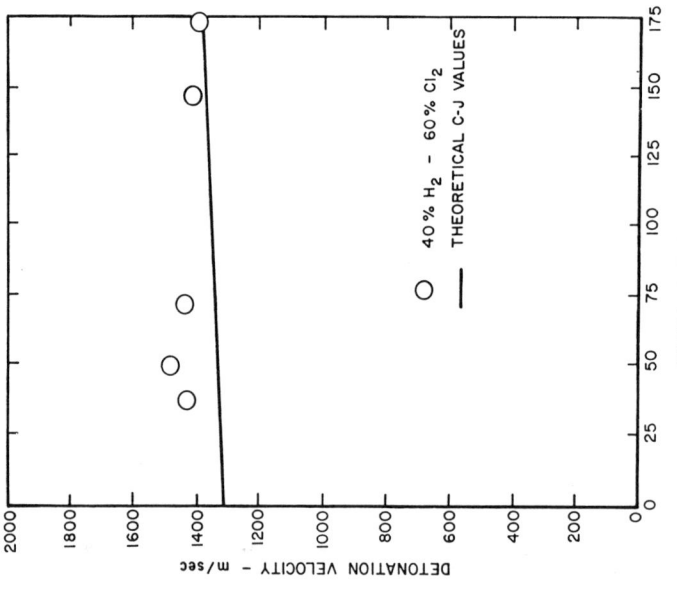

Fig. 2 Variation of detonation velocity with initial pressure for the 40% H_2-Cl_2 mixture.

DETONATION PARAMETERS

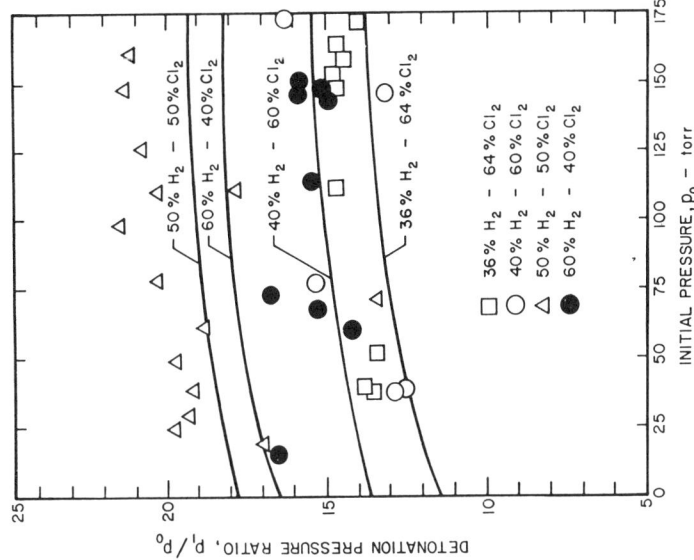

Fig. 5 Variation of detonation pressure with initial pressure in H_2-Cl_2 mixtures.

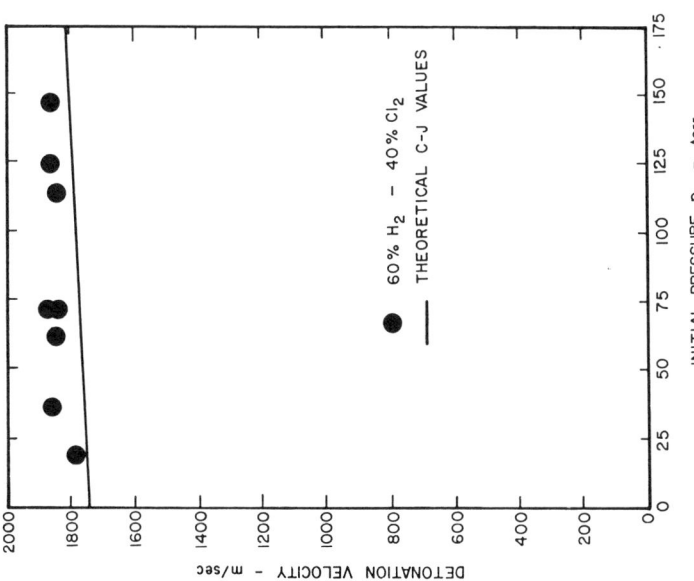

Fig. 4 Variation of detonation velocity with initial pressure for the 60% H_2-Cl_2 mixture.

nonequilibrium region giving rise to a higher than C-J velocity. For the particular case of equimolar H_2-Cl_2 at an initial pressure of 1 atm, Guenoche et al (1981) show that such an overshoot exists and leads to an increase in the detonation velocity of 6.1%. This prediction is in accord with the present study in that the experimentally measured detonation velocity is superior to the C-J value by a margin of 5-8%. The results of Akyurtlu (1975) exhibit the same superior trend, except that his experimental values are generally much higher than the C-J one, corresponding to an excess of 8-50%.

Similar predictions by Guenoche et al. are made for the detonation pressure. In Fig. 5, the comparison between the theoretical C-J detonation pressure and experimental values are shown for the four mixtures studied. In general, it is difficult to measure the detonation pressure and the experimental results show larger fluctuations than those for the velocity measurements. Again, particularly for the cases of 36% H_2-Cl_2 and 50% H_2-Cl_2, the experimental data appear to confirm the prediction of Guenoche et al. that the detonation pressure is about 12% higher than the corresponding C-J value (for 50% H_2-Cl_2 at 1 atm), which is twice the velocity increase. However, the large scatter in the data for the 40% H_2-Cl_2 and 60% H_2-Cl_2 mixtures makes it difficult to determine any definitive trend of the results for the detonation pressure.

Thus, from both the velocity and pressure data, the experiments are in accord with the theoretical model of Guenoche et al (1981). It is of particular interest to note that in spite of the fact that the theoretical model is based on one-dimensional Zel'dovich-von Neumann-Doring (ZND) structure, which is incompatible with the real three-dimensional cellular structure, the prediction of such fine details of the structure as the influence of the heat release function can be manifested experimentally.

Measurements of the detonation cell size were carried out using the standard smoked foil technique. Hydrogen-chlorine mixtures do not seem to affect adversely the soot-covered Mylar film used throughout the experiments. The cell patterns observed for H_2-Cl_2 detonations are similar to those in hydrocarbon-oxygen mixtures. The cell structure is fairly regular and the cell size can be measured rather easily in general. Typical results in the form of U-shaped curves for the variation of cell size with fuel concentration are shown in Fig. 6. At the stoichiometric (equimolar) concentration, the cell size is a minimum and increases toward the lean and rich limits as virtually all mixtures do. The variation away from the stoichiometric composition is very weak and the sharp increase occurs only as the limits are approached. The behavior of the cell size variation for H_2-Cl_2 is very similar to that for C_2H_2-O_2. It may be mentioned in passing that the cell size measurements in the present study are in disagreement with those of Akyurtlu (1975) by roughly an order of magnitude. The trend of the results in terms of the relative detonation sensitivity implicit in cell size measurements is also at variance with the trend observed by Akyurtlu.

The variation of cell size with initial pressure for the four mixtures studied is shown in Fig. 7. All the results show the standard $\lambda \sim 1/p_0$ dependence as in other hydrocarbon-oxygen mixtures. With regard to the relative sensitivity as indicated by the cell size, the present results show that equimolar H_2-Cl_2 mixtures are about as sensitive as stoichiometric ethylene-oxygen mixtures (C_2H_4 + 3 O_2). Equimolar C_2H_2-O_2 remains as the most sensitive fuel-oxygen mixture studied thus far.

Critical tube diameter measurements were carried out in the 50 mm diameter detonation tube. The critical initial pressure for successful transmission of detonation from the

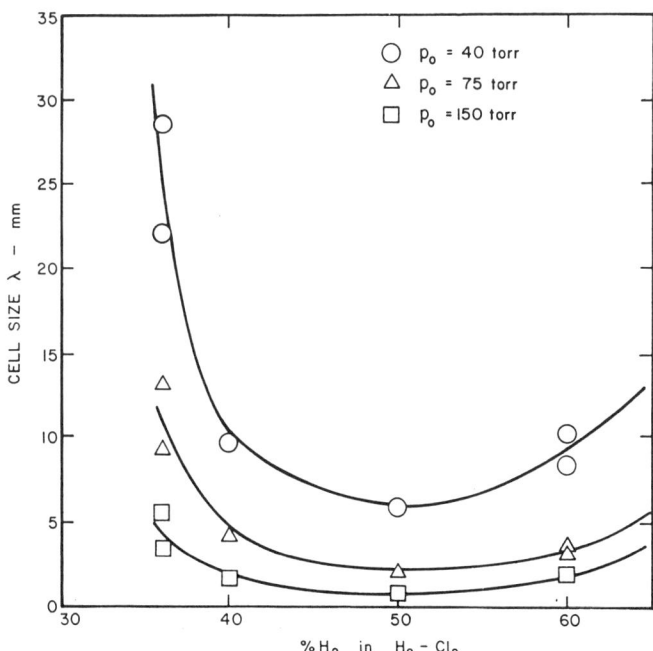

Fig. 6 Variation of detonation cell size λ with composition in H_2-Cl_2 mixtures.

tube to the chamber was determined for three hydrogen-chlorine mixtures (40, 50 and 60% H_2-Cl_2). The cell size λ, corresponding to the mixture at the critical transmission pressure, can be found from Fig. 7. The relevant critical transmission information is displayed in Table 1 where it can be seen that the d_c/λ ratio obtained varies between 12.2 and 13.5. This is well within the accuracy with which cell size can generally be measured for regular cell patterns. Although extensive experiments for the critical tube diameter in H_2-Cl_2 have not been carried out, the present data are felt to be sufficient to support the conclusion that $d_c \simeq 13\ \lambda$ is valid for the H_2-CL_2

Table 1 Transmission of detonation

Mixture	d_{crit}, mm	$p_{o_{crit}}$, torr	λ, mm	d_c/λ
40% H_2 − 60% Cl_2	50	84	4.0	12.5
50% H_2 − 50% Cl_2	50	58	3.7	13.5
60% H_2 − 40% Cl_2	50	76	4.1	12.2

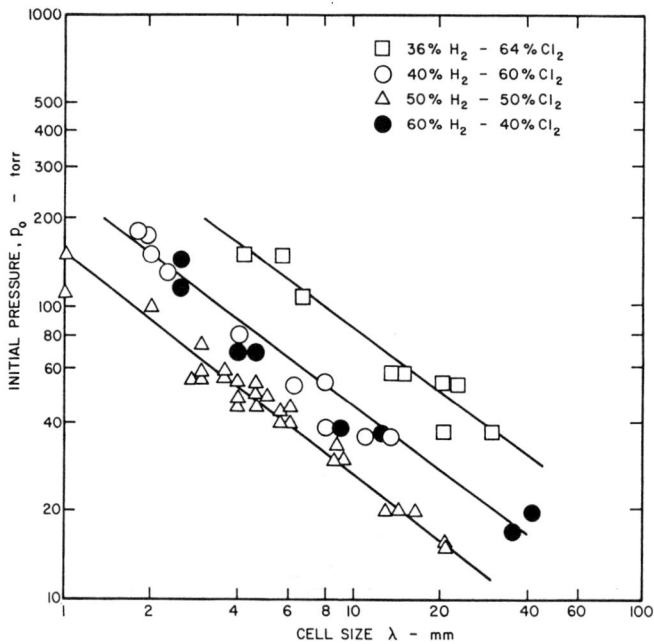

Fig. 7 Variation of detonation cell size λ with initial pressure in H_2-Cl_2 mixtures.

system as well. Thus, in spite of the fact that H_2-Cl_2 detonations also seem to present an abnormal behavior (pathological detonations according to Guenoche et al.) due to their peculiar energy release function, the $d_c = 13\lambda$ correlation applies.

Finally, the detonation cell size λ measured in the present experiments can be used to predict the critical energy for direct initiation E_c via the surface energy model [Guirao et al. (1972); Lee (1984); Benedick et al. (1986)]. That is,

$$E_c = \frac{2197}{16} \pi \gamma_o P_o I M_{CJ}^2 \lambda^3$$

where γ_o and p_o are the specific heat ratio and pressure pertaining to initial conditions, I the energy integral,

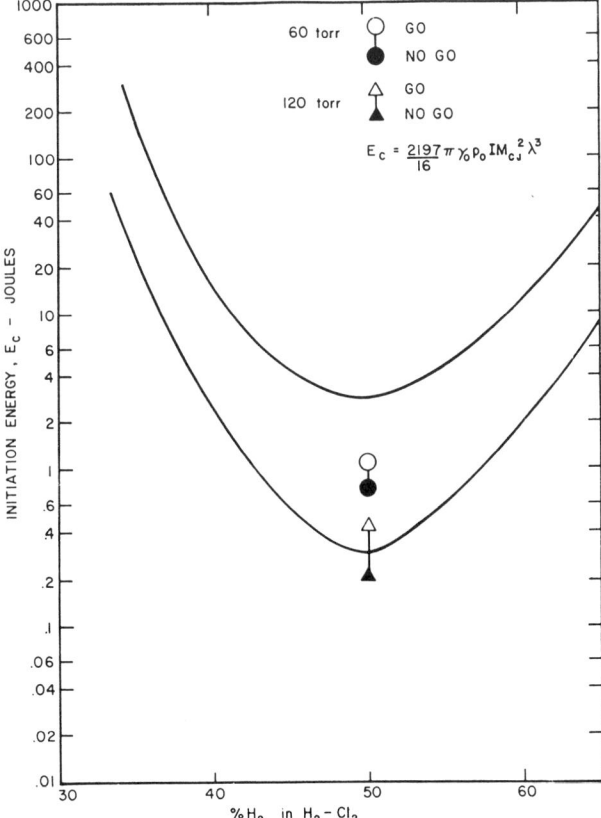

Fig. 8 Comparison between theoretical and experimental results for initiation energy in H_2-Cl_2 mixtures.

and M_{CJ} the C-J detonation Mach number. Two such predicted curves are plotted in Fig. 8 together with experimental results for initiation energy that can be deduced from our previous study [Lee et al. (1972)]. The direct initiation experiments were carried out in stoichiometric H_2-Cl_2 mixtures using electric spark initiation at two initial pressure, p_0 = 60 and 120 Torr. Although go/no-go conditions were established in both mixtures for a known condenser bank and charge voltages, the detailed discharge history necessary to determine the true source energy and hence E_c had to be inferred from similar but monitored discharges in other experiments [Lee et al. (1975)]. It has been shown [Knystautas and Lee (1976)] that the true initiation source energy corresponds to the electrical energy dissipated in the spark plasma up to the first quarter cycle of the typically underdamped oscillatory discharge. At p_0 = 60 Torr in H_2-Cl_2, the condenser bank had a capacitance C = 0.1 μF for charge voltages of 20 and 24 kV corresponding to unsuccessful and successful initiation, respectively. For the p_0 = 120 Torr case, the condenser bank had capacitances of C = 0.05 μF and C = 2 x 0.05 μF at a charging voltage of 20 kV for both, again corresponding to unsuccessful and successful initiation, respectively. The appropriate values of E_c corresponding to these initial source conditions are plotted in Fig. 8. The agreement for the p_0 = 120 Torr case with the surface energy model prediction is excellent and this also corresponds to the more reliable estimate, since direct discharge histories under virtually identical conditions were available in other experiments [Knystautas and Lee (1976)]. The agreement at p_0 = 60 Torr is not as good, even though it is of the same order of magnitude--which in itself is significant. The discrepancy can most likely be attributed to the more approximate estimate that can be made for the true or effective E_c based on the best available spark discharge information.

Conclusions

A series of experiments were carried out to measure a number of key detonation parameters in the hydrogen-chlorine system. The parameters measured were detonation velocity, pressure, cell size, and critical tube diameter as a function of mixture composition and initial pressure. Four mixture compositions were studied ranging 36-60% H_2-Cl_2 at initial pressures p_0 = 10-180 Torr. Detonation velocity and pressure in H_2-Cl_2 are invariably higher than the calculated equilibrium C-J values. The excess values fall in the range predicted theoretically by Guenoche et al. (1981), who deduced that the energy release function in H_2-Cl_2 mixtures has an overshoot unlike the monotonic behavior in other detonable mixtures. The consequence of this is that the predicted detonation states fall above the C-J levels.

Detonation cell size λ was measured directly from smoked foils inserted near the exit of the detonation tube. The cell size results show the characteristic U-shaped behavior as a function of composition. To this extent, the trend is similar to other detonable systems and is at variance with the findings of Akyurtlu (1975). The most detonable mixture is the stoichiometric (equimolar) H_2-Cl_2 mixture and its relative detonation sensitivity is comparable to that of stoichiometric ethylene-oxygen. The cell size results also show the characteristic inverse dependence on initial pressure for a given composition. Critical tube diameter measurements in H_2-Cl_2 mixtures confirm the $d_c = 13\lambda$ law. Based on cell size results, the critical energy for direct initiation can be deduced from the surface energy model. The present study indicates that such predictions are in good agreement with the available measurements.

Acknowledgments

The authors are indebted to Ms. G. Riopelle for her assistance in the experimental measurements. The work was supported by the Natural Sciences and Engineering Research Council of Canada (NSERC) under Grants A-3347 and A-7091 and by the Defense Research Establishment Suffield (DRES) under Contract 8SG84-00057.

References

Akyurtlu, A. (1975) An investigation of the structure and detonability limits of hydrogen-chlorine detonations. Ph.D. Thesis, University of Wisconsin, Madison.

Benedick, W.B., Guirao, C.M., Knystautas, R., and Lee, J.H. (1976) Critical charge for the direct initiation of detonation in gaseous fuel-air mixtures. Progress in Astronautics and Aeronautics: Dynamics of Explosions; Vol. 106, pp. 181-202. AIAA, New York.

Guenoche, H., Le Diuzet, P., and Sedes, C. (1981) Influence of the heat-release function on the detonation states. Progress in Astronautics and Aeronautics: Gasdynamics of Detonations and Explosions; Vol. 75, pp. 387-407. AIAA, New York.

Guirao, C., Knystautas, R., Lee, J.H., Benedick, W., and Berman, M. (1982) Hydrogen-air detonations. Nineteenth Symposium (International) on Combustion, pp. 583-590. The Combustion Institute, Pittsburgh, PA.

Knystautas, R. and Lee, J.H. (1976) On the effective energy for direct initiation of gaseous detonations. Combust. Flame 27, 221-228.

Lee, J.H. (1984) Dynamic parameters of gaseous detonations. Ann. Rev. Fluid Mech. 16, 311-336.

Lee, J.H., Knystautas, R., and Guirao, C.M. (1975) Critical power density for direct initiation of unconfined detonations. Fifteenth Symposium (International) on Combustion, pp. 53-67. The Combustion Institute, Pittsburgh, PA.

Lee, J.H., Knystautas, R., Guirao, C., Bekesy, A. and Sabbagh, S. (1972) On the instability of H_2-Cl_2 gaseous detonations. Combust. Flame 18, 321-325.

Hydrazine Vapor Detonations

M. D. Pedley,* C. V. Bishop,† F. J. Benz,‡ C. A. Bennett,§
R. D. McClenagan,¶ and D. L. Fenton**
NASA Johnson Space Center, Las Cruces, New Mexico
and
R. Knystautas,†† J. H. Lee,‡‡ O. Peraldi,§§ and G. Dupre¶¶
McGill University, Montreal, Quebec, Canada
and
J. E. Shepherd***
Rensselaer Polytechnic Institute, Troy, New York

Abstract

The propagation of detonation in pure hydrazine vapor was experimentally investigated. Detonation cell widths were measured in a 150 mm heated tube. The relation between cell width and critical tube diameter was also determined. The detonation velocity in pure hydrazine was within 5% of the calculated Chapman-Jouguet (C-J) velocity. Detonations in hydrazine vapor were observed at pressures from 120 kPa down to the lowest initial test pressure of 0.5 kPa. The cell width at ambient pressure

---

Copyright © 1988 by the American Institute of Aeronautics and Astronautics, Inc. No copyright is asserted in the United States under Title 17, U.S. Code. The U.S. Government has a royalty-free license to exercise all rights under the copyright claimed herein for governmental purposes. All other rights are reserved by the copyright owner.
*Advanced Systems Engineering Specialist currently with Lockheed-ESC, Johnson Space Center, Houston, TX.
† Director currently with McGean-ROHCO, Cleveland, OH.
‡ Project Manager, NASA, White Sands Test Facility.
§ Supervisor of Technical Services currently with Lockheed-ESC, High Energy Laser Test Facility.
¶ Project Engineer, Lockheed-ESC, White Sands Test Facility.
** Associate Professor currently with Department of Mechanical Engineering, Kansas State University, Manhattan, KS.
†† Professor, Department of Mechanical Engineering.
‡‡ Professor, Department of Mechanical Engineering.
§§ Currently with Department of Environmental Safety, Rhone Poulenc, Lyon, France.
¶¶ Currently with CNRS, Orleans, France.
*** Assistant Professor, Department of Mechanical Engineering, Aeronautical Engineering, and Mechanics.

and 393 K was estimated by extrapolation to be 1.8 mm. Tests were conducted at initial temperatures of 308-393 K. At constant initial pressure, the cell width for hydrazine vapor detonations increased with decreasing initial temperature. The detonation sensitivity of pure hydrazine vapor was similar to that for stoichiometric propane-oxygen mixtures. The critical tube diameter for propagation of a detonation into an unconfined environment was found to be 15 cell widths, in satisfactory agreement with the empirical relation of 13 cell widths observed for other systems. The detonation cell width measurements were interpreted using the Zel'dovich-von Neumann-Doring model with a detailed reaction mechanism for hydrazine decomposition. Excellent agreement with the experimental data for pure hydrazine vapor was obtained using the empirical relation that detonation cell width was equal to 29 times the kinetically calculated reaction zone length.

Introduction

Hydrazine (N_2H_4) and its methyl derivatives are used extensively as fuels in aerospace applications. Hydrazine itself is an extremely convenient fuel because it can act as a monopropellant [Pfeffer (1976)]. Most satellites use the exothermic decomposition of hydrazine over a catalytic bed of the iridium-based proprietary Shell 405 catalyst to provide thrust for maneuvering [Sayer and Southern (1974)]. The decomposition of hydrazine on the Shell 405 catalyst is also used to drive power units for underwater applications and the auxiliary power units of the Space Shuttle Orbiter [Schmidt (1984)].

A major drawback of using hydrazine in aerospace applications is its instability. Although incidents have been rare, occasional violent explosions have caused severe structural damage. A dramatic example was the explosions that occurred in two of the three Space Shuttle auxiliary power units during landing of the STS-9 mission [Brownfield and Korb (1984)]. Other violent explosions have occurred while venting heated hydrazine vapor [Benz and Pedley (1986)], during rapid compression of liquid hydrazine [Benz et al. (1984); Briles and Hollenbaugh (1978); Vander Wall et al. (1971)], and while studying runaway reactions in liquid hydrazine [Fritchman and Benz (1980)].

An understanding of the reason for these explosions is essential for improving the design and safety of hydrazine systems. The damage inflicted by the explosions and the speed at which they propagated suggest that the hydrazine may have detonated. The combustion properties of hydrazine were studied extensively in the 1950's and 1960's and numerous papers were published on decomposition and oxidation flames [Adams and Stocks (1953); Gilbert (1958); Gray and Holland (1970); Gray and Lee (1955), (1959); Gray et al. (1957); Gray and Spencer (1963); Maclean and Wagner (1967)], flammability limits [Furno et al. (1962); Perlee et al. (1962)], spontaneous ignition [Furno et al. (1968); Gray and Lee (1954), (1955)], thermal decomposition [Eberstein and Glassman (1965); Sawyer and Glassman (1967)], and pyrolysis and oxidation behind shock waves [Genich et al. (1974); McHale et al. (1965); Meyer et al. (1969); Michel and Wagner (1965); Moberly (1962)]. However, very few studies dealt with the detonation of hydrazine. Standard propellant explosibility tests gave negative results for liquid hydrazine [Schmidt (1984); Scott et al. (1949); Vander Wall et al. (1971)]. Detonation of pure hydrazine vapor was observed in the exploratory studies of W. Jost (1962), A. Jost et al. (1963), and Heinrich (1964). In all cases, the experiments were conducted in small tubes (diameter no more than 25 mm) and the measurements were limited to detonation velocities. Heinrich also detected evidence of the three-dimensional spinning structure of the detonation front in his streak records.

Detonation velocities depend on the thermodynamics of the chemical reactions in the detonation front and have little direct relation to the detonation sensitivity of an explosive medium. The detonation sensitivity can be characterized only by dynamic detonation parameters [Lee (1984)], which depend on the rate of chemical reaction in the detonation front. The most important experimentally measurable chemical length scale in detonative combustion is the scale of the three-dimensional cellular structure characterized by the detonation cell width λ [Knystautas et al. (1984); Lee (1984)].

The aim of this study was to measure experimentally the detonation velocity and cell widths for hydrazine decomposition over a wide range of temperatures and pressures. Additional experiments were conducted to determine whether the empirical relationships between cell width and confinement established for oxidative detonation [Mitrofanov and Soloukhin (1964); Knystautas et al. (1981); Lee et al. (1982)] also hold for hydrazine monopropellant decomposition. The approximate empirical relation

$$d_c = 13 \lambda$$

for the critical tube diameter for propagation of a detonation from a tube to an unconfined environment has been found to hold for hydrogen and hydrocarbons in tubes with air or oxygen [Lee et al. (1982)]. However, the relation is empirical and not necessarily valid for the very different hydrazine monopropellant decomposition detonation. Kinetic modeling of the chemical reactions in the detonation front was also conducted and the results compared to the experimental data.

Experimental Details

The detonation studies were conducted in a stainless steel schedule 40 pipe 1.75 m long, with an internal diameter of 150 mm and wall thickness of 7 mm. Because hydrazine is a liquid at ambient temperature (boiling point 387 K at standard atmospheric pressure), the tube was heated by an array of heating tapes bonded to the outer surface of the tube and the end flanges. The array consisted of nine equispaced 400 W heating tapes on the tube and one 300 W heating tape on each end flange. The temperature of the tube was monitored by five chromel-alumel thermocouples and controlled by a temperature controller to within ±2 K. The entire tube was insulated with fiberglass lagging, 50 mm thick.

Two methods of detonation initiation were used in these experiments. For initial studies, an oxygen-acetylene driver detonation was used to initiate hydrazine detonation. Detonation of the driver gas was initiated by a commercial electric exploding bridge wire (energy roughly 5-8 J). The driver section attached to the main detonation tube was 0.5 m long. A short length of Shchelkin spiral made up of a coiled stainless steel tube was inserted in the ignition end of the driver to assist the initiation of detonation. A manually operated ball valve with a 150 mm i.d. separated the driver from the main tube. In a typical experiment, the ball valve would be opened and the tube plus driver would then be filled with a stoichiometric oxygen-acetylene mixture at the required test pressure by partial pressure mixing. The gases would then be mixed thoroughly by a propeller fan mounted on the inside of the end flange of the main tube. The mixing fan was operated at 1750 revolutions per minute (rpm) for approximately 15 min. The ball valve was then closed and the main tube reevacuated prior to introduction of the hydrazine vapor. This method worked well at pressures close to ambient, but initiation of hydrazine vapor detonations became increasingly difficult as the pressure was reduced.

In later experiments the driver section and ball valve were removed and direct initiation was achieved by discharging a high-voltage condenser bank through an exploding bridge wire mounted in the end flange of the detonation tube. For these experiments, the Shchelkin spiral was located at the ignition end of the detonation tube. The condenser bank could be charged to energies up to 3 kJ. Discharge was triggered using either a simple three-electrode switch triggered by a 30 kV pulse or a commercial switch triggered by a pulser driving a trigger transformer. The pulser used a small capacitor charged by a 900 V battery pack and triggered by a thyratron. A significant charge remained on the condensers after initiation. Nevertheless, enough energy went into the hydrazine vapor to initiate detonation even at the lowest test pressures used. This initiator proved faster, more convenient, and more effective than the driver detonation. The only problems encountered were in protecting the detonation velocity data acquisition system from the discharge. As a result, detonation velocities were measured only in the early experiments with the oxygen-acetylene driver and in the later experiments on propagation of detonations from a tube to an unconfined environment.

Hydrazine vapor was generated in a cylindrical 1 liter stainless steel vaporizer heated to a maximum temperature of 400 K and insulated with fiberglass. Liquid hydrazine (up to 45 ml) was injected into the vaporizer with a syringe. The connecting line between the vaporizer and the detonation tube was heated to prevent condensation of hydrazine.

The pressure of hydrazine vapor in the detonation tube was measured directly with a mercury manometer, one arm of which was topped with heated Dow Corning 705 silicone oil. The hydrazine vapor contacted only the surface of the heated silicone oil meniscus in the manometer tube. The manometer was calibrated for the asymmetrical load of the silicone oil prior to each run.

Detonation velocities and peak pressures were measured by four PCB 113A24 dynamic pressure transducers at 0.4 m intervals along the tube. The transducer signals were recorded by a NEFF digital data recording system at 3 μs intervals. Detonation cells were recorded on soot-coated aluminum foils, typically 81 cm long, 36 cm wide, and 0.5 mm thick.

For the studies of detonation propagation from a tube to an unconfined environment, an additional section of heated tube 0.5 m long and 150 mm in diameter was attached to the flange at the far end of the tube from the exploding bridge wire. The critical tube orifice was a short section of thick-walled aluminum tubing welded to an aluminum disk approximately 12 mm thick. The aluminum disk was sandwiched between the flanges joining the two sections of tube. The two pressure transducers closest to the initiation source were relocated so that one was midway along the new section of tube and the other was mounted on the end flange at the far end of this section from the orifice. These transducers measured the velocity of the detonation as it came out of the propagation tube and allowed a determination of whether the detonation had failed. Three orifices with internal diameters of 24.4, 45.9, and 73.5 mm were used. After problems with reignition were experienced in the initial tests, the tubes were remachined so that they were flush with the aluminum disk on the side downstream from the main tube.

Hydrazine Vapor Detonation Modeling

Theoretical Chapman-Jouguet (C-J) detonation states for the hydrazine decomposition reaction were calculated using the NASA Lewis, Gordon-McBride numerical code [Gordon and McBride (1976)]. The code predicts that the thermodynamic equilibrium products of hydrazine vapor detonation are almost entirely nitrogen and hydrogen,

$$N_2H_4(v) \longrightarrow N_2(g) + 2H_2(g) \quad \Delta H = -94 \text{ kJ/mole}$$

At ambient temperature, the principal products of hydrazine decomposition are nitrogen and ammonia,

$$N_2H_4(v) \longrightarrow 1/3\ N_2(g) + 4/3\ NH_3(v) \quad \Delta H = -155 \text{ kJ/mole}$$

but the ammonia is almost entirely dissociated at the temperature of the detonation.

The kinetics of the hydrazine vapor detonation was studied using the Zel'dovich-von Neumann-Doring (ZND) model described by Shepherd (1986). The detailed reaction mechanism and rate constant values of Miller et al. (1983) were used. This mechanism contains 98 reactions between 23 species and describes ammonia and hydrazine decomposition and oxidation. The number of species and reactions actually used was much smaller, because all oxygen-containing species were absent.

Temperature and species profiles are given in Fig. 1-3 for a hydrazine vapor detonation at 400 K and 1 atm pressure. The temperature profile shows a pronounced maximum and a long endothermic "tail" (Fig. 1). This unusual structure is associated with the production of ammonia, an exothermic process, and the dissociation of ammonia to form nitrogen and hydrogen, an endothermic process. The role of ammonia and the temperature maximum in the reaction zone has been noted by Heinrich (1964). A. Jost et al. (1963) reported that significant concentrations of ammonia were observed for long periods behind the detonation front in hydrazine vapor detonations. The slow approach to the equilibrium C-J state is clearly seen in Fig. 1-3.

The atypical reaction zone profile suggests that the standard phenomenology for estimating cell width, critical tube diameter, and initiation energy may not be valid for this type of detonation. Calculated reaction zone lengths Δ for hydrazine detonations at initial temperatures of 308-393 K and pressures of 0.67-100 kPa are shown in Fig. 4. The reaction zone lengths were determined by the point at which the Mach number (in a shock-fixed frame) reaches 0.75 [Shepherd (1986)].

The pressure dependence shown suggests that $\Delta \approx p^{-0.8}$. This is consistent with an overall reaction that is approximately second order in hydrazine. This is the case for the initiation reaction

$$N_2H_4 + M \longrightarrow 2NH_2 + M$$

which controls the initial decomposition if hydrazine alone is considered as a reactant, i.e., $M = N_2H_4$. Michel

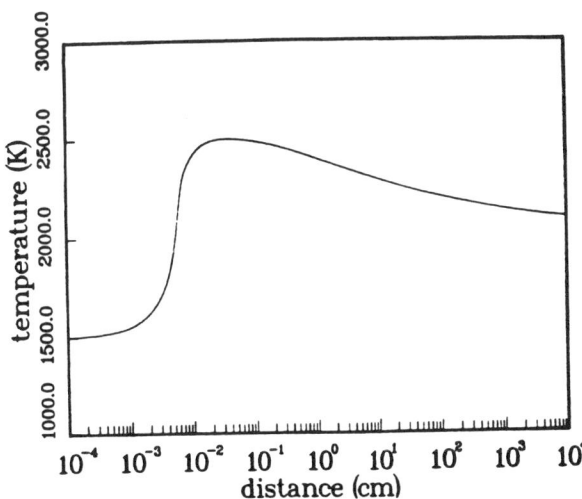

Fig. 1 Temperature profile computed with the ZND model for a hydrazine vapor detonation at initial conditions of 400 K and 101.3 kPa.

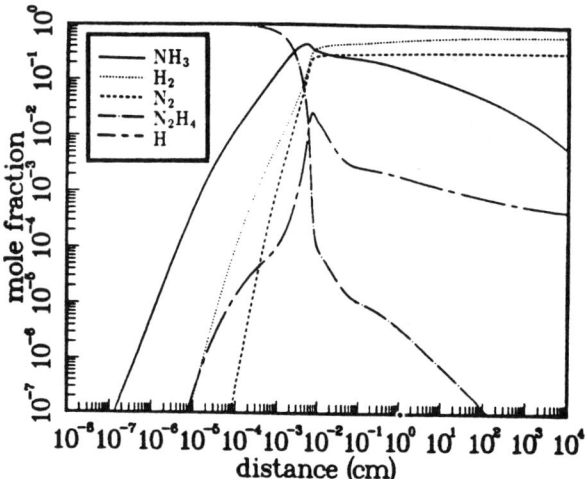

Fig. 2 Species profiles computed with the ZND model for a hydrazine detonation at initial conditions of 400 K and 101.3 kPa.

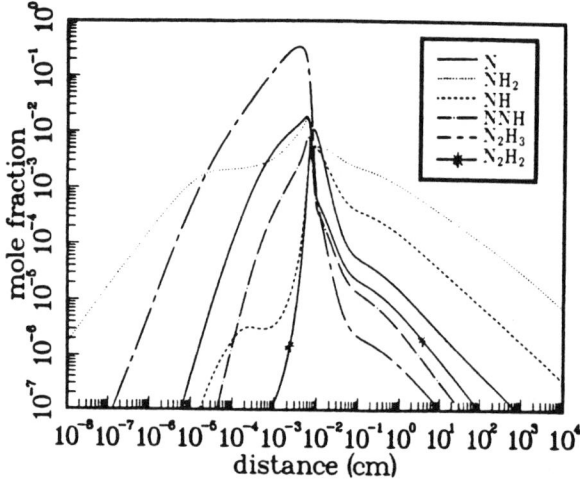

Fig. 3 Species profiles computed with the ZND model for a hydrazine detonation at initial conditions of 400 K and 101.3 kPa.

and Wagner (1965) and McHale et al. (1965) investigated the decomposition reaction under highly dilute conditions. Michel and Wagner concluded that, at a fixed reactant density, the rate constant for this reaction could be represented by

$$k = 10^{13} \exp(217,000/RT) \text{ s}^{-1}$$

indicating an activation energy of 217 kJ/mole.
By combining this rate constant with a first order
dependence on the total concentration, an approximate
induction time formula can be deduced for pure hydrazine
decomposition at high pressure

$$\tau = [N_2H_4]^{-1} 1.2 \times 10^{-17} \exp(217,000/RT) \text{ s}$$

where $[N_2H_4]$ is in mole/cm^3.
The induction times and reaction zone lengths computed
with this approximate formula were compared to the results
of the detailed kinetic computations. The approximate
induction time formula did not appear to be reliable.
Reaction zone lengths were overestimated by a factor of 2-
5. Computation of the effective activation energy from the
detailed kinetic mechanism yielded a value of 217 kJ/mole
at low pressure, in excellent agreement with the value
determined by Michel and Wagner. However, the effective

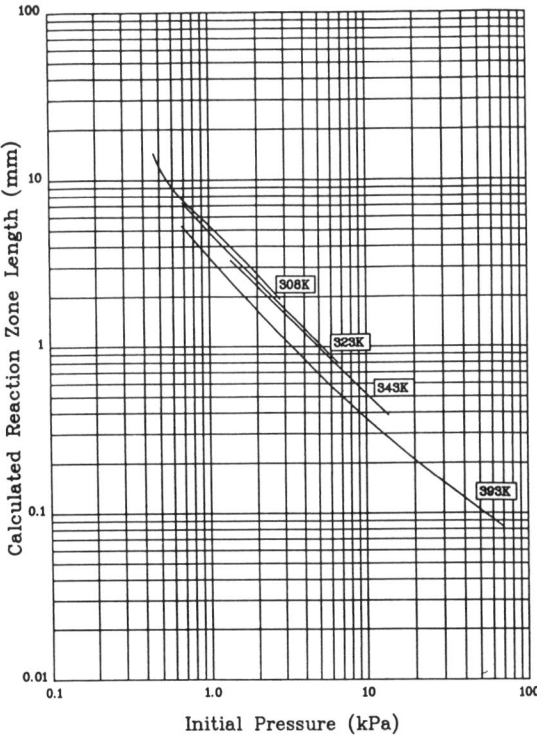

Fig. 4 Calculated reaction zone lengths for pure hydrazine vapor as a function of initial pressure.

activation energy decreases with increasing pressure; the value at 1 atm is 157 kJ/mole.

Results and Discussion

Experimental results were obtained for detonation velocities and peak pressures, detonation cell widths, and the number of cells required for detonation propagation from a tube to an unconfined environment. The results are compared with the theoretical calculations and discussed in terms of experience of explosions in hydrazine systems.

Detonation Velocity and Overpressure

Detonation velocities and overpressures were measured only in the early experiments using the oxygen-acetylene driver and the later experiments on detonation propagation. All measurements were conducted on hydrazine vapor at an initial temperature of 393 K. The experimental data were in reasonably good agreement with the calculated C-J values; values obtained using the oxygen-acetylene driver are given in Table 1. Detonation velocities were within 10% of the calculated values. W. Jost (1962) and Heinrich (1964) observed significant velocity deficits (8-21% below C-J), probably because boundary-layer effects were significant in their small tubes (less than 25 mm in diameter). A. Jost et al. (1963) measured detonation velocities within 4% of C-J in a larger 84 mm diameter tube, in good agreement with the present results.

The measured peak overpressures showed considerably more scatter, probably because of interference from the pressure transducers ringing at their natural frequency. Nevertheless, the peak overpressures generally lay somewhere between the calculated pressure for the C-J detonation and the calculated pressure for the von Neumann spike.

Table 1 Calculated and measured detonation velocities and overpressures for hydrazine vapor decomposition

Initial pressure kPa	P_{CJ} MPa	P_{v-n} MPa	P_{max} MPa	V_{CJ} m/s	V_{meas} m/s	V deficit, %
60.0	1.64	3.49	1.5-2.3	2513	2406	-4.3
32.8	0.89	1.89	1.2-3.1	2512	2524	+0.5
20.0	0.55	1.16	0.5-1.1	2511	2395	-4.6

Detonation Cell Widths

Detonation cells were obtained for hydrazine vapor at initial pressures from 67 kPa down to the lowest test pressure of 0.46 kPa and initial temperatures from 393 K down to 308 K. The cell structure was highly regular. Single head spin was not reached even at the lowest test pressure.

The results are shown in Fig. 5. The curves drawn through the data are obtained from the reaction zone length Δ, kinetically calculated using the ZND model; the empirical relation

$$\lambda = A \Delta, \text{ where } A = 29$$

was selected to give the best fit to the experimental data. The value of 29 for the empirical constant A is in reasonable agreement with past experience. The value of A calculated using the ZND model appears to increase slowly as the sensitivity to detonation (as measured by reaction length or cell width) increases. For stoichiometric hydrogen-air mixtures, A is found experimentally to be about 22 [Shepherd (1986)], and hydrazine vapor appears to be more sensitive than stoichiometric hydrogen-air.

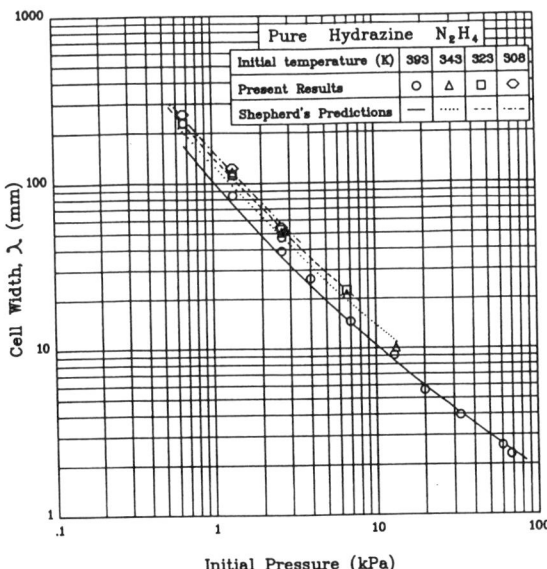

Fig. 5 Variation of detonation cell width, with initial pressure/vapor for pure hydrazine at different initial temperatures.

Interestingly, the kinetic calculations of Westbrook and Urtiew (1983) using the constant volume model give a value for A of 29 for hydrogen and many hydrocarbons in air or oxygen. However, these authors determined the reaction length by the point of maximum temperature gradient in the reaction zone; the ZND model predicts that this length is significantly shorter than that based on a Mach number of 0.75 [Shepherd (1986)].

Using A = 29, the agreement between experiment and theory is excellent over the entire range of initial temperatures and pressures tested. The theory correctly predicts that the cell width increases with decreasing pressure or decreasing temperature. The measured cell widths may be starting to drift above the theoretical curve at pressures below 2.5 kPa. The small deviation possibly results from inaccuracies in the kinetic model, but an alternative explanation is that the hydrazine vapor is not pure because decomposition before detonation initiation is significant at these very low pressures. The results show that the ZND model is valid for hydrazine vapor detonations despite the atypical reaction zone profile.

The measured cell widths show that hydrazine vapor is extremely sensitive to detonation. The cell size data for

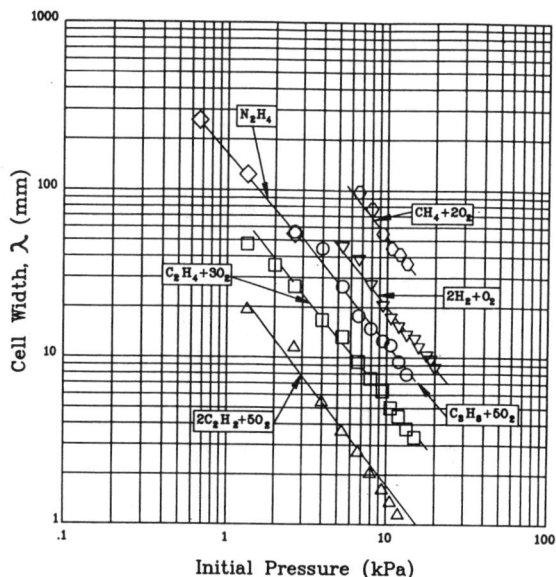

Fig. 6. Relative detonation sensitivity of hydrazine and fuel-oxygen systems.

hydrazine decomposition at 308 K are compared to data for fuel-oxygen systems at ambient temperature in Fig. 6. The results indicate that pure hydrazine vapor is more sensitive to detonation than stoichiometric hydrogen-oxygen mixtures and roughly similar in sensitivity to stoichiometric propane-oxygen mixtures.

The approximate critical energy for direct detonation of an unconfined hydrazine vapor cloud can be calculated using the equation

$$Ec = 431 \rho_o V_{CJ}^2 I \lambda^3$$

where ρ_o is the initial density of the unburned hydrazine, V_{CJ} the C-J detonation velocity, and I the energy integral for hydrazine vapor [Guirao et al. (1982)].

The energy integral I for pure hydrazine vapor was calculated to be 0.4973 using the similarity solution of Taylor (1950) for the behavior of blast waves initiated by a strong explosion. The critical energy for direct detonation of hydrazine vapor at 393 K was calculated using this energy integral, the measured cell size, and the calculated C-J velocities. The calculated critical energy ranged from approximately 8 J at a pressure of 1 atm to roughly 100 kJ at 0.67 kPa.

Detonation Propagation from Tube to Unconfined Environment

The detonation propagation experiments were limited in scope and designed only to determine if the 13 λ relation holds for hydrazine vapor detonations. All tests were conducted at an initial temperature of 393 K; the initial pressure of hydrazine vapor was up to 120 kPa. Experiments were conducted with the three critical tube orifices mounted inside the detonation tube. Failure of a detonation to propagate into the unconfined environment was clearly indicated by a significant decrease in the velocity of the detonation front recorded by the pressure transducers downstream from the orifice. Successive tests were conducted at lower pressures until the detonation failed to propagate; additional tests were then conducted to accurately determine the minimum pressure for propagation. The number of cell widths in the critical tube orifice at this minimum pressure was estimated from the orifice internal diameter and the cell width data in Fig. 5.

The results are shown in Fig. 7. The minimum critical tube diameter for detonation propagation was found to be

15.0 λ for the two smaller propagation tubes and 8.4 λ for the largest tube. The results for the two smaller tube agree with the $d_c = 13\ \lambda$ relation to within the experimental uncertainties inherent in this type of study (mostly in the measurement of cell width λ). The value of 8.4 λ for the largest tube is significantly lower than in the other tests. A probable explanation is that the ratio of the diameters of the main detonation tube to the propagation tube was approximately two. The rule of thumb for rigid confinement is that the chamber to critical tube diameter ratio must be at least three to eliminate the possibility of a false apparent detonation transmission by wall-induced reflected reinitiation. This ratio was exceeded for the two smaller tubes. Thus, in the largest tube at tube widths less than 13 λ, the detonation may have failed as it came out of the propagation tube, but reignited as it wrapped around and contacted the walls of the main tube. The time delay for reignition would have been too short to affect the measured detonation velocities.

Thus, with this one caveat, it appears that the empirical 13 λ relationship holds for hydrazine decomposition. This result combined with those from other studies strongly suggests that the relation is geometric,

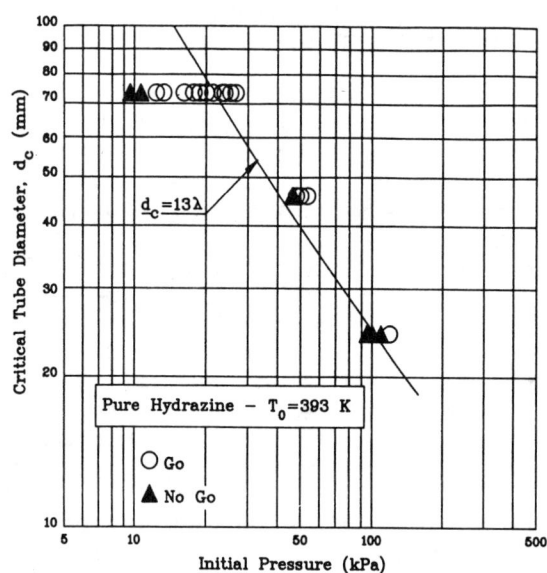

Fig. 7 Correlation of critical tube diameter d_c with the empirical relation $d_c = 13\ \lambda$ for pure hydrazine.

as is the λ/π relation for propagation in detonation tubes, and is not related to the chemistry of the detonation reaction other than by cell width. It is also highly probable that the other empirical relationships [critical diameter = 10 λ for propagation out of a confined channel with a square cross section; critical diameter = 6.5 λ for a totally unconfined detonation; critical diameter = 5 λ for a detonation confined by a single planar surface (Lee et al. 1982)] hold for hydrazine vapor detonations.

Hydrazine System Safety

The experimental results show that hydrazine vapor detonates readily and indicate that the critical energy for detonation is small at ambient pressure and above. They suggest that the violent explosions sometimes observed in hydrazine systems may result from vapor detonation or from transition to detonation. The structural damage observed cannot usually be explained by the C-J pressure for the initial system hydrazine vapor pressure, but could be explained by a transition mechanism in which the hydrazine vapor pressure is well above the initial pressure when detonation occurs.

The detonation cell width for hydrazine vapor at ambient pressure and 393 K is quite small; the value extrapolated from Fig. 5 is 1.8 mm. Hydrazine vapor can be subjected to pressures orders of magnitude above ambient in aerospace systems. For example, rapid deceleration of moving liquid hydrazine can generate pressures above 70 MPa. The cell width for vapor bubbles compressed to such pressures should be extremely small and the critical energy and dimensions for detonation should also be small. Thus, it is quite possible that the violent explosions observed in hydrazine systems result from the high sensitivity of the vapor to detonation.

Conclusions

Detonation studies were conducted on pure hydrazine vapor. The results show that hydrazine vapor is very sensitive to detonation. Detonation velocities and pressures are in acceptable agreement with those calculated from the one-dimensional Chapman-Jouguet theory.

Detonation cell widths λ were measured and found to correlate very well with chemical reaction zone lengths Δ calculated using the three-dimensional ZND model, if an empirical value of 29 for A is used in the relation

$\Lambda = A \lambda$. This empirical value is in reasonable agreement with values of A for detonation of hydrogen and hydrocarbons in air and oxygen. Detonation cell widths for hydrazine vapor were similar to those for stoichiometric propane-oxygen mixtures. The cell widths increase with decreasing temperature, as predicted by the ZND model.

The limited experiments on critical tube diameters for propagation of a detonation into an unconfined space showed that the 13 λ relationship holds reasonably well for hydrazine vapor detonations. This result suggests that all the empirical relations between confinement and cell size are purely geometric and should hold for all gas-phase detonations regardless of the nature of the reacting gases.

The results of these studies suggest that hydrazine vapor detonations may be responsible for the violent explosions that occasionally cause severe damage to hydrazine systems.

References

Adams, G.K. and Stocks, G.W. (1953) The combustion of hydrazine. Fourth Symposium (International) on Combustion, pp. 239-248. Williams and Wilkins, Baltimore, MD.

Benz, F.J., Long, T.L., and Weary, D.P. (1984) Explosive hydrazine decomposition due to rapid gas compression (adiabatic heating). JANNAF Safety and Environmental Protection Subcommittee Meeting, p. 21, CPIA Publ. 408. NASA/White Sands Test Facility, Las Cruces, NM.

Benz, F.J. and Pedley, M.D. (1986) A comparison of the explosion hazards of hydrazine and methylhydrazine in aerospace environments. Paper presented at JANNAF Propulsion Meeting, New Orleans.

Briles, O.M. and Hollenbaugh, R.P. (1978) Adiabatic compression testing of hydrazine. AIAA Paper 78-1043.

Brownfield, C.D. and Korb, L.J. (1984) APU injector tube failure analysis (STS-9). Rockwell International.

Eberstein, I.J. and Glassman, I. (1965) The gas-phase decomposition of hydrazine and its methyl derivatives. Tenth Symposium (International) on Combustion, pp. 365-374. The Combustion Institute, Pittsburgh, PA.

Fritchman, T.T. and Benz, F.J. (1980) The exothermicity of liquid hydrazine exposed to various auxiliary power unit materials. NASA TR-226-001.

Furno, A.L., Imhof, A.G., and Kuchta, J.M. (1968) Effect of pressure and oxidant concentration on autoignition temperatures of selected combustibles in various oxygen and nitrogen tetroxide atmospheres. J. Chem. Eng. Data 13(2), 243-249.

Furno, A.L., Martindill, G.H., and Zabetakis, M.G. (1962) Limits of flammability of hydrazine-hydrocarbon vapor mixtures. J. Chem. Eng. Data 7(3), 375-376.

Genich, A.P., Zhurnov, A.A., and Manelis, G.B. (1974) Decomposition of hydrazine behind reflected shock waves at high pressures. Zh. Fiz. Khim. 48(3), 728-729.

Gilbert, M. (1958) The hydrazine flame. Combust. and Flame 2(2), 137-148.

Gordon, S. and McBride, B.J. (1976) Computer program for calculation of complex chemical equilibrium compositions, rocket performance, incident and reflected shocks, and Chapman-Jouguet detonations. NASA SP-273 Interim Rev. NASA Lewis Research Center, Cleveland, OH, March 1976. N78-17724.

Gray, P. and Holland, S. (1970) Effect of isotopic substitution on the decomposition flame of hydrazine. Combust. and Flame 14(1), 203-215.

Gray, P. and Lee, J.C. (1954) The combustion of gaseous hydrazine. Trans. Faraday Soc. 50, 719-728.

Gray, P. and Lee, J.C. (1955) Explosive decomposition and combustion of hydrazine. Fifth Symposium (International) on Combustion, pp. 692-699. Reinhold, New York.

Gray, P. and Lee, J.C. (1959) Recent studies of the oxidation and decomposition flames of hydrazine. Seventh Symposium (International) on Combustion, pp. 61-67. The Combustion Institute, Pittsburgh, PA.

Gray, P., Lee, J.C., Leach, H.A., and Taylor, D.C. (1957) The propagation and stability of the decomposition flame of hydrazine. Sixth Symposium (International) on Combustion, pp. 225-263. Reinhold, New York.

Gray, P. and Spencer, M. (1963) Studies of the combustion of dimethyl hydrazine and related compounds. Ninth Symposium (International) on Combustion, pp. 148-157. The Combustion Institute, Pittsburgh, PA.

Guirao, C.M., Knystautas, R., Lee, J.H., Benedick, W., and Berman, M. (1982) Hydrogen-air detonations. Nineteenth Symposium (International) on Combustion, pp. 583-590. The Combustion Institute, Pittsburgh, PA.

Heinrich, H.J. (1964) Propagation of detonations in hydrazine vapor. Z. Phys. Chem. 42, 149-165.

Jost, A., Michel, K.W., Troe, J., and Wagner, H.G. (1963) Detonation and shock tube studies of hydrazine and nitrous oxide. ARL 63-157, Wright Patterson AFB, OH.

Jost, W. (1962) Investigation of gaseous detonation and shock wave experiments with hydrazine. ARL 62-330, Wright Patterson AFB, OH.

Knystautas, R., Lee, J.H., Moen, I.O., Guirao, C.M., Urtiew, P., Bjerketvedt, D., Rinnan, A., and Fuhre, K. (1981) Determination of critical tube diameter for acetylene-air and ethylene-air mixtures. Rept. CMI 8034030-3, Chr. Michelsen Inst., Fantoft, Norway.

Knystautas, R., Guirao, C., Lee, J.H., and Sulmistras, A. (1984) Measurements of cell size in hydrocarbon-air mixtures and predictions of critical tube diameter, critical initiation energy and detonability limits. Progress in Astronautics and Aeronautics: Dynamics of Shock Waves, Explosions, and Detonations (edited by J.R. Bowen, N. Manson, A.K. Oppenheim, and R.I. Soloukhin). Vol. 94, pp. 23-37. AIAA, New York.

Lee, J.H.S. (1984) Dynamic parameters of gaseous detonations. Ann. Rev. Fluid Mech. 16, 311-336.

Lee, J.H., Knystautas, R., Guirao, C.M., Benedick, W.A., and Shepherd, J.E. (1982) Hydrogen-air detonations. Proc. of Second Intl. Workshop on the Impact of Hydrogen on Water Reactor Safety, (edited by M. Berman), pp. 961-1005. SAND82-2456, Sandia National Laboratories, Albuquerque, NM.

Maclean, D.I. and Wagner, H.G. (1967) The structure of the reaction zones of ammonia-oxygen and hydrazine-decomposition flames. Eleventh Symposium (International) on Combustion, pp. 871-879. The Combustion Institute, Pittsburgh, PA.

McHale, E.T., Knox, B.E., and Palmer H.B. (1965) Determination of the decomposition kinetics of hydrazine using a single-pulse shock tube. Tenth Symposium (International) on Combustion, pp. 341-351. The Combustion Institute, Pittsburgh, PA.

Meyer, E., Olschewski, H.A., Troe, J., and Wagner H.G. (1969) Investigation of N_2H_4 and H_2O_2 decomposition in low and high pressure shock waves. Twelfth Symposium (International) on Combustion, pp. 345-356. The Combustion Institute, Pittsburgh, PA.

Michel, K.W. and Wagner, H.G. (1965) The pyrolysis and oxidation of hydrazine behind shock waves. Tenth Symposium (International) on Combustion, pp. 353-364. The Combustion Institute, Pittsburgh, PA.

Miller, J.A., Smooke, M.D., Green, R.M., and Kee, R.J. (1983) Kinetic modeling of the oxidation of ammonia in flames. Combust. Sci. Tech. 34, 149-176.

Mitrofanov, V.V. and Soloukhin, R.I. (1964) The diffraction of multifront detonation waves. Sov. Phys.-Dokl. 9(12), 1055-1058.

Moberly, W.H. (1962) Shock tube study of hydrazine decomposition. J. Phys. Chem. 66, 366-368.

Perlee, H.E., Imhof, A.C., and Zabetakis, M.G. (1962) Flammability characteristics of hydrazine fuels in nitrogen tetroxide atmospheres. J. Chem. Eng. Data 7(3), 377-379.

Pfeffer, H. A. (1976) Hydrazine monopropellant technology for satellite auxiliary propulsion. Rept. 98037 (N77-10158), European Space Research Organization, Noordwijk, Netherlands.

Sawyer, R.F. and Glassman, I. (1967) Gas-phase reactions of hydrazine with nitrogen dioxide, nitric oxide and oxygen. Eleventh Symposium (International) on Combustion, pp. 861-869. The Combustion Institute, Pittsburgh, PA.

Sayer, C.F. and Southern, G.R. (1974) The effect of ammonia on the starting characteristics of hydrazine/Shell 405 thrusters. Rept. RPE-TN-256, Rocket Propulsion Establishment, Wescott, England.

Schmidt, E.W. (1984) Hydrazine and Its Derivatives: Properties, Applications, Wiley-Interscience, New York.

Scott, F.E., Burns, J.J., and Lewis, B. (1949) Explosive properties of hydrazine. Rept. Inv. 4460, U.S. Bureau of Mines.

Shepherd, J.E. (1986) Chemical kinetics of hydrogen-air-diluent detonations. Progress in Astronautics and Aeronautics: Dynamics of Explosions (edited by J.R. Bowen, J.C. Leyer, and R.I. Soloukhin). Vol. 106, pp. 263-293. AIAA, New York.

Taylor, G.I. (1950) The formation of a blast wave by a very intense explosion, I: Theoretical discussion. Proc. R. Soc. London A201, 159-174.

Vander Wall, E.M. et al. (1971) Propellant-material compatibility. Final Rept. AFRPL-TR-71-41 (AD 736464) Aerojet Liquid Rocket, Sacramento, CA.

Westbrook, C.K. and Urtiew, P.A. (1983) Chemical kinetic prediction of critical parameters in gaseous detonations. Nineteenth Symposium (International) on Combustion, pp. 615-623. The Combustion Institute, Pittsburgh, PA.

Applicability of the Inverse Method to the Determination of C-J Parameters for Gaseous Mixtures at Elevated Pressures

P. A. Bauer,[*] P. Vidal,[†] N. Manson,[‡] and O. Heuzé[§]
Laboratoire d'Energétique et de Détonique, Poitiers, France

Abstract

In order to provide data on the main detonation parameters, such as pressure, density, and sound velocity, without the use of the equation of state of the detonation products, a preliminary study was undertaken. It was aimed at the relevance of the inverse method in a field of initial pressures - i.e., within a range of 40 - 80 bars - where, to date, it had never been applied. The suitability of this method required that the basic assumptions of the Chapman-Jouguet theory (i.e., a one-dimensional, stable, and self-sustained detonation propagating) should meet. These conditions together with the validity of the experimental data are discussed. It presents the comparison between results provided by an a priori calculation using the QUATUOR code based, for the present study, on the Percus-Yevick equation of state and those obtained by the inverse method. The latter involves data on the thermophysical parameters of the initial state that were obtained in an analytical way, using the Redlich-Kwong real-gas equation of state.

Copyright © 1988 by the American Institute of Aeronautics and Astronautics, Inc. All rights reserved.
[*] Associate professor
[†] Research Engineer
[‡] Professor
[§] Research Engineer

Introduction

So far, numerous equations of state (EOS) of real gases as well as thermodynamic data at high temperature (i.e., enthalpy, entropy, equilibrium constants of ideal gas) are available allowing an a priori calculation by means of the Chapman-Jouguet (C-J) theory of the detonation velocity D in both gaseous and condensed explosives. For high initial pressures, the Percus-Yevick EOS (Bauer 1985; Heuzé 1985) yields this parameter and it turns out to be in agreement with experimental data over a range of 5-150 bars. This EOS is part of the QUATUOR thermochemical code described in previous papers (see, for example, Heuzé et al. 1985).

However, to date, unlike the detonation velocity D that may be measured in an accurate way, the detonation pressure p does not lend itself to a reliable measurement over such a range of pressures. This eliminates the opportunity of checking more thoroughly the validity of the Percus-Yevick EOS by comparing the a priori calculation, not only for D but for p, density ρ, etc., to available experimental data.

Nonetheless, the validity of the a priori calculation may be checked when comparing its results with the data provided by the inverse method (Manson 1958, 1960; Wood and Fickett 1963).

Such a procedure had been successfully undertaken in the case of gaseous explosives at atmospheric or subatmospheric pressures (Brossard and Manson 1958; Pujol 1968; Brochet et al. 1966). However, the work of Brossard (1970), as well as that of Brochet et al. (1970), on divergent spherical detonation raised questions about which experimental velocities should be taken into account. Should it be that obtained when extrapolating $D(\phi^{-1})$ toward $\phi^{-1} = 0$ (see Appendix) or that D_s provided by a measurement in spherical geometry once the detonation attained its constant velocity within the uncertainty of the measurement (0.4%)? Since the latter may not be obtained at elevated initial pressures, would the results be coherent when derived from velocities measured in an appropriate i.d. tube?

The present paper provides results obtained by the use of this method in the case of an isometric mixture $(2 - x)$ $CH_4 + 0.5(1 + x) C_2H_6 + 0.5x\ H_2 + 27.35$ air at ambient temperature and initial pressures of 40-80 bars.

In what follows: 1) we specify the prominent conditions under which experimental data were obtained as well as present the numerical procedure used to calculate the thermophysical properties of the reactive mixture, 2) we describe the experiments and present their results, and 3) we present the results obtainable using the Percus-Yevick

EOS with standard thermodynamic data for the a priori calculation. These results and their comparison with those of the a priori calculation and those derived from the inverse method are discussed in the final part of the paper.

Data Required for Inverse Method

The determination of the C-J detonation parameters of a homogeneous explosive by means of the inverse method requires knowledge of the variation of the detonation velocity D and of several thermophysical properties as a function of two parameters (Manson 1960; Pujol et al. 1966) that specify its initial state, namely the initial pressure p_o, temperature T_o, density ρ_o, and a parameter x that defines the composition of this explosive in an isometric mixture. If, for instance, the initial state is defined by T_o, p_o, and x, as was done in this study, then:
1) D should be measured as a function of p_o and x for a constant T_o, whence

$$n_{Tx} = \frac{\partial \text{Log } D}{\partial \text{Log } p_o})_{T,x}$$

$$k_{TP} = \frac{\partial \text{Log } D}{\partial \text{Log } x})_{T,p}$$

2) Density ρ_o, should be expressed in terms of the volumetric expansion coefficient α_o, sound velocity a_o, specific heats ratio γ_o, as well as the coefficients,

$$\beta_x = -\frac{1}{\rho_o}\frac{d\rho_o}{\partial x})_{p,T} \text{ and } h_x = \frac{\partial h_o}{\partial x})_{T,p}$$

where h_o is the enthalpy.

In order to determine the latter, the Redlich-Kwong EOS (Kemp et al. 1975) was used. Its validity in the case of air was proved by these authors. We assumed that it could be applied to multicomponent mixtures provided that a mixing law is known. For that purpose, we assumed the following rule:

$$P_c = \sum_i X_i P_{c_i}$$

$$\frac{T_c}{P_c} = \sum_i X_i \frac{T_{ci}}{P_{c_i}}$$

where p_{ci} and T_{ci} are the critical pressure and temperature of each component i in molar proportions X_i in the mixture. These data are provided by Kemp et al. (1975). This rule may be regarded as reliable since the major component in our mixtures was air in a molar fraction of more than 90%. The whole set of data in addition with the equations of the C-J theory and of the inverse method allow to solve the problem, thus providing the detonation parameters of the mixture with exception of the temperature.

Experimental Conditions

The main purpose of our experiments, according to the concepts presented in the previous section, was the measurement of the detonation velocities in isometric mixtures at initial pressures of $40 < p_o < 80$ bars. Furthermore, in order to collect data on mixtures that have a low ability to detonate at a lesser initial pressure (Bauer et al. 1986b), we chose isometric mixtures involving methane, i.e.,

$$(2 - x)CH_4 + 0.5\left((1 + x)C_2H_6 + x\ H_2\right) + 27.35\ air$$

In view of the relevance of the C-J theory and the inverse method, discussed in the Appendix, the experiments were performed in a 20 mm i.d. tube that provides greater safety with respect to mechanical stresses (Presles et al. 1985).
The experimental apparatus and procedure have been described previously (Bauer et al. 1984b, 1986b).
Test mixtures were prepared by the use of a weighing system that has an uncertainty of 0.1 g up to 70 kg. A tank with an appropriate mixing device was filled with the components. The composition and homogeneity of the mixtures were checked by chromatography prior to each set of experiments. The latter were performed at ambient temperature, $T_o = 290 \pm 3$ K. A 2 m long and 20 mm i.d. tube (see Fig. 1) was equipped with two sets of measurements — i.e., three successive ionization gages at a distance of $L_1 = L_2 = 500 \pm 0.5$ mm one from the other as shown on Fig. 1. Each was connected to the gates of two chronometers. The first pickup was located at a distance of 1 m from the

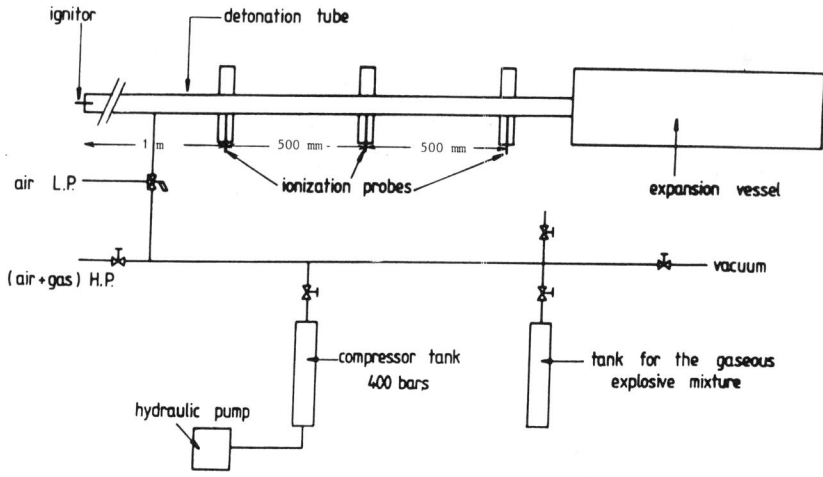

Fig. 1 Experimental device.

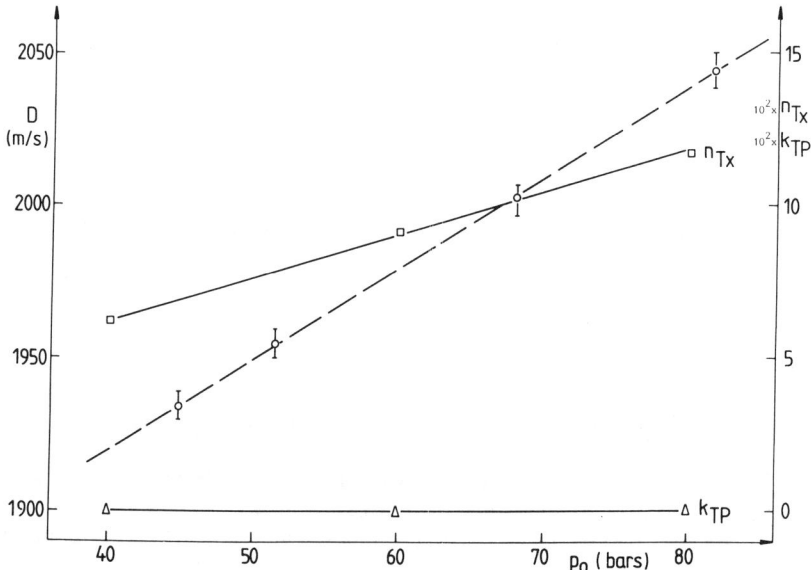

Fig. 2 Detonation velocities and derivatives n_{Tx} and k_{TP} for mixture $2\ CH_4 + 0.5\ C_2H_6 + 27.35$ air ($x = 0$).

ignition device, namely, detonator 8. The onset of the detonation is, thus, expected to occur in a straightforward way. The comparison of the two successive values of D measured along L_1 and L_2 allowed a check of the stability of the wave. We observed that the velocity was constant over the propagation distance since, both values were very

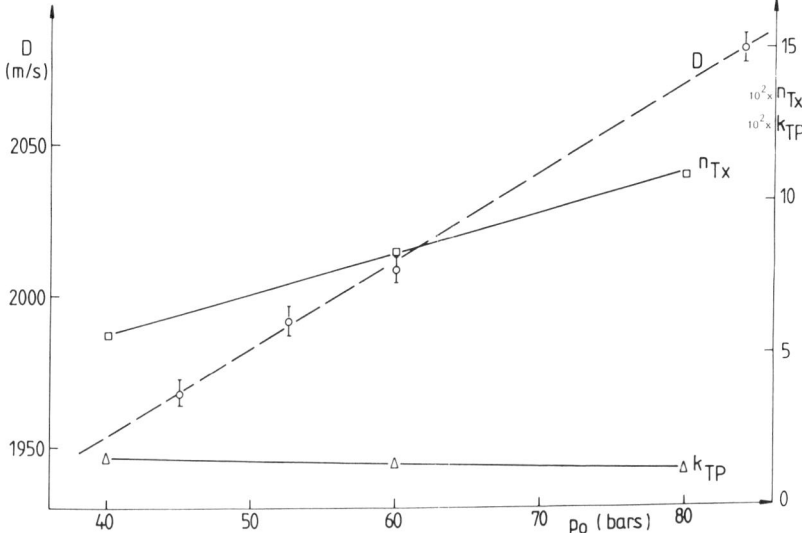

Fig. 3 Detonation velocities and derivatives n_{Tx} and k_{TP} for mixture $1.5\ C_2H_6 + H_2 + 27.35$ air (x = 2).

close and within the uncertainty of the measurements - i.e., 0.4% or so.

As to the numerical aspect of the problem, the parameters that were required for solving the equations were derived from a set of experimental detonation velocities measured at initial pressures of 40-80 bars corresponding to x = 0 (i.e., a LNG-like mixture) and x = 2. These two sets of values, together with the corresponding n_{Tx} and k_{Tp}, are presented on Figs. 2 and 3 respectively.

A linear analysis based on the least square method checks the behavior of the data. Both sets of detonation velocity data are fairly close but, nonetheless, the discrepancy between them is of the order of the threefold experimental uncertainty.

Results and Discussion

Comparison with the a priori Calculation

The comparison between the experimental results and those derived from a C-J calculation based on the Percus-Yevick EOS is reported in Tables 1 (x = 0) and 2 (x = 2).

Table 1 Experimental (in a 20 mm i.d. tube) and a priori
Calculation of Detonation Velocity in $2\ CH_4 + 0.5\ C_2H_6 + 27.35$ air ($x = 0$)

p_o, bar	45	51.5	68	81.5
Experimental D, m/s	1935	1955	2005	2045
Calculated D_{CJ}, m/s	1960	1975	2020	2055
$(D_{CJ}-D)/D_{CJ}$, %	1.3	1.0	0.7	0.5

Table 2 Experimental (in a 20 mm i.d. tube) and a priori
Calculation of Detonation Velocity in $1.5\ C_2H_6 + H_2 + 27.35$ air ($x = 2$)

p_o, bar	45	52.5	60	84
Experimental D, m/s	1970	1995	2010	2080
Calculated D_{CJ}, m/s	1985	2005	2025	2090
$(D_{CJ}-D)/D_{CJ}$, %	0.8	0.5	0.7	0.5

It obviously appears that within this level of initial pressure, and more precisely beyond p_o = 50 bars, the discrepancy between calculated and experimental data remain close to the uncertainty limits. Although this does not allow a physical interpretation of the trend of the variation of $|D_\infty - D_{CJ}|$ when the initial pressure increases, it may highlight the approximation $D_\infty \cong D_{CJ}$. Actually, this may, to some extent, be explained by the diameter effect that might weaken at elevated initial pressures.

This result has been reported in previous papers (see, for example, Bauer 1985). In the case where methane-ethane and air are involved and, unlike previous results (Bauer et al. 1986b) where the same type of mixture led to different values of the velocity on the two sets, the present experiments turned out to yield coherent values on both sets.

These results confirm our previous conclusions (Heuzé et al. 1985) about the ability of the Percus-Yevick EOS for the evaluations of at least one detonation parameter, namely, the velocity.

Here again, a question raises as to the validity of the experimental data since the evolution of D vs ϕ^{-1} as discussed in the Appendix exhibits a maximum value of the order of 1-1.5% greater, at $p_o \leq 0.5$ atm than that of the spherical mode. This maximum, according to Desbordes et al. (1983) should be located at $\phi^{-1} = 20\ m^{-1}$. Since this corresponds to a somewhat greater tube diameter than ours (50 m^{-1}), one may expect that the data obtained in our tube

C-J PARAMETERS FOR GASEOUS MIXTURES

Table 3 Pressure Density and Sound Velocity provided by a priori calculation based on the Percus-Yevick EOS and inverse method

Composition parameter x	0	0	2	2
Initial pressure, bar	40	80	40	80
p, bar				
Inverse method				
Without approx.	710	1585	734	1626
Approx. $1_{vx}=0$	751	1612	780	1661
a priori calc.	763	1618	789	1675
ρ, kg/m^3				
Inverse method				
Without approx.	76	152	76	152
Approx. $1_{vx}=0$	79	154	80	154
a priori calc.	78	152	78	152
a, m/s				
Inverse method				
Without approx.	1169	1254	1188	1266
Approx. $1_{vx}=0$	1123	1240	1137	1248
a priori calc.	1126	1227	1141	1244

lie not too far from those corresponding to a spherical mode. Therefore, we are justified in presuming that the inverse method may be applied to the present case.

Application of Inverse Method

The results obtained when applying the inverse method are reported in Table 3. They are related to both the initial pressures at each side of the range that was studied in this paper. Their values lend themselves in a satisfactory way to comparison with the a priori calculation. Indeed, the detonation pressure derived from this experimental procedure is in good agreement with the corresponding calculated value. Use of the ideal-gas EOS instead of that of Redlich-Kwong turned out to yield less satisfactory results, since it caused the discrepancy to become larger by a factor of 4%.

The discrepancy never exceeds 3% in the case where the highest initial pressure (i.e., p_o = 80 bars) is involved. For p_o = 40 bars, the detonation pressure is in a lesser agreement. This initial pressure might be regarded as the threshold of the diameter effect if one refers to the result of the comparison between the experimental and calculated C-J detonation velocities. Nonetheless, these

Table 4 Pressure Density and Sound Velocity in a
Stoichiometric C_2H_6/Air Mixture

P_o, bar	40	50	60
p, bar			
Inverse method[a]	808	1030	1258
a priori calc.	794	1008	1228
Rel. difference, %	2	2.2	2.4
ρ, kg/m³			
Inverse method[a]	82	102	122
a priori calc.	81	100	119
Rel. difference, %	1.2	2	2.5
a(m/s)			
Inverse method[a]	1141	1171	1201
a priori calc.	1130	1157	1183
Rel. difference, %	1	1.2	1.5

[a] Approximation $l_{vx} = 0$.

Table 5 Pressure Density and Sound Velocity in a C_2H_4 +
12.74 Air Mixture

P_o, bar	20	30	40
p, bar			
Inverse method[a]	410	625	851
a priori calc.	406	620	840
Rel. difference, %	1	1	1
ρ, kg/m³			
Inverse method[a]	42	62	82
a priori calc.	41	61	80
Rel. difference, %	2	2	2
a, m/s			
Inverse method[a]	1111	1139	1170
a priori calc.	1109	1138	1166
Rel. difference, %	0.2	0.1	0.3

[a] Approximation $l_{vx} = 0$.

results provide a reliable set of data and one may expect that the use of an EOS other than Percus-Yevick would lead to a greater discrepancy, since the calculated detonation velocities would not be as close to the experimental values as they are at present.

At that stage, one may make a comment on the simplification that was proposed by Manson (1960). This author discussed the validity of the assumption according to which $l_{vx} = (\partial LnD/\partial LnT)_{vx} = 0$ - i.e., detonation velocity D depending solely upon initial density ρ_o. Actually, it appears that such an assumption may be proposed in the present study. In any case, in addition to

its simplicity, this approximation allows the use of the inverse method in the case where nonisometric mixtures are involved.
These results are reported in Table 4 for ethane-air and Table 5 for ethylene-air. The calculation carried out with the use of the inverse method is based on the detonation velocities measured in a 55 mm i.d. tube. The same calculation based on the detonation velocities measured in a 20 mm i.d. tube is not yielding significantly different results. To some extent, this shows that the detonation velocities used in this study may be regarded as reliable. In the case of isometric mixtures, the results with this approximation are presented in Table 3. Luckily, they turn out to be closer to the a priori calculation. One must, however, be cautious not to raise a too rapid conclusion, since this better agreement may be explained by a greater level of uncertainty when operating without approximation since it requires twice as many experiments or the unknown still remaining as to the validity of the EOS.

Conclusion

Present study shows that, as expected, the inverse method may be applied in a different range of density than used in the past. However, this requires accurate knowledge of the parameters of the initial state of the mixture and evidence that the C-J assumption is valid - in other words, whether or not two-dimensional effects such as a tube diameter are the stumbling block of this method. At this stage of our preliminary investigations, the following conclusions may be drawn:
1) Detonation velocities measured in a 20 mm i.d. tube are in close agreement with the calculated values derived from the QUATUOR code based on the Percus-Yevick EOS. This confirms previous results.
2) Detonation parameters evaluated by the inverse method are in a satisfactory agreement with the results of a priori calculations. In other words, unlike when the detonation products are assumed to behave as ideal gases, the lack of an appropriate EOS should not form an obstacle for providing these data at an elevated range of initial pressures.
3) One may reasonably state that the diameter effect would presumably play a moderate role beyond p_Q = 40 bars. The evidence for such a statement requires that the experiments be conducted in larger and smaller size tubes.
4) The method involving an analytical procedure used here to evaluate the required thermophysical parameters of the initial state may be regarded as satisfactory.

The simplified formulation (i.e., $1_{vx}=0$) turns out to be very useful since it allows the calculation of the detonation parameters in nonisometric mixtures. Obviously, it may be regarded as an efficient tool for the assessment of the detonation parameters on the basis of only detonation velocity measurements as a function of the initial pressure only. This provides wide applications of the inverse method.

Appendix

As in the C-J theory, the basic assumptions of the inverse method are that the detonation front is self-sustained, that is, it has "forgotten" the initiation process, propagates in a stable way at a constant velocity, and is one-dimensional.

One must point out that the comparison between calculated and experimental detonation velocities requires that the latter should be a function only of initial pressure and temperature. In other words, we assume a negligible effect of the confinement in terms of diameter of the tube, wall roughness, etc.

In all our experiments, the detonation velocity, irrespective of the initial pressure and temperature of the mixture, was observed to be constant within the experimental accuracy - i.e., $\pm 0.4\%$.

As to the effect of the confinement, it is admitted that it vanishes only in the case where the detonation is spherical (Brochet et al. 1970). Actually, prior to the 1970's, in the case of atmospheric or subatmospheric initial pressure, most of the calculated values were compared to those D_{∞} derived from an extrapolation toward $\phi^{-1} = 0$ in the D vs ϕ^{-1} plot (ϕ being the tube diameter). It is acknowledged that such a procedure is not reliable since, according to Brossard (1968), this value of D_{∞} is slightly (1.5-3%) greater than that observed in spherical detonations of the same mixture. Also, Desbordes et al. (1983) showed that, for $p_o < 1$ bar, D vs ϕ^{-1} exhibits a maximum value.

Thus, strictly speaking, only detonation velocities measured in the spherical mode should be compared to calculated values. However, the experimental difficulty encountered with spherical detonations drastically increases when the initial pressure is raised.

When such high pressures are involved, safety concerns require the use of high-pressure tubes in the experiments. However, it must be emphasized that the higher the initial pressure, the thinner the structure of the detonation front

(Bauer et al. 1986a) and the less the value of the slope D vs ϕ^{-1}, which means that the difference between the values of D_s observed in tubes and the corresponding values of D_{max} in the spherical configuration provided by Desbordes et al. (1983) at an initial pressure lesser than 0.5 atm seems to decrease.

The latter point, namely the variation of the slope D vs ϕ^{-1} was confirmed by Bauer (1985) and by Bauer et al. (1984a).

Acknowledgments

The authors are indebted to H.N. Presles for his valuable comments on this paper and his contribution in the design of the experimental facilities. They also wish to express their thanks to D. Falaise, who did the gas analysis.

References

Bauer, P.A., Brochet, C. and Presles, H.N. (1984a) Detonation study of gaseous mixtures at initial pressures reaching 10 MPa, Arch. Combust. 4(3), 191-196.

Bauer, P.A., Brochet, C., Heuzé, O. and Presles, H.N. (1984b) Generation of high dynamic pressures by means of gaseous explosive mixtures, J. Phys. 45(11), 297-299.

Bauer, P.A. (1985) Contribution à l'étude de la détonation de mélanges explosifs gazeux à pression initiale élevée, Thèse de Doctorat d'Etat, University of Poitiers, France.

Bauer, P.A., Presles, H.N., Heuzé, O. and Brochet, C. (1986a) Measurement of cell lengths in the detonation front of hydrocarbon, oxygen and nitrogen mixtures at elevated initial pressures, Combust. and Flame 64, 113-123.

Bauer, P.A., Presles, H.N., Heuzé, O., Fearnley, P.J. and Boden J.C. (1986b) Influence of hydrocarbon additives on the detonation velocity of methane-air mixtures at elevated initial pressures, Progress in Astronautics and Aeronautics: Dynamics of Explosions, edited by J.R. Bowen, J.C. Leyer, and R.I. Soloukhin, Vol. 106, pp. 321-328. AIAA, New York.

Brochet, C., Manson, N., Pujol, Y. and Stainnack, P. (1966) Application de la méthode inverse à la détermination des caractéristiques de la détonation dans les explosifs gazeux, Communication presented at 36th Congress (International) of Industrial Chemistry, Brussels, Belgium.

Brochet, C., Brossard, J., Manson, N., Cheret, R. and Verdès, G. (1970), A comparison of spherical, cylindrical and plane detonation velocities in some condensed and gaseous explosives, 5th Symposium (International) on Detonation,

Naval Surface Weapons Center, White Oak, Silver Spring, Maryland, USA.

Brossard, J. and Manson, N. (1958) Détermination comparée des caractéristiques des ondes explosives dans les mélanges gazeux, CRAS Paris 247, 2105.

Brossard, J. (1970) Contribution à l'étude des ondes de choc et de combustion sphériques divergentes dans les gaz, Thèse de Doctorat d'Etat, University of Poitiers, France.

Desbordes, D., Manson, N. and Brossard, J. (1983) Influence of walls on pressure behind self-sustained expanding cylindrical and plane detonations in gases, Progress in Astronautics and Aeronautics: Shock Waves, Explosions, and Detonations, edited by J.R. Bowen, J.C. Leyer, and R.I. Soloukhin, Vol. 87, pp. 302-317. AIAA, New York.

Heuzé, O. (1985) Contribution au calcul des caractéristiques de détonation de substances explosives gazeuses ou condensées, Doctorat d'Université, University of Poitiers, France.

Heuzé, O., Bauer, P.A., Presles, H.N. and Brochet, C. (1985) Equations of state for detonation products and their incorporation into the QUATUOR code, 8th Symposium (International) on Detonation, Naval Surface Weapons Center, White Oak, Silver Spring, Maryland, USA.

Kemp, M.K., Thompson, R.E. and Zigrang, D.J. (1975) Equations of state with two constants, J. Chem. Educ. 52 (12), 802-803.

Manson, N. (1958) Une nouvelle relation de la théorie hydrodynamique des ondes explosives, CRAS Paris 246, 2860.

Manson, N. (1960) Détermination par la méthode inverse des caractéristiques des ondes explosives. Publ. 366, Ministère de l'Air, Paris.

Presles, H.N., Bauer, P.A., Heuzé, O. and Brochet, C. (1985) Investigation on detonation of gaseous explosive mixtures at very high initial pressure, Combust. Sci. Technol. 43, 315-320.

Pujol, Y., Brochet, C., and Manson, N. (1966) Relation entre les dérivées partielles par rapport aux variables caractéristiques de l'état initial de l'explosif, de la célérité de détonation, CRAS Paris 263, 1160.

Pujol, Y. (1968) Contribution à l'étude des détonations par la méthode inverse, Thèse de Doctorat d'Etat, University of Poitiers, France.

Wood, N.W. and Fickett, W. (1963) Investigation of the Chapman-Jouguet hypothesis by the "inverse method", Phys. Fluids 6, 648.

Safe Gap Revisited

H. Phillips*

Health and Safety Executive, Buxton, Derbyshire, United Kingdom

Abstract

The earlier analysis of 'safe gaps' failed to predict the safe gap for mixtures enriched with oxygen or for high explosion pressure. These deficiencies are now overcome by improvements in the treatment of heat transfer in the flange gap and entrainment into the emerging jet of hot gas. Global activation energy is estimated from the data for hydrogen and methane and related to Fenn and Calcote's values.

Nomenclature

a	= see Eq. 8
B	= rate constant
d	= flange gap size
E	= activation energy
E'	= activation energy (Fenn and Calcote's)
L	= length, from gap entry
m	= mass
m'	= mass flow rate
M	= molecular weight
M'	= momentum flux
R	= gas constant
t	= time
T	= temperature
ΔT	= temperature drop by heat transfer, see Eq. 5
v	= velocity
z	= see Eq. 3

Copyright © 1988 British Crown Copyright. Published by the American Institute of Aeronautics and Astronautics, Inc. with permission.
*Explosion Specialist; Explosion and Flame Laboratory.

η   = efficiency, see Eq. 1
ψ   = reaction rate function, see Eq. 6

Subscripts

0   = initial
f   = flame
j   = jet
v   = vortex
w   = wall

The safe gap

A flameproof enclosure is one method of preventing accidental ignition of a flammable atmosphere by sparking in an electrical apparatus. Transmission of the explosion from inside the apparatus to the external atmosphere is prevented if the gap between plain parallel flange surfaces is less than the maximum experimental safe gap (MESG). In practice, flange gaps permitted in electrical apparatus are less than the MESG to allow a factor of safety. To specify electrical apparatus for use in a specific flammable atmosphere requires knowledge of the MESG for that atmosphere.

MESG's for a wide range of flammable gases and vapours have been determined in a number of different apparatus [Phillips (1986)]. Experience has shown that experimental apparatus should resemble the electrical apparatus in use in a factory in that the external atmosphere should be at essentially ambient pressure. With this constraint, the MESG is found to be insensitive to quite large changes in apparatus, although there are important exceptions.

One apparatus, a 20 ml sphere with an equatorial flange joint, is now accepted as standard by the International Electrotechnical Commission, although results with the earlier 8 liter sphere are similar. It is in this 20 ml apparatus that the MESG is now measured. The results of experiments in other apparatus or at other ambient conditions are referred to as a "safe gap".

Because of the apparent insensitivity to apparatus changes, there have been suggestions that MESG is a fundamental parameter for a gas or vapour, rather like its burning velocity or quenching distance. This opinion might be accepted for many gases or vapours; however, with the more reactive mixtures or in atmospheres contaminated with additional oxygen, the MESG is dependent on the apparatus.

The mechanism for the operation of the safe gap was presented by Phillips (1963, 1971, 1972a, 1972b, 1973), but since then new experiments beyond the range of the older data have indicated deficiencies in the analysis. The

SAFE GAP REVISITED

problems are resolved in this "revisitation" of the safe gap. The earlier work is summarised and its deficiencies noted. Then follows the revised analysis required to resolve the problems. Data from the 8 liter spherical vessel was used to establish constants because the data encompassed a wider range of explosion pressure than that for the 20 ml vessel.

Background to theory

Experimental observation has guided the development of theory. For transmission of an explosion through a flange gap, the limiting gap is about half of the quenching distance and, generally, the external ignition was favoured when the internal ignition was close to the flange gap. This implied that the pressure in the flange gap at the time of external ignition was close to ambient and that the flame was quenched within the gap. The mechanism of explosion transmission is one of reignition of flammable gases by the jet of hot gases emerging from the flange gap (Fig. 1). Fig. 1a shows a critical ignition with only 50% of repeat experiments resulting in external ignition. A wider gap

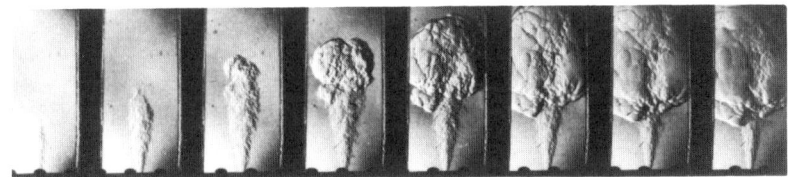

a) ignition.

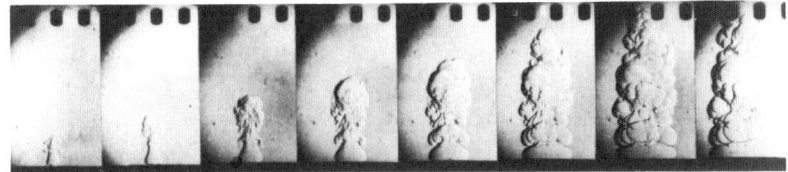

b) strong ignition.

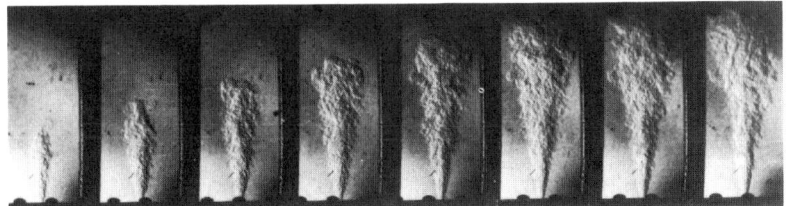

c) nonignition.
Fig. 1 Schlieren photographs of ignitions (1000 frames/s).

gives a strong ignition (Fig. 1b) and a smaller gap, failure of ignition (Fig. 1c). The emerging gases leading to a critical ignition are idealised as a planar jet surmounted by a cylindrical vortex in which the reignition is initiated (Fig. 2).

The critical processes effecting ignition are entrainment of flammable gas into the jet and the rate of burning of the entrained gases. Entrainment cools the jet, while burning tends to increase its temperature. If, by entrainment, the temperature drops so low that burning is reduced to a negligable rate, the external gas is not ignited. Other factors have an influence on this process, such as heat transfer as the hot gas passes through the flange gap and the influence of the internal explosion pressure on the speed of the ejected gas and, hence, on the rate of cooling by entrainment.

Where

$$\eta = (T - T_o)/(T_f - T_o) \qquad (1)$$

the rate of change of temperature was expressed as

$$1/\eta \cdot d\eta/dt \qquad (2)$$

The rate of entrainment for a jet was

$$1/m'_{jo} \cdot am'_j/at = z/t \qquad (3a)$$

and for a vortex,

$$1/m_{vo} \cdot dm_v/at = z/t \qquad (3b)$$

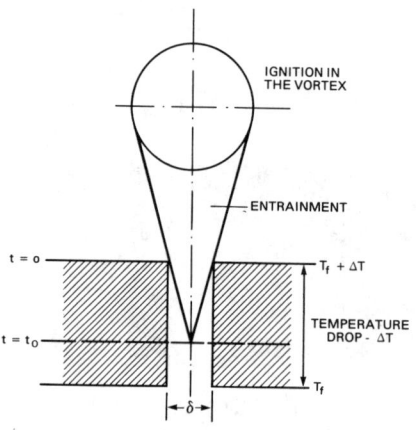

Fig. 2 Model of the hot jet.

where z was a constant, 1/3 for a planar jet and 8/9 for the cylindrical vortex at the head of the transient jet (Phillips 1972a).
An energy balance led to

$$1/\eta \cdot \partial\eta/\partial t + 1/m \cdot \partial m/\partial t = \psi \qquad (4)$$

The jet emerged from the flange gap with a temperature defined by ΔT where

$$\Delta T = (T_f - T_j)/(T_f - T_o) \qquad (5)$$

so that the rate of burning in the jet depended on the relative proportion of entrained gas and the degree to which it was already burned, as

$$\psi = \frac{BPM}{TK\eta} \frac{a/f}{(1 + a/f)} \left\{ 1 - \eta + \frac{m_o}{m} \Delta T \right\}^2 exp(E/RT) \qquad (6)$$

and

$$m_o/m = (t_o/t)^z \qquad (7)$$

Activation energy E was taken from Fenn and Calcote (1953). They had found that activation energy (in cal/g·mole) was 16 times the flame temperature at the lower limit for downward propagation of flame. The main advantage of their data was that activation energy was available for a wide range of fuels.

t_o was related to the flange gap by

$$a = avt_o \qquad (8)$$

where a is a constant related to the cone angle of the jet, 0.1 was used in subsequent calculation although Phillips (1973) indicated 0.05 to be a more appropriate value.

Heat transfer in the flange gap was assumed for fully developed laminar flow, i.e., the Nusselt number was 7.6. with a flame temperature inside the vessel of the adiabatic flame temperature at a constant pressure T_f. This followed from the assumption that external ignition was most likely early in the development of the explosion while the pressure was still low.

For ease of calculation, z was assumed to be unity. Variation in the values of a and z had no effect other than to introduce a constant of proportionality into Eq. (8). The values were chosen to be close to their predicted values and to give a unity constant of proportionality.

Solution of Eq. (4) by a Runge-Kutta method resulted in a plot of temperature against time that could be assymptotic to either ambient temperature, denoting a failure to ignite, or flame temperature, denoting an external ignition. A critical value of t_0 was found as a boundary between ignition and failure to ignite. There was an optimum velocity v in Eq. (8), which provided a minimum value of critical gap. The optimum was small for methane, but almost sonic for hydrogen. It was found through a dimensional analysis that the optimum velocity could be related to the burning velocity of the fuel, vessel volume, open area of the flange gap, and position of the internal ignition. The correlation was based on the experimental data for methane and hydrogen, but was used also for other fuels.

Using these solutions, the safe gaps calculated for a wide range of fuels and explosion vessels were compared with the experimental determinations (Fig. 3). The MESG could be correlated against the maximum value of ψ in Eq. (6) with ΔT equal to zero, see Fig. 4. This formed the basis of a method for predicting the MESG.

Problems

Whilst the earlier models have served to predict many of the experimental observations on safe gaps, such as the

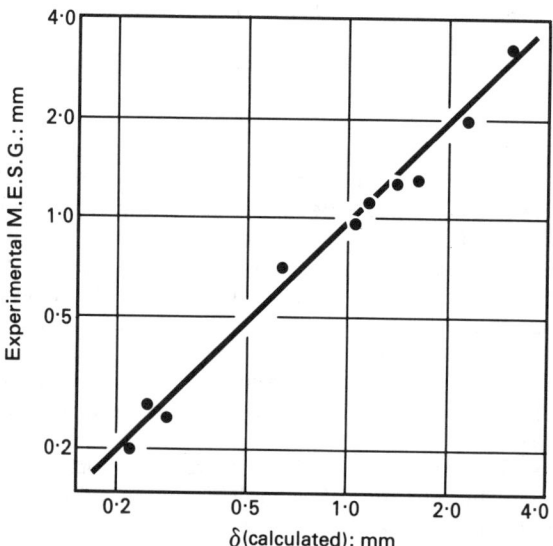

Fig. 3 Correlation between calculated values of the safe gap δ and experimental values of MESG.

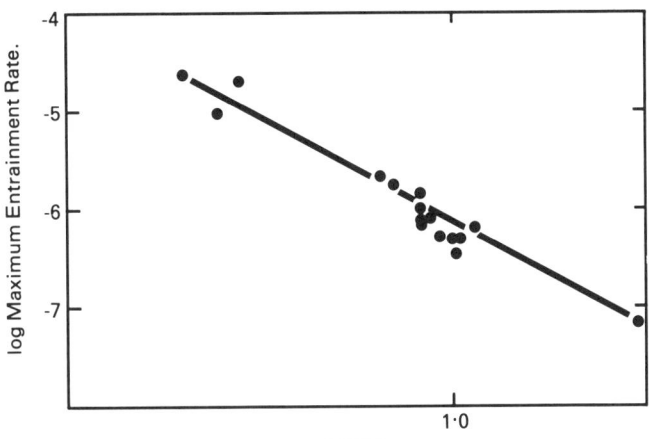

Fig. 4 Maximum of reaction rate function ψ vs MESG in the 8 liter vessel [from Phillips (1963)]

effects of a change in fuel and flange breadth, there are a number of shortcomings and a failure to predict some recent observations. This has prompted a new appraisal of the physics of the safe gaps. The basic concepts outlined earlier remain unchanged, but a closer examination of the details results in significant improvement in the accuracy of predictions of safe gaps. The problem areas are described in the following sections.

MESG of Oxygen Enriched Mixtures

Lunn (1984) reported the determination of MESG's in stoichiometric mixtures of hydrogen/nitrogen/oxygen and methane/nitrogen/oxygen in both the 8 liter and 20 ml standard explosion vessels. In each case there was a discontinuity in the plot of MESG vs nitrogen content. Below the discontinuity, with a low nitrogen content, the MESG was much less than predicted (Fig. 5). Lunn attributed this to the onset of choking flow in the discharge of hot gas from the flange gap and he postulated an "s" shaped curve to describe the change of safe gap with internal explosion pressure (Fig. 6). With side ignition, hot gas first emerges with a small explosion pressure. As pressure increases, the safe gap falls to a minimum where ignition is most likely to occur. A further increase in pressure is accompanied by an increase in the safe gap, with no further possibility of external ignition until the maximum of the curve is passed and pressure is so high that

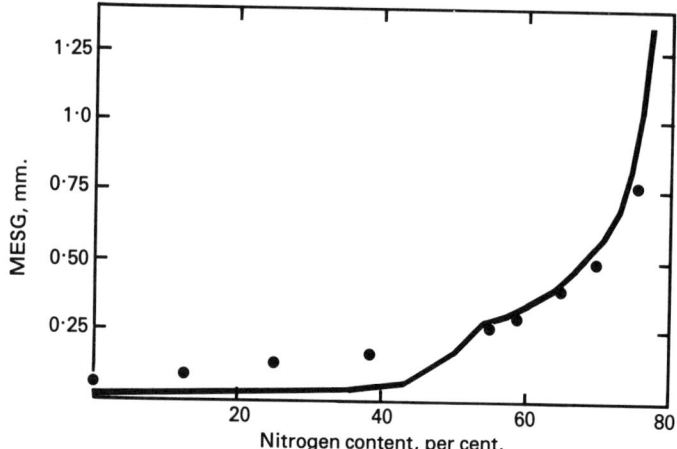

Fig. 5 Calculated and experimental safe gaps for hydrogen/oxygen/nitrogen (points are calculated values).

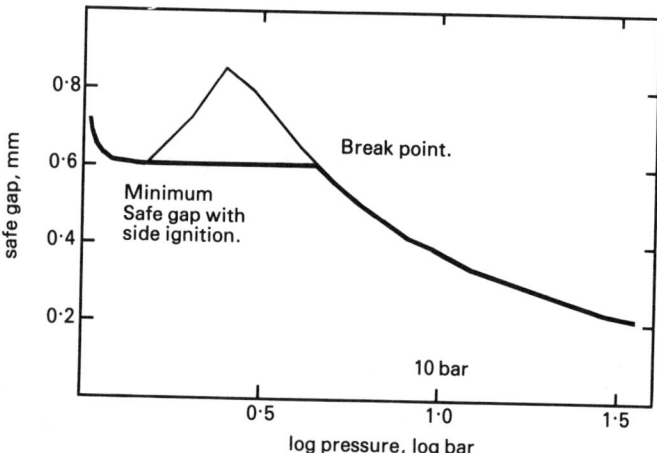

Fig. 6 "S" curve showing a minimum in the safe gap at 1.5 bar and a break point at 4.6 bar.

the safe gap has dropped below its earlier minimum. With central ignition, the hot gas first emerges at a high pressure and ignition, if it occurs, is at that pressure. The discontinuity occurs when the safe gap at high pressure becomes less than the minimum at low pressure.

The problem has commercial significance for the provision of flameproof enclosure for batteries that might discharge a mixture of hydrogen and oxygen. Experiments with an empty enclosure to simulate the battery enclosure suggested a safe gap of about 0.075 mm, yet a test on a proposed commercial battery enclosure yielded explosion transmission through a gap of 0.025 mm (unpublished test reports).

Safe Gap for Hydrogen with a High Explosion Pressure

For a hydrogen/nitrogen/oxygen mixture, the break point occurred close to a stoichiometric hydrogen/air mixture in the 8 liter vessel at a safe gap equal to the MESG. If the explosion pressure were to be increased above the maximum pressure recorded in the 8 liter vessel, the safe gap might be expected to fall below the MESG.

The hypothesis was tested with an tubular apparatus, 50 mm in diameter and 450 mm long (Fig. 7). Ignition was at one end and the 19 mm broad flange gap across a diameter at the other end. The acceleration of a hydrogen/air flame was encouraged by a spiral of wire inside the apparatus. Close to the flange gap, a pressure of 27 bar was recorded and the safe gap was 0.15 mm (Dickie 1982) compared with the MESG of 0.28 mm and the gap permitted in British Standard 5501: Part 5 of 0.15 mm for 25 mm wide spigot joints. A similar reduction in MESG was noted with acetylene but not with ethylene or propane, possibly because such high explosion pressures were not then achieved.

Jet Velocity and Temperature

The dimensional analysis relating jet velocity to the volume of the explosion vessel, flange opening and burning velocity helped to correlate the data, but did not give any insight into the physical mechanism involved. Extrapolation beyond the range of experimental data is not wise.

The assumption of laminar flow for heat transfer is probably incorrect at the higher pressure and velocity where turbulent flow is more likely.

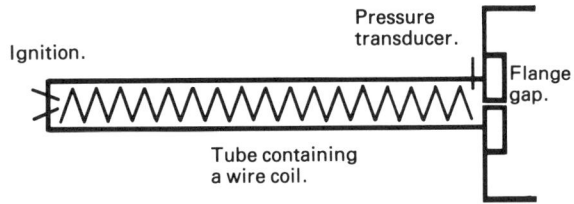

Fig. 7 Dickie's (1982) experiment: safe gap of hydrogen at high explosion pressure.

Ignition in the Jet

With side ignition, the external atmosphere is ignited while the explosion pressure is low, at the minimum of the "s" curve. With central ignition, the external atmosphere is ignited at the maximum explosion pressure. The existing model of entrainment does not explain why ignition does not occur at the minimum when the explosion pressure is falling.

The revised analysis

The objective of the re-examination of the mechanism of ignition was to offer explanation of these experiments. At the same time, it was hoped that the analysis might lead to a method of predicting the safe gap under conditions of high explosion pressure, although the uncertainty of some of the constants involved (such as entrainment rate and activation energy) might interfere with this objective.

Heat transfer in the flange gap and entrainment into the emerging transient jet were given close attention and the new results compared with the experimental data for methane and hydrogen in the 8 liter explosion vessel. The relevant data were as follows:

1) Methane: MESG for side ignition, 1.15 mm, for central ignition, 1.75 mm with a maximum explosion pressure of 1.3 bar absolute (Wheeler 1940).

2) Hydrogen: MESG for both side and central ignition, 0.28 mm. For central ignition, the maximum explosion pressure was 7.1 bar absolute (Wheeler 1939), (Smith and Blackwell 1961) with Smith (1986) providing data on the explosion pressures.

Having established parameters that allow these two sets of data to be reproduced by calculation, the correlation against data for some other fuels can be examined.

Heat transfer

Calculation of the temperature and velocity of gases emerging from the flange gap were calculated using a finite difference code with the channel between the flanges divided into 20 equal steps. The boundary conditions involved the explosion temperature and pressure at the inlet and the ambient pressure at the exit. Standard engineering equations for the heat transfer in each step were used, as found in any textbook such as Hsu (1963).

Explosion Temperature

Ignition occurs shortly following the first appearance of the jet from the flange gap, when the flame first makes contact with the internal wall of the explosion vessel. The temperature of the hot gases at that point can be calculated, assuming that the gas is first compressed and then burned at constant pressure.

Entry Loss

There is a loss of kinetic energy as the gas enters the flange gap. The loss is $0.25\rho v^2$ which is converted into heat (Spiers 1977).

Boundary Condition at Entry

At each time step, the pressure, density, and velocity are calculated at the entry. A limit is placed on this pressure, which must not exceed the explosion pressure nor be so low that the flow exceeds sonic velocity. With the corrected pressure, the velocity and density are estimated from the equations for a nozzle and the finite difference scheme continues with these new values.

Critical Reynolds Number

The flow throughout the gap is assumed turbulent if the Reynolds Number exceeds a critical value in the first grid cell.

Heat Transfer

McAdam's revision of the Colburn equation was used for turbulent heat transfer,

$$Nu = 0.02 \ Re^{0.8} \ (T/T_w)^{0.08} \quad (9)$$

and for laminar flow the Sieder and Tate equation was used,

$$Nu = 1.24 \ (Pe \ D/L)^{1/3} \ (T/T_w)^{0.08} \quad (10)$$

with the limitation that Nu does not drop below 7.6 (Kutateladze and Borishanskii 1966).

Boundary Condition at Exit

At exit from the flange gap the flow is either subsonic at ambient pressure or sonic at greater than ambient pressure. Other variables are recalculated so that the finite difference solution can continue.

Expansion outside the Flange Gap

If the exit pressure is greater than ambient, the jet expands without entrainment to achieve ambient pressure, sonic velocity, and increased area. Birch et al (1984, 1987) have examined that expansion. With sonic flow, temperature, and velocity are unchanged, but for a slot, the expanded gap width is inversely proportional to pressure. It is the expanded gap that becomes the source for subsequent entrainment.

Entrainment

The transient jet might be considered as a steady planar jet surmounted by a cylindrical vortex. In the earlier treatment the jet and vortex were both thought to entrain gas from the atmosphere with the result that the entrainment rates, $1/m \cdot dm/dt = z/t$, were not the same. This resulted in confusion over which rate characterised the ignition process. A value of z equal to unity, close to entrainment into the vortex, gave the best fit with experiment using Fenn and Calcote's (1953) activation energies.

In the revised treatment, the jet is assumed to gain gas from the atmosphere, but the vortex gains gas only from the jet.

Consideration of similarity show that at any time 28.4% of the gas is contained in the vortex, the remainder still in the jet and that the entrainment rates are the same with $z = 1/3$ for a planar jet (Phillips 1971).

Energy balances in the jet and vortex allow Eq. (4) to be replaced. For the jet, the composition of gas entering the vortex can be found from

$$z/t + 1/\eta_j \cdot d\eta_j/dt = 0.726 \, \psi \qquad (11)$$

and for the vortex from

$$(1 - \eta_j/\eta_v) \, z/t + 1/\eta_v \cdot d\eta_v/dt = \psi \qquad (12)$$

where ψ has the same significance as in Eq (6).

The starting time t_o is related to the apparent gap. From Ricou and Spalding (1961), the velocity in a planar jet is related to the distance from a point origin by

$$v = 1/K \cdot (M'/\rho r)^{1/2} \qquad (13)$$

The time taken for the jet to grow from a point until it emerges from the apparent gap is found by integrating Eq. (13) as

$$t_o = 0.4 \, d/v_o \qquad (14)$$

see Phillips (1973).

Solution of Equations

The heat transfer equations were solved to provide a look-up table of temperature, velocity, and apparent gap size for initial values of combustion temperature rise, explosion pressure, and gap size.

Using the look-up table Eqs. (11) and (12) are solved together using a Runge-Kutta method. For each value of pressure, the value of flange gap is located that defines a boundary between ignition and nonignition. Ignition is identified as a rising temperature in the vortex and failure to ignite by a falling temperature at the end of 60 steps.

Results

The main uncertainty in applying these equations to the estimation of safe gaps is in the use of Fenn and Calcote's (1953) values of activation energy. They proposed a simple means of estimating global activation energy (in cal/g·mole) as 16 times the flame temperature (in K), at the lower limit of downward flame propagation. The data referred to laminar flame propagation, but experiments in stirred reactors (for example, Longwell and Weiss 1955) suggested much higher values when the reaction was influenced by turbulence. The approach adopted was to estimate activation energies and rate constants from the data for methane and hydrogen and then to relate them to Fenn and Calcote's values.

Critical Reynolds Number

For hydrogen, Fig. 5 suggests that with central ignition the safe gap is on the right hand branch of the "s" curve (Fig. 6) and that with side ignition is at a minimum at a lower pressure. The safe gap was plotted assuming that heat transfer was turbulent with a suitable combination of activation energy E and rate constant B such that the line passed through the safe gap, 0.28 mm at the explosion pressure, 7.2 bar (Fig. 8). With the same rate data, the plot was repeated assuming laminar heat transfer. The 0.28

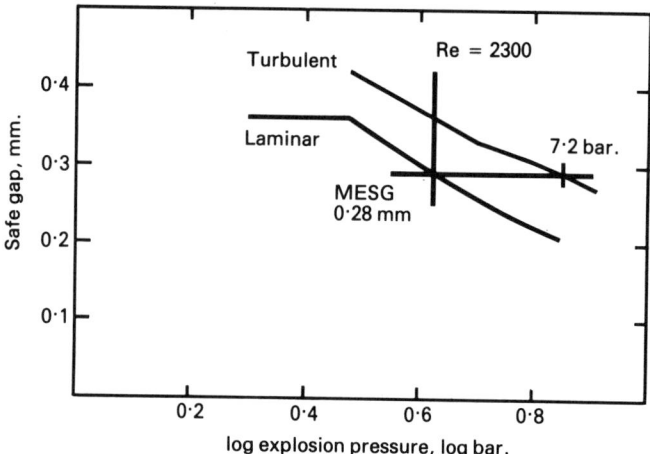

Fig. 8 Determination of critical Re from the safe gap for hydrogen.

mm safe gap was seen at a pressure of 4.3 bar, which coincided with a Reynolds number at the entry to the flange gap of 2300. This compares favourably with the usual value of Re = 2200. Changes in activation energy and rate constant had no influence on this conclusion. In the heat transfer calculation the critical Reynold's number was taken to be 2300.

Activation Energy

Burning proceeds through a system of many reactions at varying rates. Hautman et al (1981) and Peters and Lee (1987) have shown how reaction rates can be simulated by a scheme containing only four steps. Simulation of the reaction by a single step is a gross simplification but is valuable provided it is recognised that the apparent global reaction rate constant and activation energy relate only to the specific experiment and method of analysing that experiment. Fenn and Calcote (1953) used the concept of global activation energy for flame propagation and found that activation energy (cal/g·mole) was 16 times the flame temperature (K) at the lower concentration limit of downward flame propagation. In the present analysis of safe gaps global activation energy is found from the experiments with methane and hydrogen and related empirically to Fenn and Calcote's values.

The plot of the safe gap vs explosion pressure for methane was "s" shaped. The amplitude of the curve depended on activation energy and the minimum safe gap at low pressure depended on the rate constant. It was only with a specific combination of activation energy and rate constant that the safe gap for both side and central ignition in an 8 liter explosion vessel be simulated (Fig. 9). The values thus determined were rate constant $B = 0.6708 \cdot 10^9$ and Activation energy $E = 22,720$ cal/g·mole.

With the same rate constant the activation energy for hydrogen was found to be $E = 14,800$. The relation between the activation energies thus determined and those of Fenn and Calcote (E') is

$$E = 0.72E' + 3340 \text{ (cal/g·mole)} \quad (15)$$

permitting safe gaps to be estimated for a wide range of fuels, for both side and central ignition, or over a range of explosion pressures.

The plot of safe gap vs explosion pressure for hydrogen is seen in Fig. 10. A minimum occurs at the critical Reynolds number, $Re = 2300$, which denotes the MESG with side ignition. The safe gap is again 0.28 mm with central ignition in an 8 liter spherical vessel but at an explosion pressure of 27 bar the safe gap is reduced to 0.17 mm which is close to the gap found by Dickie (1982).

The plot of safe gap vs explosion pressure for methane is seen in Fig. 11. The minimum occurs at a very low explosion pressure, 1.04 bar. This is because at low pressure the heat transfer is large relative to the gas flow and the jet temperature at still lower pressures is not sufficient to maintain the reaction in entrained gas. With

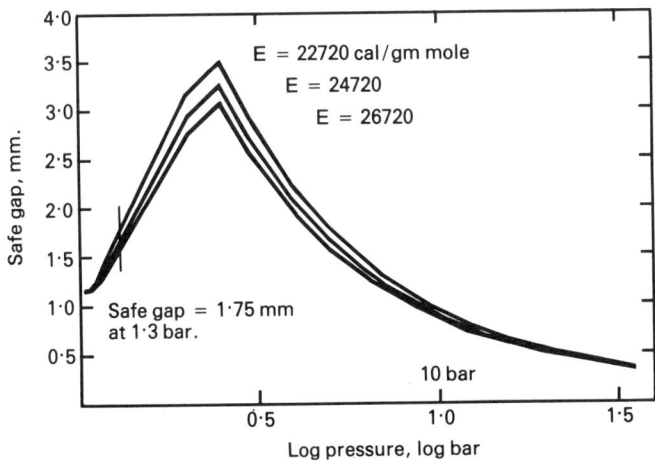

Fig. 9 Determination of activation energy from the safe gap for central ignition of methane.

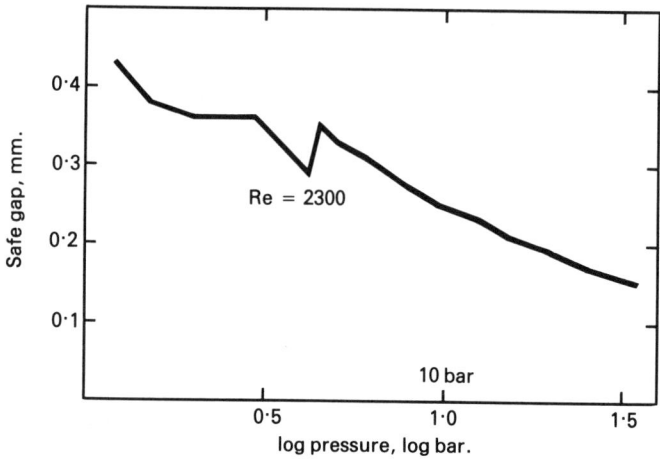

Fig. 10 Safe gap for hydrogen up to 35 bar explosion pressure.

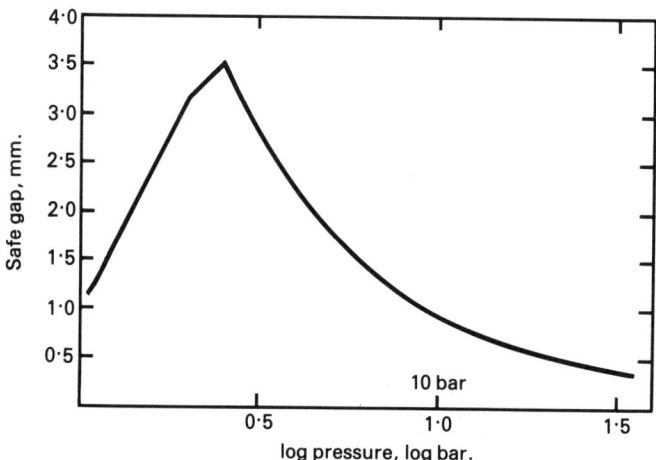

Fig. 11 Safe gap for methane up to 35 bar explosion pressure.

central ignition and the large vent offered by the flange gap in the 8 liter vessel, the explosion pressure is only 1.3 bar and the safe gap is 1.75 mm. At higher pressures the safe gap rises to 3.5 mm at 2.5 bar. In large vessels, with ignition remote from the flange gap and with only a small open gap area, the safe gap can approach this value (Thibault et al 1982).

In the 20 ml explosion vessel, which is the International Electrotechnical Commission standard, explosion pressures are sufficiently high to maintain the same MESG values at the minima of Figs. 10 and 11.

Other Fuels

Carbon disulphide has an MESG of 0.2 mm with central ignition in the 8 liter vessel (Wheeler 1939). With side ignition, the safe gap is 0.27 mm. The explosion pressures are found by interpolating Smith's (1986) data, giving 6.5, and 4.8 bar respectively.

Using the correlation for activation energy of Eq. (15) yields predictions of 0.3 and 0.36 mm for central and side ignition. Carbon disulphide is unusual in that the safe gap for side ignition is larger than the MESG with central ignition. The calculation predicts this fact, but overestimates the magnitude of the safe gaps.

In the 20 ml vessel the explosion pressure is below the pressure at the minimum and MESG in that vessel is larger than either determination in the 8 liter vessel. Helwig and Nabert (1968) found the MESG of carbon disulphide in a 20 ml vessel to be 0.34 mm.

Safe gaps were plotted for a range of activation energies (Fenn and Calcotes values of 14,000-26,000 in steps of 2,000, and 26,750) for a flame temperature of 2200 K. The two mechanisms for producing a minimum are apparent in Fig. 12. At low pressure with fuels of low reactivity (for example, methane), a minimum can occur when the rate of reaction is insufficient to compensate for cooling by entrainment - even when entrainment is slow. With higher

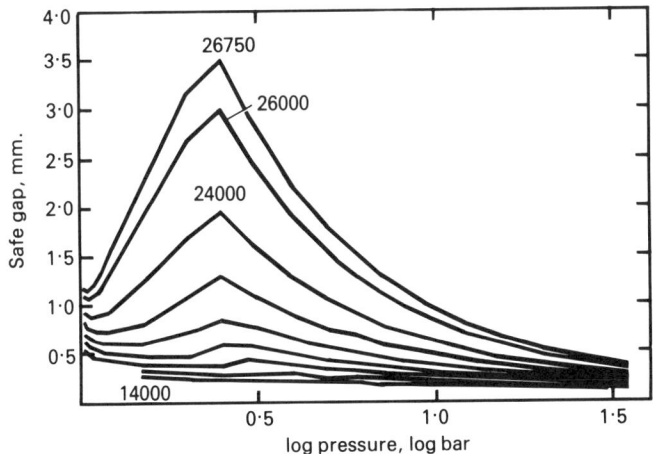

Fig. 12 Safe gap for a range of activation energies, 14,000-26,750 cal/g·mole.

reactivity, that minimum moves to a higher pressure and eventually disappears, to be replaced by a second minimum when the flow in the flange gap changes from laminar to turbulent, as for hydrogen in Fig. 10.

Location of Ignition

With the proposed new mechanism of entrainment, ignition takes place in the vortex at the head of the advancing jet. If the jet of hot gas is established at high pressure with a gap too small for reignition to take place, a subsequent reduction in pressure does not permit ignition to occur. In that event, ignition has to be in the jet and then the critical gap for ignition is larger than for the jet/vortex combination. Thus, with central ignition, the external reignition does not take place as the explosion pressure decays and the critical safe gap reduces towards its minimum.

Conclusions

The essential features of the safe gap and the MESG have been reproduced by a closer attention to heat transfer in the flange gap of flameproof electrical apparatus. Heat transfer has been calculated from conventional engineering principles and combined with a revision of the earlier analysis of safe gaps [Phillips (1963, 1971, 1972a, 1972b, 1973)]. The activation energy for the reactions have been estimated from safe gap data for hydrogen and methane and related to Fenn and Calcote's (1953) data. There must be some doubt over such a simple model for the reaction rate and this is apparent in estimating the safe gap over a range of pressures for carbon disulphide. In spite of this, the relationships between the safe gaps for carbon disulphide for side and central ignition in the 8 liter vessel and in a 20 ml vessel have been reproduced.

The earlier analysis of safe gaps was valid only for stoichiometric mixtures. There is no evidence to show that the use of a simple global reaction mechanism with the revised analysis will allow the prediction of MESGs for other than stoichiometric mixtures.

Since it appears that two mechanisms are responsible for the minimum in the plots of safe gap vs explosion pressure it is not reasonable to consider the MESG as a fundamental combustion parameter.

The drop in the safe gap as explosion pressure is increased above that normally experienced in an empty 8 liter spherical vessel with equatorial flanges open to the

MESG, for the more reactive fuels, sounds a note of warning. In large vessels with flange gaps opened over only a short length and many internal components giving rise to enhanced explosion pressures, external ignition might be possible with small flange gaps, comparable with those permitted in current standards such as IEC 79-1 (1971).

References

Birch, A. D., Brown, D. R., Dodson, M. G., and Swaffield, F. (1984) The structure and concentration decay of high pressure jets of natural gas. Combust. Sci. Technol. 36, 249-261.

Birch, A. D., Hughes, D. J. and Swaffield, F. (1987) Velocity decay of high pressure jets. Combust. Sci. Technol. 52, 161-171.

Dickie, P. (1982) The effect of pressure piling on the maximum experimental safe gap of certain gases. Internal Rep. IR/L/FL/82/15, Health and Safety Executive, Buxton, U.K.

Fenn, J. B. and Calcote, H. F. (1953) Activation energies in high temperature combustion. Fourth Symposium (International) on Combustion, pp. 231-239. Williams and Williams, Baltimore.

Hautman, D. J., Dryer, F. L., Schug, K. P., and Glassman, I. (1981) A multiple-step overall kinetic mechanism for the oxidation of hydrocarbons. Combust. Sci. Technol. 25, 219-235.

Helwig, N., and Nabert, K., (1968) Inter-relationships of the characteristics of equipment protected against explosions. PTB-Mitt. 78, 287-293.

Hsu, S. T., (1963) Engineering Heat Transfer. Van Nostrand, New York.

International Electrotechnical Commission (1971) Method of test for ascertainment of maximum experimental safe gap. Appendix to IEC 79-1.

Kutateladze, S. S. and Borishanski, V. M., (1966) A Concise Encyclopedia of Heat Transfer, p. 106. Pergamon, London, U.K.

Longwell, J. P. and Weiss, M. A. (1955) High temperature reaction rates in hydrocarbon combustion. Ind Eng Chem. 47, 1634-1643.

Lunn, G. A. (1984) The maximum experimental safe gap: The effects of oxygen enrichment and the influence of reaction kinetics. J. Haz. Mat. 8, 261-270

Peters, N. and Kee, R. J. (1987) The computation of stretched laminar methane-air diffusion flames using a reduced four-step mechanism. Combust. Flame 68, 17-29.

Phillips, H. (1963) On the transmission of an explosion through a gap smaller than the quenching distance. Combust. Flame 7, 129-135.

Phillips, H. (1971) The mechanism of flameproof protection. Res. Rep. 275, Safety in Mines Research Establishment, Buxton, U.K.

Phillips, H. (1972a) A non-dimensional parameter characterising mixing in a model of thermal gas ignition. Combust. Flame 19, 181-186.

Phillips, H. (1972b) Ignition in a transient turbulent jet of hot inert gas. Combust. Flame 19, 187-195.

Phillips, H. (1973) The use of a thermal model of ignition to explain aspects of flameproof enclosure. Combust. Flame 20, 121-126.

Phillips, H. (1986) A comparison of "standard" methods for the determination of Maximum Experimental Safe Gap (MESG). Proc. International Symposium on Explosion Hazard Classification of Vapors, Gases and Dusts, pp. 83-108, National Research Council, Washington, DC.

Ricou, F. P. and Spalding, D. B. (1961) Measurements of entrainment by axisymmetrical turbulent jets, J. Fluid Mech. 11, 21.

Smith, I. D. (1986) Explosion pressure measurement during the determination of maximum experimental safe gaps. Internal Paper IR/L/FL/86/21, Health and Safety Executive, Buxton, U.K.

Smith, P. B. and Blackwell, J. R. (1961) Flameproof enclosures: Redetermination with hydrogen air mixtures of the maximum safe gap for one inch relief flanges. Tech. Rep. D/T 117. Electrical Research Association, U.K.

Spiers, H. M. (1977) Technical Data on Fuels. British National Committee World Power Conference, London, U.K.

Thibault, P., Liu, Y. K., Chan, C., Lee, J. H., Knystautus, R., Guirao, C., Hyertager, B., and Fuhre, K. (1982) Transmission of an explosion through an orifice. Nineteenth Symposium (International) on Combustion. pp. 599-606, The Combustion Institute, Pittsburgh, PA.

Wheeler, R. V. (1939) Flameproof electrical apparatus: Maximum experimental safe gap for flanged joints in atmospheres containing carbon disulphide vapour. Tech. Rep. G/T 91, Electrical Research Association, U.K.

Wheeler, R. V. (1940) Flameproof electrical apparatus: Redetermination of the maximum experimenal safe gap for flanged joints with methane-air mixtures. Tech. Rep. G/T 112, Electrical Research Association, U.K.

Chapter II. Detonation Transition and Transmission

Concentration and Temperature Nonuniformities of Combustible Mixtures as Reason for Pressure Waves Generation

Y. B. Zel'dovich,* B. E. Gelfand,† S. A. Tsyganov,‡
S. M. Frolov,§ and A. N. Polenov¶
USSR Academy of Sciences, Moscow, USSR

Abstract

One of the basic principles of classical combustion theory is the assumption that the pure combustible mixture is uniform and temperature or concentration gradients are negligible. However, one may notice such a fundamentally important process as spontaneous origin of nonhomogeneous patterns due to probabilistic velocity distribution of molecules in a completely premixed matter. In effect, the existence of concentration and temperature nonuniformities (CTN) is usual for the overwhelming majority of technical devices. Imperfect operation of injector elements, heat fluxes, etc., are reasons for continuous periodic, or stochastic generation of regions with CTN. The classical approach to this type of engineering application does not seem to work well for investigation of combustion instability. Practical devices for some conditions, temperature, geometry, fuel-air ratio, etc., exhibit onset of undesirable pressure waves and subsequent oscillatory behavior. An analysis of some theoretical models is performed to take into account CTN in the reactive system. It is shown that the approach proves to be extremely helpful for understanding 1) the onset of high-frequency rocket combustion instability; 2) the "knock" phenomenon in

---
Copyright © 1988 by the American Institute of Aeronautics and Astronautics, Inc. All rights reserved.
*Academician; Head of Department, Institute of Physical Problems.
†Senior Researcher, Institute of Chemical Physics.
‡Head of Laboratory, Institute of Chemical Physics.
§Junior Researcher, Institute of Chemical Physics.
¶Researcher, Institute of Chemical Physics.

the internal combustion engine; 3) the generation of strong pressure waves in communicating vessels containing a reactive mixture; and 4) pressure-wave generation in vented explosions. The results demonstrate that CTN may play an important role in combustion phenomena.

Introduction

As the time scales of chemical and gasdynamic processes are essentially different, the motion of matter resulting from thermal expansion is often not taken into account in the thermal explosion theory (Frank-Kamenetski 1967). However, in conditions close to the self-ignition of a reactive mixture, new phenomena may come into existence because the chemical and gasdynamic time scales are of the same order of magnitude.

Let us consider the evolution of a local volume of the reactive mixture preconditioned to self-ignition. Self-ignition occurs in the first instance in spots where there is an optimal combination of temperature and reactant concentration. The rest of the volume is burnt thereafter in conformity with the laws of propagation of chemical reaction waves. Self-ignition is accompanied by the generation of pressure waves that heat up and mix the pure mixture with the combustion products. The process is described by at least two time scales: chemical ($\tau_c = L/U_c$) and gasdynamic ($\tau_g = L/U_g$), where L is the typical dimension of the local volume, U_c is the typical velocity of the chemical reaction front, and U_g is the typical gasdynamic velocity. The scale τ_c is typical of the time taken for burnout of the volume after self-ignition, whereas τ_g is typical of the time required for the pressure to become equally distributed over the field of interest. In laminar flame theory $\tau_c \gg \tau_g$, inasmuch as the normal velocity of the flame is $U_n = U_c \ll U_g$. The time scales of the phenomenon may be separated, and the gasdynamic effects may be set aside. In nonstationary thermal explosion theory $\tau_c \ll \tau_g$, because fast uniform burnout of the mixture is assumed to occur throughout the volume. The condition $\tau_c \approx \tau_g$ is applicable to the propagation of a detonation wave through a reactive mixture. The flow in the detonation wave is an example of the coupled motion of a shock wave and a chemical reaction front.

In general, to meet the condition $\tau_c \approx \tau_g$, the relationship $U_c \approx U_g$ must be insured. If this takes place in a local volume, the arrival of a weak pressure wave at a fixed location may coincide with the beginning of fast ignition of the mixture. In this case, pressure-wave

CONCENTRATION AND TEMPERATURE NONUNIFORMITIES 101

amplification may be expected. The situation is possible when the pressure wave is subject to a succession of "synchronization" acts with the moments of local mixture ignition. It may give rise to the rapid, progressive amplification of the wave. At some stage of the process the pressure wave may become strong enough to cause vigorous chemical activity in a pure mixture and the mixture will detonate. Thus, under certain conditions that insure the coupled motion of the pressure wave and the wave of chemical energy release, the spontaneous generation of strong shock waves and detonation may be expected.

Zeldovich and Kompaneets (1955) have suggested the idea of carrying into effect the combustion process with $U_c \gg U_g$. The idea is to control artificially or naturally the rate of propagation of the chemical release front. For example, an explosive gas mixture could be ignited with a sequence of electric sparks, or it could be preconditioned for successive self-ignition of neighboring layers. In this way the combustion process could be stimulated with arbitrarily high apparent burning velocity. According to Zeldovich and Kompaneets (1955), to generate the Chapman-Jouquet (CJ) detonation wave in the reactive mixture, the ignition source should be moved at the rate of a self-sustaining detonation. The pressure rise in a detonation is larger than any of the other rates of successive ignition. Hence, to set up the combustion mode with $U_c \approx U_g$ in the local volume, the conditions for nonsimultaneous self-ignition of the neighboring layers of the mixture should be met. (For instance, in such a manner when the layers react independently according to the laws of the adiabatic thermal explosion or branching-chain processes). If that is the case, the apparent propagation of the chemical release front that Zeldovich et al. (1980) call "spontaneous propagation" appears in the volume.

The coupling mechanism between the waves of chemical release and the pressure waves is closely connected with the gradients of the local parameters which affect the rate of a chemical reaction. It follows that fluctuations are the cause of various wave patterns at conditions close to self-ignition. The spontaneous fluctuations that arise as a consequence of possible bulk thermodynamic states of a macroscopic system and the fluctuations generated by the variations of initial and boundary conditions should be distinguished. The former may promote the spontaneous origin of chain carriers in the mixture and give an impetus to the local self-acceleration of the reaction (Zeldovich 1981). Considerable deviation from a mean value of a thermodynamic parameter may occur only in a very small spatial domain of the system. The variation of initial and boundary conditions may also result in the raising of extended portions of the mixture with spatially nonuniform distribution of local parameters (temperature, concentration, etc.). For instance, if the initially uniform pressurized, gaseous explosive mixture is suddenly expanded by the rupture of a vessel wall, the system then undergoes the sequence of nonuniform states with proper gradients of the local parameters. In the conditions close to mixture self-ignition, the gradients may bring about a nonuniform chemical release; and as a result, a number of gasdynamic wave patterns may arise in the volume.

The spatial nonuniformities in the reactive mixture appear to be potential sources of combustion instabilities that happen sometimes in the operation of technical devices. For instance, it is known as a matter of experience that, under certain conditions of mixing the fuel and oxidizer in the liquid propellant rocket engine, the process is accompanied by powerful pressure jumps that arise abruptly during normal operations (Harrje and Reardon 1972). As a rule, a high-frequency combustion instability comes into effect. The other example is the knock phenomenon in the internal combustion engine that also seems to arise because of explosion processes (Sokolik 1934). Finally, we mention the secondary shock waves that have been detected in the reaction zone of heterogeneous detonation (Ragland et al. 1968). These waves seem to give support to the stationary propagation of detonation over long distances.

The possibility of spontaneous generation of detonation waves in a spatially nonuniform explosive medium was proved for the first time by the numerical calculations of Zeldovich et al. (1970). The effect of initial temperature distribution on the evolution of the gaseous reactive mixture was considered. Gelfand et al. (1985) studied this problem parametrically and discovered the specific features characterizing the inception and transformation of various wave patterns in the system. Barthel and Strehlow (1979) have used numerical modeling to analyze the effects that appear as a result of introducing an additive with enhanced reactivity into the fuel-air mixture. The spatial distribution of the preexponential factor in the Arrhenius type of chemical reaction law was used to calculate the gradient of self-ignition delay. Gelfand et al. (1986) have investigated the role played by nonuniform reactant distribution in the evolution of the explosive mixture preconditioned to self-ignition. In all cases under review, the calculations reveal the possibility

CONCENTRATION AND TEMPERATURE NONUNIFORMITIES   103

of spontaneous generation of shock waves, their amplification, and further transition to detonation. The phenomena in point seem to have already been observed experimentally. Borisov et al. (1970) have observed the onset of detonation after a weak shock wave passing through the cloud containing preliminary ignited kerosene droplets. The weak shock-wave-stimulated droplets break up and cause the pure mixture to mix with the combustion products, thus creating the longitudinal gradient of ignition delay. Under certain conditions the fast successive self-ignition of hot mixture portions led to generation and amplification of a secondary shock wave and finally to onset of detonation. Lee et al. (1978) have conducted experiments on photochemical initiation of detonation in mixtures of $C_2H_2-O_2$ and H_2-Cl_2. To meet the condition $U_c \approx U_g$, the flash-photolysis technique was used to create the proper distribution of active radicals in the mixture. Knystautas et al. (1979) have investigated the onset of detonation by injecting the hot combustion products into the pure $C_2H_2-O_2$ mixture through perforated plates.

In this work we analyze theoretically the types of explosion processes that may be initiated in systems with spatially nonuniform distribution of temperature and reactant concentration. Of special interest are the questions relating to "critical" dimensions of the nonuniformities and sensitivity of the phenomena to variation of initial and boundary conditions. The effects accompanying multistage ignition are discussed, and some problems of practical concern are addressed.

Formulation

Consider a volume of typical dimension D containing a multicomponent gaseous explosive mixture. For the sake of simplicity, we single out only two components: a moderately active component A and a component B with enhanced reactivity. The component B may represent an additive to the basic fuel that promotes energy release in the system or the products of preflame autoxidation such as peroxides. We assume that at the initial instant the mixture is quiescent and the pressure is uniform throughout the volume. It is also suggested that a local spatial domain of characteristic dimension ϵ (henceforth referred to as ϵ-domain) exists in which the nonuniform distribution of temperature, such as

$$T(0,R) = T_0 - \chi_T R \quad \text{in } 0 < R \leq \epsilon$$
$$T(0,R) = T_e \quad \text{in } R > \epsilon \qquad (1)$$

or the mass fraction of component B, such as

$$B(0,R) = B_0(1 - \chi_b R/\epsilon) \quad \text{in } 0 < R \leq \epsilon$$
$$B(0,R) = 0 \quad \text{in } R > \epsilon \qquad (2)$$

is given. Here R is the Eulerian coordinate originated in the center of the ϵ-domain, χ_T is the temperature gradient, $0 \leq \chi_b \leq 1$ is the coefficient whose value determines the form of the distribution, and subscript o denotes the maximum value of a parameter at the instant t=0. To simplify the problem, the component A is assumed to be uniformly distributed throughout the volume. The evolution of the system after self-ignition is assumed to be governed by the one-dimensional equations of conservation of mass, momentum, energy, and reactant species without regard for effects associated with viscosity, heat conductivity, and mass diffusion. As a consequence, the analysis is restricted to processes that come into play well before the dissipative effects gain control over the evolution. The equations can be written in terms of the Lagrangian frame as follows:

$$\gamma \, \partial U/\partial \tau + (\xi/r)^{V-1} \partial p/\partial r = 0, \quad \partial \xi/\partial \tau = U$$

$$\partial \xi/\partial r = \vartheta \, (r/\xi)^{V-1}$$

$$(\gamma - 1)^{-1} \partial \theta/\partial \tau + p \partial \vartheta/\partial \tau = \alpha_a a \, \exp[\beta_a(1 - \theta^{-1})]$$

$$+ \alpha_b b \, \exp[\beta_a - \beta_b \, \theta^{-1}]$$

$$\partial a/\partial \tau + a \, \exp[\beta_a(1 - \theta^{-1})] = 0$$

$$\partial b/\partial \tau + b \, \exp[\beta_a(1 - \beta_b \beta_a^{-1} \theta^{-1})] k_b k_a^{-1} = 0 \qquad (3)$$

where p is the dimensionless values of pressure, ϑ, specific volume; U, velocity; θ, temperature; a, the mass fraction of component A; b, the mass fraction of component B; ξ, the Eulerian coordinate; r, the Lagrangian coordinate and τ, is time. The dimensionless variables are defined as:

$$p = p/p_0, \quad \theta = T/T_0, \quad U = U/\sqrt{\gamma R_* T_0}$$

$$a = A/A_0, \quad b = B/B_0, \quad \tau = t k_a \exp(-E/RT_0)$$

$$\alpha_a = Q_a/R_* T_0, \quad \alpha_b = Q_b/R_* T_0, \quad \beta_a = E_a/R_* T_0$$

CONCENTRATION AND TEMPERATURE NONUNIFORMITIES   105

$$\beta_b = E_b/R_*T_0, \quad \xi = [Rk_a \exp(-E_a/R_*T_0)]/\sqrt{\gamma R_*T_0}$$

$$r = (V \int_0^\xi \vartheta^{-1}\xi^{V-1}d\xi)^{1/V}$$

where R_* is the gas constant for the mixture; Q is the chemical energy; E is the activation energy; k is the preexponential factor; γ is the (constant) ratio of specific heats; indices a and b denote the values related to the proper components; V=1,2,3, for plane, cylindrical, and spherical symmetry. The dimensionless temperature gradient is given by

$$\lambda_T = \chi_T\sqrt{\gamma R_*T_0} / [T_0 k_a \exp(-E_a/R_*T_0)]$$

In the center of symmetry we suppose U=0 for $\tau \geq 0$. If the reactive mixture is placed in a vessel, it will be necessary to set boundary conditions at the walls that are assumed to be impermeable. The initial conditions ($\tau=0$) are

$$p = 1, \vartheta = 1 - \lambda_T\xi, \theta = \vartheta, U = 0, a = 1$$

$$b = 1 - \chi_b\xi/\xi_0 \quad \text{for } \xi \leq \xi_0$$

$$p = 1, \vartheta = 1 - \lambda_T\xi_0, \theta = \vartheta, U = 0, a = 1$$

$$b = 0 \text{ for } \xi > \xi_0 \qquad (4)$$

where $\xi_0 = \xi(\epsilon)$.
The system of Eq. 3 is solved numerically, subject to the initial and boundary conditions, through the use of the artificial viscosity technique (Richtmyer and Morton 1967).

Results

Four Types of Explosion Processes

To study wave patterns generated by local explosion events in the spatially nonuniform reactive medium, we consider the problem of the evolution of a two-component mixture with nonuniform initial distribution of a component B [see Eq. 2]. At $\tau = 0$, the temperature is assumed to be constant throughout the volume, i.e., $T=T_0=T_e$ and $\chi_T=0$ in Eq. (1). Heat release resulting from the self-ignition of component B produces temperature gradients in the ϵ-domain and creates conditions for the possible speeding up of pressure waves generated by the subsequent explosion of component A (Zeldovich et al. 1970). Depending on the dimension of ϵ-domain, the kinetic parameters, and the reaction heats, four various types of explosion processes were revealed in volume D by numerical computations. Our computations were terminated when the elapsed time was as large as the typical values of diffusion or conduction time scales. The numerical values for the parameters in Eqs. 3 and 4 have been chosen as typical for combustion chambers:

$$\gamma=1.2, \quad 800 \leq T_0 \leq 1400°K,$$

$$k_a = k_b = 10^{10} s^{-1}, \quad 5 \leq \alpha_a \leq 8,$$

$$3 \leq \alpha_b \leq 5, \quad 15 \leq \beta_a \leq 20,$$

$$12 \leq \beta_b \leq 15, \text{ and } 0.2 \leq \epsilon \leq 100 mm.$$

The four types of explosion processes that may be generated spontaneously in the system are described in the following discussion.

<u>First Type</u>. If $\epsilon \to D$ and $\chi_b \to 0$, then the constant volume explosion takes place in the system with almost uniform pressure rise throughout the volume. For the given numerical values of reaction heats, the pressure has been increased by no more than a factor of 3, compared with the initial pressure.

<u>Second Type</u>. If $\epsilon_* < \epsilon \leq D$ and $\chi_b=1$, the range of basic parameters exists in which the onset of detonation inside the ϵ-domain occurs and further stationary propagation of the detonation wave takes place in the initially uniform medium. The temporal evolution of such a flow is presented in Fig. 1, which shows pressure P against distance ξ for several values of τ. The mixture is defined

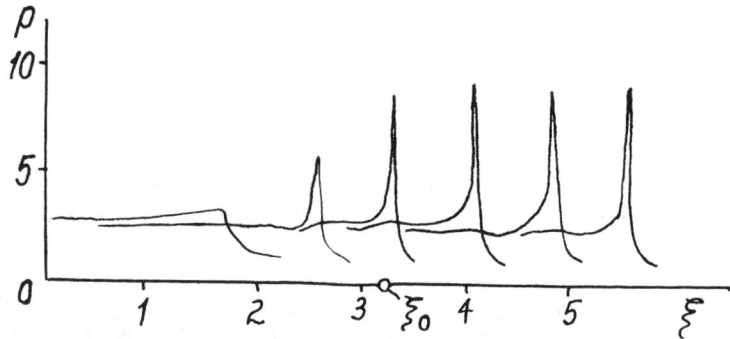

<u>Fig. 1</u> The onset of detonation in the system with distributed additive B. Shock pressure-time history in dimensionless time interval $\Delta\tau_*=0.27$.

by the following values of the basic parameters: β_a=16.7, β_b=12.5, α_a=8, α_b=3, T_0=1000°K, and ϵ_*=50 mm. The features of the evolution are the same as described by Zeldovich et al. (1970). The pressure rise behind the detonation wave is more than eight times greater that the initial pressure. This type of explosion process is important for the following analysis.

Third Type. If $\epsilon_{**} < \epsilon < \epsilon_*$, then in a certain range of basic parameters the detonation wave was generated outside the ϵ-domain with further stationary propagation in uniform mixture. The temporal evolution of the process may be broken into the following main points: a) self-ignition of component B and fast subsequent ignition of component A contained in the ϵ-domain; b) decay of the precursor pressure wave generated by the local explosion; c) a secondary explosion of component A and formation of a secondary shock wave-reaction front complex behind the decaying precursor shock wave; and d) a secondary shock-wave reaction-front complex, which overtakes the precursor shock wave, triggering detonation.

The third type of explosion process is notable for the precursor shock wave that is precipitated by the microexplosion in the ϵ-domain. The precursor wave creates the conditions for secondary wave amplification resulting from a time-phased local energy release. Figure 2 shows the temporal evolution of the mixture defined by the following values of the basic parameters: β_a=16.7, β_b=12.5, α_a=8, α_b=3, T_0=1200°K, and ϵ=5 mm.

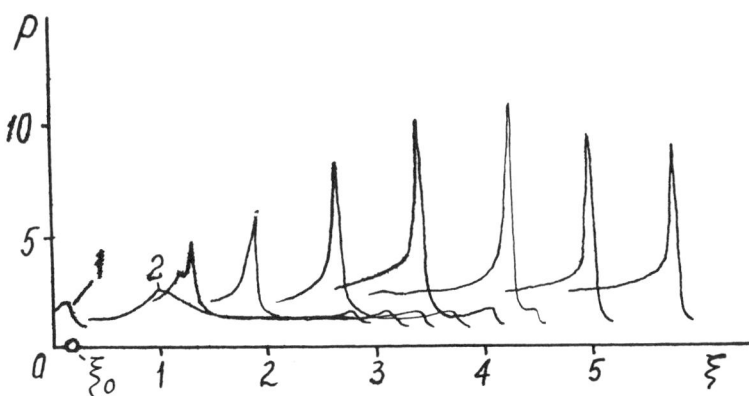

Fig. 2 The onset of detonation in the system with small spatial nonuniformity of reactant concentration ($\Delta\tau_*$=0.27). Curves 1 and 2 are drawn for τ = 0.326 and τ = 2.49, respectively.

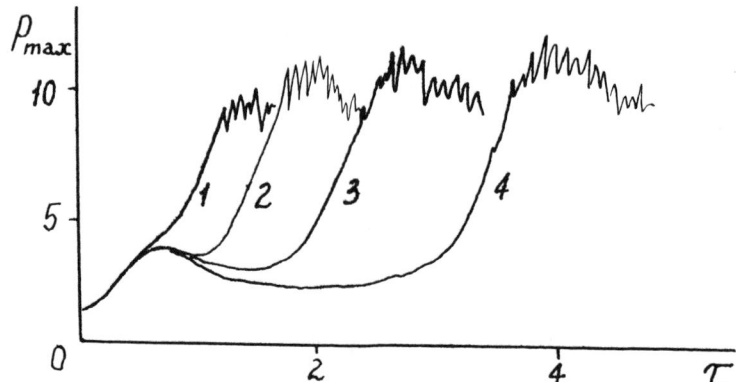

Fig. 3 The temporal evolution of pressure in a system with small nonuniformities of reactant concentration. Curve 1: $\epsilon=50$ mm; curve 2: $\epsilon=10$ mm; curve 3: $\epsilon=3$ mm; and curve 4: $\epsilon=2$ mm. The kinetic parameters are the same as for Fig. 2.

Figure 3 displays the history of peak pressure in the volume. We point to the fact that the smaller the $\epsilon < \epsilon_*$ (other things being equal), the larger the distance from the origin required for the onset of detonation. The oscillations of P_{max} arise in the numerical study as a consequence of the longitudinal instability of a one-dimensional detonation wave. The temporal shift at the instant detonation occurs is explained by the decay of the precursor shock wave and hence by the diminishing overall temperature level behind it. For $\epsilon=\epsilon_{**}\approx 2$ mm and D=100 mm, the detonation wave appeared on the rim of the vessel wall. For $\epsilon < \epsilon_{**}$, a nonstationary explosion wave propagates throughout the volume. The pressure rise in such a wave is greater than for the case of the constant volume explosion. For $\epsilon < \epsilon_{**}$, the explosion processes of the first type have been observed again.

<u>Fourth Type</u>. This type of explosion process is termed the oscillatory one. The computations revealed that in a certain range of basic parameters, the wave pattern shown in Fig. 4 was realized. The figure is drawn for the mixture defined as follows: $\beta_a=19.8$, $\beta_b=14.8$, $\alpha_a=7.9$, $\alpha_b=2.96$, $T_0=1010°K$, and $\epsilon=50$ mm. The specific feature of such processes is the existence of the decaying phase in the history of a shock wave that initially gained momentum. Figure 4 represents the case when the detonation wave is initiated in close proximity to ϵ-domain boundary, is switched off immediately, and passes to the outer region because of relatively low temperature of the region. The decaying shock wave serves as a precursor (see 3rd type)

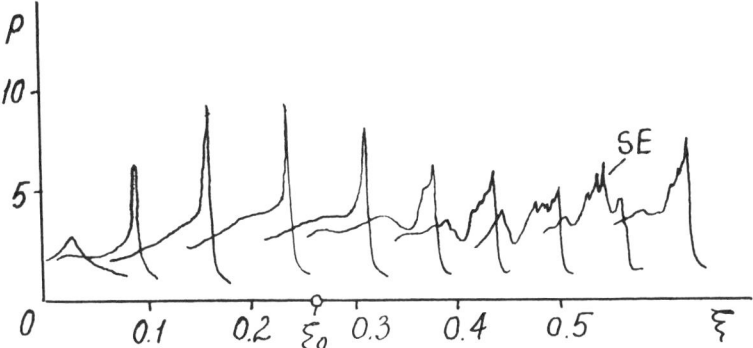

Fig. 4 Oscillatory explosion process. SE, second explosion. The time interval is $\Delta\tau_* = 0.027$.

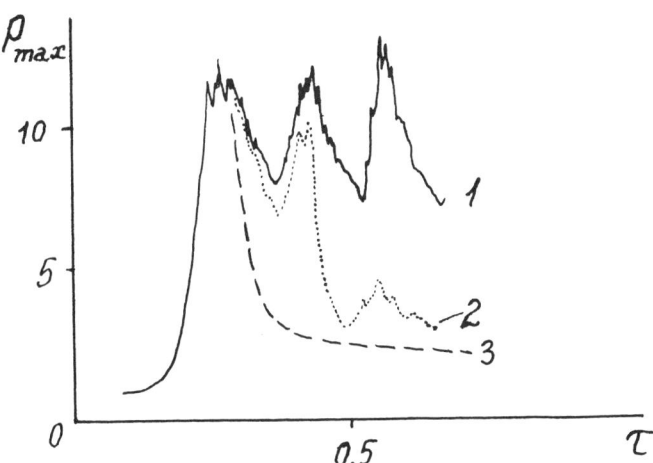

Fig. 5 The temporal evolution of pressure for oscillatory explosion processes. Curve 1 is representative of "galloping" detonation ($T_0 = 1013°K$, $\beta_a = 19.75$, $\beta_b = 14.8$, $\alpha_a = 7.9$, and $\alpha_a = 2.96$). Curve 2 is indicative of detonation fail ($T_0 = 1012°K$, $\beta_a = 19.8$, $\beta_b = 14.9$, $\alpha = 7.9$, and $\alpha_b = 2.96$). Curve 3 is representative of shock-wave decay in chemically inert medium.

and creates the conditions for amplification of a secondary shock wave.

Figure 5 shows the typical plots of peak pressure in the system against dimensionless time for this type of explosion process. Curve 1 corresponds to the "galloping" detonation that arises in a very narrow range of basic parameters. Dashed line 3 reflects the history of shock-wave decay in a chemically inert medium. It can be seen from Fig. 5 that strong shock waves are periodically being generated and decaying in the volume. The interval between subsequent secondary explosions is governed by ϵ and kinetic parameters of the mixture. Depending on the initial condition, the detonation wave or a series of shock waves reaches the wall.

Of great importance is the sensitivity that the explosion processes under investigation have to initial temperature variations. A temperature increase from $T_0=1010°K$ to $T_0=1020°K$ for the mixture whose evolution is presented in Fig. 4 led to the onset of stationary detonation like that shown in Fig. 1. For temperature near $T_0=1015°K$, the wave pattern was strongly dependent on T_0 and even on parameters of the numerical scheme (e.g., the coefficient of artificial viscosity). However, this does not mean that the chosen finite difference scheme is inadequate. There is evidence that in proximity to $T_0 \approx 1015°K$, the solution with the second type of explosion process becomes globally unstable.

Thus, our calculations show that spontaneous generation of shock and detonation waves may occur in spatially nonuniform reactive media. Fuel droplets in heterogeneous mixtures may have an effect on the propagation of an explosion wave inasmuch as temperature and vapor mass fraction gradients are, as a rule, inherent in the neighborhood of them. However, the small dimensions of gas-phase nonuniformities required for initiation of rather strong shock waves are indicative of a mechanism capable of providing some insights about the onset of nonlinear disturbances in both homogeneous and heterogeneous reactive systems.

The other question is the critical dimension of local nonuniformity for a given system. Although our attention has been restricted to isolated local spatial nonuniformity of temperature and reactant concentration, calculations were made with provision for two or more such local nonuniformities in the one-dimensional geometry. The situation is particularly interesting when the portions of additives, each taken separately, do not promote spontaneous generation of strong pressure waves. Our study shows that, under certain conditions and at the proper distance between the centers of local nonuniformities, the strong shock and detonation waves may arise spontaneously, proceeding from the hot spots formed by collisions of the precursor shock waves. These trends were also observed when the local spatial nonuniformities of temperature and reactant concentration were displayed near the wall or a contact surface of different gases. Thus, we conclude that

CONCENTRATION AND TEMPERATURE NONUNIFORMITIES   111

the presence of distributed spatial nonuniformities of chemical activity, contact surfaces, and boundaries makes for easier initiation of strong nonlinear pressure disturbances in explosive media.

Initial Temperature Gradient

The case of nonuniform initial temperature in the volume while species gradients are absent [i.e., $B_0=0$ in Eq. (2)] has been considered by Zeldovich et al. (1970) and Gelfand et al. (1984). In the following discussion the criterion for the spontaneous generation of strong pressure waves is derived. Self-ignition of the explosive mixture is assumed to occur well before the spreading or deformation of the temperature profile because of heat conduction and convection. The condition $U_c \approx U_g$ can be written in the form

$$dR/dt_i \approx \sqrt{\gamma R_* T_e} \qquad (5)$$

where t_i is the local induction time of the mixture. The derivative dR/dt_i defines the velocity of the spontaneous propagation of the chemical release front. The term dR/dt_i can be evaluated assuming that $t_i \approx t_{ad}$, where t_{ad} is the adiabatic ignition delay. The use of the well-known expression for t_{ad} (Frank-Kamenetski 1967) leads to

$$dR/dt_i \approx [(\gamma-1)/\gamma][Qk/\chi_T R_*]\exp(-E/RT) \qquad (6)$$

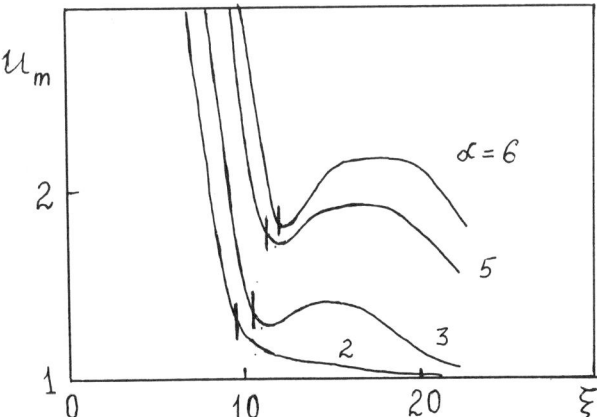

Fig. 6 Calculated velocity of peak pressure point against distance to the origin of temperature nonuniformity redrawn from [Gelfand et al. (1985)]. Vertical bars on the curves give the position predicted by Eq. (7).

Equations 5 and 6 together show that the chemical and gasdynamic time scales will be the temperature given by

$$T_c = (E/R_*)/\ln[(\gamma-1)Qk/(\gamma\chi_T R_*\sqrt{\gamma R_* T_e})] \qquad (7)$$

Amplification of a pressure wave generated by local explosion is possible in the range $T_e<T_c<T_0$. If $T_c<T_e$, then the wave patterns are very weak and almost uniform burnout of the mixture takes place throughout the volume. For temperatures approaching T_c decoupling of the motion of pressure waves and chemical release front may be expected. Hence, the critical value of temperature gradient $\chi_T = \chi_T^*$ exists such that, for $\chi_T > \chi_T^*$, the phase of pressure-wave amplification that results from interlocking with the local ignition event is absent. Thus, we conclude that the generation of strong pressure waves may be expected in the range $T_e<T_c<T_0$ for $\chi_T<\chi_T^*$. For $\chi_T>\chi_T^*$, within the limits $T_e<T_c<T_0$ as well as for $T_c>T_0$ the local explosion with further flame front formation may be expected.

Figure 6 taken from Gelfand et al. (1986) shows the dependence of the peak pressure point velocity U_m in the flow from distance to the origin. The numerical values that have been used for the parameters in the problem are $\chi_T = 2.5 \cdot 10^4$ K/m, $T_0 = 2000$ K, $K = 10^{10}$ s^{-1}, $V=1$, $\gamma=1.2$, and $E/RT_0 = 10$.

The vertical bars on the curves give the positions $R=R_c$ corresponding to the values of T_c that have been calculated on the basis of Eq. 7. The lower curve represents the case $\chi_b = \chi_T = 0$. Thus, it stands to reason that Eq. 5 reflects the main features of the phenomena that may occur in conditions close to self-ignition.

Influence of Additive

The effects produced by an additive with enhanced reactivity placed into ϵ-domain in a stepwise distribution are now considered. Thus, we assume that $\chi_b = \chi_T = 0$ in Eqs. 1 and 2 and that $T_0 = T_e$. We will show that in this situation not only explosion processes of the first and third types may make an appearance, but also processes of the second type may arise with the onset of detonation inside the ϵ-domain. The analysis under development may prove to be useful for understanding the mechanism of knock phenomenon in the internal combustion engine and for clearing up the possible combustion modes that may occur inside fuel-air clouds. In this instance the rarefaction wave propagates through the ϵ-domain after explosion-like self-ignition of component B, producing the conditions for rapid amplification of the pressure wave generated by subsequent explosion of component A. The analysis is based on Eq. 5.

Fast self-ignition of component B serves to increase the temperature to $T_b \approx T_e+(\gamma-1)Q_b/R_*$. For the first stage of ignition process $\alpha_b \approx 1.0$; i.e., the reaction heat is relatively small. This is especially true for both the additives promoting combustion and the cold-flame stage of low-temperature self-ignition of complex explosive mixtures.

As a result of thermal expansion of ϵ-domain, a weak pressure wave is generated in the outer region and a rarefaction wave propagates inside the domain. Temperature reduction at the contact surface may be easily evaluated from the elementary theory of shock tubes (Gaydon and Hurle 1963). With the acoustic approximation gives the temperature reduction is

$$\Delta T \approx \frac{\gamma-1}{\gamma} T_b (\sqrt{T_b/T_e} - 1)$$

The typical temperature gradient formed in the system before self-ignition of component A may be written as

$$\chi_T \approx \Delta T/\epsilon \tag{8}$$

To obtain the condition of pressure-wave amplification after subsequent self-ignition of component A, χ_T [Eq. 8] is equated to the temperature gradient given by Eqs. 5 and 6:

$$(T_b/\epsilon)(\sqrt{T_b/T_e} - 1) = (Q_a/R_*)[k_a \exp(-E_a/R_* T_b)]/\sqrt{\gamma R_* T_b} \tag{9}$$

After transforming and taking natural logarithms of both parts of Eq. 9 and neglecting small terms, the following condition of pressure-wave amplification can be obtained:

$$T_b \approx E_a/R_* \ln(Q_a k_a \epsilon/R_* T_e \sqrt{\gamma R_* T_e})$$

or, when rewritten in dimensionless quantities:

$$\theta_b \approx \beta_a/\ln(\alpha_a k_a \epsilon/\sqrt{\gamma R T_e}) \tag{10}$$

On the one hand, Eq. 10 allows the evaluation of the minimum spatial dimension $\epsilon=\epsilon_c$ of an additive portion with the enhanced reactivity required for triggering of the strong shock waves. On the other hand, Eq. 10 allows the determination of the heat of a subsidiary reaction Q_b, whose value is required to switch on the phenomenon

$$\alpha_b = (\delta-1)/(\gamma-1) \qquad (11)$$

where

$$\delta = \beta_a/\ln(\alpha_a k_a \epsilon/\sqrt{\gamma R T_e})$$

If the additive is absent (i.e., $\alpha_B=0$) and the mass fraction of component A is initially zero outside the ϵ-domain, then Eq. 10 allows the evaluation of initial temperature T_e, for which the explosion type of self-ignition of component A will be accompanied by amplification of the generated shock wave. This temperature may be given by the following transcendental equation:

$$1 + \beta_a^{-1} \approx \delta \qquad (12)$$

where the value of typical preignition heating $R_* T_b^2/E$ is used for initial heating of the mixture. In Fig. 7, Eq. 11 is represented by a dashed line. Our numerical

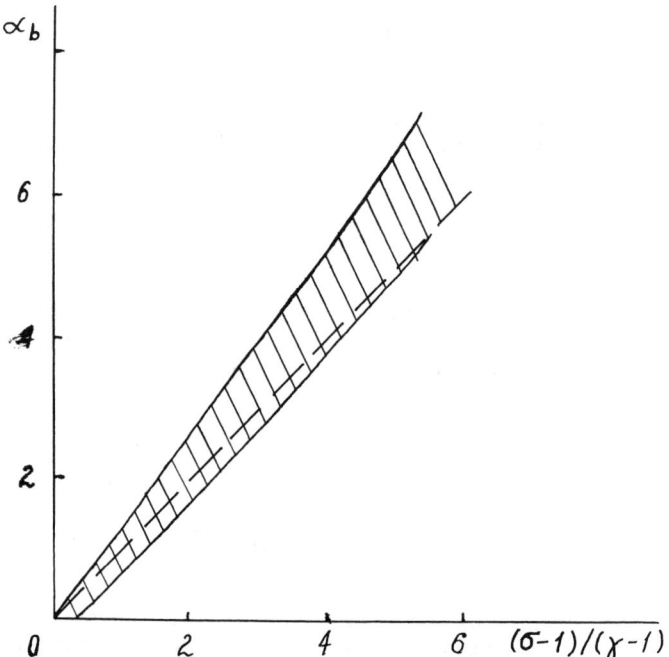

Fig. 7 Calculated dimensionless diagram that predicts the possible modes of heat release in the expanded spherical volume of gaseous reactive mixture.

calculations show that in the coordinates plotted along the axes of Fig. 7 the coupled motion of a pressure wave and a chemical release front occurs only in the dashed segment. As a matter of convenience it is assumed in calculations that the mixture outside the ϵ-domain is chemically inert. Near the lower boundary line of the segment, a shock wave lifts pressure to a maximum value at the contact surface that now may achieve the value of peak pressure in the CJ detonation wave. The upper boundary line of the segment is somewhat conditional, since it is drawn to meet condition $P_{max} \approx 1.1[1 + (\gamma-1)(\alpha_a+\alpha_B)]$. It is evident from Fig. 7 that the lower boundary line of the segment does not begin at the origin of the coordinates. The boundary lines of the segment displace only slightly under a variation of the type of symmetry.

Expansion

An analysis of a Lagrangian problem, i.e., the problem of accelerating a body by expanding combustion products is presented in the following. The main objective is to elucidate the role played by a rarefaction wave appearing in the wake of a moving body. The reactive mixture, which is confined in the volume $V_o = S\epsilon$ (where S is the cross-sectional area, and ϵ is the longitudinal dimension of a channel) between the impermeable wall from the left and a piston of mass M from the right, is assumed to undergo a one-step, first-order Arrhenius type of chemical reaction as written for component A in Eq. 3. At the initial instant the state of the mixture is uniform [i.e., $x_T=x_b=0$ in Eq. 4] and is defined by the values of pressure P_o and temperature T_o, whereas at the piston surface there is a vacuum, i.e., P=0 at the piston. For the sake of simplicity friction and heat losses are neglected when the piston is moved along the semi-infinite channel. The quantities of interest are piston velocity, maximum pressure rise, and impulse acting on the wall being associated with the motion of both the mixture and the piston.

To solve the problem, we use Eq. 3, with all terms involving the component B and the equation for component B omitted. The position of a closed-end of the channel is at R=0. The dimensionless mass of the piston is defined as

$$\mu = M\sqrt{\gamma R_* T_o} [k \exp(-\beta)]/\gamma P_o S$$

and write down the boundary conditions at the piston:

$$\xi = \xi_0: \quad dU/d\tau = P/\gamma\mu$$

Thus, the problem is formulated through the use of Eqs. 3 and 13 and initial conditions

$$\tau = 0, \quad \xi<\xi_0: \quad p = \vartheta = \theta = a = 1, \quad U = 0$$

The motion of the piston causes a rarefaction wave to propagate in the mixture. If inert gas is considered, the well-known expressions for a rarefaction wave in the time interval $0<\tau<\xi_0$ are

$$\xi - \omega = (U - \theta^{0.5})(\tau + \omega)$$

$$U = 2(1 - \theta^{0.5})/(\gamma - 1)$$

where $\omega = 2\gamma\mu/(\gamma + 1)$. The temperature distribution in an inert rarefaction wave is:

$$\theta = \{[2 - (\gamma - 1)(\xi - \omega)/(\tau + \omega)]/(\gamma + 1)\}^2$$

Let $\tau_{ad} = \gamma/(\gamma - 1)\alpha\beta$ be the adiabatic ignition delay of the reactive mixture at the initial temperature $\theta=1$. Then, for $\tau < \tau_{ad}$ and $\beta \gg 1$, the distribution given by Eq. 16 is valid for a rarefaction wave in an explosive medium.

If $\tau_{ad} < \xi_0$, then at the instant $\tau=\tau_{ad}$, a constant volume explosion takes place in the interval $(0, \xi_0 - \tau_{ad})$. For $\beta \gg 1$ the overpressure produced by such an explosion is $\Delta p = (\gamma - 1)\alpha$. The pressure jump that thus appears serves to form a compression wave in the system.

The temperature distribution of Eq. 16 results in a gradient of ignition delay increasing toward the piston and brings about nonuniform chemical energy release along the length of the channel. The process of successive self-ignition of portions of the mixture may be considered as propagation of a chemical release front.

The variation of gas temperature at a piston surface for the instant of time $\tau=\tau_{ad}$ is given by the expression

$$\Delta\theta_p = 1 - \{1 - [(\gamma - 1)/2]U_p\}^2$$

where, the subscript p denotes the quantities at the piston surface and U_p is the piston velocity. If $\Delta\theta_p$ is equal to β^{-1}, the typical scale of temperature difference for explosion processes, the conditions close to the self-ignition limit are fulfilled in the neighborhood of the

piston. Hence a distinction between the two following limits arising in the problem can be made.
The first limit is $\Delta\theta_p \ll \beta^{-1}$. In this case the velocity of the chemical release front is much greater than the typical gasdynamic velocity, and a constant volume explosion consumes the mixture with almost uniform pressure rise along the length of the channel. If we take into account that $U_p \approx \tau_{ad}/\gamma\mu$, the condition under examination may be written in the form

$$1 - \{1 - [(\gamma - 1)/2]\tau_{ad}/\gamma\mu\}^2 \ll \beta^{-1}$$

As long as $\beta^{-1} \ll 1$, $\alpha\mu \gg 1$. The same condition may be derived from the exact expression for U_p in the rarefaction wave.

Self-ignition may also occur for $\tau_{ad} > \xi_0$. At the instant $\tau = \xi_0$, the head of the rarefaction wave is reflected from the closed end of the channel. The maximum decrease in temperature in the reflected rarefaction wave is achieved at the time $\tau \approx \tau_1 = 2\xi_0 + \xi_0^2/\omega$ and is given by Eq. 17. Here, τ_1 is the time taken for the reflected wave to reach the piston in chemically inert gas. If ignition occurs at that time, Eq. 18 remains valid.

For $\tau_{ad} > \tau_1$, the wave reflected from the piston should be taken into account. Inasmuch as we consider small decreases in temperature for the time of two reflections, we may assume the stationary temperature variation in the channel. In this case the condition $\theta_p \ll \beta^{-1}$ will be given by

$$1 - (1 + \Delta\xi/\xi_0)^{-(\gamma - 1)} \ll \beta^{-1}$$

Since $\Delta\xi \approx U_p\tau_{ad}/2$, $\alpha\mu \gg \tau_{ad}/2\xi_0 > 1$ from Eq. (18).
Thus, the expression $\Delta\theta_p \ll \beta^{-1}$ is equivalent to Eq. (18). For the case in point, the value of the maximum pressure is less than or equal to $P_{max} = 1 + \Delta P$. In contrast to the Lagrangian problem with momentary energy release, the maximum piston acceleration in this instance is achieved at $\tau \geq \tau_{ad}$. Hence, the impulse acting on channel walls changes stepwise after the reaction front reaches the piston.

It should be noted that Eq. 18 contains some typical mass $\rho_0 S\sqrt{\gamma R_* T_0}/[k \exp(-\beta)]$ instead of the total mass of the mixture. Here ρ_0 is the initial density of the gas. This quantity corresponds to the mass of the mixture involved in motion by time $\tau = 1$, which is equal to the chemical time scale.

The second limit is $\alpha\mu\ll 1$. In this case, velocity of the heat release front at $\tau \geq \tau_{ad}$ is much less than the typical gasdynamic velocity. Indeed, for $\beta \gg 1$ we have

$$\alpha\mu = \gamma\mu/(\gamma-1)\tau_{ad}\beta$$

$$\approx [(\partial\xi/\partial\theta)_{\tau=\tau_{ad}}/(\partial\tau_i/\partial\theta)_{\theta=1}] = \partial\xi/\partial\tau_i \ll 1$$
$$\xi=-\tau_{ad}$$

where $\tau_i = \tau_{ad}\theta^2 \exp[-\beta(1-\theta^{-1})]$ is the induction time at temperature θ. The derivative $\partial\xi/\partial\tau_i$ determines a velocity of the heat release front as well as the typical gasdynamic velocity, where $\sqrt{\theta}\approx 1$.

In the case under study the maximum pressure rise is also not greater than $P_{max} = 1+\Delta P$ for all the time the piston is accelerated. After the self-ignition event, a decaying shock wave propagates through the mixture. The arrival of the shock wave at the piston surface leads to a change in its motion and a varying impulse acting on the channel walls. Shock-wave loading of the piston may generate inertial effects with piston velocity jumps.

In the intermediate case ($\alpha\mu\approx 1$) the amplification of a shock wave due to coupling with a local chemical release may be expected after self-ignition. In accordance with the results of section 1, the onset of detonation may occur in this case, and secondary explosions and decaying precursor shock waves may also occur. Reflection of the detonation wave from the piston may lift the pressure level to such an extent that the channel walls may burst. The additional effect on the channel walls is connected with longitudinal impulsive loading caused by the periodically accelerating motion of a body after the stroke.

Next, we will derive the condition for cessation of self-ignition. The critical condition of self-ignition at $\tau=\xi$, may be obtained by equating the rate of heat release in the mixture with the rate of cooling the mixture in the rarefaction wave. In keeping with Eq. 16 the rate of cooling in the first rarefaction wave is given by ($\tau=\xi_0$, $\xi=-\xi_0$)

$$\frac{\partial\theta}{\partial\tau} = -2(\gamma-1)/[(\gamma+1)(\xi_0+\omega)]$$

The rate of temperature rise due to chemical energy release is

$$\frac{\partial\theta}{\partial\tau} = \frac{\exp[\beta(\theta-1)]}{\beta\tau_{ad}}$$

Since $\beta(\theta-1) = -\ln(1-\xi_0/\tau_{ad})$, we have

$$\partial\theta/\partial\tau = [\beta(\tau_{ad} - \xi_0)]^{-1}$$

The critical condition may then be written in the form

$$2(\gamma-1)/[(\gamma+1)(\xi_0+\omega)] = \beta(\tau_{ad}-\xi_0)^{-1}$$

or after transformation,

$$1-\alpha\mu \geq (\xi_0/\tau_{ad})[1 + (\gamma+1)/2\beta(\gamma-1)]$$

Equation 19 is true for the first rarefaction wave and is also valid for a reflected wave ($\tau>\xi_0$). As evidenced by Eq. 19, self-ignition is possible when two conditions are met:

$$\alpha\mu \ll 1 \text{ and } \tau_{ad} \leq \xi_0[1 + (\gamma+1)/2\beta(\gamma-1)]$$

For $\alpha\mu \gg 1$ and $\beta \gg 1$, diminution of temperature in a rarefaction wave is $\Delta\theta \leq \beta^{-1}$, and self-ignition always occurs.

Thus, we may give the four possible models of system evolution distinguished by a value of the parameter $\alpha\mu$:

1) For $\alpha\mu \ll 1$ and $\tau_{ad} > \xi[1 + (\gamma+1)/2\beta(\gamma-1)]$, self-ignition does not occur, and motion of the piston is that of the case of inert gas expansion.

2) For $\alpha\mu \ll 1$ and $\tau_{ad} \leq \xi(1 +(\gamma+1)/2\beta(\gamma-1))$, a decaying shock wave propagates all along the mixture.

3) For $\alpha\mu \approx 1$, a shock wave is generated as a result of self-ignition, which accelerates because of coupling with local energy release.

4) For $\alpha\mu \gg 1$, almost uniform burnout of the mixture takes place, and shock waves are absent.

Figure 8 shows the four parametric domains. The plot is drawn on the basis of numerical solution of Eq. 3 with the initial and boundary conditions [Eqs. 13 and 14]. Along the y axis the ratio of the typical gasdynamic time scale to the typical chemical heating time scale $\Pi = \xi_0/\tau_{ad}$ is plotted. Along the x axis the ratio of the total mass of the gas to the piston mass $m/M = P_0V_0/MRT_0$ is plotted.

The boundary line separating domains 1 and 2 fits well with the equation

$$\Pi = mM^{-1}/\{[\gamma/\beta(\gamma-1)] + mM^{-1}[1 + (\gamma+1)/2\beta(\gamma-1)]\}$$

which can be obtained from the critical condition, Eq. 19. Domain 2, which is representative of the parametric region

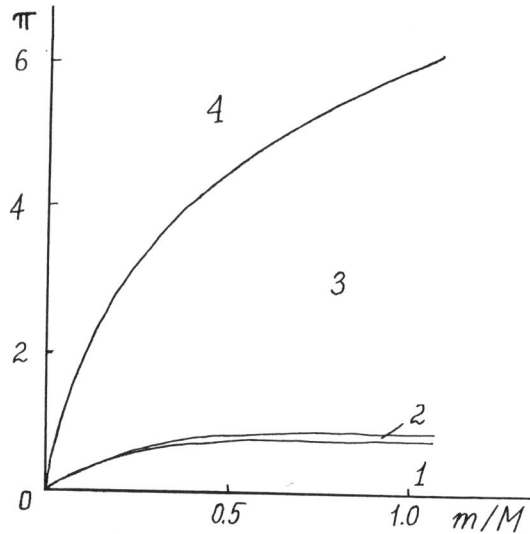

Fig. 8 Calculated dimensionless diagram predicting the flow pattern in the reactive mixture behind the moving piston.

with decaying shock waves, is narrow compared with the other domains. Such flow patterns may arise under large gradients of induction time or in conditions close to the limit of self-ignition. The boundary line separating domains 3 and 4 is conditional to some extent. In the case under consideration the boundary line is drawn through the points in which the maximum pressure is not larger than $P_{max} = 1.1(1+\Delta p)$.

The phenomena typical for domain 3 are very sensitive to small variations of temperature and slight additives that affect the rate of chemical release. It is evident from Fig. 8 that solutions fall into this domain only for a narrow range of T_0. For instance, for m/M=0.1, β=20, and T_0=1000°K, variation of the initial temperature by 50°K leads to cessation of an amplification phase in the evolution of a compression wave and the solution falls into domain 1 or 4.

The calculated time interval required for amplification of the compression wave depends of the quantities Π and m/M. For $\Pi \approx 1$ and m/M$\approx$1 it is about a hundredth of a typical time scale of chemical reaction $t_c = 1/k \exp(-\beta)$. Therefore, the processes under investigation are stable with respect to low-frequency acoustic oscillations and turbulent fluctuations whose characteristic time scales exceed the amplification time. Moreover, such oscillations and fluctuations may become a cause of shock wave generation under normal operating conditions in a compression machine operated in parametric domain 1.

The present model is indeed in qualitative agreement with experimental observations of the combustion process in an internal combustion engine. In accordance with the work

The present model is indeed in qualitative agreement with experimental observations of the combustion process in an internal combustion engine. In accordance with the work of Sokolik (1934), as the compression ratio is increasing, the knock phenomenon tends to gain momentum at first, with later attenuation and cessation. In the framework of the model in question, the knock phenomenon begins in domain 2 in Fig. 8. As temperature and pressure increase in the channel, the effect of explosion tends to increase too (see domain 3). Further increase of the compression ratio results in a pressure drop in the lead shock wave, and ultimately the phenomenon ceases (domain 4).

Figure 9 shows the calculated piston velocity history for m/M=1. Curve 1 corresponds to the case of a piston set in motion by a gaseous mixture with low reactivity ($\Pi = 10^{-5}$), during which time piston acceleration self-ignition does not occur. Curve 1 appears to be in good agreement with the analytical solution of the Lagrangian problem for the case of inert gas expansion. Curve 3 is representative

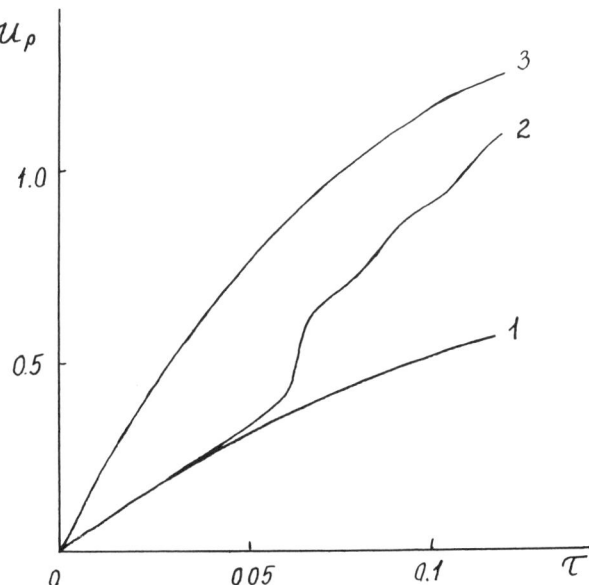

Fig. 9 Piston velocity history for various flow patterns in a channel ($t_c = 8 \cdot 10^{-3}$s)

of the piston velocity history for the gaseous mixture with high reactivity ($\Pi=10^4$). As distinct from curve 1, the greatest acceleration in the instance is achieved at the moment $\tau = \tau_{ad}>0$. For $\Pi\to\infty$ the piston velocity appears to be in good agreement with the theory of acceleration of a body by chemically inert combustion products. Curve 2 is drawn for $\Pi=2.5$, which insures that the characteristic point falls into domain 3 in Fig. 8. At the instant $\tau\approx0.062$, a stepwise acceleration of the piston occurs (seen in Fig. 9). This is a result of detonation wave reflection from the moving piston. From this point on the motion of the piston becomes oscillatory because of inertial effects of the gaseous volume.

The point of particular interest is the maximum piston velocity (for $\tau\to\infty$), when m/M becomes constant for different flow patterns in the channel. Our calculations show that the maximum piston velocity increases as Π increases, and the greatest value is achieved for the momentary heat release at $\tau=0$ ($\Pi\to\infty$).

Conclusion

The present analysis has demonstrated that the spontaneous generation of shock and detonation waves may occur in conditions close to self-ignition of the reactive mixture. A variety of factors studied may give an impetus to the onset of pressure waves. Among them are the temperature nonuniformities and/or portions of the mixture with distributed reactant concentration, boundaries and contact surfaces, and variation of initial and boundary conditions. It is obvious that the results obtained are valid not only for the initially quiescent reactive mixture but also hold true for a stream. An additional factor that may give rise to the phenomena mentioned above is the nonuniform residence time distribution of mixture portions in a combustion chamber. The results may provide insight into the onset of nonlinear combustion instability in various technical systems as well as the transition to detonation in accidental explosions.

References

Barthel, H. O. and Strehlow, R. A. (1979) Direct detonation initiation by localized enhanced reactivity. AIAA Paper 79-0286.

Borisov, A. A., Gelfand, B. E., Gubin, S. A., Kogarko, S. M., and Podgrebenkov, A. L. (1970) Amplification of weak shock waves in a burning two-phase liquid-gas system. Z. Pricl. Mech. Tech. Fiz., 1, 168-173.

Frank-Kamenetski, D. A. (1967) Diffusion and Heat Transfer in Chemical Kinetics, Nauka, Moscow, USSR.

Gaydon, A. G. and Hurle, I. R. (1963) The Shock Tube in High Temperature Chemical Physics. Chapman & Hall, London, UK.

Gelfand, B. E., Polenov, A. N., Frolov, S. M., and Tsyganov, S. A. (1985) On the onset of detonation in ununiformly heated gaseous mixture. Fiz. Goreniya Vzryva, 21, 118-123.

Gelfand, B. E., Frolov, S. M., Polenov, A. N., and Tsyganov, S. A. (1986) The onset of detonation in systems with ununiform distribution of temperature and reactant concentration. Khim. Fiz., 5, 1277-1284.

Harrje, D. T. and Reardon, F. H. (1972) Liquid Propellant Rocket Combustion Instability, NASA SP-194, Washington, DC.

Knystautas, R., Lee, J. H., and Moen, I. (1979) Direct initiation of spherical detonation by a hot turbulent gas jet. 17th Symposium (International) on Combustion, Pittsburgh, PA, 1235-1245.

Lee, J. H., Knystautas, R., and Yoshikama, N. (1978) Photochemical initiation of gaseous detonation. Acta Astronaut., 5, 971-982.

Ragland, K. W., Dabora, E. K., and Nicholls, J. A. (1968) Observed structure of spray detonations. Phys. Fluids 11, 2377-2382.

Richtmyer, R. D. and Morton, K. W. (1967) Difference Methods for Initial-Value Problems, 2nd ed, Interscience, New York.

Sokolik, A. S. (1934) Combustion and Detonation in Gases, Gostechteorizdat, Moscow, USSR.

Zeldovich, Y. B. (1955) Theory of Detonation, Gostechteorizdat, Moscow, USSR.

Zeldovich, Y. B., Librovich, V. B., Mahviladze, G. M., and Sivashinski, C. I. (1970) On the onset of detonation in ununiformly heated gas. Z. Pricl. Mech. Tech. Fiz., 2, 76-81.

Zeldovich, Y. B., Barenblatt, G. I., Librovich, V. B., and Machviladze, G. M. (1980) Mathematical Theory of Combustion and Explosion, Nauka, Moscow, USSR.

Zeldovich, Y. B. (1981) Fluctuations of induction period of branching chain reaction. Dokl. A kad. Nauk SSSR, 257, 1173-1174.

Heat Evolution Kinetics in High-Temperature Ignition of Hydrocarbon/Air or Oxygen Mixtures

A. A. Borisov,*, V. M. Zamanskii,† V. V. Lisyanskii,‡
G. I. Skachkov,§ and K. Y. Troshin¶
USSR Academy of Sciences Moscow, USSR

Abstract

Ignition delays and heat-release times in mixtures of hydrocarbons (CH_4, C_2H_4, C_2H_6, C_3H_8, and C_7H_{16}) with air or oxygen of various composition were determined using pressure and luminosity records. Experiments were carried out in a shock tube. The characteristic heat-release times in the explosion are shown to be only slightly dependent on the initial temperature (the effective activation energy of about 6 kcal/mole) and pressure (pressure exponent no more than 0.3). Their values range between 10^{-5} and 10^{-4} s for different mixtures. Unlike the ignition delays, the heat-release times are not appreciably affected by promoters. At the detonation limits in tubes the heat-release times become commensurate with ignition delays, whereas within the limits they are always greater than the induction periods. Approximate expressions are derived for modeling the heat-release rate during the induction period and explosion time from thermal explosion theory with an effective reaction order n. The value of n varies from 2 to 4.5 for different mixtures.

Copyright © 1988 by the American Institute of Aeronautics and Astronautics, Inc. All rights reserved.
* Institute of Chemical Physics.
‡ Institute of Chemical Physics.
§ Institute of Chemical Physics.
¶ Institute of Chemical Physics.

Introduction

In studying ignition processes at practical conditions, one usually assumes that the ignition delay τ_i (or induction period) is much greater than the time required for the major part of the reaction heat to be released (or the explosion time τ_e). This is true for self-ignition at very high dimensionless activation energies, E/RT_0 (where E is the effective activation energy of the reaction, and T_0 is the initial temperature of the gas mixture), i.e., for the processes occurring at relatively low temperatures. The total heat evolution time, the sum $\tau_i+\tau_e$, in this case consists virtually of τ_i only. As a consequence, in practical calculations the error in describing the heat-release curve during τ_e and in determination of τ_e itself can be neglected.

Numerous experimental data obtained in static conditions, rapid compression machines, and shock tubes at temperatures 900 and 1500°K and pressures no higher than a few bars show that as a rule τ_i is at least several times greater than τ_e. However, shock tube data at sufficiently high temperatures suggest that the relation between τ_i and τ_e may reverse and that the heat-release kinetics in an exothermic reaction may look quite different from the kinetics characteristic of the thermal, or chain-thermal, explosion.

Inasmuch as the heat release kinetics in detonations follows the adiabatic self ignition laws (at least in the major part of the reacting mixture), and since the chemical kinetics is one of the principal factors controlling the detonation wave decay at the limits, deflagration-to-detonation transition, and detonation initiation, it is important to elucidate the extent to which the heat evolution history behind the detonation front corresponds to that inherent in the chain-thermal explosion. Moreover, the detonation instability problems cannot be solved based solely on the induction periods without invoking the data on the entire heat-release profile.

While the data on ignition delays in the temperature, and pressure range characteristic of detonations, are widely available in the literature [see, e.g., Kogarko and Borisov (1960) and Steinberg and Kaskan (1955)], information on the values of τ_e and heat-release kinetics at the stage of fast acceleration of the reaction is scarce. Even detonation calculations based on detailed chemical reaction schemes [Westbrook (1982); Atkinson and Bull (1981)] cannot reliably predict heat-release rates for large burned fractions. The very high temperatures, presence of oxidative-pyrolysis products, and a large number of radicals may substantially alter the set of elementary reactions that control the heat-release rate at these stages compared with the reaction schemes based primarily on the analysis of chemical reactions during the induction period. The expressions for heat-release rates derived from the data on ignition delays were shown to be inadequate by Bishimov, et al. (1969) who introduced heat-release stages with strong and weak dependences of their rates on temperature.

In the present work we have studied experimentally the heat-release function behavior in gaseous mixtures during the entire reaction course. Our main attention was focused on the estimation of temporal characteristics of the two reaction stages, rather than on the determination of the heat-release profiles. This is because hot spot formation becomes important in ignition of energetic mixtures behind shock waves. The pressure, or luminescence, time histories recorded in the experiment can hardly insure an accurate assessment of the heat-release curve inherent in a homogeneous reaction, although the characteristic time intervals, as shown by the comparison of ignition delay data (Borisov 1970), are measured with good accuracy. Moreover, precise techniques for measuring heat-release rates characteristic of detonation processes are not developed; therefore approximate ones must be employed. And finally, it is sufficient (and frequently even desirable) for many detonation calculations to use a "brutto" (global) model for heat-release kinetics. For such an equation the characteristic times are used as the parameters for fitting the suggested heat-release curves to the experimental ones.

Experiments

Experiments were conducted behind reflected shock waves in a shock tube. Details of the measurement technique can be found elsewhere (Borisov and Skachkov 1972). The gas parameters behind reflected shock waves were calculated based on the conservation laws from the known gas enthalpies and measured velocities of incident shock waves. Pressure and light absorption (or emission) at 306 nm were monitored behind the shock-wave front.

Ignition of mixtures of CH_4, C_2H_4, C_2H_6, C_3H_8, C_4H_{10}, and C_7H_{16} with air or oxygen was studied at temperatures from 1250 to 1700°K and pressures in the range of 0.6 to 8 bar. The composition of the mixtures investigated is presented in Table 1.

Table 1 Composition of the mixtures studied

N	Type of fuel RH	Concentration RH	O_2	I
1	CH_4	6	12	(82)
2		10	20	(70)
3		9.5	19	71.5
4	C_2H_4	5	15	(80)
5	C_2H_6	6	20	74
6		4.5	16	(79.5)
7		7.8	19.6	72.6
8		4.3	21.3	74.4
9	C_3H_8	3.3	16.6	(80.1)
10		2.6	20.4	77
11		2	10	88
12		1.5	32.5	(66)
13	C_7H_{16}	1.5	16.5	(82)

Concentrations are in vol %. Diluant,I, is either N_2 or A_r. Values in parentheses indicate concentration of Ar for mixtures diluted with Ar; values without parenthesis indicate the diluant is N_2.

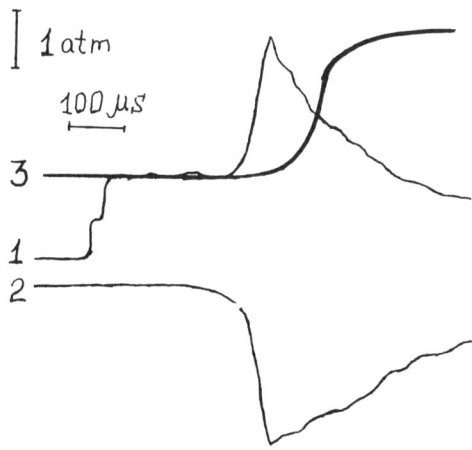

Fig. 1 Typical oscillogram for ignition of a gaseous mixture behind a reflected shock wave. Mixture: 3, T_0 = 1669°K, p_0 = 2,6 bar. Line 1: pressure-time history; line 2: light absorbtion at λ = 306nm. Line 3 is calculated using Eq. (6).

Figure 1 shows pressure and luminosity records behind a reflected shock wave. These records suggest that the heat-release curve can be divided into two parts: the induction period τ_i, during which the pressure and luminosity rises are negligible, and heat-release (or explosion) time τ_e. The characteristic times τ_i and τ_e

determined from both the pressure and luminosity traces are practically identical. The following procedure of τ_i and τ_e determination is thus adopted: τ_i was measured from the moment of the reflected shock arrival at the pressure gauge to the sharp pressure rise (this latter moment is well-defined at relatively low temperatures); τ_e is the time interval between the beginning of the steep pressure rise to the pressure maximum. At high temperatures the moment of the sharp pressure rise is not as well defined; therefore, τ_i is measured as a time cutoff on the time axis by the line tangent to the p(t) curve (where p is pressure, and t is time) at the inflection point, and τ_e is the time interval between this intersection point and that of the tangent line and the line $p = p_{max}$ (where p_{max} is maximal pressure). The two definitions of τ_e underestimate their values because they do not take into account pressure losses due to expansion of the gas behind the shock wave, which makes the $\Delta p(t)$ curve steeper than that for the constant-volume explosion and shifts the maximum of this curve toward time zero. However, since the pressure rise rates near the Δp maximum are usually very high, this time shift is believed to be small (at least less than the measured τ_e); hence, the main explosion heat is released during the τ_e. The values of τ_e measured from $\Delta p(t)$ records and light emission curves agree within the experimental error (±30%).

Figure 2 shows the Arrhenius plots (the reciprocal initial temperature) of τ_i and τ_e for self-ignition of stoichiometric mixtures of various hydrocarbons with air or

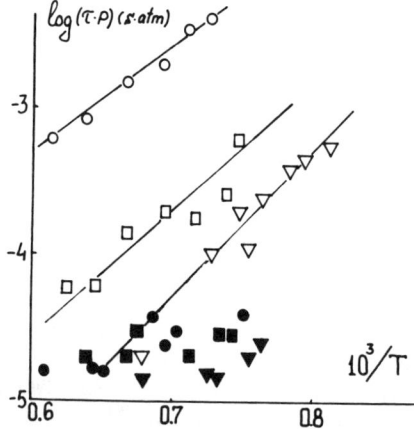

Fig. 2 Arrhenius plots of τ_i (open symbols) and τ_e (filled symbols) for various fuels. O, Mixture 1; □, mixture 9; ▽, mixture 13.

O_2 and Ar. These results indicate that τ_i depends strongly on the nature of the hydrocarbon and temperature, whereas τ_e practically does not vary from one fuel to another. The effective activation energy of τ_e is much lower than that of τ_i and is about 6 kcal/mole. A low effective activation energy of τ_e does not necessarily mean that temperature-insensitive elementary reactions (e.g., recombination reactions) become rate-controlling, since its actual value should be estimated not at the initial temperature but on the varying temperature during the explosion. At temperatures characteristic of explosions the effective activation energy undoubtedly is much higher than 6 kcal/mole.

A relatively large scatter of the data on τ_e, despite its weak temperature dependence, is presumably due to the stochastic evolution of the process after local ignition and to chaotic interactions of the compression waves generated by the exothermic reaction. The reaction occurs under the semiconfined conditions; therefore, at the initial slow heat-release stage, the pressure recorded by the gage rises slower than the temperature. The pressure rise becomes nearly proportional to the temperature rise only at the fast reaction stage, when the process occurs almost under confined conditions.

It is well-known that the induction periods τ_i for hydrocarbon/oxygen mixtures are dependent on pressure. This pressure dependence is slightly weaker than the proportionality to $(p)^{-1}$. For example, experiments performed with propane/air mixtures at pressures in the

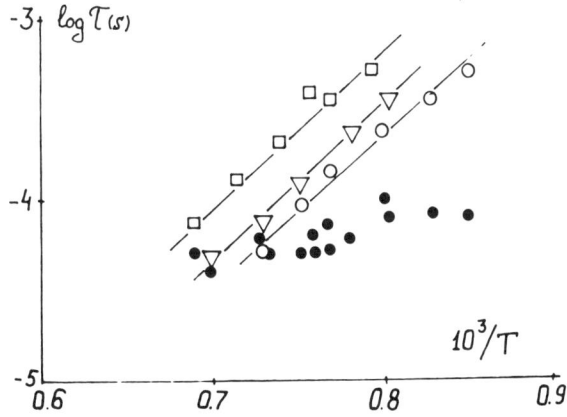

Fig. 3 Arrhenius plots of τ_i (open symbols) and τ_e (filled symbols) for mixture 8. □, p_0 = 1 bar; ▽, p_0 = 2 bar; O, p_0 = 5 bar.

range of 1 to 5 bar (Fig. 3) yield the dependence $\tau_i \propto p^{-0.7}$. A similar relation is also obtained for methane/air mixtures (Borisov et al. 1983). The measured τ_e values presented in the same figure demonstrate that τ_e dependency on pressure is much weaker. The order is difficult to determine because the experimental error is too large and the pressure range is too small, but at least the order does not exceed 0.3.

Thus, for all the hydrocarbons studied τ_i becomes equal to τ_e in the range of several tens to 100 μs at temperatures that depend on the type of hydrocarbon and pressures. The higher the pressure the lower the temperature at which $\tau_i = \tau_e$. At higher temperatures the total heat evolution time is bounded from below by the value of τ_e, and τ_i with its high pressure and temperature sensitivity can no longer be used in ignition and gas-dynamic calculations in this region of the parameters.

It was interesting to study the effect of promoters on the value of τ_e. In our experiments we employed isopropyl nitrate, methyl nitrate, cyclohexylnitrate, N_2F_4, and NO_2 as promoters. The previous studies (Borisov et al. 1985) have demonstrated that, when added to the fuel in amounts of about 15% and lower, the promoters may reduce τ_i by up to an order of magnitude in the range of τ_i below 10^{-3}s. The measurements displayed in Fig. 4 show that τ_e is only slightly affected by the additives.

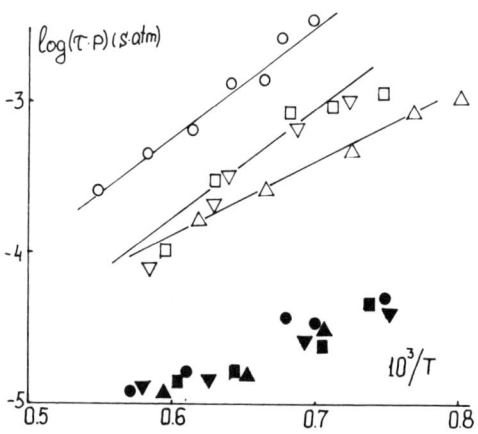

Fig. 4 Arrhenius plots of τ_i (open symbols) and τ_e (filled symbols) for mixture 1 (O) with a promoter added in various concentrations: □, 5% CH_3ONO_2 (with respect to CH_4); △, 15% N_2F_4. Black points designate heat-release times (τ_e).

Thus, we can summarize the results as follows. Ignition delays can be considered as characteristic heat-release times in hydrocarbon/air mixtures only when they are higher that 10^{-4} s. At higher heat-release rates the explosion process degenerates and τ_e becomes dominant.

Discussion

Increased temperature, pressure, and reactant concentrations may reverse the ratio of τ_e to τ_i. This changeover is quite plausible in detonation waves, especially at atmospheric initial pressure, because they are characterized by temperatures and pressures behind the lead shock front higher than 1400°K and 20 bar. Therefore, it is interesting to compare quantitatively the values of τ_i and τ_e in detonation waves.

The times τ_i and τ_e behind the lead shock wave of Chapman-Jouquet detonations in ethane-air mixtures of different composition are presented in Fig. 5a. The dashed lines indicate the concentration limits in these mixtures measured by Borisov and Loban (1977). The τ_e vs 1/T curve is plotted using the expression

$$\tau_e = 7.8 \times 10^{-6} p^{-0.3} s \, \exp(6000/RT)$$

found in this work based on the experimental data. Here the activation energy is in calories per mole, and the pressure is in bars. The equation for τ_i was borrowed from the work of Burcat et al. (1972). The results for other hydrocarbon/air mixtures are quite similar (see, e.g., Fig. 5b, where the data for ethylene/air mixtures are shown).

Although, in a detonation wave with a cellular structure a variety of characteristic reaction times exist, we can infer from Fig. 5 that, for the major fraction of the cross-sectional area of the detonation front within the detonation limit, heat evolution kinetics is determined by τ_e rather than by τ_i. This implies a degenerative explosion with a lower temperature sensitivity of the heat-release profile than in a normal explosion. This circumstance will certainly enhance stability of the detonation wave with respect to gasdynamic disturbances when the mixture composition departs from the concentration limits. The detonation limit of air/hydrocarbon mixtures per se arises apparently when the temperature sensitivity of the heat-release time changes abruptly (i.e., when $\tau_i \approx \tau_e$). The heat losses in the induction zone may vary its length dramatically because of a high effective activation energy of the induction period and thus result in local decay of the detonation front that eventually can cause

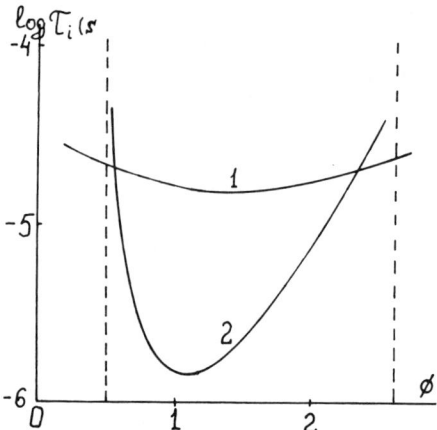

Fig. 5a τ_e and τ_i as functions of the air equivalence ratio in detonation waves C_2H_6/air mixtures behind the plane shock wave propagating at Chapman-Jouquet velocity. Dashed lines indicate the concentration detonation limits from Borisov and Loban (1977). The relationships for τ_i are from Burcat et al. (1972), and τ_e are measured in the present work.

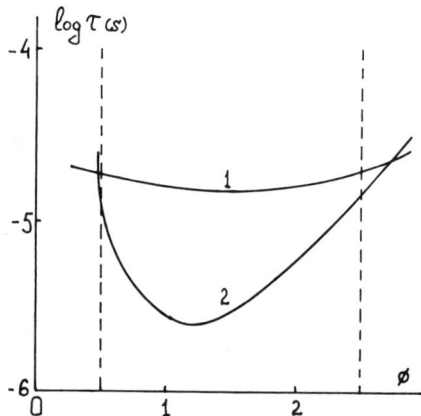

Fig. 5b τ_e and τ_i as functions of the air equivalence ratio in detonation waves in C_2H_4/air mixtures behind the plane shock wave propagating at a Chapman-Jouquet velocity. Symbols as in Fig. 5a.

detonation failure. The same losses are unable to appreciably change the heat-release profile under the conditions when $\tau_e > \tau_i$.

The conclusion that limiting detonation conditions are achieved when the temperature-sensitive induction period becomes dominant is confirmed by the data on initiation of a spherical detonation. Radial shock-velocity profiles for an explosion of a high explosive charge in an unconfined C_3H_8/air mixture (Edwards et al. 1976) are presented in Fig. 6. The estimates based on our data on τ_i and τ_e show that for a critical charge size (below which no initiation of gaseous detonation occurs) the values of τ_i and τ_e are comparable at the shock velocity (D) that corresponds to the minimum of the D(r) where r is the radial distance from the ignition source curve. This implies that as soon as the heat-release curve acquires the shape typical of thermal explosion, i.e., as soon as the temperature-sensitive induction stage of the reaction becomes dominant, the wave becomes unstable and decays. More stable regimes of degenerative explosions result in reinitiation of detonation after the wave velocity goes through the minimum, with a much higher probability than in the case of ideal chain-thermal explosion.

Thus, the concept of characteristic τ_i and τ_e offers a reasonable explanation of some marginal features of detonations. These quantities can also be utilized in detonation calculations as parameters characterizing individual stages of the reaction. The heat-release profile can be reconstructed based on these parameters. We will try to derive a single expression for the heat-release rate that can 1) be applied in a wide temperature range, 2) describe the temperature rise and reactant concentration variations during both the induction period and energy release stage, and 3) reflect more or less reasonably the chemistry of the explosion reaction.

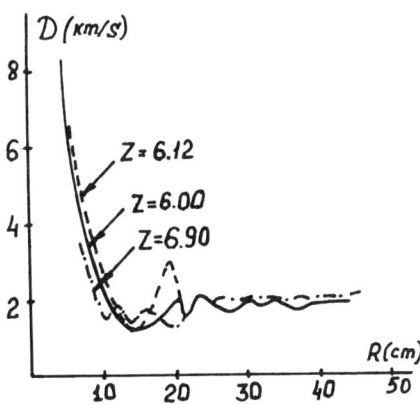

Fig. 6 Observed variations of detonation velocity with radial distance in mixtures $C_3H_8+5O_2+ZN_2$. Initiating source 2.5 g of tetryl, and p_0 = 1 bar..

When deriving this expression, we employ the relation between τ_i and the reaction rate that follows from the theory of thermal explosion (Frank-Kamenetzky 1967)

$$\tau_i = \frac{C_p[M]RT_0^2}{QEW_0} \tag{1}$$

Here C_p is the specific heat, $[M]$ the total concentration of molecules, T_0 the initial temperature, Q the reaction heat per fuel mole, E the effective activation energy, and W_0 the reaction rate at time zero. We assume that the reaction rate W_0 in Eq. (1) is just the initial value of the reaction rate $[W = Z_0(a)^n(b)^m \exp(-E/RT)]$ that describes the behavior of the reaction during the entire heat evolution period, i.e., during the induction period (when concentrations $a \approx a_0$, $b \approx b_0$, and $T \approx T_0$) and the explosion period. Thus,

$$W = W_0(a/a_0)^n(b/b_0)^m \exp[-(E/R)(1/T - 1/T_0)] \tag{2}$$

Combining Eqs. (1) and (2) yields

$$-da/dt = \frac{C_p[M]RT_0^2}{EQ\tau_i}(a/a_0)^n(b/b_0)^m \cdot$$

$$\exp\{[-(E/R)[1/T - 1/T_0]\} \tag{3}$$

According to the definition of Q

$$a_0 - a = \frac{C_p[M]}{Q}(T-T_0) = [(T-T_0)/(T_b-T_0)]a_0 \tag{4}$$

where T_b is the maximal explosion temperature. A similar equation can be derived for b

$$b_0 - b = \beta[(T-T_0)/(T_b-T_0)]b_0 \tag{5}$$

where β is the stoichiometric coefficient equal to unity when one molecule of a consumes one molecule of b. Without restricting the generality, one can assume $\beta = 1$ and $b_0 = a_0$. Then,

$$-\frac{d(a/a_0)}{dt} = \frac{C_p[M]RT_0^2}{EQ\tau_i a_0}(a/a_0)^{m+n}]\} \cdot$$

$$\exp\{-(E/RT_b)/[1-(T_b-T_0)(a/a_0)/T_b] + E/RT_0\} \tag{6}$$

This equation yields the correct value of τ_i, since it was derived from the integral expression in which a was assumed to be constant and equal to a_0. It should also describe the later stages of the reaction. There is one parameter in this equation available to fit the measured value of τ_e: an unknown total reaction order m+n. We adopted the following procedure of m+n determination from the data on τ_i and τ_e. The heat release curve is represented schematically as a portion with zero temperature rise (τ_i) and a portion with a linear temperature rise (τ_e). The slope (tgα) of the straight line T(t) is assumed to be equal to the slope of the line tangent to the real heat-release curve at the inflection point, i.e.,

$$\tau_e = (T_b - T_0)/\text{tg}\alpha \qquad (7)$$

To find tgα, we replace the variable T in Eq. (6) for a and after that differentiate the resulting equation with respect to t. Equating d^2T/dt^2 to zero yields the inflection point temperature

$$T_{infl} = [E/2R(m+n)]\{\sqrt{1 + [4R(m+n)T_b/E]} - 1\} \qquad (8)$$

When substituting this temperature into the equation for dT/dt derived from Eq. (6), we take into account that for hydrocarbons the m+n probably cannot exceed 5, so that the quantity [4R(m+n)/E] T_b is not greater than 2-3. This yields the following approximate expression for tgα = $(dT/dt)_{infl}$

$$\text{tg}\alpha = [RT_0^2/E\tau_i] \; \{1-[E/2R(m+n)-T_0]/(T_b-T_0)\}^{m+n} \cdot$$
$$\exp\{-(E/RT_b)/[1-(T_b-T_0)(a/a_0)/T_b] + E/RT_0\} \qquad (9)$$

Combining Eqs. (4), (7) and (9) results in

$$[T_b/(T_b-T_0)][1-E/2RT_b(m+n)] =$$
$$\{[(T_b-T_0)E/RT_0^2](\tau_i/\tau_e) \cdot$$
$$\exp[-(E/RT_0)/(1+1.1T_0/T_b)]\}^{1/(m+n)} \qquad (10)$$

Eq. (10) was solved graphically with respect to m+n. The results of the solution of Eq. (10) for stoichiometric mixture C_3H_8/air are presented in Fig. 8 for two T_0. Intersection of curves A and B yields the sought-for value of (m+n)≈3 for this mixture, and (m+n) drops with rising temperature. Similar calculations for ethylene/air

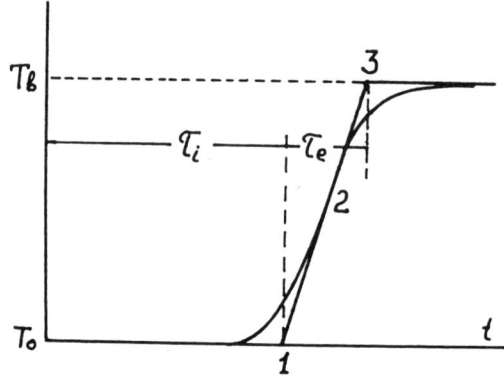

Fig. 7 Temperature rise-time history in a thermal explosion. T_0, initial temperature; T_b, maximal explosion temperature.

mixtures also yield (m+n) ≃ 3. For air and oxygen mixtures of methane of various composition, (m+n) was found to be 4.5.

Although such high values of the reaction order seem unrealistic, they simply reflect the fact that the stage of deep fuel burnout is controlled by secondary reactions with lower activation energies. Therefore, reactant consumption

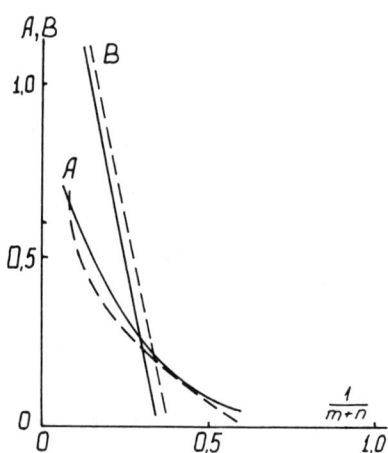

Fig. 8 Graphic solution of Eq. 10 for a stoichiometric C_3H_8/air mixture. $A = T_b[1-E/RT_b(m+n)]/(T_b-T_0)$; $B=\{((T_b-T_0)/RT_0^2]E(\tau_i/\tau_e) \cdot \exp[-(E/RT_0)/(1 + 1.1(T_0/T_b))]\}^{1/m+n}$. Solid lines, $T_0 = 1250°K$; dashed lines, $T_0 = 1430°K$.

HEAT EVOLUTION KINETICS IN HIGH-TEMP. IGNITION 137

Table 2 Calculated and measured total times

Mixture	T, °K	m+n	$(\tau_i + \tau_e)_{exp}$, s	$(\tau_i + \tau_e)_{cal}$, s
3	1430	4.5	$6.4 \; 10^{-3}$	$8.1 \; 10^{-3}$
	1669	4.5	$5.2 \; 10^{-4}$	$7.5 \; 10^{-4}$
	1785	4.5	$2.1 \; 10^{-4}$	$2.7 \; 10^{-4}$
5	1430	3.0	$9.5 \; 10^{-5}$	$1.6 \; 10^{-4}$
	1670	3.0	$3.0 \; 10^{-5}$	$3.4 \; 10^{-5}$
8	1250	3.0	$1.7 \; 10^{-3}$	$3.5 \; 10^{-3}$
	1430	3.0	$2.4 \; 10^{-4}$	$5.0 \; 10^{-4}$
	1669	3.0	$3.1 \; 10^{-5}$	$7.4 \; 10^{-5}$

Calculations were performed with Eq. (6) at p_0 = 1 bar

affects the global heat release rate more strongly than at the stage of small burnt fractions.

To verify the validity of Eq. (6) with the estimated magnitude of the effective reaction order over a wide temperature range between T_0 and T_b, we integrated it numerically. The integration with respect to a was performed from a_0 up to the concentration at the maximum of the reaction rate, i.e., to point 2 in Fig. 7. This choice of integration limit was dictated by the fact that the concentration a at the maximum explosion temperature is not well-defined. Ideally the maximum temperature is attained when a=0, i.e., when t→∞; however, Δp_{max} on the experimental pressure curve used for τ_e estimation is attained at a finite value of a, which is determined by the pressure gain due to reaction and pressure losses that are caused by gas expansion. The inflection point on the $\Delta p(t)$ curve is believed to be much better defined and closer to the inflection point on the ideal T(t) curve. Calculations of the pressure profile using Eq. (6), under an assumption that the explosion takes place in a closed vessel, are presented in Fig. 1.

Values of $\tau_i + \tau_e$ estimated using the procedure discussed earlier of the numerical integration of Eq. (6) are compared with measured ones in Table 2. As seen from Table 2, Eq. (6) describes fairly well the variation of the effective reaction time $\tau_i + \tau_e$ with temperature.

Conclusions

1) The heat-release time in an explosion (τ_e) is an important characteristic of the ignition process that is a measure of the total reaction time under certain conditions.

2) When detonation propagates in hydrocarbon-air mixtures that are supercritical with respect to

concentrations of the fuel or oxidizer, the rate of the heat release is governed by the explosion stage, i.e., by τ_e. The concentration detonation limits are characterized by an approximate equality $\tau_i \approx \tau_e$.

3) An approximate formula is suggested for the global reaction rate in a wide temperature range convenient for gasdynamic calculations. The formula describes with reasonable accuracy the experimental data on ignition delays and heat-release times and substantially simplifies the chemical kinetics equations in detonation waves.

References

Atkinson, R. and Bull, D. C. (1981) Kinetic Modeling of Ethane/Air Detonability, Shock Waves, Explosions and Detonations edited by J. R. Bowen et al. Vol. 87 of AIAA Progress in Astronautics and Aeronautics, AIAA, New York, 318-332.

Baker, J. A. and Skinner, G. B. (1972) Shock-tube studies on the ignition of ethylene-oxygen-argon mixtures. Combust. Flame, 19, 347.

Borisov, A. A. (1970) Selfignition and detonation of gases and two-phase systems, Doctoral thesis, Moscow, USSR, p.289.

Bishimov, E., Korobeinikov, V. P., and Levin, V. A. (1969) Strong Explosion in a Combustible Gas Mixture. Astronautica Acta 15, 267-274.

Borisov, A. A., Dragalova, E. V., Lisyanskii, V. V., Skachkov, G. I., and Zamanskii, V. M. (1983) Kinetics and mechanism of methane oxidation at high temperatures. Oxid. Commun. 4, 45.

Borisov, A. A., Gelfand, B. E., Dragalova, E. V., Zamanskii, V. M., Lisyanskii, V. V., Skachkov, G. I., and Tsyganov, S. A. (1985) Selfignition of hydrocarbons in the presence of promoting additives. Sov. J. Chem. Phys. 2, 1377-1394.

Borisov, A. A. and Loban' S. A. (1977) Detonation limits of hydrocarbon/air mixtures in tubes. Fiz. Foreniya Vzryva, 13, 719.

Borisov, A. A. and Skachkov, G. I. (1972) Selfignition of nitrous oxide. Kinet. Katal. 13, 42-47.

Burcat, A., Crossley, R. W., Scheller, K., and Skinner, C.B. (1972) Shocktube investigation of ignition in ethane-oxygen-argon mixtures. Combust. Flame, 18, 115.

Edwards, D. H., Hooper, G., and Morgan, J. M. (1976) An experimental investigation of the direct initiation of spherical denonations. Acta Astron. 3, 117.

Frank-Kamenetskii, D. A. (1967) Diffusion and Heat Transfer in Chemical Kinetics, Moscow, USSR, p.491.

Kogarko, S. M. and Borisov, A. A. (1960) On ignition delay measurements at high temperatures. Izv. Akad. Nauk 8, 1348.

Steinberg, M. and Kaskan, W. E. (1955) The ignition of combustible mixtures by shock waves. 5th Symposium (International) on Combustion, The Combustion Institute, Pittsburgh, PA, p. 664.

Westbrook, C. K. (1982) Chemical kinetics of hydrocarbon oxidation in gaseous detonation. Combust. Flame, 46, 191.

Fluid Dynamic Effects on the Transition to Detonation from Turbulent Flame in Unconfined Gas Mixtures

Shiro Taki* and Yuji Ogawa†
Fukui University, Fukui, Japan

Abstract

The mechanisms of the noticeable acceleration of propagating turbulent flame and the transition to detonation have been investigated in a space where combustible gas mixtures are slightly confined by a soap bubble. Five comb-shaped grids are inserted repeatedly facing to the flame front in the gas mixture over a distance of 20 cm. In the gas mixture of stoichiometric ethane-oxygen slightly diluted by nitrogen and helium, where, for example, initial flame speed is about 64 m/s when oxygen is diluted by 20%, a transition from deflagration to detonation can take place. The framing or streak photographs are taken and the time histories of the pressure and light emission at a point are measured. It is ascertained that the positive feedback mechanism to accelerate the turbulent flames works well even in unconfined space. A lot of blast waves are observed in highly accelerated turbulent flames. Some are made by fast burning of the unburned gases in eddies behind grid wires where flamelets, trapped before the arrival of the flame front, heat the gases; others are made by burning of the segregated unburned gases surrounded by flames. In the final period close to the onset of detonation, reaction kernels are observed between the leading shock and main flames in schlieren photographs. Numerical simulations on the transient phenomena as the initiation of detonation are carried out using simple two-step chemical reaction model to show how the exothermic reaction kernels are generated by shock interactions and how they make blast waves. The results of numerical simulations suggest that these processes depend mainly on the fluid dynamic effects rather than the chemical kinetics.

Copyright © 1988 by the American Institute of Aeronautics and Astronautics, Inc. All rights reserved.
*Associate Professor, Department of Mechanical Engineering.
†Technical Assistant, Department of Mechanical Engineering.

TRANSITION TO DETONATION FROM TURBULENT FLAME

Introduction

Transition from deflagration to detonation (DDT) in a premixed gas is promoted by obstacles. The Schelkin (1963) spiral is a well-known assist to the initiation of detonation. Because most of the experiments to date on DDT have been carried out in a tube, it is still not clear what conditions in unconfined space are needed to cause DDT. In the presence of obstacles, turbulence is produced by the induced flow. The turbulence will increase the transport of mass and energy and the area of flame front, leading to the increase in the rate of combustion reactions and, hence, to enhancement of the pressure waves. An explosion of an unburned gas pocket trapped behind the flame front is considered to be another mechanism of flame acceleration. All these efforts to increase the burning rate depend on the induced flow velocity, which in turn depends on the burning rate itself. This relation is able to form a positive feedback system. This idea of positive feedback mechanism is proposed by Moen et al. (1980), who showed the large acceleration of cylindrically expanding methane-air flames when repeated obstacles were appropriately placed in the flame path. Their experiments, however, were carried out in confined space. Experiments in partially confined space were later carried out by their group (Chan et al. 1983), but such large accelerations could not be seen. Based on their idea, Stock et al. (1985) tried to accelerate a propagating flame by an explosion behind the flame front, but a transition to detonation did not occur even though a fairly strong blast was used.

It should be remembered that the transition to detonation is initiated by explosions in shock-heated unburned gas mixtures, as shown by beautiful experiments of Urtiew and Oppenheim (1966) over 20 years ago. They also found that a lot of small explosions occurred preceding the detonative explosion. Taki and Fujiwara (1985) got numerical solutions, although only one-dimensional space was considered, showing that an explosion, occurring behind a leading shock where the temperature of gas mixture is high enough to ignite easily, does not directly form a detonation; rather several explosions, induced by the initial explosion occur before the onset of detonation.

Ogawa et al. (1985) found that the propagating flame can be easily accelerated even in fully unconfined gas mixtures when comb-shaped grids are placed repeatedly in the flame path, based on the idea of the positive feedback mechanism. With repeated grids of appropriate sizes and separations, a detonation occurs. In the present paper, we investigate the mechanisms of the transition to detonation in unconfined space, mainly based on the records of streak and framing schlieren photographs and pressure waves. A numerical simulation on the onset of detonation is carried out to help in understanding the experiments.

Experimental Apparatus

The combustible gas mixture column is settled in atmosphere confined by soap bubble films, which are supported by steel wires as shown in Fig. 1. The size of the mixture column is about 500 mm long and 75 x 75 mm^2 in cross section. The confinement by soap bubble walls has little affect on the pressure waves, so we can consider the gas mixture column to be unconfined (Leyer et al. 1974).

Turbulences are generated by grids placed repeatedly facing the flame front in the gas mixtures. The grid is made of straight rods arranged in parallel into a comb shape. The clearances between the rods are the same as the diameters of rods. Two kinds of rods are used in the present experiments, one 3 mm in diameter and the other 6.4 mm. Five grids are inserted into the gas mixture column at the positions of ① 140, ② 190, ③ 240, ④ 290, and ⑤ 340 mm distant from the igniter.

The gas mixtures used in the present experiments are composed of the stoichiometric ethane-oxygen and the diluent nitrogen-helium. Helium is used to reduce the mass density of mixtures in order to float the soap bubbles in the atmosphere. The mixtures used can be written as $C_2H_6 + 3.5(O_2 + aN_2 + bHe)$. In the present paper, we call the diluent ratio as $d = a + b$. When diluent ratio $d = 0.25$, we use a mixture of $a = 0.17$ and $b = 0.08$ and, when $d = 0.2$, $a = 0.07$ and $b =$

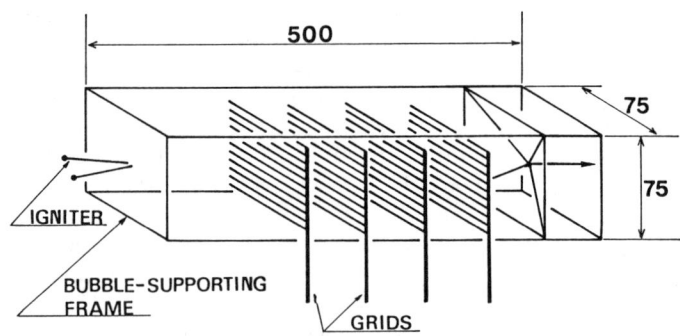

Fig. 1 Steel wire frame to support combustible gas mixtures in soap bubble films and the arrangement of grids and a igniter.

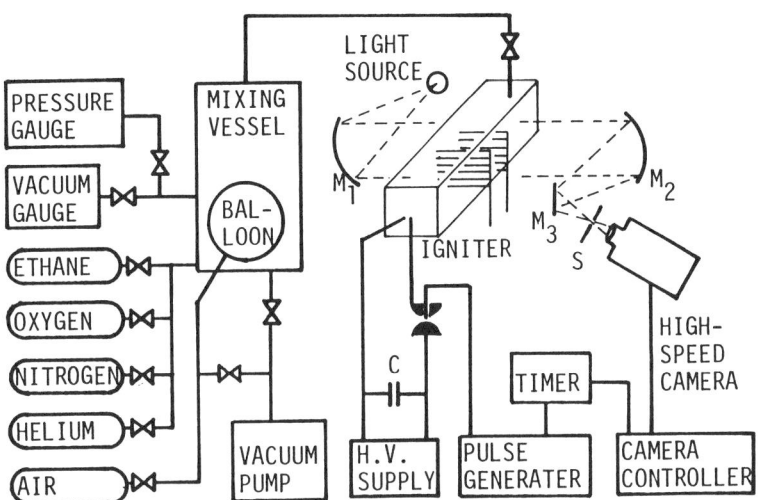

Fig. 2 Schematic diagram of the whole system of experimental apparatus.

0.13 in the present experiments. A spark igniter is inserted into the mixtures to ignite them. Without the grids, a flame propagates in the mixture column with an almost constant speed of about 49 m/s when $d = 0.25$ and of about 64 m/s when $d = 0.2$.

The propagations of the flames and shock waves are observed by schlieren framing or streak photographs, using a high-speed camera. The brightness of the flames is also observed at a point where the light emissions are measured by a photocell system, the space resolution of which is less than 10 mm. It should be noted that weak emissions such as those from cool flames can not be measured. Ultraviolet rays are cut by the lens system. Two kinds of filters were used: one passes waves shorter than about 520 nm and the other longer than about 540 nm. However, except for the signal strength, the differences in the obtained results are so small that the filter was not used for most of the experiments to get clear signals. The transient pressure waves are measured by P.C.B. piezoelectric pressure transducers, which are set just outside of the soap bubble. A schematic diagram of the whole system of the experimental apparatus is shown in Fig. 2.

Experimental Results and Discussions

History of Pressure Waves and Light Emission at a Point

A pressure transducer is placed at a point 100 mm from the center axis of the gas mixture column between the grids

④ and ⑤. A photocell system to measure the light emission is also aimed at the center axis between grids ④ and ⑤. The measured histories of the light emissions and pressure waves of two typical examples are shown in Figs. 3 and 4 for runs 694 and 701, respectively. All the pressure waves show the characteristic feature of the blast waves. In unconfined space, whole pressure pressure increase by a piston effect

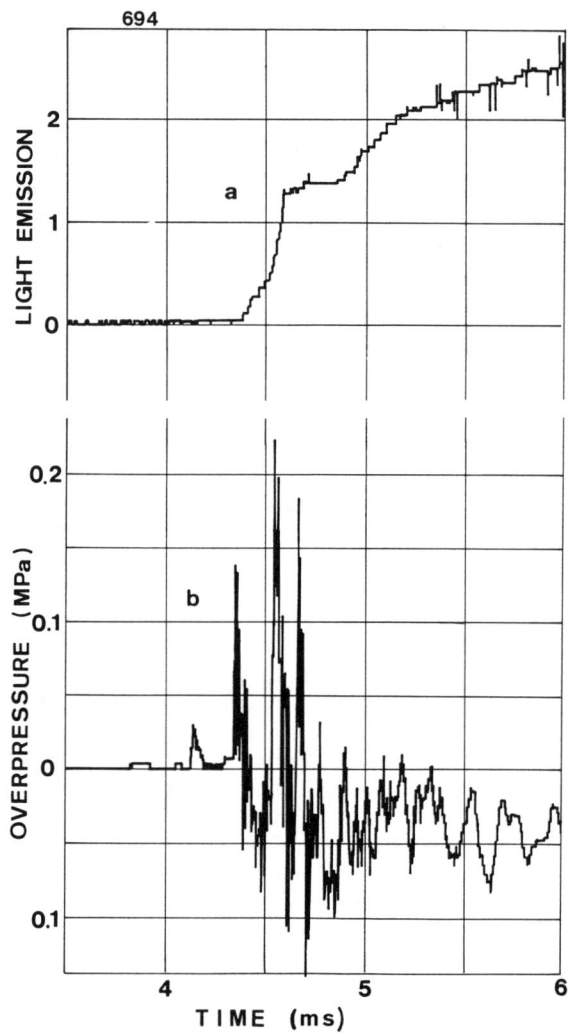

Fig. 3 History of the strength of the light emission (a) and overpressure (b), measured at between grids ④ and ⑤ for the non-detonation case of run 694 when diluent ratio $d = 0.25$ (time is measured from ignition).

TRANSITION TO DETONATION FROM TURBULENT FLAME   145

of burned gas could not be expected. For the case of Fig. 3b, when $d = 0.25$, a detonative explosion does not occur, although a lot of strong pressure waves can be seen; on the other hand when $d = 0.2$ for the case of Fig. 4b, a detonative explosion follows pressure waves similar to those of

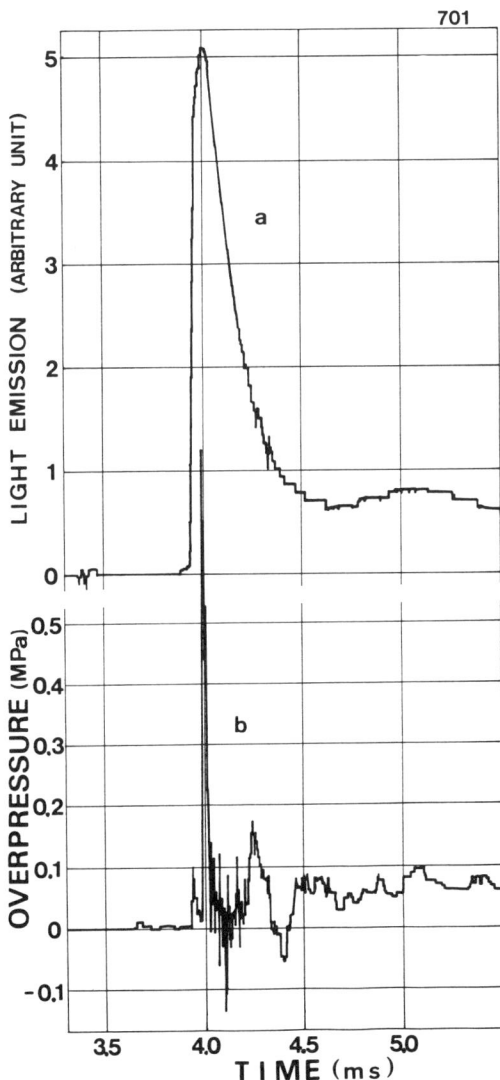

Fig. 4 History of the strength of the light emission (a) and overpressure (b), measured at between grids ④ and ⑤ for the detonation case of run 701 when diluent ratio $d = 0.20$ (time is measured from ignition).

Fig. 3b. The time histories of the light emissions from the flames show the large differences between these two cases. Figure 4a, for the detonation case, shows that after a slight emission of light triggered by a blast, a very strong emission appears, corresponding to a detonative blast wave in Fig. 4b, followed by weak emissions. For the case of nondetonation, as shown in Fig. 3a, the light emission is initiated by a blast similar to the case of Fig. 4a and increased suddenly by the strongest blast; however, after that, the light emission still continued to increase gradually for long time, contrary to the detonation case.

Streak Schlieren Photographs

Figures 5 and 6 show the streak schlieren photographs of runs 695 and 702, respectively. The experimental conditions of run 695 are same as those of run 694, which is a nondetonation case, while those of run 702 are same as those of run 701, which is a detonation case. The position at which the photographs are taken is the center line of the gas mixture column. We can see both shock waves and flames as dark regions on these pictures. A lot of traces of shock waves can bee seen ahead of the flames--that is, many explosions occur in front of the flames. The gradients of the traces of shock waves to the time axis show the shock strengths, the randomness of which suggests the explosions of unburned gas pockets made by strong turbulence. The pictures of the regions in front of the flames are getting dark with time, the reasons of which are discussed in the following experimental results.

Schlieren Framing Photographs

Framing photographs by the high-speed schlieren system are shown in Fig. 7, where the transition from turbulent flame to detonation occurs. The time written on the left of the pictures have an arbitrary origin. The grids in this run are made of bigger steel wires of 6.4 mm in diameter compared with those 3 mm in diameter in the runs above shown, for the bigger steel wires make larger eddies, which are more clearly seen in framing pictures. In Fig. 7, a very dark region moving from left to right corresponds to high-temperature gases. We can see standing eddies behind the grid wires as small dark circles, which are considered to be burning already. It is interesting that in schlieren pictures the gas mixtures, between grids ③ and ④ before the flame arrives, are getting dark inhomogeneously, which may consist of a lot of exothermic reaction kernels. A detona-

TRANSITION TO DETONATION FROM TURBULENT FLAME 147

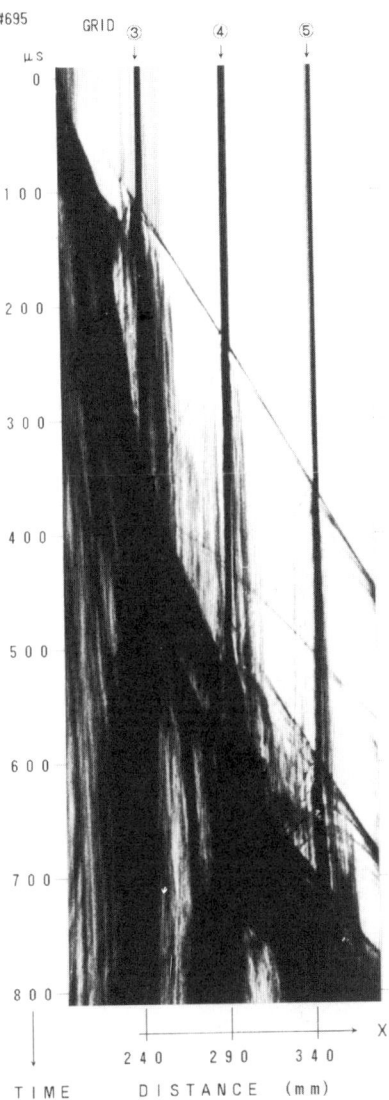

Fig. 5 Streak schlieren photograph of the accelerating flames and the pressure waves; run 695, when diluent ratio $d = 0.25$, a non-detonation case (origin of time is arbitrary).

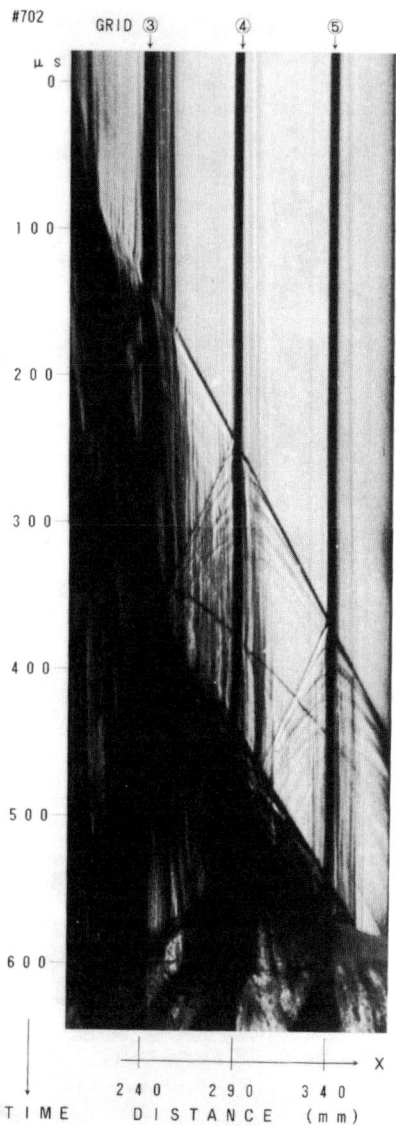

Fig. 6 Streak schlieren photograph showing the transition to detonation from accelerating flames and the traces of pressure waves; run 702, when diluent ratio $d = 0.20$, a detonation case (origin of time is arbitrary)

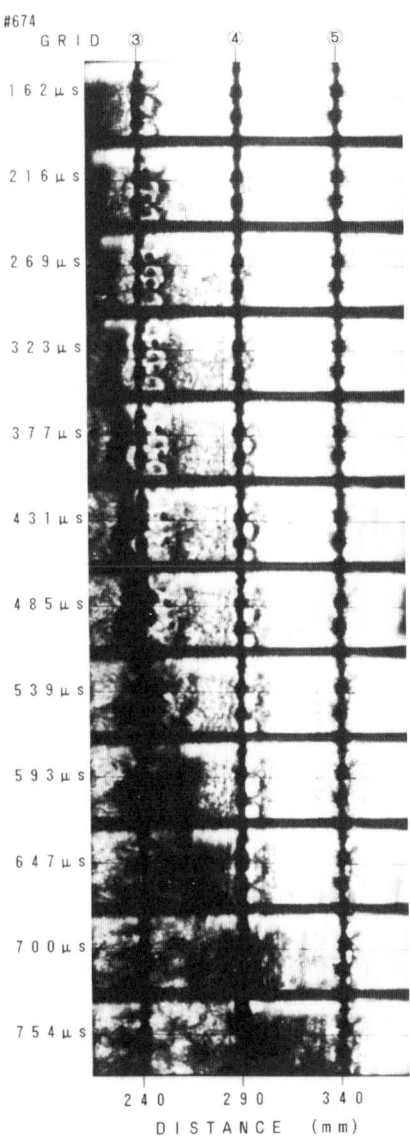

Fig. 7 Successive schlieren photographs showing the transition to detonation from accelerating flames. Five grids are separated by 50 mm, each grid having steel rods with diameters of 6.4 mm and clearances of 6.4 mm, arranged into a comb shape. Run 674, when diluent ratio $d = 0.20$, is a detonation case (origin of time is arbitrary).

tive explosion occurs at around 750 μs between grids ④ and ⑤. To help to understand these phenomena, a numerical simulation is tried as described in the following section.

Numerical Analysis

Results obtained by numerical simulations are shown in Fig. 8, where isobars are drawn by solid lines. Flame fronts are expressed by broken lines and the burned gas regions are shaded in Fig. 8b. The two boundaries are not unconfined but walls of a channel.

Model

Two-dimensional phenomena with a simple model are considered. As an initial condition, a plane shock propagates in a channel filled with a combustible gas mixture. To generate a blast wave, some gas mixtures behind the shock are forced to ignite as shown in Fig. 8a at time = 0.0. The chemical reaction model adopted is two-step one, i.e., the induction step and the recombination step, which is same as used by Taki and Fujiwara (1978). The rate equations for α, the induction reaction progress parameter, and β, the recombination reaction progress parameter, are written as follows.

Induction reaction

$$w_a = d\alpha/dt = -k_a \rho \exp\{-E_a/RT\} \qquad (1)$$

Recombination reaction

$$w_b = d\beta/dt$$
$$= \begin{cases} -k_b p^2 [\beta^2 \exp\{-E_b/RT\} - (1-\beta)^2 \exp\{-(E_b+Q)/RT\}] \\ \qquad\qquad\qquad\qquad\qquad\qquad \text{when} \quad \alpha \leq 0 \\ 0 \qquad\qquad\qquad\qquad\qquad\qquad \text{when} \quad \alpha > 0 \end{cases} \qquad (2)$$

The fundamental equations solved are Navier-Stokes equations with conservation equations on chemicals, but the transport coefficients are assumed to be constant. All the values are nondimensionalized by three values, i.e., the heat value Q, initial density of mixtures ρ_0, and induction reaction length of Chapman-Jouguet detonation l^*.

Finite-Difference Calculations

The explicit MacCormack-FCT method (Taki et al. 1981) with second-order accuracy for both time and space is used

TRANSITION TO DETONATION FROM TURBULENT FLAME 151

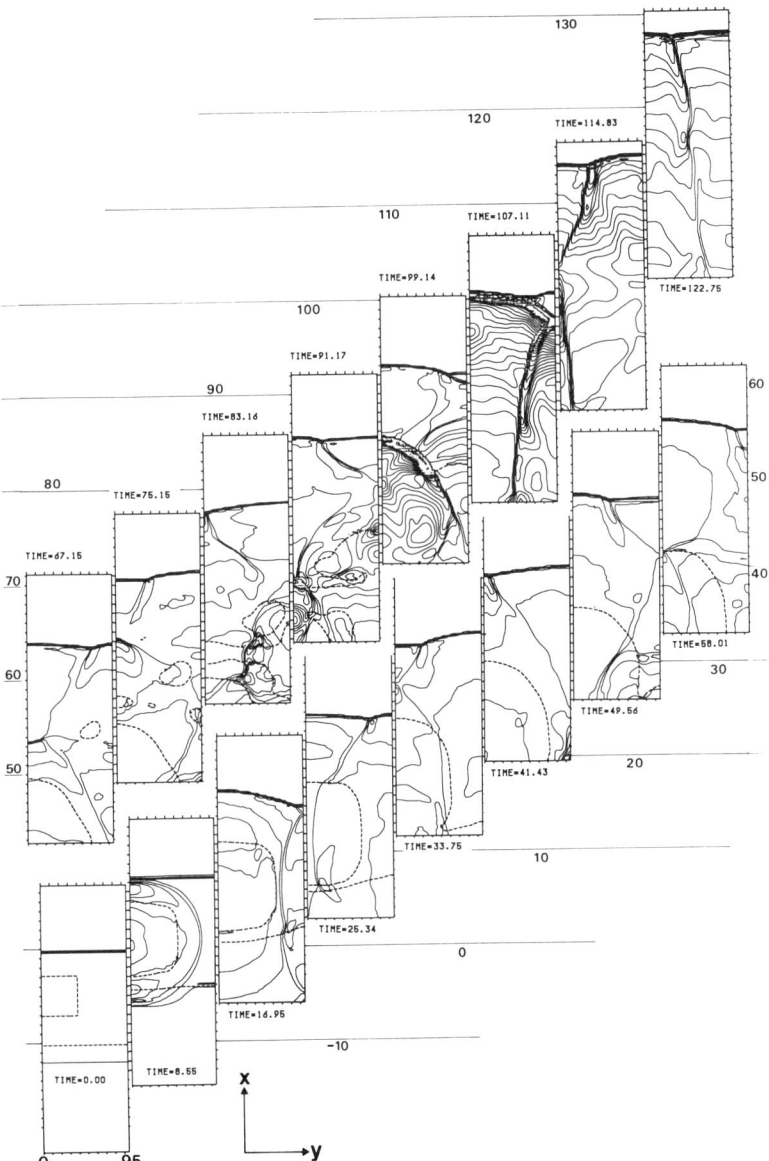

Fig. 8a Numerical results shown by time sequences of isobars, the intervals of which are 2.5 p_0. The leading shock is propagating in a two-dimensional channel with the width of 9.5 l^*. Flame fronts are shown by broken lines.

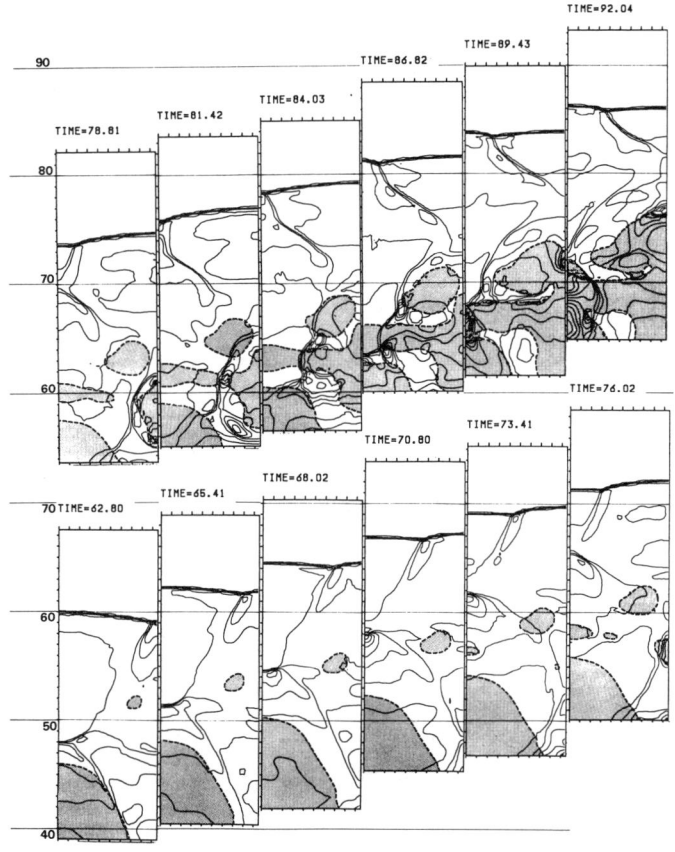

Fig. 8b Some parts of the same results as Fig. 8a, showing that the exothermic reaction kernels are growing. Solid lines are the isobars, the intervals of which are 2.5 p_0. The burned gases are identified by shaded areas.

to solve the unsteady two-dimensional equations mentioned above. The following values of parameters are used in the present simulations:

$Q/RT_0 = 20.0$, $E_a/Q = 1.7$, $E_a/E_b = 4.9$
$k_a/k_b p_0 Q^2 = 20.0$, $\gamma = 1.4$, $Re^* = 10,000$
$\Delta x = \Delta y = 0.211 \ l^*$, width of channel = $9.5 \ l^*$.

Numerical Results and Discussion

Time sequences of isobars after initial explosion are shown in Fig. 8a. The initial blast wave creates a triple-shock structure on the leading shock wave, but this structure attenuates because of the separation of flames to shocks. The initial blast wave is followed by many explosions, at (x,y;t) = (0, 9.5; 14), (14, 9.5; 16), (8, 0; 27), (18, 9.5; 40), (38, 0; 40), (40, 0; 57), and so on. Then, a lot of blasts interact with each other, which will make many hot spots in a very complicated manner. A new flame spot appears at (x,y;t) = (55, 6.5; 67) as shown in Fig. 8b, where the burned regions are shaded. At time = 75, three other flame spots can be seen, one of which becomes a center of a rather strong explosion at (x,y;t) = (55, 9.5; 75). This explosion, however, is not a detonative explosion, but induces one at (x,y;t) = (67, 0; 90). The present numerical result is somewhat similar to the experimental results, although the space is not open in numerical model.

Conclusions

A detonation transition from deflagration is observed in almost unconfined space filled with slightly diluted ethane-oxygen gas mixtures. Propagating flames are easily accelerated by comb-shaped grids made of steel wires. It is ascertained that the positive feedback mechanism proposed by Moen et al. (1980) works well to accelerate the turbulent flames by using repeated grids. In highly accelerated turbulent flames a lot of blast waves are generated in turbulent burning regions, where the segregated unburned gases surrounded by flames burn explosively.

The positive feedback mechanism, however, is not enough to initiate a detonation. In the final period close to the onset of detonation, reaction kernels are observed between the leading shock and main flames in framing schlieren photographs. Numerical simulations of such transient phenomena as the initiation of detonation, using a simple two-step chemical reaction model, show how the exothermic reaction kernels are made by shock interactions and how they generate blast waves. The results of numerical simulations suggest that these processes depend mainly on the fluid dynamic effects rather than the chemical kinetics.

References

Chan, C., Moen, I. O., and Lee, J. H. (1983) Influence of confinement on flame acceleration due to repeated obstacles. Combust. Flame 49, 27-39.

Leyer, J. C., Guerraud, C., and Manson, N. (1974) Flame propagation in small spheres of unconfined and slightly confined flamable mixtures. Fifteenth Symposium (International) on Combustion, pp. 645-653, The Combustion Institute, Pittsburgh, PA.

Moen, I. O., Donato, M., Knystautas, R., and Lee, J. H. (1980) Flame acceleration due to turbulence produced by obstacles. Combust. Flame 39, 21-32.

Ogawa, Y., Kondou, S., and Taki, S. (1985) Experiments on transition to detonation from an accelerating flame in unconfined gaseous mixtures. JSME B-51, 3421-3425 (in Japanese)

Schelkin, K. I. and Troshin, Ya. K. (1963) Gasdynamics of Combustion. NASA TT F-231.

Stock, M., Schildknecht, M., Geiger, W. (1985) Flame acceleration by a postflame local explosion. Progress in Astronautics and Aeronautics: Dynamics of Shock Waves, Explosions, and Detonations: (edited by J. R. Bowen, N. Manson, A. K. Oppenheim, and R. I. Soloukhin), Vol.94, pp. 491-503, AIAA, New York.

Taki, S. and Fujiwara, T. (1978) Numerical analysis of Two-dimensional nonsteady detonations. AIAA J. 16, 73-77.

Taki, S. and Fujiwara, T. (1981) Numerical simulation of triple shock behavior of gaseous detonation. Eighteenth Symposium (International) on Combustion, pp. 1671-1681, The Combustion Institute, Pittsburgh, PA.

Taki, S. and Fujiwara, T. (1985) Numerical simulations on the establishment of gaseous detonation. Progress in Astronautics and Aeronautics: Dynamics of Shock Waves, Explosions, and Detonations: (edited by J. R. Bowen, N. Manson, A. K. Oppenheim, and R. I. Soloukhin), Vol.94, pp. 186-200, AIAA, New York.

Urtiew, P. A. and Oppenheim, A. K. (1966) Experimental observations of the transition to detonation in an explosive gas. Proc. R. Soc. London A295, 13-28.

Numerical Simulations of the Development and Structure of Detonations

E. S. Oran,* K. Kailasanath,† and R. H. Guirguis‡
Naval Research Laboratory, Washington, DC

Abstract

Multidimensional time-dependent numerical simulations have been used to calculate the evolution of the instability that leads to the cellular structure of detonations and to study the detailed behavior of the interacting shock waves and reaction zones forming the detonation wave. The simulations consist of two-dimensional, time-dependent solutions of the convective transport of mass density, momentum density, and energy coupled to models for chemical energy release. We discuss the the behavior of the multidimensional structure of a detonation and how it depends on the differences of the thermodynamic properties in the induction zones behind the Mach stem and the incident shock. The role of turbulence on detonation ignition and propagation is discussed.

Introduction

In this paper, we describe some of the properties of the the structure of multidimensional propagating detonations in liquids and gases that we have learned from numerical simulations. A major strength of a simulation is that it is an excellent vehicle for looking at fundamental interactions in a systematic way. Thus, we can isolate the fundamental controlling processes and study their interactions in idealized environments. In this paper, we summarize several important aspects of detonations that we have used simulations to study and then discuss the role of turbulence in detonations.

Numerical simulations of gas-phase detonations are based on solutions of the compressible, time-dependent, conservation equations for total mass density ρ, momentum density ρv, and energy E,

$$\frac{\partial \rho}{\partial t} = -\nabla \cdot \rho \mathbf{v} \qquad (1)$$

$$\frac{\partial \rho \mathbf{v}}{\partial t} = -\nabla \cdot (\rho \mathbf{v} \mathbf{v}) - \nabla P \qquad (2)$$

This paper is the work of the U.S. Government and is not subject to copyright protection in the United States.

$$\frac{\partial E}{\partial t} = -\nabla \cdot (E\mathbf{v}) - \nabla \cdot (\mathbf{v}P) \tag{3}$$

where $\mathbf{v}$ is the fluid velocity and P the pressure. In a multispecies fluid in which chemical reactions affect transformations among the species, we also need individual species number densities $\{n_i\}$

$$\frac{\partial n_i}{\partial t} = -\nabla \cdot n_i \mathbf{v} + Q_i - L_i n_i, \quad i = 1, ..., N_s \tag{4}$$

where the $\{Q_i\}$ and $\{L_i\}$ are chemical production and loss terms, respectively, for species i. The effects of molecular diffusion have been omitted from these equations because they are generally insignificant on the time scales of interest for detonations. There is a constraint that defines the total number density N,

$$N = \sum_{i=1}^{N_s} n_i \tag{5}$$

where N_s is the total number of different kinds of species present. The total energy is a sum of the kinetic and internal energy,

$$E = \frac{1}{2}\rho \mathbf{v} \cdot \mathbf{v} + \rho \epsilon \tag{6}$$

where ϵ is the specific internal energy.

An ideal-gas equation of state is used for the gas-phase calculations,

$$P = NkT = \rho RT \tag{7}$$

Also, we assume

$$\rho \epsilon = \sum_i \rho_i h_i - P = \rho h - P \tag{8}$$

$$h_i = h_{io} + \int_{T_o}^{T} c_{pi} \, dT \tag{9}$$

so that the properties of the individual species are taken into account. Here, k is Boltzmann's constant, $\{h_i\}$ the enthalpies of each species i, $\{h_{io}\}$ the heats of formation, and $\{c_{pi}\}$ the specific heats.

We also describe calculations of detonations in liquid nitromethane, for which we use a HOM equation of state [see Mader (1979)] for both the condensed fuel and the products. The equation of state for the condensed phase is based on the Walsh and Christian technique (1955). The equation of state of the gas products is constructed by solving the equilibrium Chapman-Jouguet (C-J) detonation problem using the BKW equation of state for the final products. The details of the equation of state are given in Guirguis and Oran (1983).

In all of the two-dimensional calculations described below, the full details of the chemical reactions are not included in the model. Instead, we use the induction parameter model that reproduces the essential feature of the chemical reaction and energy release process. In this model, three quantities are tabulated as a function of temperature, pressure, and stoichiometry: the chemical induction time, time of energy release, and final amount of energy released. These may be obtained by integrating the full set of elementary chemical reactions, as we have done for hydrogen-oxygen combustion, or they may be gathered from experimental data, as we have done for liquid nitromethane. Then, a quantity called the induction parameter is defined, which is convected with the fluid in a Lagrangian manner. This parameter records the temperature history of a fluid element and, when the element is heated long enough, energy release is initiated. This model for including the properties of a chemical reaction mechanism in a numerical simulation was described originally by Oran et al. (1981) and has been been developed further by Kailasanath et al. (1985a,b) and Guirguis et al. (1986).

The crux of being able to simulate the multidimensional structure of detonations is solving coupled continuity equations for density, momentum, and energy with enough accuracy [Oran and Boris (1987a,b)]. The numerical solution of continuity equation, which is by far the most difficult part of solving Eqs. (1)–(4), has been the subject of zealous arguments for the last 20 years. The problem occurs because both numerical diffusion and numerical dispersion arise from the Eulerian finite-difference formulation, and these errors must be minimized. In addition, there are unavoidable Gibbs errors that arise simply because we are using a finite representation.

The most straightforward finite-difference approaches do not work for solving the continuity equation. The fundamental problem was stated originally by Gudonov (1959): a linear algorithm cannot be monotone unless it is first order so that numerical diffusion damps the unphysical oscillations due to numerical dispersion. For example, a donor-cell algorithm is first order and very diffusive, but it is monotone. The standard Lax-Wendroff algorithm is second order, but can oscillate wildly at steep gradients. To solve shock and detonation problems, higher-order accuracy is needed, but we cannot tolerate the oscillations introduced by numerical dispersion. More complicated, nonlinear approaches are required, but they impose the physical requirements of positivity and monotonicity on the finite-difference algorithm itself. Flux-corrected transport (FCT) [see, for example, Boris (1971); Boris and Book (1976)] was the first algorithm to do this and, since its appearance, a number of algorithms have emerged based on these same principles. The FCT algorithm was used in all of the calcualtions presented below.

The Development of Cellular Structure

The front of a self-sustained propagating detonation in an energetic gas and, certainly in some energetic liquids, is not uniform. Its structure is complex and multidimensional, involving several shocks continuously interacting with each other and the boundaries of the region through which the detonation is moving. The triple points formed at the intersection of these shocks, designated as incident shock, Mach stem, and transverse wave, trace patterns that are called detonation cells. These structures may be degenerate for certain material conditions or if the detonation is heavily overdriven, but they can occur in all self-sustained gas-phase detonations. Extensive experimental data [see, for example, Strehow (1984) and Fickett and Davis (1979)] have shown that the size and regularity of this cell structure is characteristic of the particular combination of initial material conditions, such as composition, density, and pressure.

Planar detonation fronts may be unstable to perturbations in the transverse directions. One result of this is that given an initially planar front, a perturbation causes it to depart from a one-dimensional configuration. Consider Fig. 1, which shows a series of pressure contours from a numerical simulation of the early effects of perturbing a planar detonation front [Kailasanath et al. (1985a)]. This simulation was initialized by placing an elliptical pocket of hot, unburned gas behind a planar detonation. The pocket, which was merely used as a device for initiating the perturbation, burns slowly and sends out pressure disturbances in all directions. These waves interact with the incident shock front, causing the front to curve outward. The pressure waves also reflect from the side walls of the channel and move transverse to the incident shock front, as seen in frame 2 of Fig. 1. These pressure waves are strengthened due to collisions with each other and further increase the curvature of the incident shock front, as can be seen in frame 7. After a short time, a portion of the incident shock reflects from the side walls of the channel. Frame 8 shows a Mach reflection that has been formed between the pair of triple points. The reflected shock waves, which are the transverse waves, are initially weak but are later strengthened due to collisions with each other and the walls.

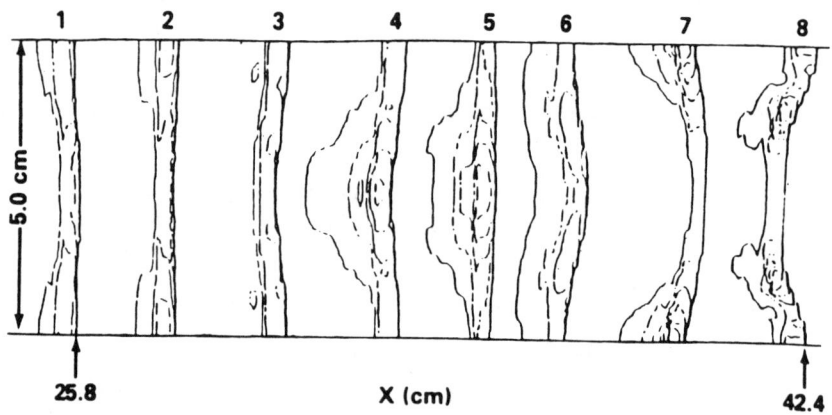

Fig. 1 Composite of eight "snapshots" of the pressure contours near the detonation front at intervals of 10 μs for the early stages of the formation of a pair of triple points. The detonation is propagating in a 65 Torr (8.66 kPa), 298 K stoichiometric hydrogen-oxygen mixture diluted with 60% argon (Kailasanath et al. 1985a).

DEVELOPMENT AND STRUCTURE OF DETONATIONS

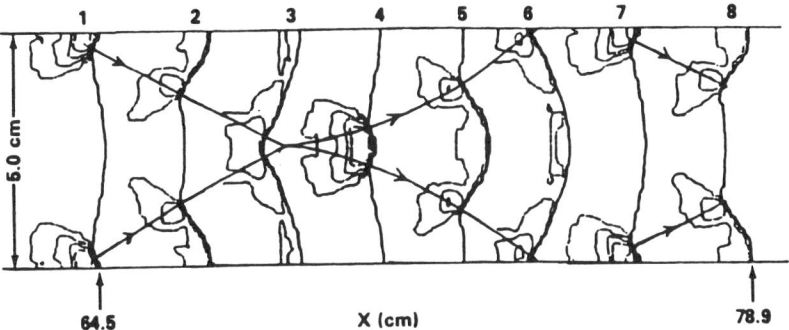

Fig. 2 Composite of pressure contours at later times than those shown in Fig 1. The lines with arrows show the movement of the triple points (Kailasanath et al. 1985a).

The further evolution of the pair of triple points is shown in Fig. 2. The path and the direction of movement of the triple points are indicated by the lines with arrows. In the first frame, the transverse waves are moving toward each other and away from the wall. By the fourth frame, they have collided with each other and are moving away from the center of the channel. In frame 7, they are again moving towards the center after colliding with the walls. Frames 7 and 8 are similar to frames 1 and 2, respectively, showing that an equilibrium configuration detonation has been established.

In the numerical simulations, the multidimensional structure can be initiated in many different ways. Calculations by Taki and Fujiwara (1981) describe a system with a number of exothermic spots located across the path of the detonation. These spots perturb the planar detonation wavefront and the system evolves into a group of triple points. A number of ways have been used to initiate the detonation, including perturbations in front and behind planar detonations, oblique shocks that transition to detonations, and explosions in corners. For example, we have taken two main approaches. The first is to set up a propagating planar detonation with some kind of a perturbation either behind or in front of it [e.g., Taki and Fujiwara (1981); Kailasanath et al. (1985a,b)]. The second approach is to set off a shock or explosion in a homogeneous detonable mixture, and watch the evolution to a detonation [Oran et al. (1982)].

The most widely explored way to initate multidimensional structure is by perturbing a planar front, because this approach gives the cleanest transition to triple points at the detonation front. In general, we find that the *eventual* detonation cell structure formed is apparently independent of the initial perturbation, as long as the transverse perturbations are strong enough. The number of triple points *initially* formed depends on the number and symmetry of the obstacles perturbing the planar flow. *However, the calculations seem to indicate that the system relaxes toward a final state that is independent of the initial conditions.*

For example, the calculations by Kailasanath et al. (1985b) show that the system prefers to relax to a final state that is symmetric about the centerline of the tube. A typical evolution to a symmetrical solution is shown in the set of calculations by Kailasanath et al. (1985b) that modeled detonations in 5 and 10 cm tubes. The initial configurations studied are shown in Fig. 3. Configurations a and b were the same, except that in case b the disturbance was asymmetric. The evolution of the perturbed detonation started very differently. Figure 2 shows the evolution of configurtion a and Fig. 4 shows that of configuration b. However, case b goes through a gradual change in the number and pattern of triple points and eventually becomes the symmetric cell structure seen for case a.

The effect of changing the channel width is studied by simulating planar detonations interacting with hot spots shown in cases b and c in Fig. 3. Here, case b is half of case c and if the wall is truly a line of symmetry, the solution of b should be the upper half of c. The result of case c was a symmetrical pattern with a full cell centered in the tube and partial structures above and below it. In case b, the triple point structure started out looking like the upper half of case c, but soon deviated and went to a structure corresponding to case a. The conclusion is that, for the dilute $H_2/O_2/Ar$ mixture studied and with the restriction to two dimension and a fixed channel width, the wave structure tries to go to a natural mode of the system. For some other chemical systems, this could take an inordinately long time and the results might appear chaotic.

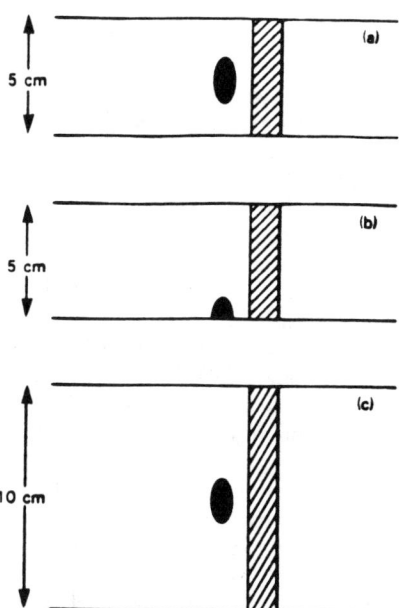

Fig. 3 Schematic of the initial conditions used to initiate the multidimensional structure of detonations (Kailasanath et al, 1985a). a) symmetric perturbation behind a planar front in a 5 cm tube; b) an asymmetric initiation; c) a symmetric initiation in a 10 cm tube.

The triple-point pattern in the 5 cm tube is shown qualitatively in Fig. 5a. We see that the triple point structure does not appear to reflect immediately when it hits the wall and a complete detonation cell has not been formed. A time and space gap appears at the walls as the structure reforms. (The formation of this incomplete cell is discussed further below.) Increasing the width of the channel to 7 cm, as shown in Fig. 5b, considerably reduces the gap in the path of the triple points near the walls. Finally, the locus of the triple points for a 9 cm wide channel,

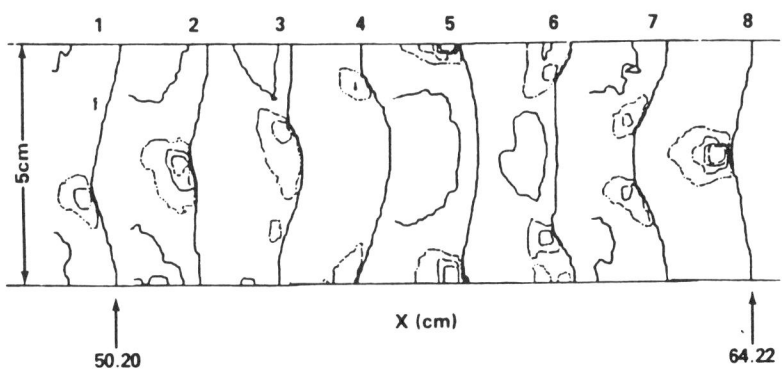

Fig. 4 Pressure contours for a detonation in a 5 cm wide channel initiated using the asymmetric pocket (Kailasanath et al. 1985b).

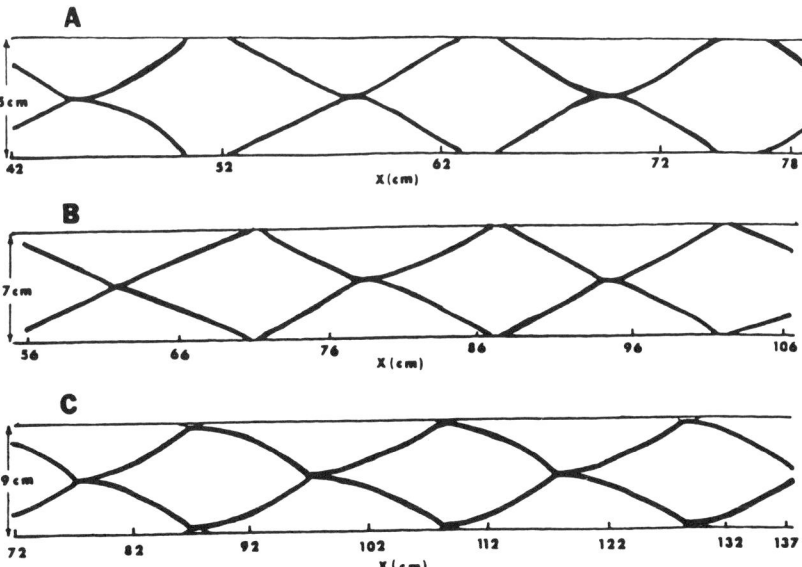

Fig. 5 Calculated paths of triple points for detonation propagation in (A) 5 cm wide channel, (B) 7 cm channel, and (C) 9 cm channel.

shown in Fig. 5c, forms a complete detonation cell and what appears to be partial structures above and below it. From the figure, we estimate the cell width and length to be about 8.5 and 19.6 cm, respectively.

In calculations for wide channels, for example, channels whose widths are at least several cell sizes, we have found that a number of stable or metastable states can exist which correspond to different numbers of cells. Additional perturbations to one stable pattern can make the system alternate between several states, or transition from one metastable state to a more stable state. Experiments and high-resolution calculations (such as those by Guirguis for hydrogen and nitromethane systems) both find detailed substructures, which we believe are of similar origin. The experiments have a certain inherent noise level and some loss processes that can generate noise. The high-resolution calculations resolve more high-frequency sound waves. In both cases, this noise creates the perturbation necessary to trigger substructures.

The Importance of Transverse Waves

One surprising feature of the calculations shown in Fig. 5 is the flattened shape of the cell when the channel width is less than what appears to be a natural cell size. To better understand the factors that determine the shape of the cell and thus the calculated cell size, we must look at the curvature of the transverse wave as it reflects from the wall or from another transverse wave. For the purposes of this explanation, we call that portion at and close to the triple point the "head" of the transverse wave and we call the "tail" that region of the transverse wave extending back toward the burned material, as shown schematically in Fig. 6. This is based on the temperature, density and pressure contours from calculations for the 5 cm tube. First, the transverse waves are moving toward each other (a) and the distance between the two heads is larger than the distance between the two tails. It appears the tails collide with each other earlier than the heads. After the collision (c), the curvature of the transverse wave is reversed and the heads are closer

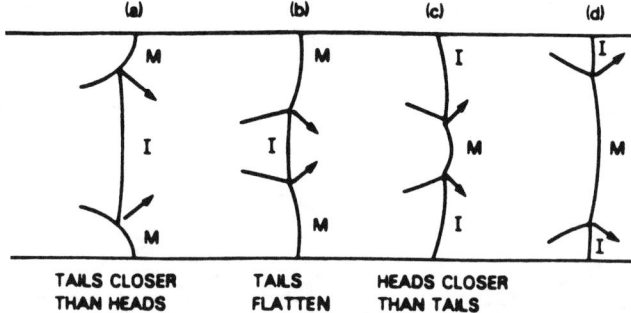

Fig. 6 Schematic diagram corresponding to selected contours from the calculation of the 5 cm wide channel showing the structure of the transverse waves at the detonation front.

to each other than the tails. Comparing Figs. 6b and 6c, we see that the curvature of the transverse wave is reversed at or around the time of collision. Figure 6d shows the situation in which the tails collide with and reflect from the walls earlier than the heads. Such a reflection causes a larger pressure difference around the tail than across the unreflected front segment. This high-pressure region pushes the incident shock front forward and results in the flattened detonation cell seen in Fig. 5.

As the width of the tube is increased, the transverse wave can travel faster and become weaker before collision; thus, the inclination of the transverse wave to the channel walls is less. If the channel width is increased so that it is slightly larger than the detonation cell width, we expect new triple points to be generated in the shocked material in the direction ahead of the transverse waves. Explosion can occur only in compressed unreacted regions. This produces a new pair of transverse waves that propagate toward the two transverse waves already present in the system. The new and old waves then collide with each other, rendering the transverse wave spacing equal to the detonation cell width. Therefore, in a channel slightly larger than the detonation cell width, we would observe four transverse waves at certain periods in the detonation cell cycle as well as a complete detonation cell within the channel. This is what is seen in Fig. 5c.

In the 9 cm channel shown in Fig. 5 we also observed the presence of unburned gas pockets near the walls behind the detonation front. Such unburned gas pockets have been observed earlier both in experiments and in numerical simulations [Subbotin (1975); Oran et al. (1982); Hiramatsu et al. (1984)]. The origin of these pockets in this calculation can be explained by extending the argument used above on the inclination of the transverse wave. Consider a case for which the channel width is slightly larger than the detonation cell width. In this case, a transverse wave moving toward the walls, which does not encounter another transverse wave moving in the opposite direction, continues to propagate, although it is considerably weakened. The incident shock is also considerably weakened and hence the reaction zone ahead of the transverse wave also falls way behind. If the head of the transverse wave reflects earlier than the rear segment, the portion of the gas near the head of the transverse wave burns first, effectively cutting off a gas pocket. The formation of unburned gas pockets has been discussed in detail elsewhere [Oran et al. (1982); Kailasanath et al. (1985b)] for the same gas mixture at the same initial temperature and pressure used in these calculations.

Irregularity in the Structure of Detonation Waves

Most of the experimental results showing detonation cells do not show the regular structure we have seen in the calculations shown above, but show irregular structure in which the size of a cell is not clearly determined. In the experiments, the regular cell patterns appear in the highly diluted argon mixtures of hydrogen and oxygen, as we have predicted in the calculations. If the mixture is less dilute, or argon is replaced by molecular nitrogen, the structure again becomes irregular.

Some insight into the mechanism causing the irregularity of the structures was provided by the recent work of Guirguis et al. (1986, 1987a) that studied detonation cell structure in liquid nitromethane. In this case, the parameters for the induction parameter model were taken from experimental data, as opposed to being derived from a detailed chemical reaction mechanism. Guirguis defined two quantities, the induction parameter, defined in the form $\tau_i = A^i exp(E^i/RT)$, and an energy release time, which also had an Arrhenius form $\tau_r = A^r exp(E^r/RT)$. The initial calculations were performed with the experimentally derived values of τ_r and τ_i and typical results are shown in Fig. 7. The outstanding feature of this figure is that the structure appears very irregular. The solid horizontal lines drawn through the pressure contours outline the movement of the stronger triple points and the dashed lines indicate weak triple points.

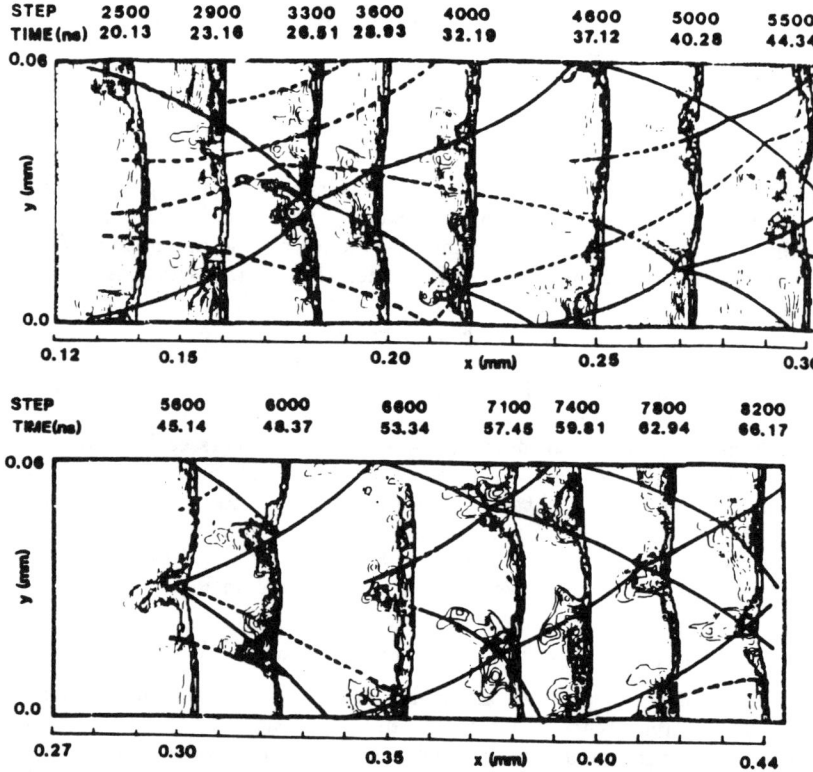

Fig. 7 Pressure contours of a detonation wave propagating in a channel filled with liquid nitromethane. Solid traces are loci of main triple points and dashed traces loci of secondary triple points. Induction time and energy release parameters derived from experimental data (Guirguis et al. 1986, 1987).

DEVELOPMENT AND STRUCTURE OF DETONATIONS 165

Numerical tests in which the timestep and computational mesh spacing were varied proved that these results are a physically correct property of the calculation, and not numerically induced. The next question is what is the origin and what controls this structure. Figure 8 shows the results of increasing the value of E^i in the expression for the induction time. This has the effect of increasing the difference of the size of the induction zones behind the Mach stems and reflected shocks. The structure is now quite regular and qualitatively similar to the dilute hydrogen-oxygen calculations described above. An interesting feature of this calculation is the formation of large unreacted pockets behind the propagating detonation front, which was not really seen in the calculations represented in Fig. 7. The formation of such a pocket is shown in Fig. 9, which shows how they form as two triple points collide.

Similar studies were carried out which varied the energy release parameters. From these, we found that detonation structure is also affected by the energy release parameters, A_r and τ_r. Instantaneous energy release leads to a one-dimensional structure. Very fast energy release results in less regular structures and very slow energy release results in large pockets and highly curved fronts. In this last case, the detonation may die out.

The obvious question now is what can we say about irregularity in the gas phase. To answer these questions, we have been doing calculations of detonations in hydrogen-oxygen diluted with argon, as above, and compared these to calculations of hydrogen-oxygen diluted with nitrogen, which we know gives irregular structures [Guirguis et al. (1988)]. The calculations indicate that the nitrogen dilution has a tendency to give irregular structures. This result indicates that the multidimensional structure of a detonation depends on the differences of the thermodynamic properties in the induction zones behind the Mach stem and the incident shock. Whereas the chemical induction times for equivalent nitrogen and argon dilutions are the same, the rate and amount of energy

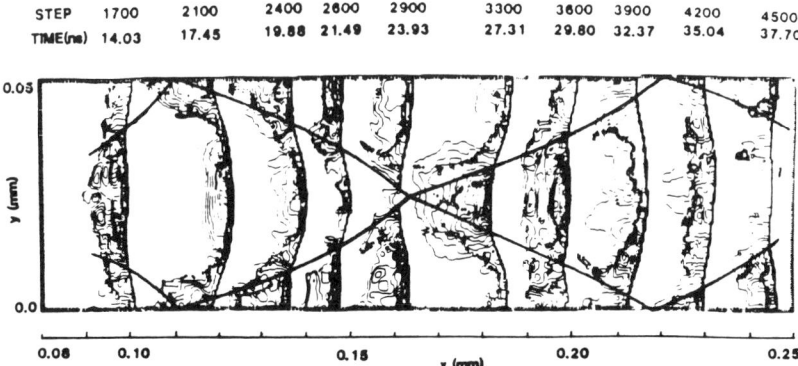

Fig. 8 Pressure contours of a detonation wave propagating in a channel filled with liquid nitromethane with modification to the temperature dependence of the experimentally derived induction time (Guirguis et al. 1986, 1987).

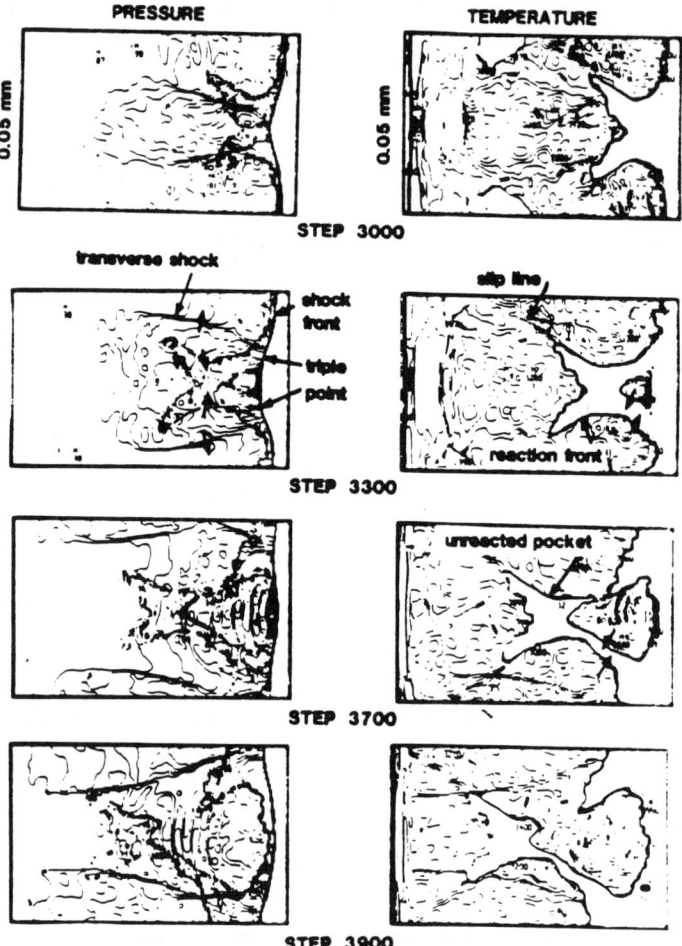

Fig. 9 Pressure and temperature contours behind the detonation wave, showing the collision of two triple points at the center of the channel (Guirguis et al. 1986, 1987).

release is different, and so the relative sizes of the induction zones behind the Mach stem and incident shock are different for argon and nitrogen dilution.

The Role of Turbulence in Detonations

Recently the role of turbulence in detonations has become a subject of some discussion. Turbulence could be important in 1) the shock-to-detonation transition, 2) the deflagration-to-detonation transition, and 3) detonation propagation. Here we breifly discuss each of these and leave a more detailed discussion to a future paper.

First consider the shock-to-detonation transition. A shock propagating in an energetic medium does not leave a homogeneously heated and compressed region behind it. Both experiments and calculations have shown that hot spots are a natural result of shocking a compressible medium. They are the natural result of the presence of acoustic perturbations, and these may be enhanced by the presence of a nonequilibrium, reacting medium. In many cases, these spots are hot enough to cause ignition, or at least to generate pressure waves that spread out radially from the spot. When the spots do ignite, a reaction wave (that may itself be a detonation) propagates in the reacting medium and may eventually catch up with the shock and produce a detonation. Even before the spots ignite, they may perturb the shock front and even accelerate it, eventually causing more hot spots or directly initiating a detonation at the front. The region behind the propagating shock is thus a noisy region, with a broad spectrum of pressure fluctuations. In this case, we could say that the flow in this inhomogeneous region is turbulent and these turbulent fluctuations are the source of hot spots that lead to transition to detonation.

The fundamental physical mechanisms in detonation propagation are convection and energy release from chemical reactions. In addition to these, molecular and thermal diffusion are also fundamental mechanisms in flame propagation. In a deflagration-to-detonation transition, the characteristic energy release rates and flow velocities are fast enough that the diffusion processes must become negligible. This transition can occur through a number of mechanisms. One mechanisms is very similar to that described above in a shock-to-detonation transition: hot spots can form behind a flame front and these spots generate pressure waves that result in an increase in flame velocity, eventually leading to detonation. The pressure waves could directly accelerate the flame by increasing the pressure behind the front. They can indirectly cause the flame front to become distorted, so that its surface area is increased, more energy is released, and the flame front seems to move faster. These mechanisms are the result of pressure disturbances generated by a noisy or turbulent flow behind the flame and these disturbances directly or indirectly accelerate the flame front.

The final issue is the effect of turbulence on a propagating detonation. The question here is where is the turbulence and what can its effects be. We know that the regions behind the leading shock fronts are very noisy and that there is a spectrum of pressure fluctuations. These fluctuations occur in both the fully reacted material and the reaction zones. A major effect of such perturbations is to accelerate the initiation process. They provide a mechanism of reinitiation of triple points by causing hot spots in either reaction zones behind the Mach stem or incident shocks or by speeding up the reactions in any unreacted gas pockets cut off by transverse waves.

A source of some of these fluctuations could be Kelvin-Helmholtz instabilites at slip lines behind the shock fronts. The major effect here would not necessarily be convective mixing, which is normally associated with such shear-layer instabilities, but the noise generated by the compressible turbulent flow. It is not yet clear, however, how important these fluctuations are relative to fluctuations generated by the chemical-acoustic processes in nonequilibrium flows. For example, in the case of liquid nitromethane shown in Fig. 7, the chemical-acoustic effects are dominant and the flow is very sensitive to perturbations.

Summary

From our two-dimensional calculations of the structure of propagating detonations in liquids and gases, we have been able to conclude:

1) The multidimensional structure of a detonation depends on the differences of the thermodynamic properties in the induction zones behind the Mach stem and the incident shock.

2) The formation of unreacted pockets behind the detonation front depends on the inclination of the transverse waves and the curvature of the shock fronts. Highly curved fronts may result in large pockets.

3) The temperature dependence of the induction time is a major factor in the regularity of detonation structure.

4) Detonation structure is affected by the energy release parameters. Instantaneous energy release leads to one-dimensional structures. Fast energy release results in less regular structures. Very slow energy release results in large pockets, highly curved fronts, and the detonation may die out.

5) In large systems, that is, those in which many cells can exist, it is possible to have more than one configuration of cell patterns. Given the right perturbation, the system can alternate among metastable states or can transition to a more stable one.

6) Pressure fluctuations behind flames and shocks are important in the transition to detonation process. Pressure fluctuations behind propagating detonations provide perturbations that can accelerate ignition and generate triple points. Highly resolved calculations show more pressure fluctuations, that can act as perturbation to create more substructure.

Acknowledgments

The research has been carried out by the authors, J. Boris, T.R. Young, J.M. Picone, and C.W. Oswald, under sponsorhip from the Naval Research Laboratory through the Office of Naval Research.

References

Boris, J.P. (1971) A fluid transport algorithm that works. *Computing as a Language of Physics*, pp. 171-189. International Atomic Energy Agency, Vienna.

Boris, J.P. and Book, D.L. (1976) Solution of the continuity equation by the method of flux-corrected transport. *Methods Comput. Phys.* 16, 85-129.

Fickett, W. and Davis, W.C. (1979) *Detonation*. University of California Press, Berkeley.

Godunov, S.K. (1959) Finite difference methods for numerical computation of discontinuous solutions of the equations of fluid dynamics. *Mat. Sb.* 47, 271-306.

Guirguis, R. and Oran, E. (1983) *Reactive Shock Phenomena in Condensed Materials: Formulation of the Problem and Method of Solution*, NRL Mem. Rept. 5228, Naval Research Laboratory, Washington, DC.

Guirguis, R., Oran, E.S., and Kailasanath, K. (1986) Numerical simulations of the cellular structure of detonations in liquid nitromethane – regularity of the cell structure. *Combust. Flame* 65, 339–365.

Guirguis, R., Oran, E.S., and Kailasanath, K. (1987a) The effect of energy release on the regularity of detonation cells in liquid nitromethane. pp. 1659–1668. *21st Symposium (International) on Combustion*, The Combustion Institute, Pittsburgh, PA.

Guirguis, R., Oran, E.S., and Kailasanath, K. (1988) The regularity of cellular structure in gaseous detonations, I. the effect of thermochemical parameters. To be submitted to *Combust. Flame*.

Hiramatsu, K., Fujiwara, T., and Taki, S. (1984) Modeling of gaseous detonation limits, *Proc. Fourteenth International Symposium on Space Technology and Science*, p. 549.

Kailasanath, K., Oran, E.S., Boris, J.P., and Young T.R. (1985a) A computational method for determining detonation cell size. AIAA Paper 85-0236.

Kailasanath, K., Oran, E.S., Boris, J.P., and Young, T.R. (1985b). Determination of detonation cell size and the role of transverse waves in two-dimensional detonations. *Combust. Flame*, 61, 199.

Oran, E.S., Boris, J.P., Young, T.R., Flanigan, M., Burks, T., and Picone, M. (1981) Numerical simulations of detonations in hydrogen-air and methane-air mixtures, *Eighteenth Symposium (International) on Combustion.* p. 1641. The Combustion Institute, Pittsburgh, Pa.

Oran, E.S., Young, T.R., Boris, J.P., Picone, J.M. and Edwards, D.H. (1982) A study of detonation structure: the formation of unreacted gas pockets, *Nineteenth Symposium (International) on Combustion*, pp. 573–582. The Combustion Institute, Pittsburgh, PA.

Oran, E.S. and Boris, J.P. (1987a) Numerical Methods in Reacting Flows. AIAA Paper 87-0057.

Oran, E.S. and Boris, J.P. (1987b) *Numerical Simulation of Reactive Flow*. Elsevier, New York.

Mader, C.L. (1979) *Numerical Modeling of Detonations*. University of California Press, Berkeley, CA.

Strehlow, R.A. (1984) *Combustion Fundamentals*. McGraw Hill, New York.

Subbotin, V.A. (1975) *Fizika Goreniya Vzryva* 11, 96–102.

Taki, S. and Fujiwara, T. (1981) Numerical simulation of triple shock behavior of gaseous detonation. *Eighteenth Symposium (International) on Combustion*, pp. 1671–1681. The Combustion Institute, Pittsburgh, PA.

Walsh, J.M. and Christian, R.H. (1955) *Phys. Rev.* 97, 1544.

Transmission of Overdriven Plane Detonations: Critical Diameter as a Function of Cell Regularity and Size

D. Desbordes*
Laboratoire d'Energétique et de Détonique, Poitiers, France

Abstract

Steady overdriven detonations in $C_2H_2/O_2/Ar$ mixtures are produced by means of the shock tube method. The experimental reduction of cell spacing λ/λ_{CJ} with the degree of overdrive M_s/M_{CJ} of the detonation is well predicted by assuming a constant proportionality between the characteristic size of the structure and the chemical induction length. When the plane detonation is an overdriven wave, a drastic reduction of the critical diameter of transmission d_c is observed. d_c varies as the cell spacing according to the classical rule $d_c = 13\lambda$. As a consequence, the critical energy required for initiation of spherical expanding detonation, derived from this type of ignition, cannot have a unique value. Deviation from the $d_c = 13\lambda$ rule is observed when the degree of overdrive becomes too high and when the dilution with argon is important (a deviation up to a factor of two is observed: $d_c = 26\lambda$). This behavior is correlated with cell regularity, which seems, in addition to its mean characteristic size, to play an important role in transient phenomena in detonation.

Introduction

At the present time, transient phenomena observed in gaseous detonation (for instance, in direct initiation by shock compression produced by a point source in an infinite medium or in transmission of a plane detonation propagating in a tube into a spherical expanding detonation) are undoubtedly linked to the three dimensional structure of the detonation front. The existence of gaseous detonation depends upon the ability of the reactive medium to create

---

Copyright © 1988 by the American Institute of Aeronautics and Astronautics, Inc. All Rights reserved.
*Assistant Professor.

TRANSMISSION OF OVERDRIVEN PLANE DETONATIONS   171

and regenerate in various situation the periodical cellular structure by the collisions of transverse waves.

Particularly since the precursor work of Zel'dovich et al. (1956), the critical diameter of transmission of a detonation from a circular tube to free space is now recognized as a measure of the detonability of a gaseous mixture, and can be used to determine the critical energy for direct initiation (see Matsui and Lee (1978) and Lee (1984)).

In a general way, the critical diameter d_c includes 13 times the cell spacing λ_{CJ} of the gaseous mixture. This rule was checked for various mixtures such as hydrogen and common hydrocarbons with oxygen and air at different equivalence ratios, initial pressures, and rates of nitrogen dilution (Matsui and Lee (1978); Edwards et al. (1981); Knystautas et al. (1982); Moen et al. (1982)).

In recent experiments, Desbordes and Vachon (1986) have shown that $d_c = 13 \lambda$ remains valid for transmission of a plane detonation if it is an overdriven wave (for degree of overdrive M_s/M_{CJ} up to 1.1). So, a drastic reduction of the critical diameter of a mixture is observed, with a corresponding small increase of the degree of overdrive of the detonation.

The present paper treats the problem of the critical diameters of transmission of overdriven detonations with a higher degree of overdrive than those used in the 1986 study, and their link with cell size.

Laboratory-scale experiments are basically performed with $C_2H_2/O_2/Ar$ systems.

First of all, we examine the dependence of λ/λ_{CJ} on M_s/M_{CJ} and the consequence of the good prevision of experimental results if we admit constant proportionality between cell spacing and chemical induction length L_i, i.e., $\lambda = A L_i$.

The second objective is to determine how far in degree of overdrive the critical diameter obeys the $d_c = 13 \lambda$ rule and in that case what the overdriven detonation transmission provides about the critical energy needed for direct initiation of a reactive mixture.

Experimental Details

The experimental device used in the laboratory for the tests on overdriven detonation transmission is shown in Fig. 1. Two cylindrical, 4 m long steel tubes, 21 or 52 mm in diameter, are connected to a 38 cm i.d., 50 cm long chamber. The tube is divided into two parts separated by a thin Mylar foil, the first about 3.6 m long (the high-pressure section) and the second 0.4 m long (the low-pressure section) and protrudes directly inside the chamber. These two parts formed a kind of shock tube apparatus.

The high- and low-pressure (HP and LP) sections both contain the same mixture, but the HP section has the higher initial pressure. Direct quasisteady detonation (velocity D $\sim D_{CJ}$) is achieved at the closed end of the tube by means of an exploding wire device. The LP section is empty, with the gaseous reactive mixture tested at a chosen initial pressure p_f.

The tests were performed with premixed gases (stoichiometric C_2H_2/O_2 and $C_2H_2/O_2/Ar$ with 50 and 75% in Ar) at different initial pressures p_f of 20 - 300 Torr.

Two pressure transducers, separated by 500 mm, are mounted next to the Mylar membrane in the HP section. In the LP section, a pressure gage is located near the orifice of transmission and connected with five ionization probes (25 mm apart), providing measurement of local velocity D of the plane overdriven wave. Moreover, smoked foils are positioned at the orifice of transmission, on the axis inside the LP section of the tube and in the chamber, in order to monitor cell size λ inside the tube and at the sudden area change and the transmission or failure processes of the detonation.

In that configuration, reproducible quasisteady overdriven detonation are produced (because of the quasi-negligible expansion in the LP section of the Zel'dovich-Taylor wave produced in the driver section)

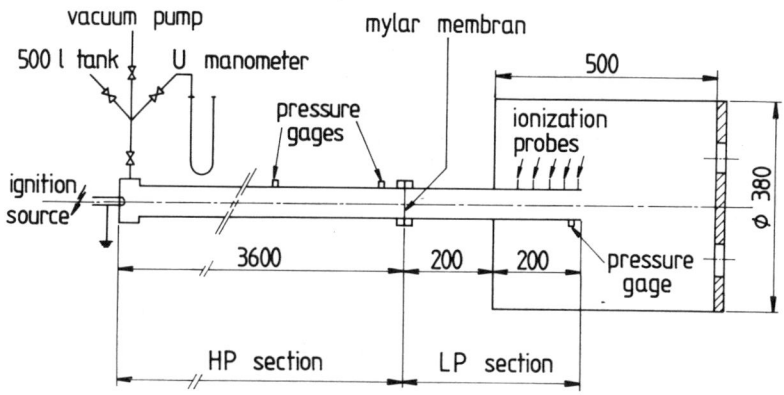

Fig. 1 Experimental apparatus.

depending, for a given p_f in the LP section, on the initial pressure of the HP section.

Overdriven detonation conditions at the tube exit of the LP section are defined in terms of its degree of overdrive M_s/M_{CJ} (or D/D_{CJ}) and its cell spacing ratio λ/λ_{CJ}. M_{CJ} and λ_{CJ} concerning stable self-sustained detonation in tubes are related to the mixture contained in the LP section. (Note: C-J = Chapman-Jouguet.)

Results and Discussion

Correlation between λ/λ_{CJ} and M_s/M_{CJ}

First of all, results of cell spacing measurements λ_{CJ} when the detonation waves are self-sustained (C-J conditions) are plotted as a function of the initial pressure p_f in Fig. 2. We have extended the range of

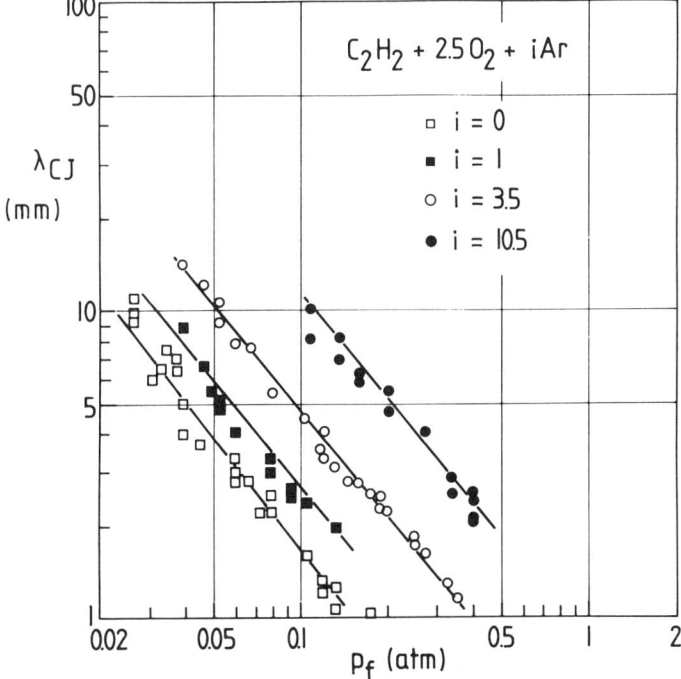

Fig. 2 Cell spacings λ_{CJ} vs initial pressure p_f for $C_2H_2 + 2.5\ O_2$ + i Ar mixtures (i = 0, 1, 3.5 and 10.5).

experimental degree of overdrive of the detonation up to 1.3 at different initial pressures p_f.

Experimental results λ/λ_{CJ} are plotted as a function of M_s/M_{CJ} in Fig. 3. Irrespective of the mixture and initial pressure, a mean curve can be deduced. The results of Vasiliev and Nikolaev (1978) and Gavrilenko and Prokhorov (1983) for $C_2H_2 + 2.5\ O_2$ mixtures at $p_f = 760$ Torr are shown for comparison.

As it has been shown previously by Desbordes and Vachon (1986) with C_2H_2/O_2 systems and by Huang and Xu (1987) in H_2/O_2 systems, the $\lambda/\lambda_{CJ} \propto M_s/M_{CJ}$ law can be simply deduced assuming that a constant proportionality exists between the cell spacing and the global chemical induction length of the mixture behind the shock wave, whatever the velocity of detonation.

We consider the simple formulation of the global chemical induction time behind shock wave of a reactive mixture expressed as (cf. Soloukhin and Ragland 1969)

$$\tau_i = K\ \rho_s^{-n}\ \exp(E_A/RT_s) \qquad (1)$$

where K is the kinetic rate factor, ρ_s and T_s density and temperature behind shock wave respectively, E_A the global activation energy of the mixture, and n a constant depending on the reactive mixture (n = 1 generally).

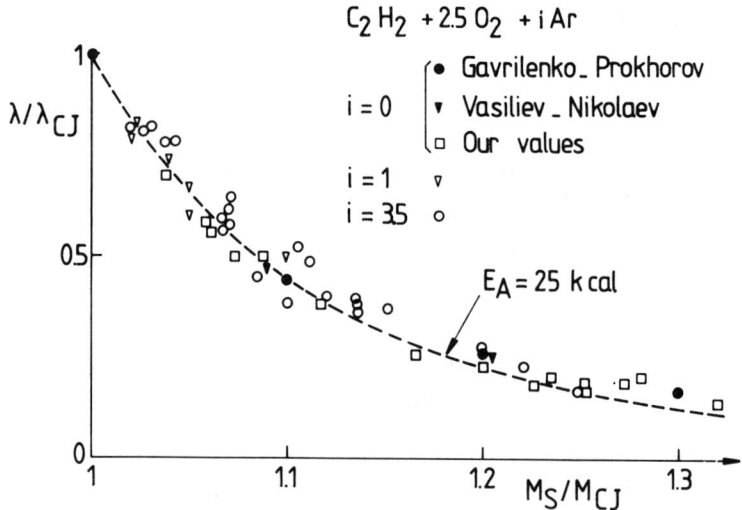

Fig. 3 Comparison of observed and calculated cell spacing ratios λ/λ_{CJ} as a function of the degree of overdrive M_s/M_{CJ} of the detonation.

Chemical induction length L_i behind a shock wave of Mach number M_s is deduced from the following relation:

$$L_i = \tau_i(D - V_s) = \tau_i D \frac{\rho_f}{\rho_s} = \tau_i M_s a_f (\frac{\rho_s}{\rho_f})^{-1} \quad (2)$$

where ρ_s/ρ_f is the density ratio across the shock and a_f the sound wave velocity in fresh gas.
Using Eq. (1),

$$L_i = K \rho_f \rho_s^{-(n+1)} M_s a_f \exp(E_A/RT_s) \quad (3)$$

If we assume that $\lambda = A L_i$ (A is a constant independent of M_s), one can obtain

$$\frac{\lambda}{\lambda_{CJ}} = \frac{M_s}{M_{CJ}} (\frac{\rho_s}{\rho_{ZND}})^{-(n+1)} \exp(\frac{E_A}{R} (\frac{1}{T_s} - \frac{1}{T_{ZND}})) \quad (4)$$

where ZND is relative to postshock conditions when $M_s = M_{CJ}$.
This relation represents quite well experimental data obtained for overdriven detonation (cf. Fig. 3) (n is chosen equal to 1).

The slope of the curve λ/λ_{CJ} (M_s/M_{CJ}) can be obtained with classical Rankine-Hugoniot relationships behind the shock wave with the approximation of constant γ (γ is the ratio of specific heats), i.e.,

$$\frac{d\lambda/\lambda}{dM_s/M_s} = -2 \left(\frac{2 + (\gamma - 1)M_s^2}{2 + (\gamma - 1) M_s^2} \frac{E_A/RT_s}{} - \frac{1}{2} \right) \quad (5)$$

This expression gives the relative change in cell size resulting from a small relative change in the Mach number of the shock wave. It represents the sensitivity of the detonation characteristic length to small changes in the shock Mach number, essentially depending on the reduced activation energy E_A/RT_s. This slope is maximum at the C-J condition (minimum velocity of detonation) and is more pronounced as E_A/RT_{ZND} increases. For stoichiometric H_2/O_2 and C_2H_2/O_2 mixtures, reduced activation energies when $M_s = M_{CJ}$ are minimum or ~ 5.5. For other hydrocarbon/O_2 and hydrocarbon/air mixtures, values are greater than 7 (see, for example, values given by Moen et al. 1986).

Nevertheless, high values of E_A/RT_s promote the level of instability in the reacting flow as has been shown by the stability criterion of Shchelkin and by other stability theories for detonations (see, for example, Fickett and Davis 1979).

The consequences are experimental observations that detonation structures present more irregularities as E_A/RT_{ZND} increases (Ulyanitskii 1981; Moen et al. 1986). Our observations in the case of overdriven detonations confirm this conclusion. As M_s/M_{CJ} increases, apparent substructure inside the cell at the C-J condition progressively disappear, according to the conclusion of the one-dimensional stability analysis of Fickett and Wood (1966) and Abouseif and Toong (1982).

The different mechanisms responsible for cell irregularity are hidden in the global term E_A/RT_s. Libouton et al. (1981) and more recently Vandermeiren (1985) attribute cell irregularities to competition between some elementary chemical processes inside the cell itself. The rate of liberation of chemical energy depending on this process is qualitatively correlated to cell regularity.

Critical Diameter in the C-J case

Successfull transmission of a planar C-J detonation through a circular orifice of diameter d into free space, where the plane detonation turns into a spherically expanding detonation, is achieved when diameter d equals or exceeds what we call the critical diameter d_c. As observed for various reactive mixture, d_c is correlated to cell spacing λ_{CJ} of the mixture by a constant of 13, first discovered by Mitrofanov and Soloukhin (1962).

Nevertheless, deviations from the $d_c = 13 \lambda_{CJ}$ rule were noticed by Edwards et al. (1981) and Moen et al. (1984) for very irregular structure systems. If, in these cases, the deviations can be attributed to difficulties in finding a good estimation of cell structure length, recent experiments reported by Moen et al. (1986) indicate that in $C_2H_2 + 2.5\ O_2 + 75\%$ Ar mixtures, where the structures are relatively regular, the $d_c = 13 \lambda_{CJ}$ rule vanishes if initial pressure is not too high. In that case, d_c is 13-26 λ_{CJ}, depending on the initial pressure.

Our experiments, performed with stoichiometric C_2H_2/O_2 mixtures undiluted and diluted by 50 and 75% argon in the case of C-J planar diffraction show (cf. Table 1) that if, for the two first mixtures, the $d_c = 13 \lambda_{CJ}$ relationship holds, then for the third one a critical diameter of 24-26

TRANSMISSION OF OVERDRIVEN PLANE DETONATIONS 177

λ_{CJ} (52 mm i.d. tube) is found. This result confirms the general trends observed by Moen et al. (1986) for very large argon dilution.

Critical Diameter in Overdriven Cases

For each tube, initial pressure p_f of the mixture contained in the chamber and in the LP section of the tube is chosen lower than its critical initial pressure p_{f_c} of transmission of the self-sustained plane detonation indicated in the Table 1.

By increasing shot by shot the initial pressure in the HP section, we can produce quasisteady overdriven detonation waves in the LP section in such a way that transmission of the detonation in the chamber is observed again. Critical transmission is defined by a minimum strength of the detonation ($M_s > M_{CJ}$) and a maximum value of detonation cell spacing λ ($\lesssim \lambda_{CJ}$).

Figures 4-6 summarize in the $\lambda - p_f$ plane the results of overdriven detonation transmission. Each tube is characterized by the critical cell width $\lambda_c = d_c/13$ (1.62 and 4 mm). For each initial pressure p_f, go/no go results are represented by the value of cell spacing measured at the tube exit.

With the 52 mm i.d. tube, it is not possible to perform experiments with high degree of overdrive because the chamber is not big enough.

In the 21 mm i.d. tube, one can notice the following:
1) In $C_2H_2 + 2.5\ O_2$ mixtures, the transmission occurs by overdriven detonation waves in the range of pressure 80 (= p_{f_c}) $\geq p_f \geq$ 30 Torr, equivalent to Mach number conditions of $1 \leq M_s/M_{CJ} \leq$ 1.16-1.17, with the $d_c = 13\ \lambda$ rule. For $p_f = 25$ Torr, the number of cells needed in the diameter for the transition increases sharply, since we

Table 1 Conditions of Critical Reinitiation of Self-Sustained Plane Detonations in $C_2H_2 + 2.5\ O_2$ + i Ar Mixtures

Mixture	d, mm	p_{f_c}, Torr	λ_{CJ}, mm	d/λ_{CJ}
i = 0	52	30	4	13
	21	80	1.6	13
i = 3.5	52	80	4	13
	21	190	1.6	13
i = 10.5	52	290	2	26

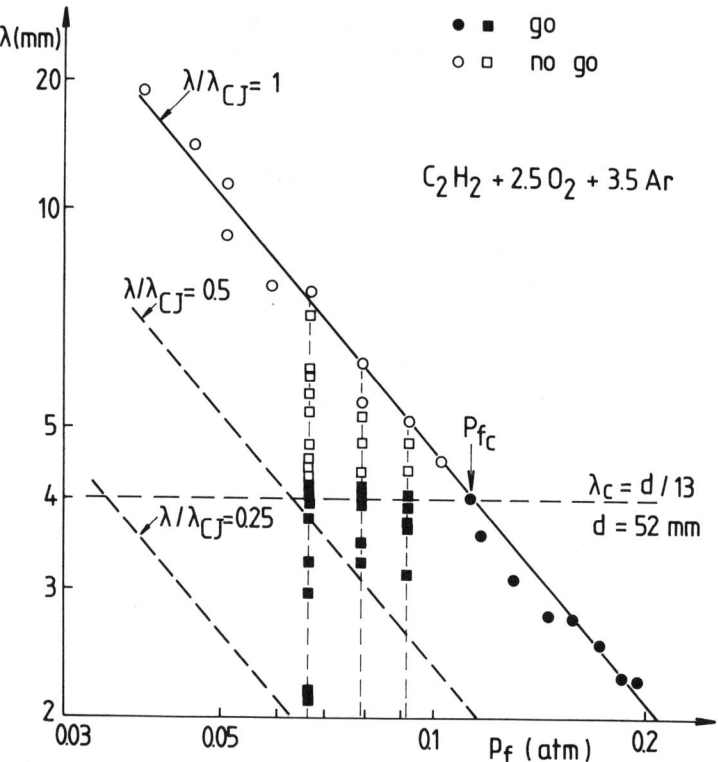

Fig. 4 Go/no go for transmission of overdriven plane detonation in $C_2H_2 + 2.5\ O_2 + 3.5$ Ar mixture at different initial pressure p_f for 52 mm i.d. tube.

find $d_c \sim 24-26\ \lambda$. For $p_f = 20$ Torr, the 25 cells spacing contained in the tube diameter does not permit observation of the transition.

2) In $C_2H_2 + 2.5\ O_2 + 50\%$ Ar mixtures, for $190\ (= p_{fc}) \geq p_f \geq 90$ Torr (i.e., for $1 \leq M_s/M_{CJ} \leq 1.10-1.11$), $d_c^{fc} \cong \overline{13}\ \lambda$ holds. For $p_f = 70$ Torr, $\bar{d}_c \cong 24-26\ \lambda$ is observed.

Thus, it can be deduced that if the overdriven degree of detonation is not too high, especially when the argon dilution is low in $C_2H_2 + 2.5\ O_2$ systems, the relationship between all critical diameters of a mixture is given by

$$\frac{d_c}{d_c(CJ)} = \frac{\lambda}{\lambda_{CJ}} \tag{6}$$

where λ/λ_{CJ} is well represented by Eq. (4).

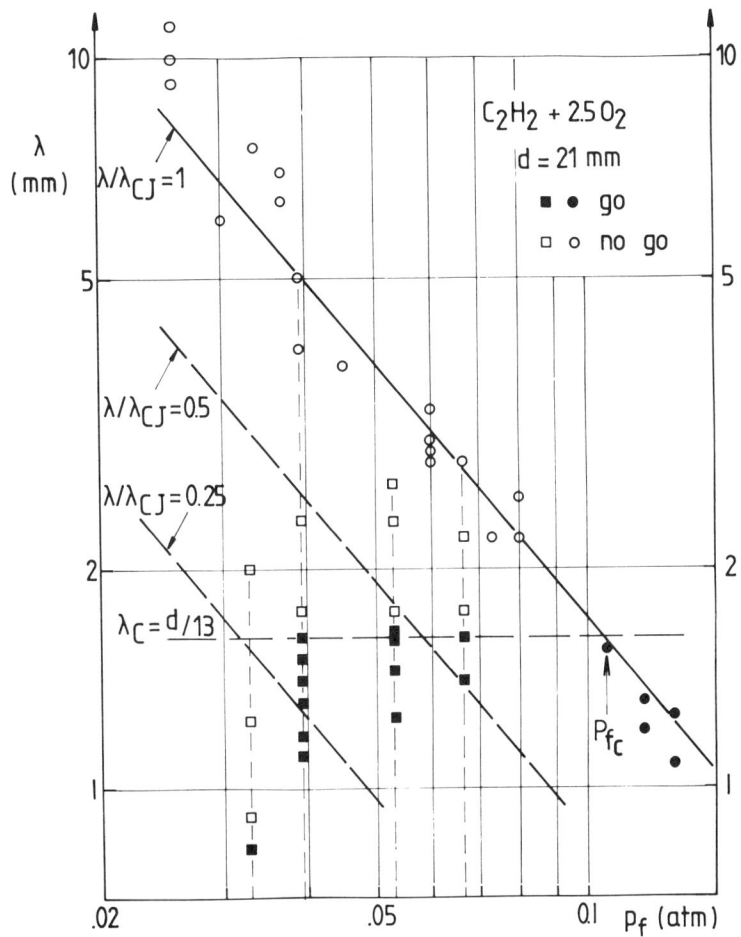

Fig. 5 Go/no go for transmission of overdriven plane detonation in $C_2H_2 + 2.5\ O_2 + 3.5$ Ar mixture at different initial pressure p_f for 21 mm i.d. tube.

This means that the critical diameter of a mixture can be reduced by an important factor (for example, by three for $C_2H_2 + 2.5\ O_2$), if plane detonation is an overdriven wave.

As has been already noted, the $d_c = 13\ \lambda$ law has been established for a variety of systems with poor and irregular cellular structure.

The deviations we observe from this law happen when the cell becomes very regular. Following Moen et al. (1986), we think that the organization of cellular structure plays a prominent role in the sustenance of the detonation wave, particularly in the transition and initiation processes.

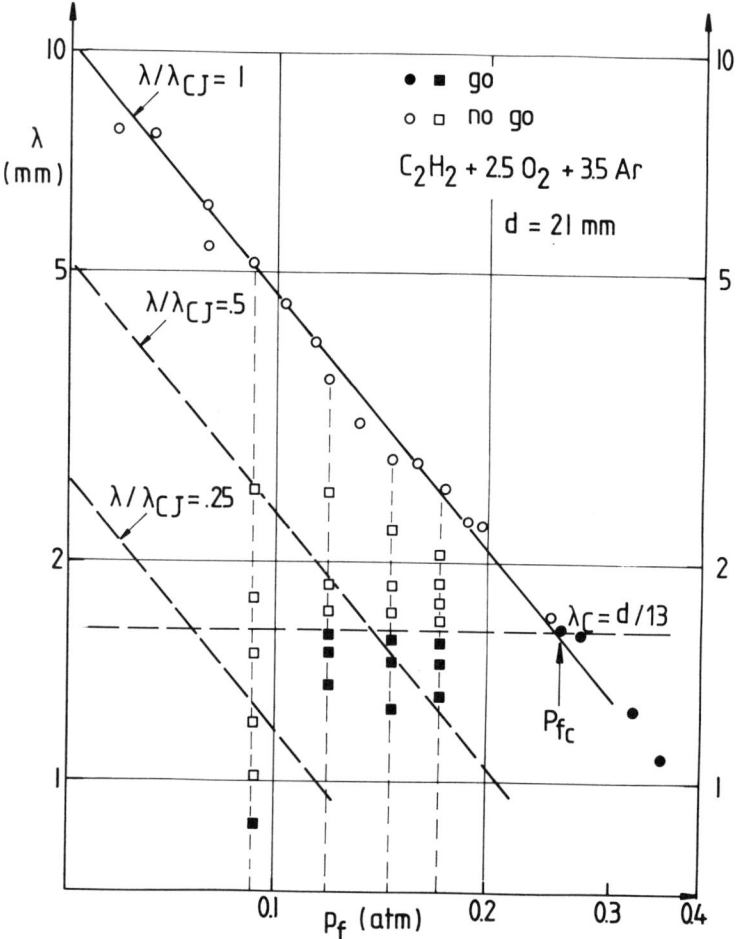

Fig. 6 Go/no go for transmission of overdriven plane detonation in C_2H_2 + 2.5 O_2 mixture at different initial pressure p_f for 21 mm i.d. tube.

The addition of large amounts of argon in a reactive gaseous mixture is widely used to obtain very regular cells. Besides, growth of the degree of overdrive of a detonation wave, as discussed above, has the same effect on the cell regularity. Thus, the combination of both effects (increase in argon dilution and/or of the strength of the detonation) causes a more regular cell structure, changing the coefficient of proportionality between d and λ. As M_s/M_{CJ} grows, a rapid increase in this coefficient is observed by a factor of two. For high dilution by argon

(75%), the overdriven effect on detonation is not needed to observe the same deviation.

Critical Initiation Energy

Because there exists an infinity of critical conditions, characterized by two parameters (strength of detonation M_s and critical diameter d_c) that ensure the transmission of the plane detonation into a spherical one in a reactive mixture, one can wonder about the problem of the critical initiation energy of a spherical expanding detonation produced through this indirect mode.

Indeed, when the strength of the plane detonation increases, the critical diameter drastically drops (we discuss only the case where $d_c = 13\ \lambda$). The critical diameter initiation of a spherical detonation has to be considered as an energy source of variable characteristics, since the plane detonation strength varies.

Considering that, at transition, the plane overdriven detonation wave propagating in the tube is quasisteady, we assume the action of a constant velocity piston behind it that supports the detonation wave at a constant Mach number M_s. If the detonation front is assumed to be a discontinuity, the burned gases behind the front have constant characteristics: velocity, pressure, sound velocity, etc. (respectively U_b, p_b, a_b), which belong to the overdriven detonation branch of the Hugoniot curve of the burned gases of the mixture.

If S is the cross-sectional area of the tube, energy given to the gas by the piston during the time t of action is

$$E = p_b\ U_b\ S\ t \quad (7)$$

We consider that, at the critical transmission of the detonation, the energy brought to the free space is that of the plane piston acting during the time $t = t_c$, where t_c is the duration elapsed when detonation just comes in the free space and the onset of spherical detonation. It corresponds approximatively to (cf. Matsui and Lee 1978)

$$t_c \simeq \frac{d_c}{2a_b} \quad (8)$$

And the cross-sectional area $S = S_c$ is given by

$$S_c = \frac{\pi d_c^2}{4} \quad (9)$$

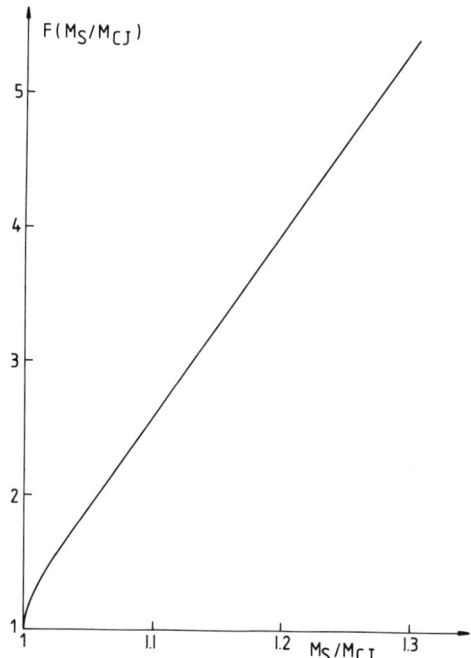

Fig. 7 Variation of $F = (p_b/p_{CJ})(u_b/u_{CJ})(a_b/a_{CJ})^{-1}$ in function of the degree of overdrive M_s/M_{CJ} of the detonation.

If we admit the $d_c = 13\ \lambda$ relationship, the critical energy can be written as

$$E_c = \frac{2197\,\pi}{8} \frac{p_b U_b}{a_b} \lambda^3 \qquad (10)$$

If we normalize E_c to its C-J value $E_c(CJ)$, we obtain

$$\bar{E}_c = \frac{E_c}{E_c(CJ)} = \frac{p_b}{p_{CJ}} \frac{U_b}{U_{CJ}} \left(\frac{a_b}{a_{CJ}}\right)^{-1} \left(\frac{\lambda}{\lambda_{CJ}}\right)^3 \qquad (11)$$

where adimensional parameters p_b/p_{CJ}, U_b/U_{CJ}, and a_b/a_{CJ} are calculated by the QUATUOR code (cf. Heuzé et al. 1985) and can be expressed in first approximation by unique relations depending on M_s/M_{CJ}, regardless of the reactive mixture.

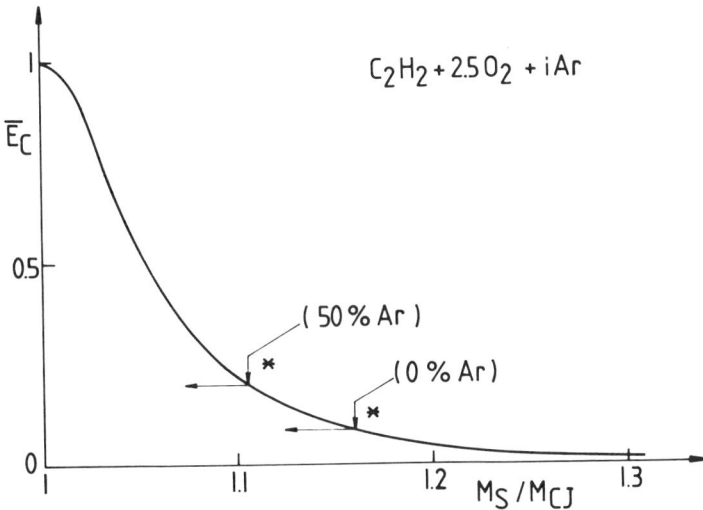

Fig. 8 Variation of the reduced energy $\bar{E}_c$ given by plane piston at critical transmission of plane detonation.

Thus, Eq. (11) can be written as a product, i.e.,

$$\bar{E}_c = F(M_s/M_{CJ})(\lambda/\lambda_{CJ})^3 \qquad (12)$$

The function F essentially represents the thermomechanical characteristics of the wave and grows monotonically with M_s/M_{CJ} (see Fig. 7).

The second part, λ/λ_{CJ} at a power of 3, including thermochemical effects of the wave, is essentially a decreasing function of M_s/M_{CJ}, except for very high Mach numbers M_s (see Eq. (4)).

The $\bar{E}_c$ value plotted in Fig. 8 for $C_2H_2/O_2/Ar$ systems is a decreasing function of M_s/M_{CJ} up to M_s^*/M_{CJ}, M_s^* being the Mach number of the wave where the $d_c = 13 \lambda$ rule breaks down.

It can be noticed that in the case of the reduction of the critical diameter by a factor of 3, the critical energy of initiation of the spherical detonation in this way diminishes by a factor of 10.

Conclusions

The present investigation has provided further evidence of the importance of the detonation structure size in transient phenomena in gaseous detonation.

Laboratory experiments on transmission of overdriven plane detonations in $C_2H_2/O_2/Ar$ systems have shown the following:

1) The cell size (for instance, λ) is very sensitive to the state of the detonation (i.e., its degree of overdrive), especially at the minimum velocity of the detonation wave ($M_s/M_{CJ} = 1$). This size varies with M_s proportionally to the one-dimensional chemical induction length of the mixture, at least for M_s/M_{CJ} up to 1.3. Nevertheless, the more λ is sensitive to M_s (the greater the value of E_A/RT_s), the more an irregular detonation structure appears.
2) The critical diameter of transmission of an overdriven plane detonation varies in the same way as its cell spacing with the classical proportion $d_c/\lambda = 13$ as long as the degree of overdrive and/or argon dilution are not too high.
3) When the degree of overdrive and/or dilution by argon is high enough, deviation from the 13λ law is unambiguously observed. According to the conclusion of Moen et al. (1986) it seems that the cell regularity plays a key role in the behavior of the detonation, not only in transient phenomena, but also in self-sustained situations.
4) The critical energy required for the initiation of a spherical detonation by this type of ignition can be reduced by an important factor when the plane detonation is overdriven.

Further investigations on overdriven detonation transmission are needed to extend the applicability of the $d_c = 13\lambda$ law at more irregular cellular systems, where generally $d_c = 13\lambda_{CJ}$ is observed.

An important effort is also required to better understand the intimate mechanisms that govern cell regularity and the onset of instabilities at higher frequencies in the detonation front.

References

Abouseif, G.E. and Toong, T.Y. (1982) Theory of unstable detonations. Combust. Flame 45(1), 67-94.

Desbordes, D. and Vachon, M. (1986) Critical diameter of diffraction for strong plane detonations. Progress in Astonautics and Aeronautics: Dynamics of Explosions, edited by J.R. Bowen, J.C. Leyer, and R.I. Soloukhin, Vol. 106, pp. 131-143. AIAA, New York.

Edwards, D.H., Thomas, G.O., and Nettleton, M.A. (1981) Diffraction of a planar detonation in various fuel oxygen mixtures at an area change. Progress in Astonautics and Aeronautics: Gasdynamics of Detonations and Explosions, edited by J.R. Bowen, N. Manson, A.K. Oppenheim, and R.I. Soloukhin, Vol. 75, pp. 341-357. AIAA, New York.

Fickett, W. and Davis, W.C. (1979) Detonation. University of California Press, Berkeley.

Fickett, W. and Wood, W.W. (1966) Flow calculations for pulsating one-dimensional detonations. Phys. Fluids 9, 903-916.

Gavrilenko, T.P. and Prokhorov, E.S. (1983) Overdriven gaseous detonations. Progress in Astronautics and Aeronautics: Shock Wave, Explosions and Detonations, edited by J.R. Bowen, N. Manson, A.K. Oppenheim, and R.I. Soloukhin, Vol. 87, pp. 244-250. AIAA, New York.

Heuzé, O., Bauer, P., Presles, H.N., and Brochet, C. (1985) The equation of state of detonation products and their incorporation into Quatuor code. VIIIth Symp. (Int.) on Detonations, Albuquerque, NM.

Huang, Z.W. and Xu, B. (1987) Cellular structure of overdriven gaseous detonations. Combust. Flame 67(2), 95-98.

Knystautas, R., Lee, J.H., and Guirao, C. (1982) The critical tube diameter for detonation failure in hydrocarbon-air mixtures. Combust. Flame 48(1), 63-83.

Lee, J.H. (1984) Dynamic parameters of gaseous detonations. Ann. Rev. Fluid Mech. 16, 311-336.

Libouton, J.C., Jacques, A., and Van Tiggelen, P.J. (1981) Cinétique, structure et entretien des ondes de détonation, Proceedings of the International Colloquium Berthelot-Vieille-Mallard-Le Châtelier, Vol. II, pp. 437-444, Bordeaux, France.

Matsui, H. and Lee, J.H. (1978) On the measure of the relation detonation hazards of gaseous fuel-oxygen and air mixtures. Seventeenth Symposium (International) on Combustion, pp. 1269-1280. The Combustion Institute, Pittsburgh, PA.

Mitrofanov, V.V. and Soloukhin, R.I. (1964) On the instantaneous diffraction of detonation front. Dokl. Akad. Nauk SSSR 159(5), 1003-1006.

Moen, I.O. et al. (1982) Diffraction of detonation from tubes into a large fuel-air explosive cloud. Nineteenth Symposium (International) on Combustion, pp. 635-644. The Combustion Institute, Pittsburgh, PA.

Moen, I.O., Funk, J.W., Ward, S.A., Rude, G.M., and Thibault, P.A. (1984) Detonation length scales for fuel air explosives. Progress in Astronautics and Aeronautics: Dynamics of Shock Waves, Explosions and Detonations, edited by J.R. Bowen, N. Manson, A.K. Oppenheim, and R.I. Soloukhin, Vol. 94, pp. 55-79. AIAA, New York.

Moen, I.O., Sulmitras, A., Thomas, C.O., Bjerketvedt, D., and Thibault, P.A. (1986) Influence of cellular regularity on the behavior of gaseous detonations. Progress in Astronautics and Aeronautics: Dynamics of Shock Waves, Explosions and Detonations, edited by J.R. Bowen, N. Manson, A.K. Oppenheim, and R.I. Soloukhin), Vol. 106, pp. 220-243. AIAA, New York.

Soloukhin, R.A. and Ragland, K.W. (1969) Ignition processes in expanding detonations. Combust. Flame 13(3), 295-302.

Ulyanstskii, V. Yu (1981) Role of "flashing" and transverse - wave structure in gases. Fiz. Goreniya Vzryva 17(2), 127-133.

Vandermeiren, M. (1985) Ondes de détonation dans les mélanges à l'acétylène : Structure et cinétique chimique. Doctorat ès Sciences, Louvain la Neuve, Belgium.

Vasiliev, A.A. and Nikolaev, Yu. (1978) Closed theoretical model of a detonation cell. Acta Astronaut. Vol. 5, 983-996.

Zel'dovich, Ya. B., Kogarko, S.M., and Simonov, N.N. (1956) Experimental study of gaseous spherical detonations. J. Phys. Technol. 26(8), 1744-1772.

Role of an Inhibitor on the Onset of Gas Detonations in Acetylene Mixtures

M. Vandermeiren* and P. J. Van Tiggelen†

Université Catholique de Louvain, Louvain-la-Neuve, Belgium

Abstract

The experimental investigation of detonations in lean $C_2H_2/O_2/Ar$ mixtures has been performed over a pressure range of 25-250 Torr. The influence of adding variable amounts of CF_3Br (1-50% with respect to the fuel) has been studied by recording the cell length L and the detonation velocity D. The presence of inhibitor does not drastically affect the stable regime of detonation. Only a slight increase of D and a simultaneous decrease of L occur when small amounts of CF_3Br are added (<5%). Furthermore, the onset of the detonation regime behind a flange with an inlet diameter of 1.1 cm exhibits a specific role of CF_3Br during the incipient stage of the detonation. This behavior is particularly obvious from soot imprints. The ignition distance behind the orifice is twice as long in the presence of only 2% of inhibitor in the mixture. The different role of CF_3Br traces on acetylene-oxygen detonation in well-sustained regimes as well as at the incipient stage of the process will be discussed. An attempt to explain the data is offered by means of chemical kinetics arguments based on a rather simple mechanism.

Introduction

Inhibitors are compounds added in small quantities to modify drastically the combustion behavior of flammable mixtures. These compounds are of great practical interest

---

Copyright © 1988 by the American Institute of Aeronautics and Astronautics, Inc. All rights reserved.

*Research Assistant, Laboratoire de Physico-Chimie de la Combustion (currently with Air Products-S.A. Belgium).

†Professor, Laboratoire de Physico-Chimie de la Combustion

for the extinction of flames and unwanted fires as well as for the suppression of detonation hazards.

Among the usual extinguishers, halogenated compounds are used to a large extent and one of the most efficient is bromotrifluoromethane CF_3Br (halon 1301). A review of the inhibiting influence of fluorocarbon compounds on flames has been reported previously (Fristrom and Van Tiggelen 1979). A comparative experimental study of their influence on the burning velocity of different flammable mixtures has already been published (Vandermeiren et al. 1981). There is a systematic reduction of the burning velocity irrespective of the fuel. Very recent measurements (Rothschild et al. 1986) indicate the inhibition effect of CF_3Br on the propagation of CH_4-air flames stabilized on a flat-flame burner. However, it has been reported (Nzeyimana et al. 1986) that CF_3Br traces barely modify the transition from deflagration to detonation. In that investigation, it has been noticed that a major step of the flame acceleration process is the development of the turbulence in the flow ahead of the flame. The turbulence is not at all sensitive to a minute change in chemical composition; as a matter of fact, the viscosity of the mixture is not affected by compounds added in traces, e.g., CF_3Br.

Furthermore, in shock tube measurements the influence of halogenated inhibitors does not seem to be so systematic. Recent studies (Hidaka et al. 1985a; Hidaka an Suga 1985b) indicate, indeed, that the influence of an additive like CF_3Cl on the induction periods depends on the nature of the fuel used: hydrogen-oxygen mixtures and ethylene-oxygen mixtures exhibit an increase of the induction time when CF_3Cl is added, although the opposite is observed for CH_4-O_2 mixtures.

A similar trend has been noticed for detonations. Libouton et al. (1975) and Dormal et al. (1983) have shown that addition of traces of halogenated compounds inhibits detonation and therefore greatly increases the cell length of the detonation structure for $CO/H_2/O_2/Ar$ mixtures. On the contrary, Macek (1963) has noticed that CF_3Br and $C_2F_4Br_2$ sensitize detonations by decreasing the initial energies required to initiate H_2/O_2 detonations. Similarly, Jacques (1978) pointed out a promoting effect of CF_3Br on H_2-O_2-Ar detonation: the cell length decreases with the amount of CF_3Br added to the mixture. Moreover, Bull (1982) mentions that CF_3Br addition to methane-vitiated air mixtures enlarge the detonation limits. Finally, modeling of the gaseous detonations performed by Westbrook (1984) suggests significant inhibiting action upon the detonation of hydrocarbon, but data of large-scale tests of critical tube diameter indicate on the contrary a slight promoting effect [Moen et al.(1984)].

It is the purpose of this work to investigate experimentally the role of CF_3Br traces on the behavior of acetylene-oxygen detonations in well-sustained regimes as well as at the incipient stage of the process where the inhibition efficiency is most prominent.

Experimental Details

The experimental setup has been described previously (Vandermeiren and Van Tiggelen 1985). Measurements of the detonation velocity D and the cell length L have been performed for detonations propagating in acetylene-oxygen mixtures with an equivalence ratio of 0.33 and diluted by 50% argon at initial pressures of 25-250 Torr. The choice of that particular lean mixture was guided by our previous experience on acetylene detonations (Vandermeiren and Van Tiggelen 1985). Indeed, according to Libouton et al.'s classification (1981), regular detonation structures have been observed in a range of lean equivalence ratios. Variable percentages of the inhibitor CF_3Br with respect to the fuel molar fraction have been added to the initial mixture: 1, 2, 3, 5, 10, and 50% CF_3Br. Since the data about the CF_3Br influence on self-sustaining acetylene detonations have already been discussed (Vandermeiren and Van Tiggelen 1986), only the most salient features of that study will be summarized here.

Figure 1a exhibits the slight increase (1.5%) of the detonation velocities for small traces of CF_3Br added. It occurs for any initial pressure with a transmission orifice of 1.69 cm. Beyond 5% of CF_3Br added, the detonation velocities return to their original values measured in the uninhibited mixtures. Furthermore, in Fig.1b, the cell lengths L and consequently the characteristic time (t_{car} = L/D) decreases significantly with the adition of CF_3Br. The minimum of L that coincides with the maximum increase of D corresponds to 60% of the values of the uninhibited mixture, irrespective of the initial pressure. The slight promoting effect of CF_3Br traces can be explained qualitatively as follows. When small amounts of inhibitor are added to the mixture, the induction time is decreased as a consequence of the high level of the von Neumann temperature in those mixtures (see later discussion), but the time of the heat release pulse is not modified. Therefore, the detonation wave (shock plus reaction zone) is better coupled and a higher velocity is observed. On the contrary, when the percentage of CF_3Br

added becomes significant (>5% with respect to the fuel), the time of the heat release pulse is enhanced due to a mere thermodynamic effect. It leads to a decrease of the quality of the coupling between shock and reaction zone, which in turn reduces the detonation velocity.

It is worth noticing that no detonation sets in for mixtures either at an initial pressure of 25 Torr when more than 1% of CF_3Br is added or at 50 Torr when 50% of CF_3Br is present. These observations lead us to assume a specific role for the inhibitors during the incipient stage of

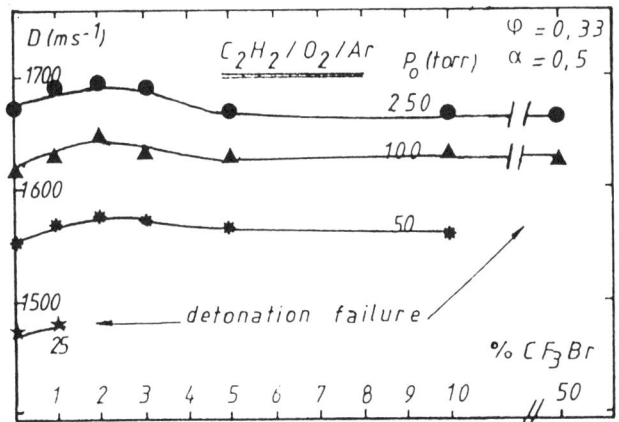

Fig. 1a Detonation velocity D vs the CF_3Br percentage for C_2H_2-O_2-Ar mixtures (ϕ = 0.33, α = 0.5).

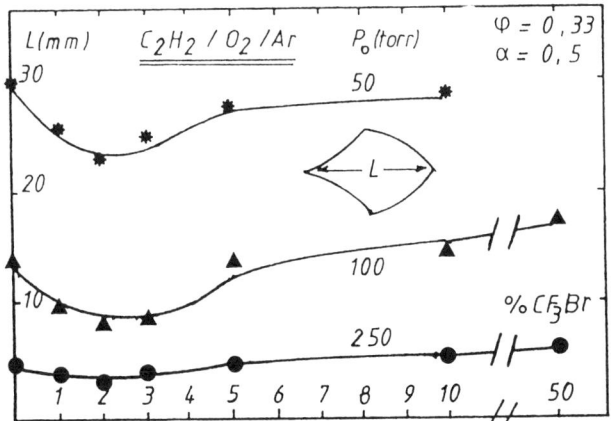

Fig. 1b Cell lengths L vs the CF_3Br percentage for $C_2H_2/O_2/Ar$ mixtures (ϕ = 0.33, α = 0.5).

detonation, i.e., at the initiation process. Therefore, a systematic investigation on the onset of detonation behind a bored flange with a variable orifice diameter has been carried out for mixtures with different amounts of inhibitor.

The test section of the rectangular tube with a cross section 3.2 x 9.2 cm is sketched on Fig.2 and is located immediately behind the bored flange. A Mylar film diaphragm (F) separates the test and the initiation section of the tube where a hydrogen-oxygen mixture at 300 Torr was detonated. The detonation velocity in the test section was measured by means of 16 ionization gages (G) located at a distance of 6 cm of each other. The signal was recorded on

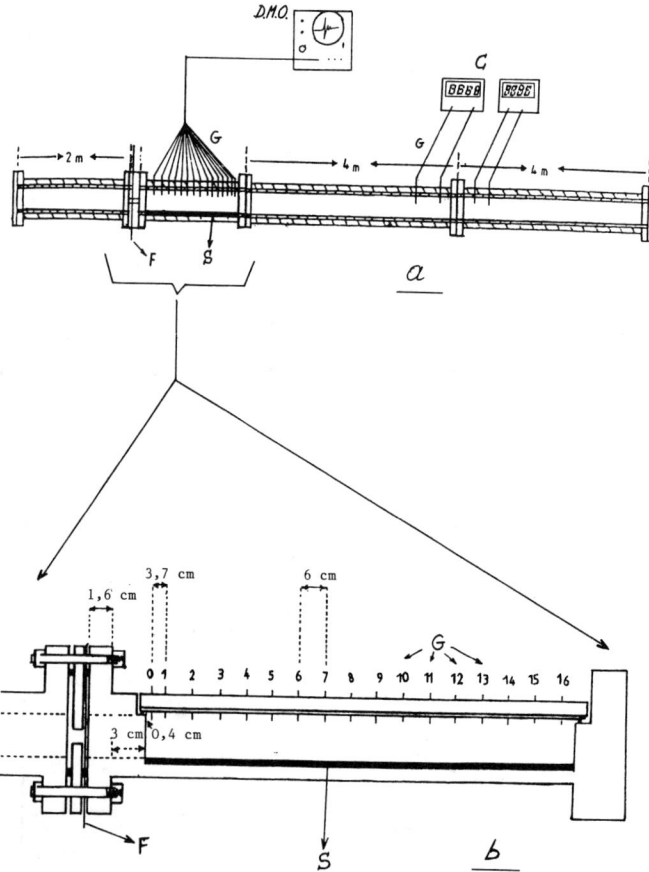

Fig. 2 Experimental set up for detonation transmition measurements :a) detonation tube. b) test section. DMO = digital memory oscilloscope, F = diaphragm, G = gages, C = time counters, S = soot record).

GAS DETONATIONS IN ACETYLENE MIXTURES 191

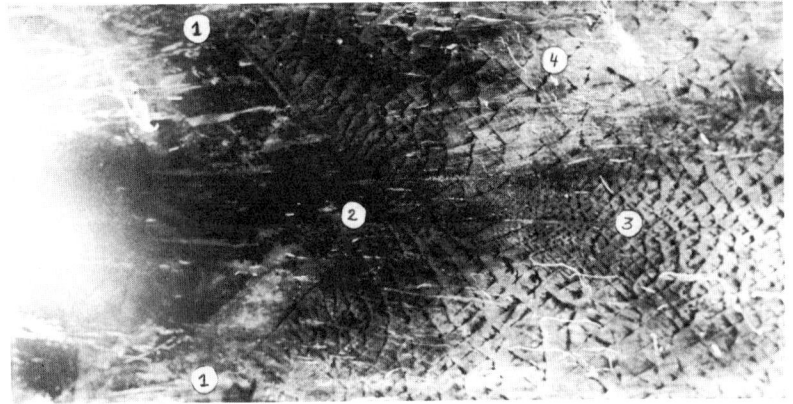

A. Soot record.

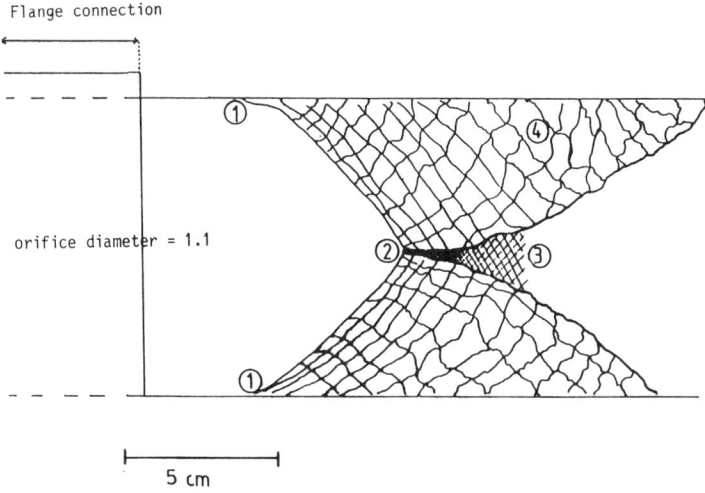

B. Sketch.

Fig. 3 Typical soot record of the onset of a detonation in an uninhibited $C_2H_2/O_2/Ar$ mixture ($\phi = 0.8$, $\alpha = 0.5$, $P = 100$ Torr, diameter of the flange orifice 1.1 cm).

a digital storage oscilloscope (Kikusui DSS6520). Soot imprints (S) were recorded simultaneously. Moreover, the velocity of the detonation wave in the well-established regime was recorded also at about 4 m after the end of the test section. These detonations were run in a lean $C_2H_2-O_2-Ar$ mixture ($\phi = 0.8$, $\alpha = 0.50$, and $P_o = 100$ Torr).

For an uninhibited mixture, a typical soot record of the onset of the detonation is shown on Fig. 3. The diameter of the orifice in the flange is 1.1 cm. The lower part of Fig. 3 is a sketch of the picture. It appears that the initiation of the detonation occurs either behind the reflection of the leading shock wave with the walls (1) or after the collision of two triple points trajectories (2). Such a collision (2) occurs around 15 cm after the bored flange orifice. The detonation wave issued from the reflection of the leading shock with the wall (1) is usually unsteady. The increase of the cell length in this zone (4) emphasizes the decaying character of the detonation originating from point 1. Such a detonation would thus probably not survive more than a few decimeters.

However, at position 2, an overdriven detonation starts and relaxes toward the steady detonation wave in the zone 3. This observation is confirmed by centerline measurements of the local detonation velocities at several stations during the onset of the detonation behind the bored flange (Fig. 4). The detonation velocity profile reaches a maximum value (2.7 km/s) which is around 1.5 times the average detonation velocity (1.86 km/s) for the mixture studied. The maximum of D is located around 45 cm behind the orifice. The velocity profile evolves then toward its average value over a distance of about 40 cm.

The process of the onset of detonation when inhibitor traces are added to the original mixture is presented on

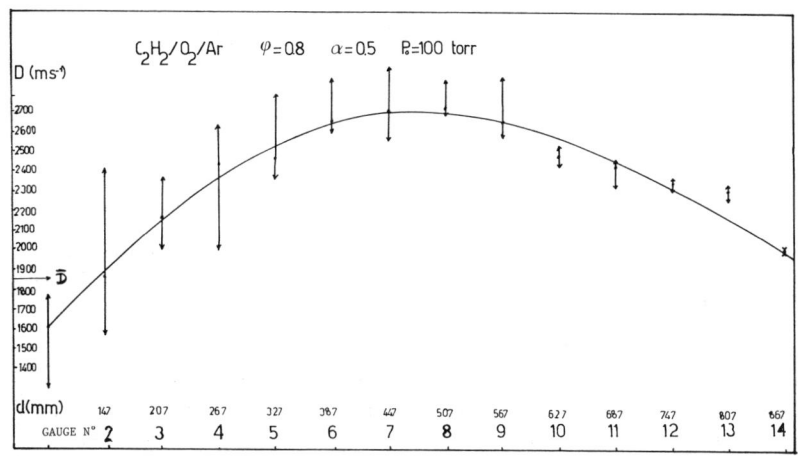

Fig. 4 Detonation velocity profile vs distance behind a bored flange during the onset of detonation of an uninhibited $C_2H_2/O_2/Ar$ mixture (ϕ = 0.8, α = 0.5, P = 100 Torr, flange orifice = 1.1 cm).

GAS DETONATIONS IN ACETYLENE MIXTURES

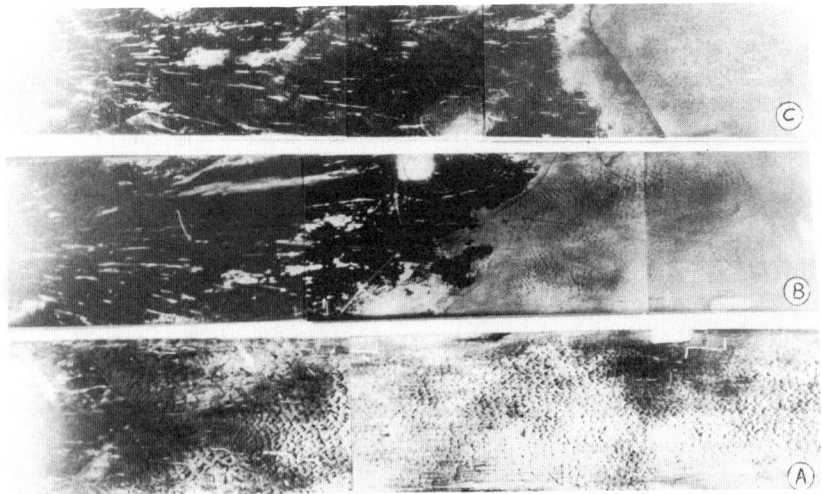

Fig. 5 Comparative soot records of the onset of uninhibited and inhibited $C_2H_2/O_2/Ar$ mixtures (flange orifice = 1.1 cm, tube width = 9.2 cm): A) ϕ = 0.8, α = 0.5, P = 100, %CF_3Br = 0; B,C) ϕ = 0.8, α = 0.5, P = 100, %CF_3Br = 2.

Fig. 5. The diameter of the orifice is 1.1 cm and the width of the soot records is 9 cm. Run A corresponds to the unhihibited mixture. Two records of the onset of detonation for mixtures inhibited by 2% of CF_3Br with respect to the fuel are shown in the middle (run B) and the upper part (run C) of Fig. 5. For mixtures with 2% of CF_3Br, the collision of triple-point trajectories seems not to be able to induce the onset of a detonation. A predetonation distance of about 40 cm is required before the appearance of the small cells characteristic of the initiation process behind a strong leading shock (Fig. 5 runs B and C). Such a delay time (or distance) has always been observed for any run in mixtures containing 2% of CF_3Br with respect to the fuel.

When percentages of CF_3Br larger than 2% were added to the original mixture, the onset of the detonation takes place even later and soot imprints exhibit no structure at all on the 1 m test section. However, a detonation sets in somewhat later, since at a distance 4 m behind the test section a detonation velocity was recorded even for mixtures containing up 50 % of CF_3Br. This experimental observation confirms the data recorded previously for well-sustained detonation in inhibited mixtures.

However, when the diameter of the bored flange is 1.69 cm, the strength of the leading shock overshadows the influence of the inhibitor on the onset of the detonation (see left-hand side of Fig. 6). When increasing quantities of CF_3Br are added a characteristic nose cone configuration 1 comes out in the early stage of the initiation. Nonetheless, for minute amounts of CF_3Br (Fig. 6b), a cell structure is still present in the nose cone region. Such a cell structure disappears completely if higher quantities of CF_3Br are added (Fig. 6c and 6d).

Two questions now arise: why is there a specific action of CF_3Br noticeable at the onset of detonation process and not during the well-sustained regime and how can these contradictory observations be reconciled with one another?

Discussion

Gaseous detonation can be depicted as an exothermic self-ignition process induced by a leading unsteady shock. The self-ignition relies ultimately upon the chemical kinetics of radical reactions with a chain branched mechanism. For hydrocarbons and most fuels containing hydrogen, the elementary reaction for chain branching is

$$H + O_2 \longrightarrow OH + O \qquad (1)$$

The coupling between the heat release zone at the end of the induction period and the leading shock provides the feedback mechanism to sustain the detonation (Dormal et al. 1981).

Therefore, any chemical or physical means that can lengthen the induction period of the self-ignition process will decrease the stability of the detonation wave. On the other hand, reducing the induction period will favor the sustenance mechanism for detonation. By adding traces of CF_3Br, the induction period is modified substantially by the appearance of two new elementary reactions,

$$CF_3Br + H \longrightarrow HBr + CF_3 \qquad (2)$$

$$HBr + H \longrightarrow H_2 + Br \qquad (3)$$

Both steps are competing with the chain branching reaction (1). The kinetics of the mechanism [reactions (1-3)] has been well documented by several mass spectrometric studies of flames (Biordi et al. 1975; Safieh et al. 1982). Its net effect is to replace active hydrogen atoms with bromine

GAS DETONATIONS IN ACETYLENE MIXTURES

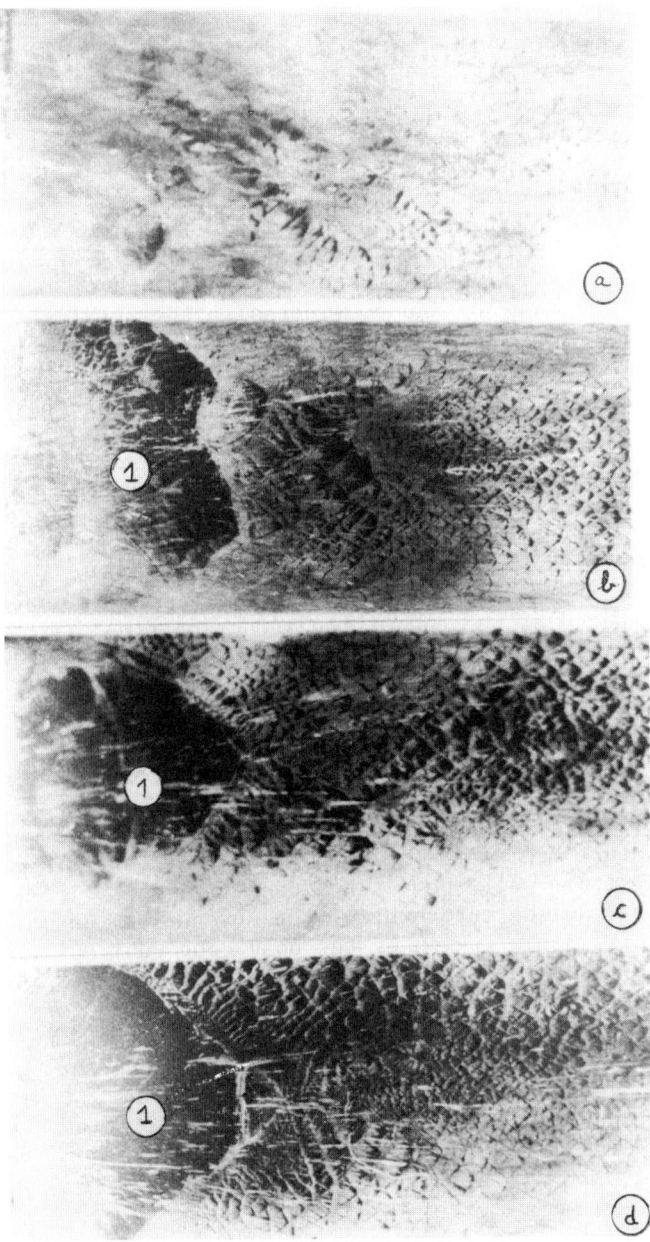

Fig. 6 Comparative soot records of the onset of uninhibited and inhibited (C_2H_2/O_2/Ar mixtures all mixtures $\phi = 0.8$, $\alpha = 0.5$, P = 100 Torr, diameter orifice = 1.69 cm):
a) 0 % CF_3Br.   c) 10 % CF_3Br.
b) 5 % CF_3Br.   d) 50 % CF_3Br.

atoms and with CF_3 radicals, which are not as efficient to multiply the number of active radicals (H, O, OH). Consequently, when CF_3Br is added to the original mixture, it increases the time required to reach a critical radical concentration and, therefore, delays the self-ignition period. Consequently, the characteristic time of the cellular structure is also increased. Libouton et al. (1975) has observed such behavior, indeed, for detonations in $CO/H_2/O_2$ mixtures.

But, at a higher temperature range such as for C_2H_2/O_2 detonations, one has to consider also the pyrolysis of CF_3Br molecules

$$CF_3Br \longrightarrow CF_3 + Br \qquad (4)$$

One may compare the reaction frequencies [reactions (2) and (4)] at several temperatures (Table 1) by using rate constants from Westbrook (1983): $k_2 = 2.2 \ 10^{14} \exp(-8900/T)$ $cm^3 \cdot mole^{-1} \cdot s^{-1}$ and $k_4 = 2.10^{13} \exp(-128000/T)s^{-1}$, and by assuming H-atom concentration of 7.10^{-9} mole·cm^3, i.e., the one computed at the end of the induction time. It should be borne in mind with this last assumption that the role of the process of reaction (2) is overestimated.

Thus, in the low-temperature range (T ≤ 1700 K) reaction (2) dominates. On the other hand, above 1700 K, the pyrolysis reaction (4) of CF_3Br takes over and consumes the inhibitor. This behavior explains the lack of inhibition for well-established detonations where the von Neumann temperature behind the leading shock is of the order of 2470 K, i.e., well above the 1700 K limit.

But, during the onset of the detonation, and more precisely in the region of the nose cone configuration (Fig. 6, zone 1), the unsteady leading shock emerging from the test section at an average velocity of 1400 m/s corresponds to a von Neumann temperature of 1650 K. In that case, CF_3Br traces play their inhibiting role through

Reaction frequencies

T/K	$k_2[H]/s^{-1}$	k_4/s^{-1}
1000	3.7×10^4	0.2
1500	1.8×10^5	9×10^3
2000	3.7×10^5	$2. \times 10^6$
2500	5.9×10^5	4.8×10^7

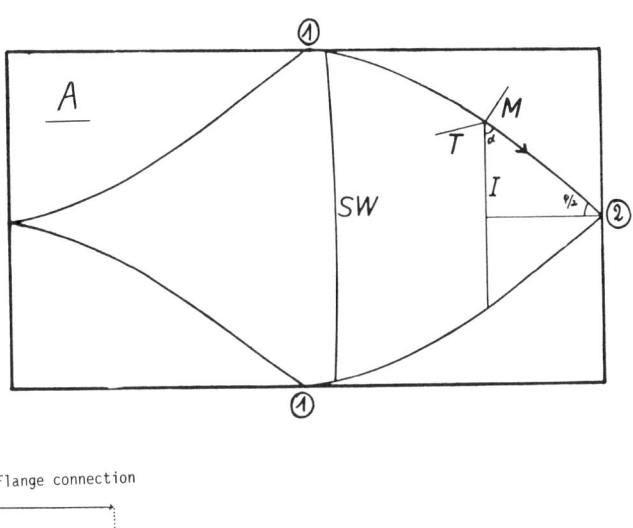

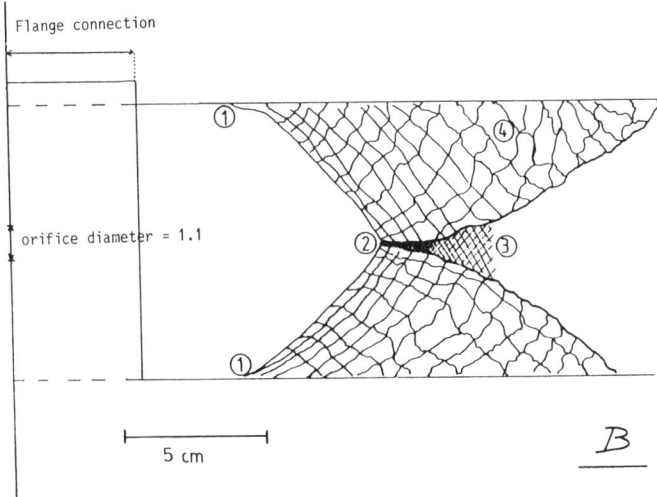

Fig. 7 Comparison of sketches of an onset of detonation and of an reinitiation process at the end of a detonation cell.

reaction (2), which induces the decoupling between the leading shock and the heat release zone clearly visible on the soots records of Fig. 6, zone 1. On the other hand, a similarity is noticeable for soot traces recorded at the end of a detonation cell (Fig. 7a) and in the nose cone configuration before the onset of detonation (Fig. 7b). It has been demonstrated previously (Dormal et al. 1981) that transverse waves propagating in a precompressed, preheated mixture, due to a decaying shock wave, can induce the self-ignition of the hot gases. The latter ignition creates a reactive shock wave, which then overtakes the leading decaying shock. So, at the end of the cell, as well as at the nose cone configuration, the temperature behind the leading shock remains much lower than the von Neumann temperature at midcell. The inhibitor influence is therefore much more drastic at the end of the cell as it was recorded in a previous paper (Dormal et al. 1983).

In the presence of trace inhibitors for a detonation in a well-established regime, the cell length decreases; whereas, at the incipient stage of a detonation the onset, is delayed, more especially when the amount of CF_3Br is large. The fact that the detonation finally sets in for some inhibited mixtures may be ascribed to a mechanism similar to the DDT process (Lee 1986) where a series of successive shock interacts. This latter consideration explains why the onset of detonation in an inhibited mixtures is sometimes even more than twice the onset distance recorded for the uninhibited mixture.

Finally, it is worth recalling here also that the complete failure of detonation in inhibited mixtures occurs only for a mixture close to the detonation limit, for instance $\Phi = 0.33$; $\alpha = 0.5$, and $p_o = 25$ Torr if % $CF_3Br \geq 2\%$, when the von Neumann temperature is quite low.

Acknowledgments

Financial support of Fonds de la Recherche Fondamentale et Collective (Belgium) is acknowledged. The first author is also indebted to Institut pour l'Encouragement de la Recherche Scientifique dans l'Industrie et l'Agriculture (Belgium) for a Ph.D. scholarship.

References

Biordi, J.C., Lazzara, C.P., and Papp, J.F. (1975) The effect of CF_3Br on radical concentration profiles in methane flames. Amer. Chem. Soc. Symp. Ser. 16, 256-291.

Bull, D.C. (1982) Towards an understanding of the detonability of vapor clouds: fuel-air explosions. SM Study, No 16, 139, University of Waterloo, Ont., Canada.

Dormal, M., Libouton, J.C., and Van Tiggelen, P.J. (1981) Reinitiation process at the end of a detonation cell. Progress in Astronautics and Aeronautics: Gasdynamics of Detonations and Explosions : (edited by J.P. Bowen, N. Manson, A.K. Oppenheim, K.I. Soloukhin), vol. 75, pp. 358-369. AIAA, New York.

Dormal, M., Libouton, J.C., and Van Tiggelen, P.J. (1983) Etude expérimentale des paramètres à l'intérieur d'une maille de détonation. Explosifs 36(1), 63.

Fristrom, R.M. and Van Tiggelen, P.J. (1979) An interpretation of the inhibitor of C-H-O flames by C-H-X compounds. Seventeenth Symposium (International) on Combustion, pp. 773-785, The Combustion Institute, Pittsburgh, PA.

Hidaka, Y., Kawano, H. and Suga, M. (1985a) Inhibition of H_2-O_2 reactions by CF_3Cl. Combust. Flame 59(1), 93.

Hidaka, Y. and Suga, M. (1985b) Additive effect of CF_3Cl on OH*, CH* and C_2* emissions : shock tube study with $C_2H_4/O_2/CF_3Cl$ and $CH_4/O_2/CF_3Cl$ mixtures. Combust. Flame 62(2), 183.

Jacques, A. (1978) Effet de la composition chimique sur la structure des ondes de détonation. Mémoire de Licence, Université Catholique de Louvain, Belgique.

Lee, J.H.S. (1986) On the transition from deflagration to detonation. Progress in Astronautics and Aeronautics: Dynamics of Explosions (edited by J.R. Bowen, J.C. Leyer, and R.I. Soloukhin), Vol. 106, pp. 3-18. AIAA, New York.

Libouton, J.C., Dormal, M., and Van Tiggelen, P.J. (1975) The role of chemical kinetics on structure of detonation waves. Fifteenth Symposium (International) on Combustion, pp. 79-86. The Combustion Institute, Pittsburgh, PA.

Libouton, J.C., Jacques, A., and Van Tiggelen, P.J. (1981) Cinétique, structure et entretien des ondes de détonation. Actes du Colloque International Berthelot-Vieille-Mallard-Le Chatelier Tome II, pp. 437-442. Bordeaux, France.

Macek, A. (1963) Effect of additives on formation of spherical detonation waves in hydrogen-oxygen mixtures. AIAA J. 1, 1915-1919.

Moen, I.O., Ward, S.A., Thibault, P.A., Lee, J.H., Knystautas, R., Dean, T., and Westbrook, C.K. (1984) The influences of diluents and inhibitors on detonations. Twentieth Symposium (International) on Combustion, p.1717. The Combustion Institute, Pittsburgh, PA.

Nzeyimana, E., Vandermeiren, M., and Van Tiggelen, P.J. (1986) Influence of chemical composition on the deflagration-detonation transition. Progress in Astronautics and Aeronautics : Dynamics of Explosions (edited by J.R. Bowen, J.C. Leyer, and R.I. Soloukhin), Vol. 106, p. 19. AIAA, New York.

Rothschild, W.G., Kaiser, E.W., and Lavoie, G.A. (1986) Effects of fuel air equivalence ratio, temperature and inhibitor on the structure of laminar methane-air flame. Combust. Sci. Technol. 47 (3+4), p.209.

Safieh, H.Y., Vandooren, J., and Van Tiggelen, P.J. (1982)
Experimental Study of Inhibition induced by CF_3Br.
Nineteenth Symposium (International) on Combustion, pp.
117-126, The Combustion Institute, Pittsburgh, PA.

Vandermeiren, M., Safieh, H.Y., and Van Tiggelen, P.J. (1981)
Action d'inhibiteurs halogénés sur la propagation des flammes.
Ist Specialist Meeting of the Combustion Institute, Bordeaux,
France, pp. 178-183.

Vandermeiren, M. and Van Tiggelen, P.J. (1985) Cellular Structure
in Detonation of Acetylene-Oxygen Mixtures. Progress in
Astronautics and Aeronautics : Dynamics of Shock Waves,
Explosions, and Detonations: (edited by J.R. Bowen, J.C.
Leyer, and R.I. Soloukhin), vol. 94, pp. 104-114, AIAA,
New York.

Vandermeiren, M., and Van Tiggelen, P.J. (1986) Structure of Gas
Phase Detonation in Acetylene Mixtures : Role of an Inhibitor.
Proceedings of the International Symposium on Intense Dynamic
Loading and its Effects, pp. 219-224. Beijing, China.

Westbrook, C.K. (1983) Numerical Modeling of Flame Inhibition by
CF_3Br. Combust. Sci. Technol.34, pp. 201-225.

Westbrook, C.K. (1984) Chemical Kinetic Factors in Gaseous
Detonations. The Chemistry of Combustion Process. Amer. Chem.
Soc. Ser. 248, pp. 175-192.

Experimental and Theoretical Investigation of the Effective Energy in a Shock Tube

Tang Mingjun and Peng Jinhua
East China Institute of Technology, Nanjing, China

Abstract

The energy and average power of an initiation source are two important parameters that investigators are very interested in, no matter when the gaseous or the spray detonations are dealt with. In this paper the two parameters of blast initiation are obtained from experiments and calculations. The experimental values were obtained from the measured trajectories of the shock waves produced by the initiation source. The theoretical values were calculated by using the shock tube relationship. Consistency is shown between the experimental and calculated results. A procedure for predicting the effective initiation energy and average power of a blast initiation source in a shock tube is illustrated. It is noteworthy that the effective initiation energy that transmitted from the driver to the driven gas is approximately 40-50% of the total energy released by the initiation source.

Introduction

During the last thirty years the shock tube has been rapidly developed as a research tool for studying processes in high-temperature gases and fast gas flows. At the same time, it is widely used to study spray and dust detonations. The shock tube usually consists of a high-pressure chamber (driver section) and a low-pressure chamber (driven section), with a diaphragm between them. After the diaphragm bursts, a compression wave is formed in the low-pressure gas, which rapidly steepens to form a shock front. Simultaneously, an expansion or rarefaction wave moves back into the high-pressure gas. This rarefaction wave will reflect when it propagates to the end of the high-pressure chamber. Then it can overtake the shock front propagating into the low-pressure gas. Dabora (1980) proposed that the energy transmission from the high-pressure gas to the low-pressure gas is only effective until the reflected rarefaction wave catches up with the shock front. If the driven gas is a reactive mixture, this energy is called the "effective initiation energy."

The purpose of this paper is to deal with the effective energy that is transmitted from the driver gas to the driven gas. This is a basis for studying the initiation of spray and dust detonations. The calculation is based on Dabora's suggestion and the properties of shock waves and rarefaction waves. The experimental data were taken from the work of Tang et al. (1986), where the effective energy was obtained from the measured trajectories of shock fronts in a vertical shock tube.

Experimental Results

The experiments by Tang et al. (1986) were conducted in a vertical shock tube 8.2 m long with a square cross section of 4.13 cm by 4.13 cm. The blast wave initiator was a detonation tube with a circular cross section of 5 cm i.d. and 1.2 m in length. It was mounted at an angle of 15 deg to the main tube. For all tests, the initiator was filled with a $2H_2-O_2$-He mixture. The detonation products of the mixture were used as driver gases. The main tube was filled with air (a nonreactive mixture) at atmospheric pressure.

Eleven pressure switches located along the vertical tube at 0.0, 1.0, 1.5, 2.0, 2.5, 3.5, 4.0, 5.0, 5.5, 6.0, and 6.5 m were used to sense the arrival of the shock front.

After the rarefaction wave is reflected from the closed end of the initiator (driver section) and catches up with the shock front, the wave can be considered a constant-energy blast wave. Then its decay can be calculated approximately by using the equation developed by Dabora (1980):

$$\frac{dM}{M} = - \frac{\alpha+1}{2}\left(1 - \frac{1}{M^2}\right)\frac{dx}{x} \tag{1}$$

INVESTIGATION OF ENERGY IN A SHOCK TUBE 203

For the vertical shock tube, $\alpha=0$ (planar case); after integrating, Eq. (1) becomes

$$\frac{x}{x_c} = \frac{M_1^2-1}{M^2-1} \qquad (2)$$

where x is the blast wave distance, x_c is the distance from the diaphragm at which the reflected rarefaction wave catches up with the shock front, M is the blast-wave Mach number, and M_1 is the initial Mach number. Tang et al. (1986) have verified that Eq. (2) is consistent with the experimental results. And beyond a certain location, indeed it is true that $x(M^2-1)$ is approximately a constant. Thus, we can obtain x_c and M_1 from the experimental curve x-t. For example, the initial mixture $2H_2-O_2-He$ in the initiator is at 1.68 atm and 298.15 K, the detonation pressure is 32.4 atm, and the sound speed a_1 in the main tube is 346 m/s. The measured arrival time t at which the shock front (blast wave) reaches distance x, and corresponding velocities of the shock front, are shown in Table 1. From Table 1, we can see that, at $x \simeq 350$ cm, $x(M^2-1)$ approaches a constant.

Table 1. The determination of x_c and M_1
(for initiator pressure 1.68 atm)

x (m)	t (10^{-6}s)	D (m/s)	M	$x(M^2-1)$ (m)
0.0	0.0			
0.50		943	2.73	3.21
1.00	1060			
1.25		779	2.25	5.09
1.50	1702			
1.75		705	2.04	5.51
2.00	2411			
2.25		654	1.89	5.79
2.50	3176			
3.00		678	1.96	8.52
3.50	4650			
3.75		675	1.95	10.52
4.00	5391			
4.50		635	1.84	10.65
5.00	6967			
5.25		611	1.77	11.07
5.50	7786			
5.75		592	1.71	11.08
6.00	8630			
6.25		574	1.66	10.95
6.50	9501			

Therefore, it can be concluded from the experiments that $x_c=350$ cm, and $M_1=2.73$ for initiator pressure 1.68 atm.
The rate of energy input to the driven gas must equal the rate of the work performed by the driver gas at the interface. Thus, the power per unit area is

$$P = \frac{2\gamma_1 M_1^2 - (\gamma_1 - 1)}{\gamma_1 + 1} \cdot \frac{2(M_1^2 - 1)}{(\gamma_1 + 1)M_1} p_1 a_1 \qquad (3)$$

where p_1, a_1, and γ_1 are the pressure, sound speed, and the ratio of specific heats of driven gas, respectively.

The energy is equal to this rate multiplied by the time over which this rate is effective. This time is taken as that required by the rarefaction wave to traverse the driver section, reflect from the driver end wall, and overtake the shock front. Therefore, the energy per unit area is

$$E = P t_c = P \frac{x_c}{D} = P \frac{x_c}{M_1 a_1} \qquad (4)$$

where D is the velocity of the shock front and t_c is the time at which the reflected rarefaction wave overtakes the shock front.

After x_c and M_1 are obtained from the measured curves of the trajectories of shock waves x-t, the average power P and the energy E per unit area can be obtained. The results are shown in Table 2.

Theoretical Calculations

Calculation of the Initial Mach Number in the Driven Section. In the experiments, the driver section and the

Table 2. Experimental results of P and E

Ini. pressure in initiator	atm	1.68	2.02	3.72	4.40	5.76	7.80
$P_4 P_1$		16.2	19.6	36.5	43.8	57.8	79.1
M_1		2.73	2.97	3.91	3.95	4.62	5.34
x_c	m	3.50	3.75	4.50	5.00	5.50	6.50
t_c	10^{-3} s	3.71	3.65	3.33	3.66	3.44	3.38
P	10^6 w/m^2	566	752	1820	1885	3095	4838
E	10^6 J/m^2	2.10	2.75	6.06	6.90	10.6	16.4

INVESTIGATION OF ENERGY IN A SHOCK TUBE 205

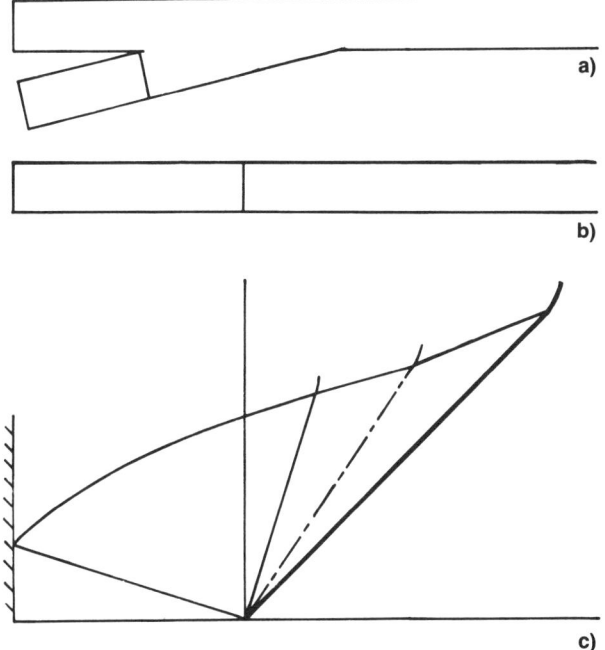

Fig. 1 The simplified model and the flowfield.

driven section are not coaxial, and the two cross-sectional areas are different. The ratio of the cross-sectional areas of the driver to driven section is 1.15 (i.e., $A_4/A_1 = 1.15$). For simplification, we assume that the two sections are coaxial (see Fig. 1).

The variables in zone 4 are the parameters after the constant-volume detonation of the gaseous mixture ($2H_2$-O_2-He); i.e., ρ_4 is equal to ρ_0. And assuming that the ideal gas equations can be applied to both unburned and burned gases, based on

$$\left(\frac{\partial \ln p}{\partial \ln v}\right)_s = C_p/C_v \left(\frac{\partial \ln v}{\partial \ln p}\right)_T$$

from Gordon and McBride (1971) and using the symbols

$$\gamma_s \equiv \left(\frac{\partial \ln p}{\partial \ln \rho}\right)_s$$

and

$$\gamma \equiv C_p/C_v$$

thus

$$\gamma = -\gamma_s \left(\frac{\partial \ln v}{\partial \ln p}\right)_T$$

Therefore,

$$p_4 \simeq p_{CJ}/2$$

$$\gamma_4 = -\gamma_s \left(\frac{\partial \ln v}{\partial \ln p}\right)_T$$

and

$$A_4 = (\gamma_s/p_4 \rho_4)^{1/2} = (\gamma_s p_4/\rho_0)^{1/2}$$

where p_{CJ} and γ_s are the Chapman-Jouquet pressure and the ratio of specific heats of the burned gas, which are computed by using the Gordon-McBride code (1971).

As seen in Fig. 1b, the driver and driven sections are coaxial, and if they have the same areas ($A_4/A_1=1$), the equation of shock tube performance for ideal gas

$$\frac{p_4}{p_1} = \frac{2\gamma_1 M_1^2 - (\gamma_1 - 1)}{\gamma_1 + 1} \left[1 - \frac{\gamma_4 - 1}{\gamma_1 + 1} \frac{a_1}{a_4}\left(M_1 - \frac{1}{M_1}\right)\right]^{\frac{-2\gamma_4}{\gamma_4 - 1}} \tag{5}$$

can be used to determine M_1. A stronger initial shock wave is expected when the effect of the area change near the diaphragm is taken into account. To calculate M_1, an "equivalent factor" g is introduced and the initial pressure ratio p_4/p_1 is multiplied by g, and the sound-speed ratio a_4/a_1 is multiplied by

$$g^{\frac{\gamma_4 - 1}{2\gamma_4}}$$

as described by Bradley (1962). For the driver gas, $\gamma_4 \simeq 1.2$, and the ratio of cross-sectional areas, $A_4/A_1 = 1.15$, a value of g is chosen as g = 1.05 and

$$g^{\frac{\gamma_4 - 1}{2\gamma_4}} \simeq 1.004$$

INVESTIGATION OF ENERGY IN A SHOCK TUBE

Hence, using the corrected Eq. (5), M_1 can be computed for every ratio p_4/p_1.

Calculation of the Average Power and the Effective Energy. As mentioned in the section on experimental results, after obtaining M_1 we can use Dabora's theory to calculate the average power and the effective energy. The rate of energy per unit area is computed by Eq. (3), and the effective energy per unit area is computed by Eq. (4). The difference between the two calculations in experimental and theoretical models is that, in the former, M_1 and x_c were obtained from experiments, whereas in the latter, M_1 and x_c were calculated from the corrected Eq. (5), and

$$x_c = M_1 a_1 t_c \tag{6}$$

where

$$t_c = \frac{\frac{2\ell}{a_4}\left[1 - \frac{\gamma_4-1}{\gamma_1+1}\frac{a_1}{a_4}\left(M_1 - \frac{1}{M_1}\right)\right]^{-\frac{\gamma_4+1}{2(\gamma_4-1)}}}{1 + \left[\frac{(\gamma_1-1)M_1^2 + 2}{2\gamma_1 M_1^2 - (\gamma_1-1)}\right]^{1/2}} \tag{7}$$

As shown, a_4/a_1 in Eq. (7) must be multiplied by

$$g^{\frac{\gamma_4-1}{2\gamma_4}}$$

The theoretical calculation results are shown in Table 3.

Table 3. Calculation results

Ini. pressure in initiator	atm	1.68	2.02	3.72	4.40	5.76	7.80
p_4/p_1		16.2	19.6	36.5	43.8	57.8	79.1
M_1		2.92	3.13	3.90	4.16	4.56	5.05
x_c	m	3.62	3.86	4.90	5.24	5.88	6.71
t_c	10^{-3} s	3.58	3.56	3.63	3.65	3.73	3.84
P	10^6 w/m^2	737	926	1869	2284	3053	4192
E	10^6 J/m^2	2.64	3.30	6.78	8.33	11.4	16.1
E_o	10^6 J/m^2	6.43	7.81	15.0	18.0	23.9	33.1
E/E_o	%	41.1	42.3	45.2	46.4	47.7	48.7

The total energy E_0 of the driver gas in the driver section is the reactive energy released in the initiator tube as a result of the constant volume combustion of the $2H_2-O_2-He$ mixture. Only a fraction of the total energy is transmitted to the driven gas. As seen in Table 3, E/E_0 values are about 40-50%.

Conclusions and Discussion

The investigation into the effective energy shows that the results from the experiments and the calculations are consistent. Thus, the procedure illustrated earlier could be used to predict the approximate initiation energy for the initiation of reactive gas spray and dust, even though the driver and the driven sections were not coaxial and had different cross-sectional areas. Of course, it does not allow for the difference between the two cross-sectional areas to be large (i.e., $A_4/A_1 \simeq 1$).

It is found from the research that when the initial pressure ratio p_4/p_1 (or M_1) increases, the P, x_c, and E will increase, but the t_c, when the reflected rarefaction wave reaches the shock front, is approximately a constant.

For the planar case, M_1 should be a constant before $t = t_c$, but in experiments it decays gradually. This result may be attributed to the fact that the driver and the driven sections are not coaxial. As seen in Fig. 1a, after the bursting of the diaphragm, the gas flows not only down to the main tube, but also upward to the top of the main tube. Thus, the rarefaction from the top can immediately catch up with the shock front propagating down to the main tube. Therefore, it should be emphasized that only when $x(M^2-1)$ approaches a constant can the above procedure be used to determine x_c.

References

Bradley, J. N. (1962) Shock waves in chemistry and physics., Methuer, London, Great Britain, p. 124.

Dabora, E. K. (1980) Effect of additives on the lean detonation limit of kerosene sprays, final rep., U.S. Army Research Office, Grant No. DAAG 29-78-G0074.

Gordon, S. and McBride, B. (1971) Computer program for calculation of complex chemical equilibrium compositions, rocket performance, incident and reflected shocks, and Chapman-Jouguet detonations. NASA SP-273.

Tang, M. J. et al. (1986) Direct initiation of detonation in a decane spray. AIAA Series Vol. 106, pp. 474-489.

Chapter III. Nonideal Detonations and Boundary Effects

Nonideal Detonation Waves in Rough Tubes

Y. B. Zel'dovich,* A. A. Borisov,† B. E. Gelfand,‡
S. M. Frolov,§ and A. E. Mailkov¶
USSR Academy of Sciences, Moscow, USSR

Abstract

Results of investigations of detonation waves in gaseous systems, with friction and heat transfer effects taken into account, are summarized. The analysis applies to detonations in rough tubes. The following flow models in the reaction zone and downstream of it are considered: (1) the bulk heat release model due to self-ignition of the gas behind the lead shock wave; (2) the model of combustion initiated near the wall due to self-ignition in shock waves multiply reflected at the roughness elements (in this case, the major fraction of the mixture burns in a turbulent flame that eventually covers the entire cross section of the tube); and (3) the model of combined ignition, both in the volume and at the tube walls. An analysis of the problem, with due regard for the two types of mixture ignition, points to the existence of a range of parameters where the detonation velocity selection is ambiguous. The experiments conducted in a rough tube with H_2-O_2-N_2 and C_3H_8-air mixtures show that in real conditions only one detonation mode is realized in rough tubes in which the two ignition mechanisms are operative. It is concluded that quasi detonations in rough tubes can be treated with a non-one-dimensional model of the detonation process. The quasi-detonation wave, with a velocity deficit confined between the values characteristic of the two ignition mechanisms in detonation waves, is shown to be unstable. The heat evolution rate in this wave is subject to large-scale fluctuations.

Copyright © 1988 by the the Institute of Physical Problems, USSR Academy of Sciences. Published by the American Institute of Aeronautics and Astronautics, Inc. with permission.
*Academician of the USSR Academy of Sciences, Head of the Department of Theoretical Physics.
†Head of Laboratory.
‡Senior Researcher.
§Junior Researcher.
¶Junior Researcher.

Introduction

In recent years nonideal detonation regimes have been extensively discussed in the literature devoted to explosion hazard problems. This is because in most real accident situations, the so-called classical detonations did not occur. However, the destruction caused by accidental explosions shows that with the intensity of the pressure waves and their ability to propagate for long distances, the nonideal explosions are comparable to normal detonations.

The difference between nonideal explosion processes and classical detonations is that the former are substantially affected by energy and momentum losses. Deceleration of the combustion products at the duct walls, and heat transfer, may change the propagation velocity and structure of the reactive wave. A rough internal duct surface may drastically alter the conditions of mixture ignition behind the shock front. For instance, if the temperature of the shocked gas is not sufficient for fast ignition, a possibility of local mixture ignition exists because of multiple reflections of the shock waves at the obstacles. Rough walls strongly turbulize the flow behind the wave front. A turbulent flame arises behind the shock front and is stabilized by continuously appearing hot spots at the walls. Unlike the induction zone in a normal detonation wave, this reaction zone pattern only weakly responds to possible gas temperature perturbations. Thus, quasi-steady explosion processes propagating at low supersonic velocities become feasible.

Hence, the losses dramatically alter the flow pattern behind detonation waves and lead to steady-state solutions that pertain to the waves propagating with a velocity intermediate between slow deflagration and normal detonation.

In the present work we studied both theoretically and experimentally the structure and parameters of detonation waves in rough tubes. The specific features of high-velocity detonations with self-ignition of a homogeneous mixture behind the lead shock (Fig. 1) and low-velocity detonations with mixture ignition in shock waves reflected at obstacles (Fig. 2) are considered. The possible mechanism of nonideal wave propagation under real conditions is discussed.

Formulation of the Problem

In the frame of reference attached to the wave front, the equations governing the one-dimensional (1-D) structure of the reaction zone in a steady-state detonation wave propagating in a rough tube are as follows (Zeldovich 1940):

$$(D-w)/v = D/v_o$$

$$d[p + (D-w)^2/v]/dx = \pi\sigma/\phi$$

$$d[(D-w)(H+(D-w)^2/2)/v]/dx = -\pi q/\phi$$

$$+ \pi\sigma D/\phi \qquad (1)$$

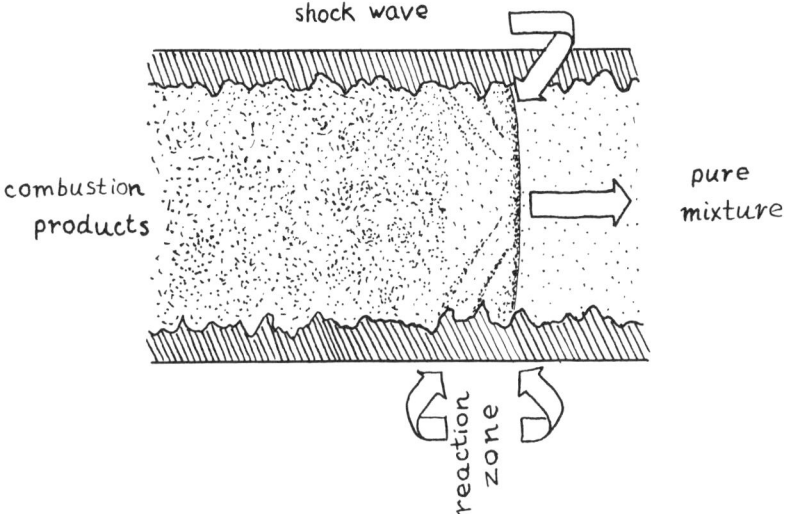

Fig. 1 Schematic of a detonation wave in a rough tube with the bulk self-ignition behind the shock wave.

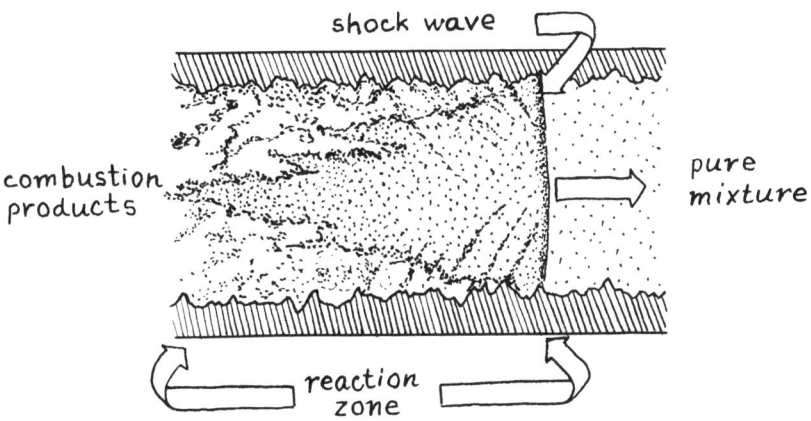

Fig. 2 Schematic of a detonation wave in a rough tube with ignition at the walls in multiply reflected shock waves and turbulent mixture burning.

where D is the wave velocity; w, v, and p are the particle velocity, specific volume, and pressure behind the shock front, respectively; π and ϕ are the tube perimeter and cross-sectional area; and the subscript 0 refers to the initial state.

The friction force per unit area of tube surface is

$$\sigma = C_f w|w|/2v \qquad (2)$$

where $C_f=C_f(x)$ is the local drag coefficient. The form of Eq. (2) was chosen to account for the possibility of a reverse of the sign of the combustion product velocity.
The heat flux to the wall is calculated by the formula

$$q = \alpha(T_{st} - T_w) \qquad (3)$$

where $\alpha=\alpha(x)=\kappa Nu/2r$ is the local heat transfer coefficient, κ is the heat conductivity of the gas, r is the tube radius, Nu is the Nusselt number, and T_{st} and T_w are, respectively, the gas stagnation temperature and wall temperature. The gas enthalpy is calculated as

$$H = Q\beta + \gamma p v/(\gamma-1) \qquad (4)$$

where γ is the ratio of specific heats, Q is the heat of the combustion reaction, and β is the unburned fraction described by the equation

$$d\beta/dx = f(\beta, T, \psi_1, \psi_2, \ldots) \qquad (5)$$

$\psi_1, \psi_2, \ldots$ are the parameters on which the reaction rate may depend.

The solution of Eqs. (1), (4), and (5) with the additional relationships of Eqs. (2) and (3) must meet the boundary conditions

$$x = 0 \quad v = v' \quad p = p' \quad \beta = 1$$
$$x \to \infty \quad v \to v_o \quad p \to p_o \qquad (6)$$

In the boundary conditions [Eq. (6)], we assume that reaction occurs without a change in the mole number. The prime quantities are calculated from the relationships at the lead shock front.

To close the problem we must define the coefficients $C_f(x)$ and $\alpha(x)$. We will neglect to the first approximation the effect of gas compressibility and longitudinal pressure gradient on the flow deceleration at the beginning of the boundary layer, and assume the velocity profile in it to be of an exponential form. One can then obtain from kinematic similarity that $C_f \sim (k_s/\delta_1)^{2n}$, where k_s is the equivalent roughness (Schlichting 1951), δ_1 is the momentum displacement thickness, and n is the exponent of the velocity distribution law. From the relationship $C_f/2 = d\delta_1/dx$, we can deduce $C_f \sim (k_s/x)^{2n/(2n+1)}$; when n=0.125, we obtain $C_f \sim (k_s/x)^{0.2}$.

The proportionality coefficient can be estimated from comparison with the interpolation formula (Schlichting 1951) $C_f = (1.89+1.62 \log x/k_s)^{-2.5}$ valid for $10^2 < x/k_s < 10^6$. For $x/k_s = 10^{-3}$ we find

$$C_f \approx 0.03(k_s/x)^{0.2} \qquad (7)$$

From Eq. (7) it follows that $C_f \to \infty$ when $x \to 0$. Here we encounter the limit of the applicability of the 1-D theory. For the sake of simplicity we consider the drag coefficient averaged over the reaction zone length ℓ (over the zone where $D-w \leq C$, C being the local acoustic velocity):

$$C_f \approx \bar{C}_f \approx 0.04(k_s/\ell)^{0.2} \qquad (8)$$

Coefficient k_s is formally determined from Schlichting's formula

$$\lambda^{-0.5} = 2 \log r/k_s + 1.74 \qquad (9)$$

where λ is the hydraulic resistance coefficient for a tube as measured from the pressure drop per unit length in an isothermal flow. For the low Reynolds number inherent in detonations, coefficient λ depends only on r and characteristic roughness size. Equations (8) and (9) permit the losses associated with gas deceleration to be estimated if the tube charateristics are given.

In the reaction zone, α appearing in Eq. (3) is calculated from the formula $Nu \approx 0.029\eta \, Re_x^{0.8}$ (Kutateladze 1979) valid for $Re_x > 5 \times 10^5$. Here $Re_x = xw/\nu$ (where ν is the kinematic viscosity) is the Reynolds number of the flow in a rough duct, and factor η accounts for the heat transfer intensification due to vortex flow near the obstacles, and for the increase in the surface area. To make the problem simpler, we assume that $Nu \sim Re_x$ and express the average heat transfer coefficient in the reaction zone of a detonation wave as

$$C_h \approx \bar{C}_h \approx 0.008 \, \eta \kappa \ell v/\nu C_p r \qquad (10)$$

Here $\bar{C}_h = \bar{\alpha} v/C_p |w|$; C_p is the specific heat of the gas.
There is a difficulty in calculating the gas deceleration and cooling downstream of the reaction zone (where $D-w>C$). On one hand, the flow here tends toward a steady state; and on the other, its velocity diminishes, which makes the flow laminar. In this situation reasonable physical approximations are needed that reflect the principal qualitative features of the phenomenon. In the case under consideration the effects of heat losses and friction on the detonation velocity are "localized" in the reaction zone. Therefore, application of one or another relations for calculating gas deceleration and cooling in the supersonic flow reign may solely affect its length, which is not of primary interest for the present consideration. For this reason the drag coefficient C_f can for convenience be assumed to be constant and equal to its averaged value in the reaction zone.

When the flow velocity is low, Nu tends toward a constant value ($Nu_0 \approx 3.66$). The flow may become totally stagnant ($v \to v_0$) when the gas temperature differs from that of the walls, and this has to be taken into account when considering the second boundary condition ($P \to P_0$). The simplest assumption that can be made for the supersonic zone with $Re_x < 5 \times 10^5$ is $Nu = Nu_0$, i.e., to ignore the region of intermittance of flow regimes and the region of laminar flow with an enhanced heat transfer.

The problem thus formulated is completely closed provided that the burning law [Eq. (5)] is known. In the following discussion we will determine the main properties of detonation waves propagating in rough tubes using various physical models of the combustion reaction. The boundary conditions [Eq. (6)] allow the calculation of not only the reaction zone structure in a detonation wave but the detonation velocity eigenvalue D as well. These relationships require no conditions for selection of the detonation velocity other than those that have already been introduced in the established theory for ideal detonations. This was first pointed out in the work of Zeldovich (1940).

When solving the problem, one has to take into account the fact that Eqs. (1) and (5) have a singularity. The two following conditions from Eqs. (1) and (5) must both be met:

$$D - w = C \qquad (11a)$$

$$-Q\frac{d\beta}{dx} = \frac{\pi}{\phi}[-9\frac{v}{D-w} - \sigma(\frac{v}{\gamma-1} - v_0+v)] \qquad (11b)$$

The second condition implies that at the section where D-w=C an unburned mixture exists (since $d\beta/dx \neq 0$). An analysis shows that the singular point (β_*, p_*, v_*) is of the generalized saddle type [subscript * labels the values at the section where Eqs. (11a) and (11b) are valid]. Consequently, the value of D must be determined by means of the trial-and-error method, i.e., by integrating Eqs. (1) and (5) at various magnitudes of D until conditions in Eqs. (11a) and (11b) are met at some flow cross-section x_*. The integral curve must eventually come back to the $v = v_0$ and $p = p_0$ state downstream of the singular point x_*.

Because of its complexity, the problem must be solved numerically. The fourth-order Runge-Kutta method was employed in the calculations. To check the correctness of the determination of the D value, the integration was performed from the singular point toward the detonation front where the following conditions are met: $v = v'$, $p = p'$, and $\beta = 1$. To improve the accuracy of calculations in the vicinity of the singular point, new variables were introduced: $\bar{\beta} = \beta-\beta_*$, $\bar{v} = v-v_*$, and $\bar{p} = p-p_*$.

Homogeneous Ignition of the Shocked Mixture

In this section we consider propagation of a detonation wave in which the mixture is ignited homogeneously as a result of its heating behind the lead shock. We will ignore the fact that some fraction of the mixture burns near the wall behind shock waves reflected from the obstacles. The chemical kinetic equation [Eq. (5)] assumes the form

$$d\beta/dx = -k\beta^m \exp(-E/RT)/(D-w) \qquad (12)$$

where k is the preexponential factor, m is the reaction order, E is the activation energy, and R is the gas constant.

The calculations were made for the following magnitudes of the main parameters. C_f was varied between 0 and 0.5, and C_h between 0 and 0.05. The r ranged from 2.5×10^{-3} to 1.5×10^{-1} m; m = 0, 1, and 2; E = 25 to 40 kcal/mol, and Q was varied from 13 to 15 kcal/mol, while k value = 10^{10} s-1 and $\gamma = 1.3$. The velocity deficit in all the calculations was no more than $\Delta D = (D_0-D) = 0.13D_0$, where $D_0 = [2(\gamma^2-1)Q]^{1/2}$ is the ideal detonation velocity.

A calculated p-v diagram for the case $C_f = 0.5$, $C_h = 0.02$, $E = 25$ kcal/mol, $n = 1$, and $D/D_0 = 0.92$ is presented in Fig. 3. In the 1-D case considered, a pressure rise is observed behind the shock front due to the work done by friction forces (when $C_f \leqslant 0.2$, the pressure hump is not seen).

Point A in Fig. 3 pertains to the states of the initial mixture behind the shock front ($\beta=1$). At point B half of the mixture has burned out ($\beta=0.5$). Point C is the Chapman-Jouguet (CJ) plane wherein conditions in Eqs. (11a) and 11b) are fulfilled ($\beta=\beta_*=6 \times 10^4$), and point G represents the final state of the stagnant and totally cooled combustion products. As seen in Fig. 3, the particle velocity changes its sign in the expansion process, and the specific volume becomes greater than 1. This effect was first explained by Zeldovich (1940).

Figure 4 shows the detonation velocity as a function of the drag coefficient for different reaction orders and acti-

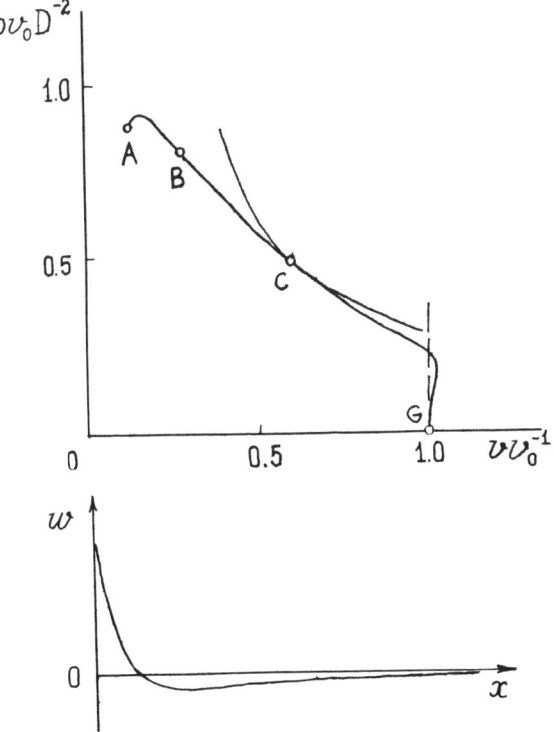

Fig. 3 Detonation Hugoniot for a wave with losses and an analog of the Rayleigh line, the bulk ignition case. A, the state behind the lead shock wave; B, the point where 50% of the mixture has reacted; C, the CJ point; and G, the complete stagnation and cooling point. Schematic of the particle velocity dependence on distance in the laboratory frame of references.

vation energies. The dotted continuation of one of the curves represents the unstable branch of the solution to Eqs. (1) and (12). From Fig. 4 it is seen that the detonation velocity drops abruptly at a particular value of C_f. Such a behavior is typical of the dependence of the experimental detonation velocity on mixture composition, initial pressure, tube diameter, blockage ratio, etc., and is associated with the detonation limits.

In accordance with the estimates made by Zeldovich and Kompaneets (1955), the detonation velocity at the limit drops by a value

$$\Delta D/D_o = \epsilon\, RT'/E$$

where T' is the temperature at the shock front and ϵ is a coefficient. The preceding expression for ΔD was derived with heat losses alone taken into account. The value of ϵ is 0.5 in this case.

Friction heats up the mixture additionally, which may extend the detonation limits. In an approach similar to that of Zeldovich and Kompaneets (1955), it can be shown that for

$$w = (\gamma+1)^2 E/[2(\gamma-1)\gamma^2 D_o^2] \gg 1, \text{ then } \epsilon = 0.5.$$

For the curves of the two families in Fig. 4, $w = 7.3$ and 6.3, respectively, and $\epsilon = 1.5$ and 1.8. Thus, for mixtures with a lower activation energy, flow deceleration results in broader detonation limits. As expected, the calculated ΔD decreases as the activation energy increases and $\epsilon \to 0.5$.

The limiting value of C_f (or λ) at which the detonation still propagates grows as the reaction order and activation energy increase. At a fixed value of C_f (or λ), the detonation velocity is higher the greater the reaction rate and the smaller the reaction order. As the activation energy increases, the influence of the reaction order on the detonation velocity deficit in a rough tube becomes weaker.

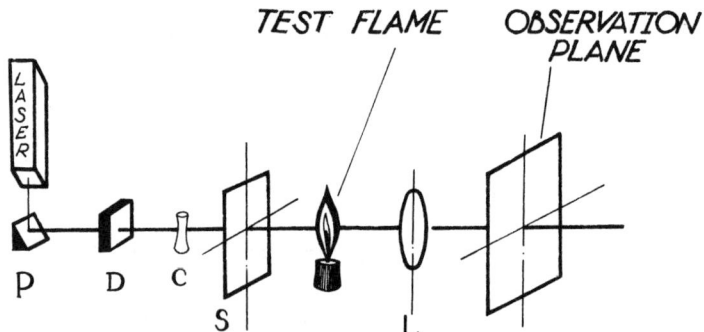

Fig. 4 Detonation velocity in a rough tube as a function of friction losses (the bulk ignition case). Solid lines are for E=25 kcal/mol, and dashed lines are for E=27.5 kcal/mol. Numbers at the curves represent reaction order.

NONIDEAL DETONATION WAVES IN ROUGH TUBES

The heat losses affect the detonation velocity only slightly and are essential solely near the limit (curves 1' and 2' in Fig. 4).

The unburned fraction at the CJ point is low in all the variants calculated and exerts virtually no effect on the detonation velocity. The highest values of β_* were obtained for m = 2. For instance, with C_f = 0.5 and m = 2, the unburned fraction is somewhat higher than 2%. As C_h and C_f grow, the unburned fraction increases. In mixtures with a higher activation energy, β_* is smaller at the same D.

An analysis shows that Fig. 4 has a more general meaning than has been indicated above. If a quantity $(C_f/r) \times 10^{-2}$ is plotted along the x axis in Fig. 4 instead of C_f, the curves will be independent of the tube radius. This is because at C_h = 0, r appears in Eqs. (1) and (12) only in the combination C_f/r. Hence, the condition C_f/r is the same for a given mixture and permits the utilization in large-scale experiments of the data on detonation propagation obtained in laboratory-scale tubes.

Figure 5 shows the detonation velocity vs the reaction heat dependencies for different values of C_f/r. The curves presented in the figure reflect to some extent the detonation velocity vs the mixture composition dependencies, and are in good qualitative consistency with the experimental data (Matsui 1981).

The Model of Detonation with Ignition at the Walls

Experiments by Schelkin (1949) demonstrate that quasi-steady explosion processes may propagate at a velocity almost one-half as high as the ideal detonation velocity D_0. According to the preceding estimates, a stable detonation with self-ignition behind the lead shock wave is impossible in these conditions.

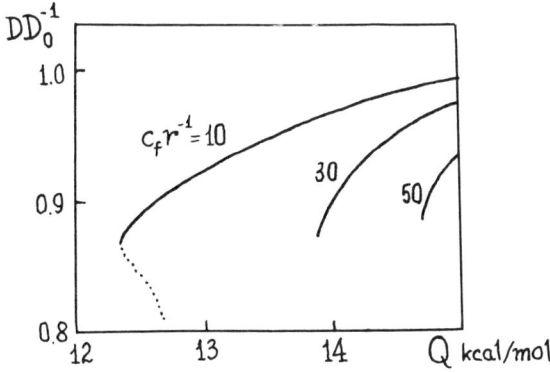

Fig. 5 Detonation velocity in a rough tube as a function of the reaction heat for various drag coefficients (the case of bulk ignition).

Zeldovich (1940 and 1955) and Schelkin (1949) have suggested what the mechanism of propagation might be for such a low-velocity detonation in rough tubes. According to this mechanism, the mixture is ignited near walls because of its compression and heating in shock waves multiply reflected from the obstacles. The further mixture combustion takes place in a turbulent flame that emanates from the wall, forming a cone.

We illustrate the aforesaid combustion mechanism using a methane-air mixture as an example. The detonation velocity as a function of the equivalence ratio α_f (solid line) is shown in Fig. 6 (Westbrook 1982). Also presented is the shock wave velocity (dashed line) at which the ignition delay in a wave that has undergone a single reflection at a rigid wall is equal to that in an incident shock wave propagating at the normal CJ detonation velocity.

The limits of the low-velocity detonation in the stoichiometric methane-air mixture in rough tubes are $0.8 < \alpha_f < 1.2$, corresponding to the Mach number interval of 2.98 to 3.45. From the data by Kogarko and Borisov (1960), it follows that the ignition delay for the strongest incident wave from this interval is $t_i = 0.79$ s. Behind the reflected shock waves at M = 2.98 to 3.45, $t_i < 10^{-3}$ s.

When simulating the detonation with wall ignition, we suppose that the law of mixture burnout [Eq. (5)] is determined by the growth of the turbulent boundary layer behind the shock front. The temperature in the region of multiply reflected shock waves is assumed to be so high that burning starts immediately behind the wave front. The burning rate

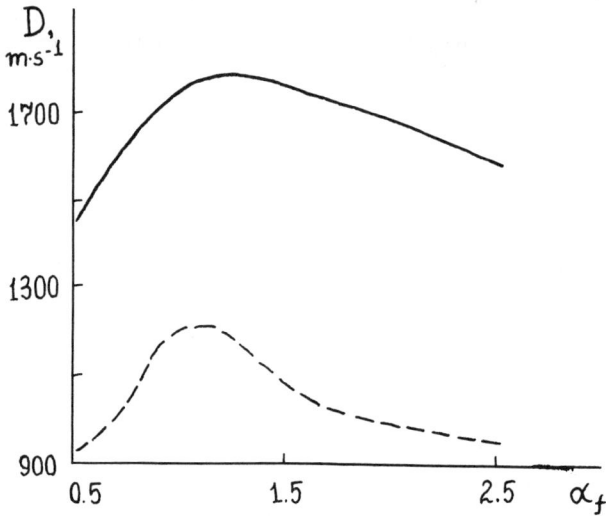

Fig. 6 The calculated detonation velocity for the two heat release mechanisms in rough tubes (for methane-air mixtures). The dashed line is obtained for the wall-ignition mechanism under an assumption that the ignition delay in the reflected shock wave equals that in the incident wave propagating at the CJ velocity; α_f is the equivalence ratio.

NONIDEAL DETONATION WAVES IN ROUGH TUBES

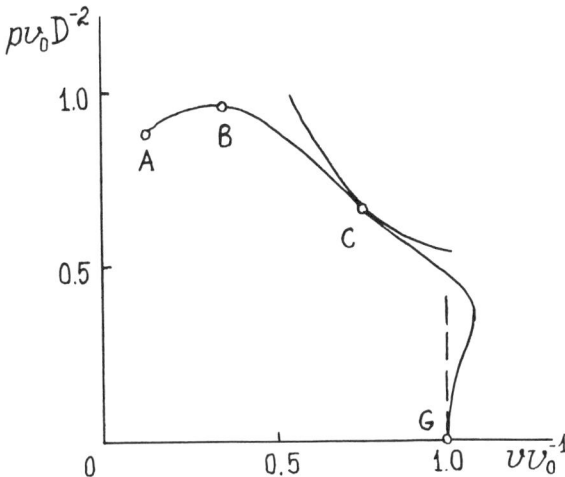

Fig. 7 Detonation Hugoniot and an analog of the Rayleigh line for low-velocity detonation in a rough tube. A, the state behind the lead shock wave; B, the state at the point with 50% mixture burnout; C, the CJ point and G, the flow stagnation and cooling point.

is practically temperature-independent and depends on the turbulence intensity in the boundary layer. The equation for the chemical reaction rate [Eq. (5)] in this case can be written as

$$\beta = \pi(r-\delta)^2/\pi r^2 \qquad (13)$$

where δ is the thickness of the boundary layer. For the sake of simplicity the rate of the boundary layer growth is assumed to be constant, i.e., $\delta = k_\delta x$. Here $k_\delta \approx 0.02$ w'/ (D-w') (Zeldovich and Kompaneets 1955).

Calculations were performed for the following values of the basic parameters: C_f was varied between 0 and 0.05, r from 2.5×10^{-3} to 1.5×10^{-1} m, Q between 13 and 15 kcal/mol, and $\gamma = 1.3$.

Figure 7 demonstrates a calculated p-v diagram for the detonation regime with wall ignition (similar to that in Fig. 3). The characteristic points have the same meaning as in Fig. 3. The unburned fraction at point C (CJ point) is $\beta_* \lesssim 0.1$. The reaction zone length is large in the case considered (5 to 10 tube radii).

The detonation velocity as a function of coefficient C_f is plotted in Fig. 8. As can be seen, the losses caused by the deceleration and cooling of the reacting gas can reduce the velocity of a self-sustained detonation by a factor of two compared with its ideal thermodynamic value.

Figure 9 presents calculated profiles of the gas temperature averaged over the tube cross section. As the blockage ratio of the duct increases, the temperature in the reaction zone changes only slightly, despite the fact that the detonation velocity drops by a factor of two. This is

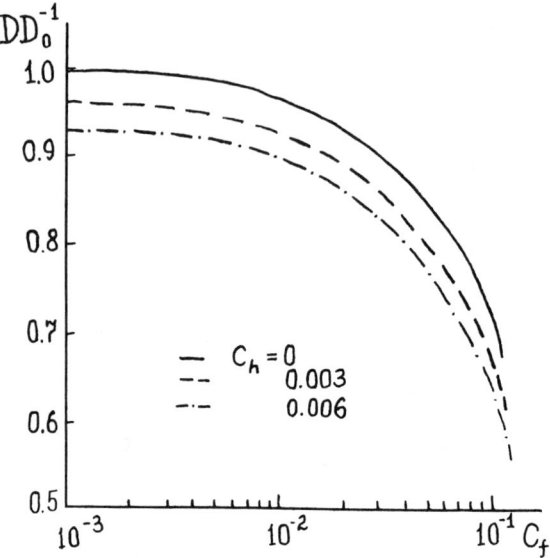

Fig. 8 Detonation velocity as a function of the momentum and heat losses (the wall-ignition case).

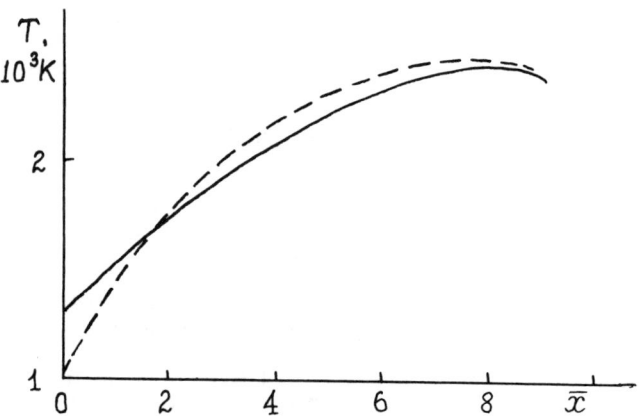

Fig. 9 Temperature in the reaction zone as a function of the dimensionless distance from the wave front ($\bar{x}=x/r$). The solid line represents the tempereature in a wave with small losses, and the dashed line is the temperature in a wave with large losses propagating at a comparably low velocity.

associated with the transformation of part of the kinetic energy into friction heat, which compensates for the temperature drop caused by lower velocities of wave propagation. Such a temperature variation pattern may be one of the reasons for low values of detonation velocities at which the wave in rough tubes still propagates steadily.

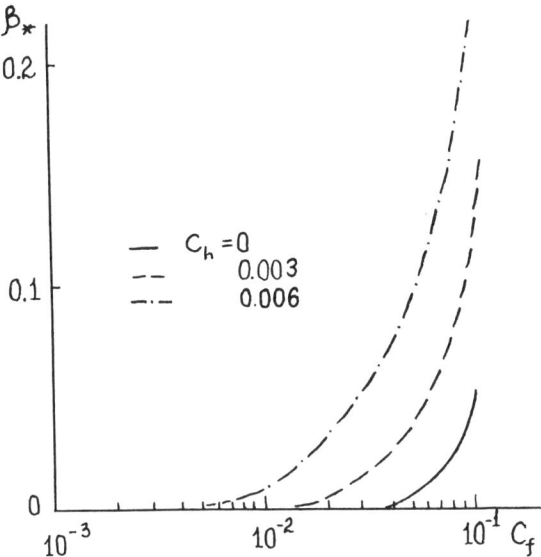

Fig. 10 The unburnt fraction as a function of the momentum and heat losses. The detonation regime with a mixed heat release.

The basic property of detonation in a rough tube is incomplete mixture burning in the reaction zone. The unburned fraction increases as the drag coefficient grows and may reach more than 20% (Fig. 10). Due to this phenomenon, thermodynamic calculations can no longer give any idea of how large the detonation velocity is in a system with losses unless the real flow pattern in the reaction zone is taken into account. Incomplete mixture burning in the CJ plane may additionally diminish the wave velocity by more than 10%. Note that occurrence of the reaction downstream of the CJ plane may have no effect on the detonation wave propagation, since the flow in this zone is supersonic.
However, incomplete burning of the mixture in the reaction zone may also bring about multifront detonations. Burning of the mixture in the supersonic zone of the flow leads in this case to appearance of infinite derivatives of the thermodynamic parameters and gas velocity; as a consequence, the continuation of the solution to the final state becomes impossible. Such a situation can be connected with generation of secondary shock waves behind the lead shock front. The unburned mixture behind the primary CJ plane may burn out behind the secondary waves; thus, several CJ planes may arise in the flow. Mixture burning behind subsequent detonation fronts is represented in the p-v diagram by the discontinuous curve.
One-dimensional calculations yield a qualitatively correct prediction of the detonation velocity drop caused by a decrease of the diameter of a rough tube. Figure 11 displays $D(r)/D_0$ curves for $C_h = 0.006$ and various C_f values. It can be seen in Fig. 11 that, when C_f = idem and $d \gg d_{cr}$

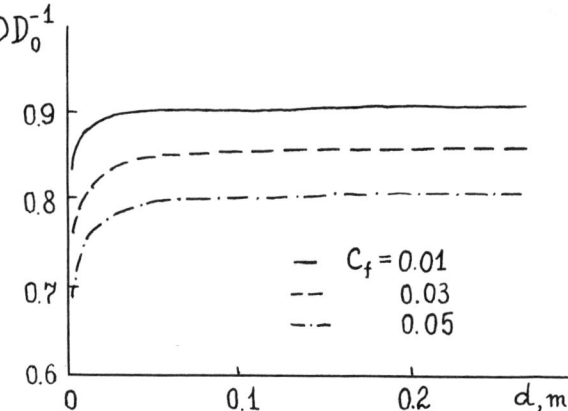

Fig. 11 Detonation velocity as a function of the tube diameter for various drag coefficients (the wall ignition model).

(d_{cr} is the critical diameter for detonation propagation in a duct), the D is virtually independent of the tube diameter. This implies that, in tubes of different diameters having geometrically similar roughness, the low-velocity detonation processes in mixtures of the same composition propagate with nearly the same velocities. This property may be considered a condition for geometric modeling of low-velocity detonations. To verify the criterion, the data of Peraldi et al. (1986) are used in Fig. 12.

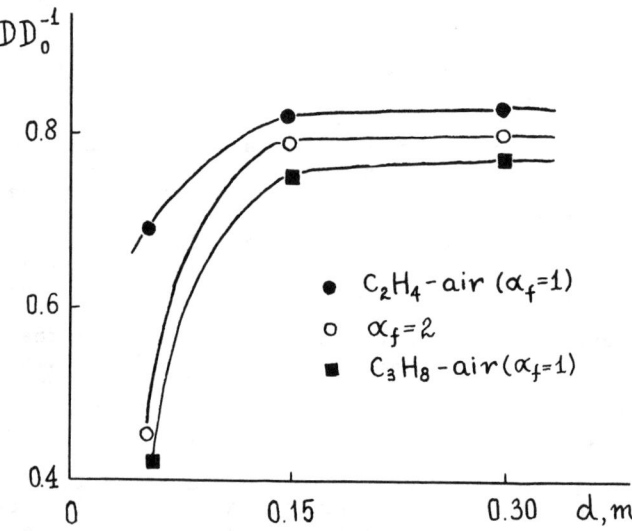

Fig. 12 Detonation velocity as a function of the tube diameter according to the data of Peraldi et al. (1986).

From a comparison of Figs. 4 and 8, it follows that for tubes of small diameters ($r \sim 10^{-2}$ m) the limiting drag coefficients $C_{f\ell}$ in the two models (of the wall and bulk mixture ignition) coincide by an order of magnitude. However, the reaction zone in the wall-ignition model is much longer than in the bulk-ignition model. This means that the high-velocity detonation regime may be realized only in tubes with relatively low λ. In tubes of large diameters $C_{f\ell}$ is high (this follows from the similarity condition C_f/r = idem). Therefore, the blockage ratio in such tubes may be high, yet the high-velocity detonation regime is still possible.

Near the limit of the high-velocity regime the detonation velocity is nonunique in the 1-D approximation. The data obtained by Lee et al. (1984) seem to confirm this nonuniqueness. When λ is below its limiting values for the high- and low-velocity regimes, which of the two regimes will be realized in practice is a priori unknown (provided that the 1-D approximation describes the real detonation wave more or less correctly).

Nonuniqueness of the Detonation Regimes in Rough Tubes

To illustrate the nonuniqueness of detonation processes in rough tubes, we considered a 1-D problem that includes the two possible ignition mechanisms.

It is assumed that the core flow ahead of the flame front expands adiabatically due to the action on it of the reacting boundary layer. In such an approach one can use the results of the 1-D calculations of the reaction zone in a detonation wave with wall ignition up to the instant when the mixture self-ignites in the core flow. As the detonation velocity increases, the self-ignition point in the core flow approximates the leading wave front and the bulk reaction becomes possible in addition to the reaction in the boundary layer.

The calculations of such a detonation wave were carried out for $r = 10^{-2}$ m, $C_f = 0.05$, $C_h = 0.005$, $E/R = 18000$ K, $m = 1$, and $\gamma = 1.3$. The calculation results are presented in Fig. 13. Characteristic points in which one of the conditions [Eq. (11)] is met were indicated in the reaction zone for each fixed value of D. Lines I and II represent locii in which D-w=C. Lines AB and GD are drawn through the points in which the condition in Eq. (11b) is met. Line III represents locii where self-ignition is observed in the adiabatic core flow. As seen from the graph, the conditions in Eqs. (11a) and (11b) are fulfilled simultaneously in two points, B and G. Point B corresponds to the low-velocity detonation with $D \approx 0.78 D_0$. The reaction zone is long ($\ell \approx 10r$) and $\beta_* \approx 0.1$.

Point G pertains to the high-velocity detonation with $D \approx 0.99 D_0$. The reaction zone is short, and the unburned fraction is small ($\beta_* < 0.005$).

The preceding example clearly shows that the simplified 1-D approach to the detonation velocity calculations (with dissociation, relaxation, and other factors being neglected) in rough tubes yields two detonation velocities in a certain region of parameters. The question of which of these

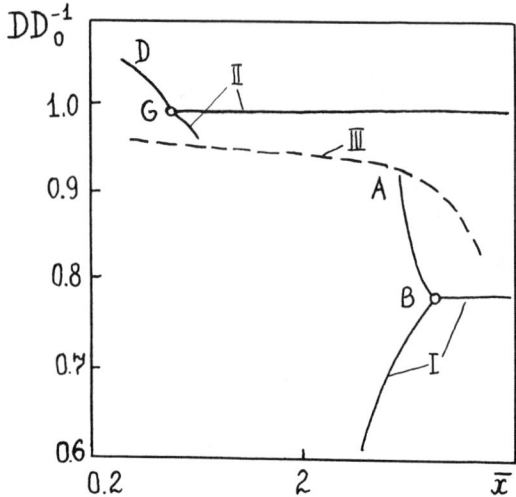

Fig. 13 Diagram illustrating a nonuniqueness of the detonation velocity in a rough tube (mixed heat release model); $\bar{x}$ is the dimensionless length of the reaction zone. Lines I and II are locii in which the total heat release and heat losses rates are equal [condition in Eq. (11b)]. III is the locus in which the mixture self-ignites in the core flow.

regimes is realized in reality can be answered after solving the nonsteady 2-D problem.

Thus we can conclude that

1) The losses diminish the detonation velocity, change the structure of the reaction zone, and may result in the onset of multifront structures and nonunique selection of detonation velocity.

2) Detonation is not possible at every level of loss: there exist limiting values of the losses at which the eigenvalue problem of detonation velocity calculation has no solution.

3) An approximate scaling of the process is possible based on the 1-D solution.

Experimental

To check the stability of detonations in rough tubes and the feasibility of the two previously mentioned detonation regimes, we have conducted experimental studies in a detonation tube. Two versions of the tube were used. In the first version, detonation was initiated by exploding a small volume of the stoichiometric propane-oxygen mixture in a smooth tube (70 mm in diam and 9 m long); then the mixture entered a flexible metal hose of 68-mm i.d. and 3.1-m long. The corrugations (obstacles) were 6-mm high and spaced 6 mm apart. Beyond the hose, the detonation wave again entered a

smooth tube of the same diameter and was monitored along
11 m of the tube. Both the average wave velocity and its
distribution along the hose were measured. In the second
version, detonation was initiated in a smooth tube 70 mm in
diam and 25-m long; it then entered a flexible metal hose
15-m long. The detonation velocity was measured at several
bases along the hose. Experiments were carried out with
propane-air, hydrogen-oxygen-nitrogen, and methane-air mixtures. Each mixture was prepared at an elevated pressure in
a mixer and then an amount of about three tube volumes was
channeled thru the tube.

Figure 14 shows the detonation velocity in the smooth
tube and the mean wave velocity in the 3.1-m-hose as functions of the fuel (propane) concentration in air. As with
other mixtures, the mean detonation velocities in the hose
are much lower than in the smooth tube. From this it follows that a detonation wave entering the rough tube rapidly
changes structure and propagates at a unique mean velocity
corresponding in the case under consideration to the wall
ignition regime.

Two more peculiarities of the above data are worth
noting. After the detonation wave passes by the rough tube
section, it recovers almost immediately in the smooth tube
but within a narrower concentration range than the detonation limits in a smooth tube. The detonation wave in the
smooth tube after it escapes the rough section has a velocity that is higher than the CJ velocity. This is an indication of the fact that transition from the wall-ignition

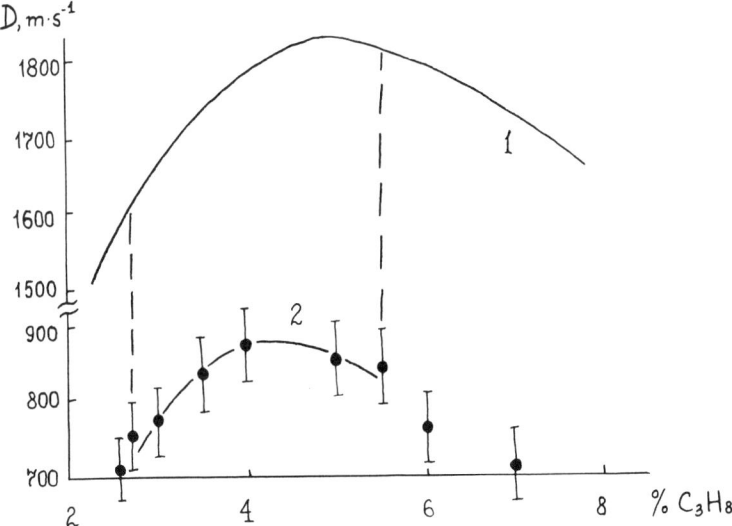

Fig. 14 Detonation velocity in the 3.1-m-long metal hose and in
the smooth tube for propane-air mixtures as a function of the
mixture composition. Curve 1 is for the smooth tube and curve 2
for the rough one. Vertical dashed lines show the limits of
detonation recovery in the smooth tube.

regime to the normal detonation occurs via "explosion in explosion" and overdriven detonation.

The second peculiarity is that the limits of low-velocity detonation are somewhat narrower than those in a smooth tube and that the detonation velocity at the limit in a rough tube drops down to very low values ($M \approx 2$). This is illustrated by a comparison of curves 1 and 2 in Fig. 14. The extreme points for the rough tube pertain to the low-velocity detonation limits, and the vertical dashed lines indicate the limiting mixture compositions at which the low-velocity detonation recovers in the smooth tube.

Lee et al. (1984) and Peraldi et al. (1986) report the data that give evidence of the nonuniqueness of the detonation velocity in rough tubes in hydrogen-air mixtures. However, our data (Zeldovich et al. 1984) seem to indicate that detonation propagates with a single velocity. A question arises in this regard: How stable is the low-velocity detonation (i.e., will it propagate at the same velocity for longer distances)? It was also interesting to note how the low-velocity detonation converts into the normal CJ detonation when the mixture reactivity is increased. Such a transition was studied in this work using hydrogen-oxygen-nitrogen mixtures of different compositions. The measurements of the mean wave velocity in the 3.1-m-long hose demonstrate that each mixture has its own detonation velocity. A comparison of curves 1 and 2 obtained in the smooth and rough tubes (Fig. 15) shows that the transition from one detonation regime to the other occurs smoothly (though there

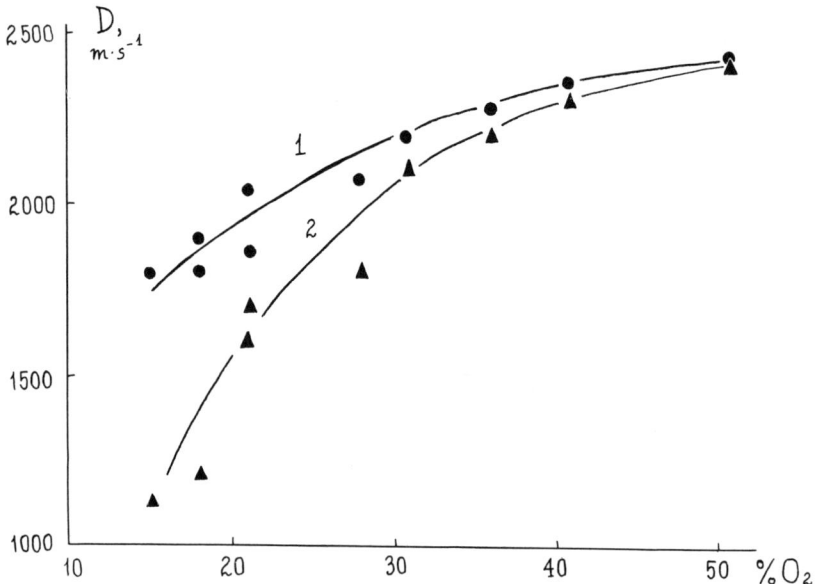

Fig. 15 Detonation velocity in the smooth (1) and rough (2) tube as a function of the composition of the H_2-O_2-N_2 mixture. The data are obtained in the 3.1-m-long hose.

exists a steep portion of the D-nitrogen content curve). The velocity deficit comes close to deficits characteristic of the bulk self-ignition mechanism when the oxygen concentration exceeds 25%. In real detonation waves that are to be described by multidimensional models, the heat may be released by a mixed mechanism, and the velocity deficit may acquire intermediate values between the two extreme cases mentioned previously. It should be emphasized here that the detonation wave "feels" the losses at the rough tube walls, even when the reaction zone is short (notice the difference between curves 1 and 2 in Fig. 15). There is no well-defined criterion for determining the conditions at which the detonation wave ceases to feel the wall roughness, and therefore the concept of "three detonation cells" proposed by Peraldi et al. (1986) is somewhat conventional.

Since the wave velocity dropped to very low values in the 3.1-m-long hose (Fig. 16), a question arose whether the low-velocity detonation was capable of further steady propagation. Indeed, for the incident waves propagating at velocities of 600-700 m/s in propane-air mixtures, the ignition delays even in reflected shock waves must be much above the millisecond range; hence such a shock wave is not likely to be capable of driving a steady detonation process.

The experiments with the 15-m hose section were conducted to answer this question. The detonation velocities as measured with pressure transducers along the hose are plotted in Fig. 16. A characteristic feature of the data shown in this figure is a deep minimum on the D-distance curve near the point of wave entry into the hose. This

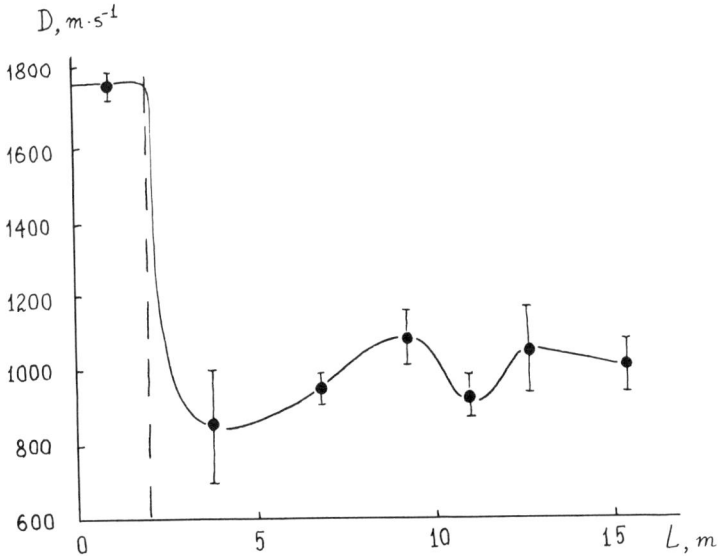

Fig. 16 Detonation velocity in the rough tube as a function of the distance from the hose front edge for the propane-air mixture.

indicates that the low wave velocities (600-700 m/s) are not characteristic of the steady detonation regime. The wave accelerates after the first 3 m of its travel in the hose and reaches a mean velocity of about 900-950 m/s. However, it should be noted that the scatter of the measured detonation velocities from run to run and in a single run is rather high. Jumpwise velocity fluctuations that attain 200 m/s are observed. This is evidence of the instability of the heat release zone behind the lead shock front, associated presumably with the strong turbulization of the flow, local combustion quenching, and self-ignition of a part of the mixture in the volume.

Smoked plates positioned radially along the rough tube gave rather inconclusive information about the wave structure. However, the smoked plate writings reveal some interesting features of the process. For instance, the flow near the obstacles has no structure characteristic of an overdriven detonation. This indicates that only "mild" ignition occurs, evidently in the space between the obstacles, without formation of detonation waves. Inasmuch as detonation waves in rough tubes have very low velocity, which is insufficient for the ignition delays to be short enough (10^{-4} s and lower), even in reflected waves the ignition near the obstacles can be expected to occur in very small volumes owing to focusing of the shock waves. This explains the detonation propagation at unreasonably low velocities.

Furthermore, inside the tube the soot prints differ from those characteristic of a turbulent flame, since against the structureless background there appear some sporadic prints that furnish evidence of a formation of local waves with a cellular structure.

Hence, from the experimental studies of detonation in rough tubes, it follows that certain phenomena exist that require for their explanation a non-1-D model, although the 1-D model gives a qualitatively correct description of the effect of losses on the velocity and structure of the wave. Among these phenomena are (1) the cellular wave structure observed in experiments, where the detonation velocity is much lower than that inherent in the waves with bulk ignition of the mixture; and (2) the strong large-scale instability of the low-velocity detonation regimes. The fact that detonation velocity goes through a minimum when the wave travels from the smooth tube to the rough one also merits special consideration.

References

Kogarko, S. M. and Borisov, A. A. (1960) On the measurement of ignition delays in the high temperature atmosphere. Izv., Akad Nauk SSSR, 1348-1353.

Kutateladze, S. S., (1979) The fundamentals of heat transfer theory, Atomizdat, 416.

Lee, J. H. S., Knystautas, R. and Freiman, A. (1984) High-speed turbulent deflagrations and transition in H_2-air mixtures to detonation. Combust. Flame 56, 227-239.

Matsui, H. (1981) Detonation of gas mixtures in tubes. Proceedings of the 19th Japan Symposium on Combustion. Tokyo, Japan, 68-73.

Peraldi, O., Kuystautas, R. and Lee, J. H. (1986) Criteria for transition to detonation in tubes. Proceedings of the 21st Symposium (International) on Combustion. Munich, Federal Republic of Germany (to be published).

Schelkin, K. I. (1949) High-speed deflagration and spinned detonation. Voenizdat, 156.

Schlichting, H. (1951) Grenzschicht-Theorie. Karlsruhe, 528.

Westbrook, C. K. (1982) Towards an understanding of the detonability of vapour clouds. Fuel-Air Explosions. Waterloo, Ontario, Canada, 139-154.

Zeldovich, Y. B. (1940) On the theory of detonation propagation in gaseous systems. Zh. Eksp. Teor. Fiz., 10. 542-568.

Zeldovich, Y. B. and Kompaneets, A. S. (1955) The Theory of Detonation. Gostechizdat, 268.

Zeldovich, Y. B., Borisov, A. A., Gelfand, B. E., Chomik, S. V. and Mailkov, A. E. (1984) Low-speed quasi detonations of fuel-air mixtures in rough tubes. Dokl. Akad. Nauk SSSR 279, 1318-1321.

Influence of Obstacle Spacing on the Propagation of Quasidetonation

L. S. Gu,* R. Knystautas,† and J. H. Lee‡
McGill University, Montreal, Quebec, Canada

Abstract

Experiments of flame propagation have been conducted in an artificially roughened 15 cm diameter tube over a wide range of mixture compositions for C_2H_2-air, C_2H_4-air, CH_4-air, C_3H_8-air, and H_2-air. Wall surface roughness is enhanced with annular obstacle plates equispaced at one half, one and two tube diameters. The blockage ratio [BR $= 1 - (d/D)^2$] remains fixed at BR = 0.5. Five combustion wave regimes are observed: quenching, weak turbulent deflagration, choking, quasidetonation, and C-J detonation. For obstacle spacings of one or two tube diameters, transition from choking to quasidetonation occurs when $d/\lambda \simeq 1$, while for a spacing of D/2 it occurs $4 < d/\lambda < 10$. In the quasidetonation regime, the flame speed is a function of the obstacle spacing, with higher values corresponding to greater spacing. For C_2H_2-air, there exists a critical value of d/λ beyond which the momentum loss has no influence on the flame speed. The previous criterion that $d/\lambda \geq 13$ for C-J detonation is not valid for some cases, since it does not predict the effect of spacing. It is proposed that the mechanisms of the combustion wave are controlled by the energy balance in the choking regime. The principle mechanism of quasidetonation is considered to be autoignition. A theoretical model is presented that accounts for the effects of heat and momentum losses, tube diameter, and mixture composition on the terminal velocity of the combustion wave. A critical tube roughness degree beyond which a quasidetonation is impossible is obtained from the

---

Copyright © 1988 by the American Institute of Aeronautics and Astronautics, Inc. All rights reserved.
* Research Assistant, Department of Applied Mechanics
† Professor, Department of Mechanical Engineering
‡ Professor, Department of Mechanical Engineering

INFLUENCE OF OBSTACLE SPACING

analysis. The theoretical conclusions agree well with experimental results.

Nomenclature

a	=stoichiometric mole number of fuel in the chemical reaction
A	=cross sectional area of the tube
b	=stoichiometric mole number of oxidizer in the chemical reaction
c	=sonic speed in the gas mixture
C_r	=concentration of the fuel
C_o	=concentration of the oxidizer
C_p	=specific heat at constant pressure
d	=internal tube diameter
d	=internal orifice diameter
D	=detonation (or quasidetonation) velocity
D_{CJ}	=Chapman-Jouguet velocity
$[D_{CJ}-D]/D_{CJ}$	=detonation velocity deficit
E	=effective activation energy
k_0	=constant in Arrhenius law
ℓ	=perimeter of the tube
P	=pressure of the mixture
$\dot{q}$	=heat loss rate per unit area through the tube wall
Q_0	=heat of reaction
Q_r	=effective heat of reaction ($Q_r=\phi Q_0$ for $\phi<1$; $Q_r=Q_0/\phi$ for $\phi>1$)
R	=gas constant
S	=spacing of the orifice plates
T	=temperature of the mixture
v	=specific volume
w	=absolute particle velocity
x	=relative distance behind the shock wave
γ	=ratio of specific heat for the mixture
λ	=detonation cell width or transverse wave spacing
μ	=friction roughness degree
ρ	=density of the mixture
ϕ	=equivalence ratio of the mixture

Subscripts

c	=critical condition
CJ	=Chapman-Jouguet condition

0   =state before the compression of the
            shock wave

Superscripts

*   =state corresponding to the Chapman-
            Jouguet condition

Introduction

One of the most fundamental problems in combustion is to determine how fast the combustion front propagates in a combustible medium. The speed of the combustion wave depends on initial and boundary conditions. There have been a number of studies performed in recent years on the propagation of flames in obstacle-filled tubes with different boundary parameters such as blockage ratio and tube diameter [Lee (1986); Peraldi et al. (1986)]. The experimental results of steady flame velocity as a function of mixture composition indicate that the final steady-state propagation can be described in terms of five regimes, categorized as self-quenching, weak turbulent deflagration, choking, quasidetonation, and Chapman-Jouguet detonation. The self-quenching regime occurs for very high blockage ratios and in relatively insensitive mixtures. Thus, the flame propagation is essentially one of successive hot-jet reignition of the gas in the compartments bounded by two orifice plates. The weak turbulent deflagration regime occurs in near-limit mixtures where the burning rate is slow. The propagation is mainly diffusion controlled and the strong turbulence can lead only to flame quenching. Of interest are the choking, quasidetonation, and the C-J detonation regimes. There appears to be a continuous spectrum of possible wave speeds from the sonic speed of the burned product to the full C-J detonation velocity that depends on the obstacle blockage ratios and tube diameters. In previous investigations, the obstacle spacing was kept constant at $S/D = 1$. Research on the effect of obstacle spacing on the propagation of combustion waves was carried out in the late 1940's by Guenoche and Manson (1949), who repeated Shchelkin's study with different mixtures in a tube 7 mm in diameter filled with a spiral ring. It was observed that steady state detonation velocities below 40% of the normal C-J values are possible in the $2C_2H_2 + O_2$ mixture. However, their study was carried out in small tubes with very sensitive fuel-oxygen mixtures. The present work reports some recent results on the influence of obstacle spacing on the propagation of quasidetonation in atmospheric fuel-air mixtures. A one-dimensional quasisteady model is presented to elucidate the influence of heat and momentum losses on the detonation velocity. Of particular interest in the theoretical study is the demonstration of the existence of a critical degree of tube roughness above which a stationary quasidetonation wave is not possible. However, the quantitative comparison to experiment is difficult due to the lack of friction and heat transfer coefficients of such obstacle-filled tubes in reacting gases.

Experimental Details

In the present study, the propagation of deflagration and detonation waves was investigated in a steel tube with an approximate internal diameter of 15 cm. The tube with circular cross-section was closed at both ends and 18m long. It is possible to judge the stationary character of the combustion front with assurance only if the tube is sufficiently long. The degree of wall roughness depends on the orifice spacing as well as the effective blockage ratio (defined as $BR = 1 - d^2/D^2$, where d and D are, respectively, the internal diameters of the orifice plate and steel tube). The orifice plates employed in the experiment have approximately the same opening diameters of 10.2 cm, yielding a blockage ratio of 0.5. The obstacle rings were joined together in an equispaced array on four threaded rods so that the spacing could be adjusted to the desired value. Orifice plate separations of two, one, and one-half tube diameter were used in the experiment. The length of each obstacle section was about 3 m and the first section started from the ignition end. Four individual section of obstacle plates were assembled before being inserted into the tube, forming a total obstacle field length of 12 m and giving a L/D ratio of 80. In most of the tests performed, the flame front was observed to accelerate to supersonic speed within 2-3 m from the ignitor and then propagate steadily downstream along the tube. Although the obstacle field was only two-thirds of the total tube length, it was long enough to ensure that the flame velocity reached a steady state.

The ignition system is a glow-wire ignitor (a tungsen filament) powered by a 12 v automobile battery. The diagnostics used to obtain the flame velocity include ionization probes employed to measure the arrival time of the flame at a given location. The bulk of the ion gages were spaced 1 m or 50 cm apart, depending on the spacing arrangement. Generally speaking, four to six obstacle rings were arranged between any two gages in order to reduce the local effect of the obstacle on the individual measured velocity. The signals from the ion gages were amplified and used to trigger time interval counters (HP52336). The wave velocities were computed by dividing the distance between any two gages by the corresponding transit time. For low subsonic flame speeds (less than 200 m/s) a digital waveform recorder (Biomation 2805) was used to determine the velocity, since the signal is too weak to trigger the time counter.

A wide variety of fuel compositions, ranging from the lean to the rich limit of flammability, were used in the present experiments. Hydrogen, acetylene, propane, ethylene, and methane of CP grade (greater than 99.5% pure) were used for this study. All tests were carried out with initial conditions of room temperature and 760 mm Hg in pressure. No attempts were made to control the daily temperature fluctuations. The experimental procedure was started by flushing the tube with compressed air. Then, the tube was sealed fully so that in any given experiment it could be evacuated to 3-4 mm Hg by two vacuum pumps (Edwards 18). The tank was refilled with fuel to the desired partial pressure and a total pressure of 760 mm Hg was achieved by adding the appropriate volume of air. Homogenization of fuel and air throughout the tube was attained by using a recirculation system that consisted of a closed loop and a circulation pump (Metal Bellows Corps MB 302 or MB 602). To ensure homogeneity, the circulation pump was operated until at least the equivalent of six tube volumes were displaced.

Result Discussions

The combustion front velocities observed in the 15 cm diameter tube, filled with obstacles equally spaced at $S/D = 1/2$, 1, and 2 (where S is the spacing of the orifice plates) are shown in Figure 1-5 for acetylene-air, hydrogen-air, ethylene-air, propane-air, and methane-air mixtures, respectively. It is found that for each fuel concentration there are five distinct regimes (quenching, weak turbulent deflagration, choking, quasidetonation, and C-J detonation) based on the final steady-state wave velocity. "Quenching" is observed in near-limit mixtures with flame speeds on the order of tens of meters per second prior to extinguishing itself. It is of interest to note that the flame propagation in the "quenching" regime is governed by two mechanisms: chemical kinetics and heat loss. The extinction mechanism near the lean or rich limit of quenching is thus associated with the reactive species

INFLUENCE OF OBSTACLE SPACING 237

limit for a given geometry of the vessel. The flame will extinguish itself if either the minimum dimension d of the vessel or the fuel-oxidizer concentration is decreased, since quenching is due to a high blockage ratio and an insensitive mixture. Thus, for a given mixture and initial conditions, the quenching diameter can be defined from experiments; for a given minimum dimension, the quenching concentration of the mixture may also be determined. In the present experiments with fixed dimension d, the quenching limit is found to be slightly different for the three spacing cases. This feature may result from the fluctuations in the laboratory temperature, since the experiments were performed at room temperatures of 8 - 30°C. The flame propagation near the quenching limits strongly depends on initial conditions (room temperature and pressure). For the composition far from stoichiometric, but slightly more sensitive than the quenching limits, a laminar flame can be obtained in which the reaction preceeds as a subsonic wave propagating into the fresh mixture by the diffusive transport of mass and energy. Steady state propagation occurs as the increase in burning rate due to "flame folding" (i.e., flame area increase) and turbulent

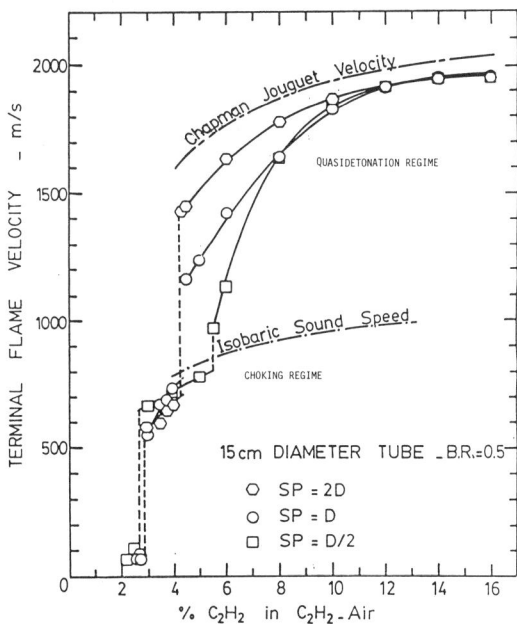

Fig. 1 Steady state flame velocity in C_2H_2-air mixtures as a function of fuel concentration.

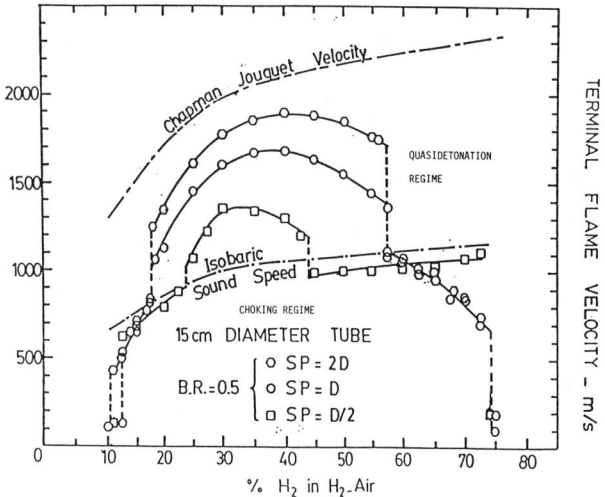

Fig. 2 Steady state flame velocity in H_2-air mixtures as a function of fuel concentration.

transport is balanced by the quenching effect of flame stretching and the rapid cooling by turbulent mixing.

As mixture compositions become progressively more sensitive than that corresponding to the region of weak turbulent deflagration, the flame speed transits to a range somewhat less than the isobaric sound speed of the burned product. This regime is referred to as the "choking" regime and is characterized by flame speeds of 600 - 1000 m/s. The sonic or choking regime is energetics controlled, as suggested by the close proximity of the deflagration wave speed to the sound speed of the burned gases. Strictly speaking, the choking regime is in essence a detonation regime where the energetics of the mixture governs the wave speed. The low velocity of the wave itself suggests that shock ignition is not the principle mechanism responsible for ignition and self-sustained propagation. It has been suggested by Zel'dovich and Kompaneets (1960) that shock reflections at the obstacles provide the high temperature necessary for autoignition. From these hot spots, the combustion then spreads toward the axis of the tube. However, the streak schlieren photographs of Brochet (1966) conclusively demonstrate that the ignition does not arise from shock reflections. On entering an obstacle field, the flame zone following the shock accelerates to catch up with the front. Figure 6 is a streak record from Wagner's (1982) work, illustrating this important point. Spark schlieren photographs taken by Chan (1985) of the wave structure in the choking regime also show the absence of autoignition by shock reflections. Thus, it may be concluded that the choking regime propagates via autoignition by turbulent mixing. The turbulent mixing between hot products and unburned explosive gases has already been demonstrated by Knystautas et al. (1978).

It has been noted that, for small spacing (i.e., $S/D = 1/2$), the wave speeds are much closer to the isobaric sound speed than in the case of larger spacing. For larger spacings, the deviation from the isobaric sound speed is more significant. This result suggests that the velocity fluctuations between obstacles with a small spacing are smaller, since the wave speeds are averaged over a number of obstacles.

As the mixture concentration approches stoichiometric mixture, the combustion front moves at a speed above the isobaric sound speed, but lower than the normal C-J velocity. This is typical of the quasidetonation mode. In Fig. 1-5, the transition is identified by the dotted vertical lines that separate the "quasidetonation" regime from the "choking" regime. Even if the velocity jump from "choking" to "quasidetonation" is not significant (e.g., about 20% higher than the "choking" speed in the case of D/2 spacing), transition can be observed. From previous

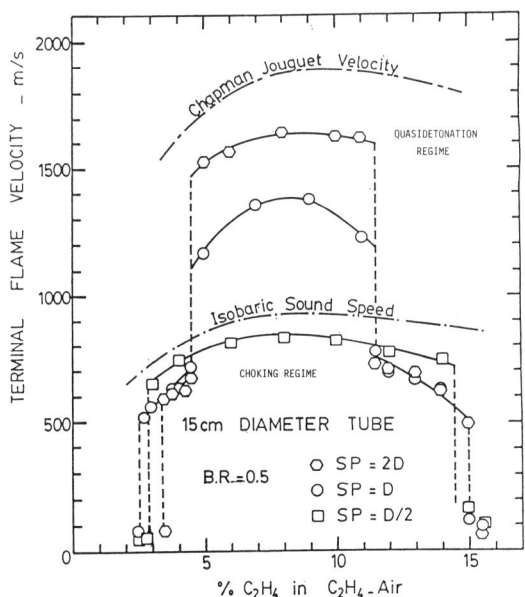

Fig. 3   Steady state flame velocity in C_2H_4-air mixtures as a function of fuel concentration.

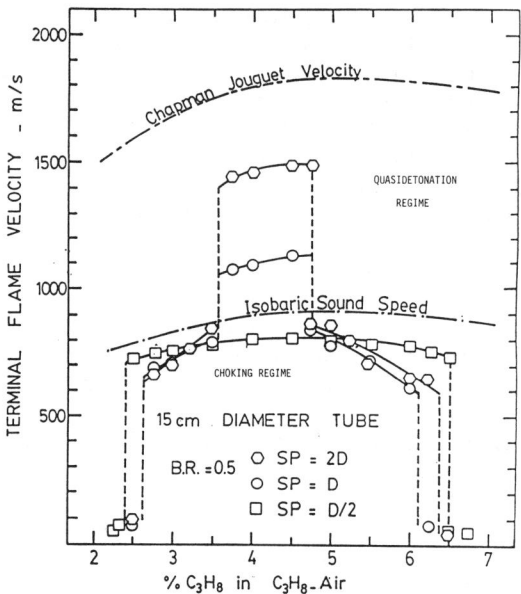

Fig. 4   Steady state flame velocity in C_3H_8-air mixtures as a function of fuel concentration.

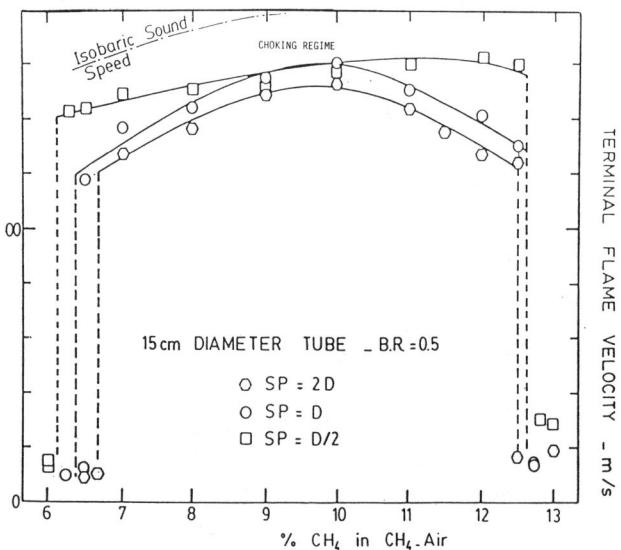

Fig. 5   Steady state flame velocity in CH_4-air mixtures as a function of fuel concentration.

studies (Lee 1986; Peraldi et al. 1986), it was proposed that in a very rough walled tube a self-sustained detonation wave can propagate only through obstacles whose dimension d is greater than the characteristic cell size λ. Present experiments show that one would expect the critical d/λ ratio for propagation to be in the 1 - 10 range, depending on the wall roughness (i.e., spacing arrangement). For spacing greater than D, the transition point remains at $d/\lambda \simeq 1$. For smaller spacing ($S/D = 1/2$), the occurrence of transition to quasidetonation corresponds to a more sensitive mixture (i.e., $d/\lambda \simeq 10$ for H_2-air and $d/\lambda \simeq 4$ for C_2H_2-air). When d/λ exceeds the required critical value, the transition to detonation does not necessarily occur. Results with C_3H_8-air and C_2H_4-air mixtures at a spacing of $D/2$ indicate no occurrence of an abrupt jump in the terminal flame velocity. Even for the two sensitive gases (C_2H_2 and H_2), the transition is relatively weak, as indicated by the small jump in the wave speed. These results may be due to the higher momentum loss and relatively lower heat of reaction. Thus, criterion of $d/\lambda \simeq 1$ cannot be used in some cases. It was found that the terminal flame velocity is a function of spacing with higher values corresponding to larger spacing. The maximum velocity deficit $[(D_{CJ}-D)/D_{CJ}]$ is the same (54%) in H_2-air and C_2H_2-air mixtures. It might be plausible to infer that the maximum velocity deficit is an important parameter for evaluating the transition. For a given boundary condition, if the mixture concentration corresponding to the maximum velocity deficit is outside the region of $d/\lambda \geq 1$, the transition boundary can be determined by $d/\lambda = 1$ since it is a necessary condition (Peraldi et al. 1986). If the mixture composition otherwise lies inside the region of $d/\lambda \geq 1$, the maximum velocity deficit (corresponding to the highest possible roughness degree) is expected to occur at transition boundary. Therefore, if kinetic data and empirical information of friction roughness and heat transfer loss coefficients are available, theoretically speaking, the transition boundary can be estimated.

Present work has shown that the amount of heat addition (effective heat release) and momentum losses play a dominant role since in the equilibrium state integral effects are particularly important. The principle mechanism of combustion wave propagation in this regime is considered to be the energetics controlled via autoignition by turbulent mixing and/or shock wave compression. Hence, for a given mixture, one can always adjust the boundary parameters (i.e., obstacle configuration, heat and momentum losses, tube diameter, etc.) to give a

242 L. S. GU ET AL.

Fig. 6   Self luminous streak photograph of the steady-state
deflagration in the choking regime of a CH_4-air flame.

quasidetonation wave moving at a speed between the isobaric sound speed and the Chapman-Jouguet value.

The normal Chapman-Jouguet detonation mode of propagation was obtained only in a C_2H_2-air mixture (Fig. 1). It was observed that, as the fuel concentration is increased further toward the stoichiometry, the velocity of the quasidetonation wave rises accordingly and progressively approaches the C-J value. It was found that when the concentration exceeds a critical value, the friction has no effect on the terminal flame velocity. The velocity jump between the quasidetonation and normal C-J detonaion regimes is not obvious, but it can be discerned from the effect of boundary conditions on the sensitivity of the velocity. Currently, this mixture composition is 12% by volume, which corresponds to minimum cell size of 0.45 cm [Peraldi et al. (1986)], giving a ratio of $d/\lambda \approx 22$. The existing criterion for a normal Chapman-Jouguet detonation is that $d/\lambda \geq 13$ [Liu et al.(1984)]. However, the present study indicates that this criterion cannot be applied for all cases, since it does not include the influence of

INFLUENCE OF OBSTACLE SPACING

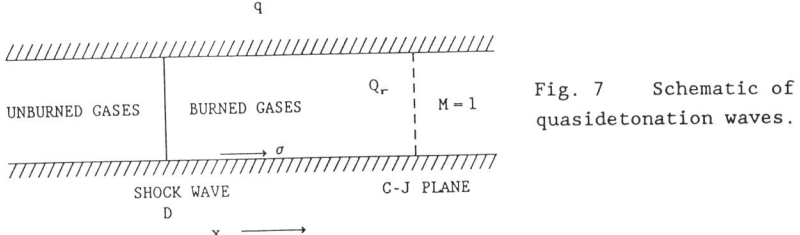

Fig. 7 Schematic of quasidetonation waves.

spacing. The experimental results also shows the normal detonation speed still has a velocity deficit of about 3-4% relative to the C-J value. This speed deficit results from the incomplete combustion (i.e., $\beta^* \neq 0$). In other words, the energy cannot be released completely at the C-J plane due to the existence of friction resistance.

Theoretical Modeling

It is of interest to elucidate the role of momentum losses on the quasidetonation waves. The schematic diagram is shown in Fig. 7.

The one-dimensional model was performed based on the following assumptions:
1) the detonation front is planar (no reflection); 2) the flow is one dimensional and quasisteady; 3) the friction resistance coefficient (μ) is constant; 4) perfect gas equation of state is applicable; 5) the heat loss is given as a function of x.

Basic Equations and Derivations

Continuity equation:

$$\frac{d(\rho u)}{dx} = 0 \qquad (1)$$

$$\rho_0 D = \rho u = \rho(D-w) \qquad (2)$$

Momentum equation:

$$\rho u \frac{du}{dx} = -\frac{dP}{dx} + \frac{\sigma \ell}{A} \qquad (3)$$

where $\sigma = (\mu/2)\rho w^2$.

Setting $d_H = A/\ell = d/4$, Eq. (3) can be integrated into Eq. (4) as

$$P_0 + \rho_0 D^2 = P + \rho(D-w)^2 - \frac{1}{d_H}\int_0^x \sigma dx \qquad (4)$$

Energy equation:

$$\rho_0 DA \frac{d}{dx}\left[\frac{\gamma}{\gamma-1}(P/\rho) + \beta Q_r + \frac{(D-w)^2}{2}\right] = \frac{-\dot{q} + D\sigma}{\ell} \quad (5)$$

where $\beta = C_r/C_{r_0}$.

Integrating Eq. (5), we have the integral form

$$\rho_0 D\left(\frac{\gamma}{\gamma-1} Pv + \frac{(D-w)^2}{2} - \frac{\gamma}{\gamma-1} P_0 v_0 - \frac{D^2}{2}\right)$$

$$= \frac{1}{d_H}\int_0^x (-\dot{q} + D\sigma)dx + (1-\beta)Q_r \quad (6)$$

From perfect gas assumption,

$$Pv = RT \quad (7)$$

and

$$\frac{d\beta}{dt} = \frac{1}{C_{r_0}}\frac{dC_r}{dt} \quad (8)$$

Assuming that the Arrhenius law applies leads to

$$\frac{d\beta}{dx} = k_0 \exp\left(-\frac{E}{RT}\right) a \frac{1}{D-w} (C_{r_0})^{a+b-1} \beta^a \left[\frac{b}{a}(\beta-1) + \frac{C_{o_0}}{C_{r_0}}\right]^b \quad (9)$$

Replacing P in Eq. (6) by using Eq. (4) results in

$$\frac{\gamma}{\gamma-1} P_0(v^* - v_0) + \frac{\gamma}{\gamma-1}\left(\frac{v^*}{v_0} + \frac{\gamma-1}{2\gamma}\right)D^2 - \frac{\gamma+1}{2(\gamma-1)}(D-w)^2$$

$$= Q_r(1-\beta^*) - \frac{1}{d_H}\int_0^{x^*} \dot{q}\,dx - \frac{1}{d_H}\int_0^{x^*}\left[\frac{\gamma}{\gamma-1}(D-W) - D\right]\sigma\,dx \quad (10)$$

Rearranging Eq. (10), one can obtain, using Zel'dovich's analogy,

$$rD^3 + sD^2 - t = 0 \quad (11)$$

where $r = \mu B$, $B = \frac{2}{d}\int_0^{x^*}\left(1 - \frac{v}{v_0}\right)^2 \frac{1}{v}\left(\frac{\gamma}{\gamma-1}\frac{v}{v_0} - 1\right)dx$,

$s = 1/2(\gamma^2 - 1)$, and $t = Q_r(1-\beta^*) - \frac{4}{d}\int_0^{x^*}\dot{q}\,dx$.

Setting $D = V - (s/3r)$, Eq. (11) becomes

$$V^3 + \delta V + \tau = 0 \quad (12)$$

where $\delta = -s^2/3r^2$, $\tau = (2s^3/27r^3) - t$.

From mathematics, there is a unique real solution for Eq. (12) if only

$$(\tau/2)^2 + (\delta/3)^3 \geq 0 \qquad (13)$$

Inequality equation (13) can be modified as

$$\mu \leq \mu_c = \sqrt{s^3/27tB^2} \qquad (14a)$$

that is,

$$\mu \leq \mu_c = \frac{\left[[1/2(\gamma^2-1)]^3/27 \, [\, Q_r(1-\beta^*) - \frac{4}{d}\int_0^{x^*} \dot{q}\, dx\,]\right]^{1/2}}{\frac{2}{d}\int_0^{x^*}(1-\frac{v}{v_0})^2 \frac{1}{v}(\frac{\gamma}{\gamma-1}\frac{v}{v_0} - 1)dx} \qquad (14)$$

Therefore, there exists a critical value μ_c for the possible quasisteady, one-dimensional quasidetonation wave.

Under the premise of inequality equation (14a), the solution to Eq. (12) is

$$V = \left[-(\tau/2) + \sqrt{(\tau/2)^2 + (\delta/3)^3}\right]^{1/3}$$

$$+ \left[-(\tau/2) - \sqrt{(\tau/2)^2 + (\delta/3)^3}\right]^{1/3} \qquad (15)$$

In summary, four parameters have been found to effect the quasidetonation velocity: tube roughness, tube diameter, heat of reaction, and heat-transfer loss. An decrease in either tube roughness degree or heat-transfer loss results in an increase in quasidetonation velocity; an increase in the quasidetonation velocity is also the result of a greater heat of reaction and larger tube diameter.

Conclusions

The propagation of a detonation was considered to be independent of boundary conditions in the classical C-J theory. Results of the previous studies [Lee (1986); Peraldi et al. (1986)] and the present work have shown that the detonation velocity is a function of the orifice configuration (blockage ratio and spacing) and the tube diameter. Theoretical analysis confirms the experimental result that there exists a critical tube roughness degree beyond which a stationary propagation of a quasidetonation wave is impossible. The results also indicate that the transition to the quasidetonation mode depends on the obstacle spacing as well as the blockage ratio and tube diameter. Therefore, the minimum quasidetonation velocity, corresponding to the critical tube roughness degree, is suggested to be the criterion for the transition phenomenon in addition to $d/\lambda \simeq 1$. The existing criterion of $d/\lambda \geq 13$ for Chapman-Jouguet detonation [Liu et al. (1984)] is invalid for some cases, since it does not account for the spacing influence. It is proposed that the mechanism in the quenching regime is diffusion controlled (diffusion of heat and free radicals), while in the choking regime it is controlled by the energetics. Autoignition plays a dominant role in the quasidetonation and normal detonation modes.

Acknowledgments

The authors are grateful to Dr. O. Peraldi for his assistance in the experiments. The work was supported by the Natural Sciences and Engineering Research Council of Canada under Grants A-7091 and A-3347 and by the Defense Research Establishment Suffield (DRES) under Contract 8SG84-00057.

References

Brochet, C. (1966) Contributions a l'etude des detonations instables dans les melanges gazeux. Doctoral Theses, University of Poitiers, Poitiers, France.

Chan, C. K. (1985) Private communication.

Guenoche, H. and Manson, N. (1949) Inflence des conditions aux limites transversales sur la propagation des ondes de Choc et de combustion. Rev. Inst. Fr. Pet. 2, 53-69.

Knystautas, R., Lee, J.H., Moen, I.O., and Wagner, H.Gg. (1978) Direct initiation of spherical detonation by a hot turbulent gas jet. Seventeenth Symposium (International) on Combustion, pp. 1235-1245. The Combustion Institute, Pittsburgh, PA.

Lee, J.H. (1986) On the transition from deflagration to detonation.

Liu, Y. K., Lee, J. H., and Knystautas, R. (1984) Effect of geometry on the transmission of detonation through an orifice. Combust. Flame 56, 215.

Peraldi, O., Knystautas, R., and Lee, J.H. (1986) Criteria for transition to detonation in tubes, Twenty-first Symposium (international) on Combustion, pp. 1629-1637. The Combustion Institute, 1986, Pittsburgh, PA.

Peraldi, O. (1986) Private communication.

Wagner, H.G. (1982) Some experiments about flame acceleration. First International Specialist Meeting on Fuel-Air Explosion, SM Study 16. University of Waterloo Press, Waterloo, Ontario, Canada.

Zel'dovich, Ya. B. and Kompaneets, A. S. (1960) Theory of Detonation. Academic Press, New York and London.

Propagation of Detonation Waves in an Acoustic Absorbing Walled Tube

G. Dupré*
Centre National de la Recherche Scientifique, Orléans, France
and
O. Peraldi,† J. H. Lee,‡ and R. Knystautas§
McGill University, Montreal, Quebec, Canada

Abstract

The present study attempts to demonstrate the essential role of transverse waves in the propagation of a detonation by eliminating them. A cellular detonation was generated in a smooth tube and propagated through a short section where an acoustic absorbing material was lined on the tube wall. The absorbing material consisted of a metallic grid rolled up several times to vary its thickness. The influence of grid parameters on the propagation of the detonation downstream of the grid was analyzed for a series of fuel-oxygen mixtures, at various initial pressures, in tubes of different diameters.

Streak records enabled the evolution of the detonation wave to be observed and the velocity to be measured over a field of view upstream and downstream of the grid. The patterns drawn on downstream smoked foils could confirm the presence or disappearance of transverse waves at the exit of the grid section. The reflected pressure from the grid layers was compared to that from a rigid wall, via pressure transducers.

If only initial pressure is varied for a given mixture, there exists a very distinct critical pressure below which the detonation suffers a total lack of cellular structure downstream of the grid and an abrupt drop in velocity of 40% - 60% of the upstream near the

---

Copyright © 1988 by the American Institute of Aeronautics and Astronautics, Inc. All rights reserved.
*Chargée de Recherche, Research Center on the Chemistry of Combustion and High Temperatures (C.R.C.C.H.T., C.N.R.S.).
† Research Associate, Department of Mechanical Engineering, Shock Wave Physics Laboratory.
‡ Professor, Department of Mechanical Engineering, Shock Wave Physics Laboratory.
§ Professor, Department of Mechanical Engineering, Shock Wave Physics Laboratory.

Chapman - Jouguet (C-J) detonation velocity. A deflagration emerges from the grid. The efficiency in the damping of transverse waves slightly depends on the mixture and tube diameter, but is strongly dependent on grid parameters. All the results demonstrate that heat and momentum losses through the grid section are negligible and the main mechanism for velocity deficit and lack of cells downstream is the damping of transverse waves across the detonation front itself. They confirm the necessity of transverse waves in any detonative combustion.

Introduction

That transverse waves are essential to the propagation of a detonation wave has never been clearly demonstrated. Perhaps the most convincing experiment that indicates their important role is the work of Evans et al. (1955). They studied the transition to detonation in a tube lined with an acoustic damping material (i.e., porous sintered bronze) and found that, in such a porous walled tube, transition is delayed or prevented. It was also shown by Meyer (1957) that a normal shock wave reflecting from a thin plastic movable wall attenuates. Thus, the porous wall in Evans' experiments would delay the amplification of the transverse waves to the level at which a self-sustained detonation wave can be formed. In other words, the porous wall would cause an increase of the transition distance to detonation.

The present study attempts to demonstrate the primary role of transverse waves in the propagation of a detonation, by eliminating them from the wavefront. A cellular self-sustained detonation wave is first generated in a smooth tube and then allowed to propagate through a short length of tube lined with an acoustic absorbing material. The reflection of the transverse waves from the acoustic absorbing material attenuates the waves and may cause their destruction. In the latter case, after traveling this short length of tube, the smoked foil records are characterized by the absence of transverse waves. The conditions and consequences of either attenuation or loss of transverse waves on the propagation of detonation are examined and compared to those of momentum and heat losses through the absorbing tube section.

Experimental

A series of experiments with fuel-oxygen mixtures ($H_2/0.5O_2$, $C_2H_2/4O_2$, $C_2H_4/3O_2$, and $C_3H_8/5O_2$) were carried out at various initial pressures P_0 and ambient temperature in three Plexiglass tubes, 3.8 m long, and of different inner diameters (d = 50.8, 76.2, and 101.6 mm). The ignition was produced by using an exploding wire and discharging into it a capacitor of 100 μf charged under 4000 V. The initiation energy available was maintained constant for all experiments, that is 800 J.

To study the influence of transverse waves on the propagation of detonation, we used, as absorbing material, a metallic grid, 50 mesh, made of stainless steel with a porosity of 0.56 (where the porosity is defined as the ratio of open area to total area). The grid was inserted inside the tube at about 2.8 m downstream from the ignition end, along the inner wall, and rolled up several times to vary the thickness of the grid section between 1 and 15 layers (n = 1 to 15). In most experiments, the grid length L was maintained constant (equal to 3d). A few tests were carried out to study the influence of grid length (L/d = 1 to 10).

The main diagnostic consisted of a streak camera to observe the wave propagation and measure the velocity over a 1.35 m field of view upstream and downstream of the grid. In some experiments, smoked foils were used to determine the cell size upstream and downstream of the grid section. The foils were soot-covered Terphane sheets that were rolled up on the inner tube wall. They show the evolution of the three-dimensional detonation structure and eventually the destruction of the transverse waves once the detonation passed through the grid section.

To measure the damping due to the grid layers, additional experiments were carried out to study the reflection of a detonation wave from the grid layers. Three and 15 layers were superimposed to cover the end plate of the detonation tube and two pressure transducers were mounted, one at the downstream end of the tube, and the other on the side wall at 12 mm upstream. The reflected shock strengths from the grids were compared with the normal reflections from the rigid, non-grid-covered wall.

Results and discussion

For a given mixture, tube diameter, length of grid-lined section, and number of grid layers of a given porosity, it is found that there exists a critical value of the initial pressure P_c, above which the detonation wave propagates through the grid section without any observable change in velocity or cell structure. Below

DETONATION IN AN ABSORBING WALLED TUBE

this critical pressure, the combustion wave suffers a total disappearance of any cellular structure downstream of the grid and an abrupt drop in velocity of 40% - 60%. This abrupt velocity drop generally occurs within a pressure range of 1 to 2 Torr, i.e., a few percent of the initial pressure. Figure 1 shows two typical streak records for a $C_2H_2/4O_2$ mixture in a 50.8 mm diameter tube, with a grid length of 152 mm (L/d = 3) with three layers of grids, at an initial pressure of 17 Torr (upper picture) and at 16 Torr (lower picture), that is above and below the critical pressure, respectively. It can be observed that at 17 Torr, except in the immediate vicinity of the grid, the detonation velocity remains constant throughout. For an initial pressure only 1 Torr below that

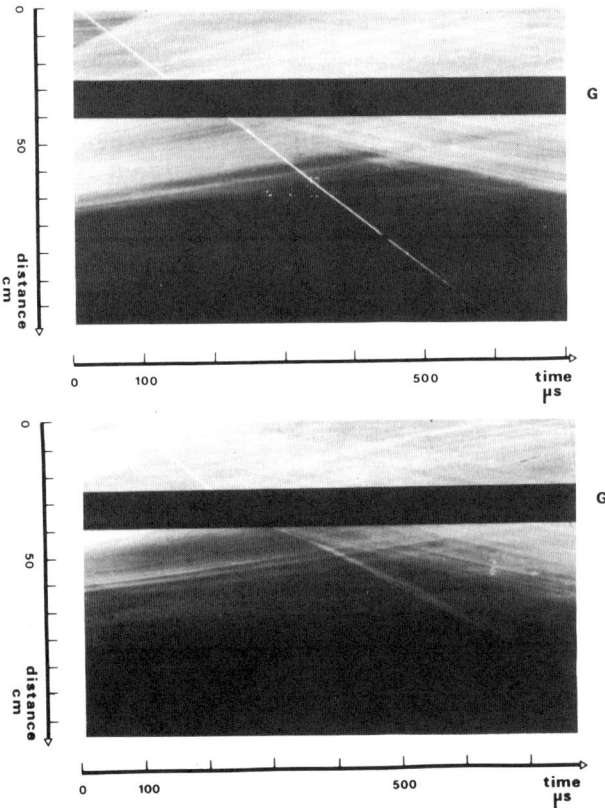

Fig. 1 Typical streak records obtained for P_0 = 17 Torr (upper picture) and 16 Torr (lower picture), that is above and below the critical pressure, respectively, for the $C_2H_2/4O_2$ mixture. G = position of the grid section. d = 50.8 mm; L/d = 3; n = 3.

for the previous case, a deflagration emerges from the grid section. Its velocity is typically 60% of the near C-J detonation velocity upstream of the grid and remains approximately constant for the last 0.85 m of travel to the end of the tube.

From the smoked foil records as well as from the velocity data, we can see that the critical pressure is very distinct. Above the critical pressure, almost no difference between the cellular patterns can be observed before and after the grid. Below the critical pressure, there is a complete absence of cells downstream of the grid section. Thus, all transverse waves are damped. At around the critical value itself, it is sometimes possible to observe some enlarged cells as the detonation first emerges from the grid section. However, the normal cell size of the mixture is rapidly restored for a few cell lengths of travel thereafter.

Different fuels (H_2, C_2H_2, and C_2H_4) in different tubes (d = 50.8, 76.2, and 101.6 mm) have been studied. For a given length of grid section (L/d = 3) and given number of grid layers (n = 3), the dependence of the critical pressure on tube diameter is shown in Fig. 2 for the different mixtures. Note first that the critical pressure depends somewhat on the mixture for a given diameter, and second, that it decreases with increasing diameter in general. For a $C_2H_4/3O_2$ mixture, P_C varies approximately as 1/d. Since the transverse waves spacing or cell size λ is inversely proportional to initial pressure, the almost linear dependence of P_C with 1/d suggests a quasiconstant critical value of d/λ.

If we use the cell size data as measured previously by Knystautas et al. (1982), a plot of the critical value of d/λ for the different mixtures and three different tube diameters is shown in Fig. 3. The results do not indicate the existence of a constant value of $(d/\lambda)_C$, whatever the mixture is. For a $C_2H_4/3O_2$ mixture, an almost linear increase of $(d/\lambda)_C$ with tube diameter can be pointed out, whereas for $C_2H_2/4O_2$ and $H_2/0.5O_2$ mixtures, no constant or linear dependence of $(d/\lambda)_C$ on diameter is observed. It would have been fortuitous to find a universal value of $(d/\lambda)_C$ as for the case of the critical tube diameter problem (Knystautas et al., 1982). Further extensive experiments are needed before definitive conclusions can be made regarding the influence of the tube diameter. However, it seems clear that the damping of transverse waves by a stack of grid layers depends on the properties of the mixture, i.e., the transport coefficients and sound speed. For C_2H_4/O_2 and C_2H_2/O_2 mixtures where the sound

DETONATION IN AN ABSORBING WALLED TUBE

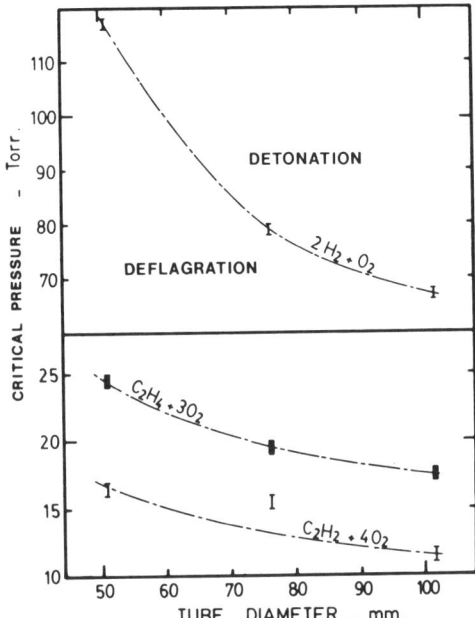

Fig. 2 Influence of tube diameter on the critical pressure for three different mixtures: $H_2/0.5O_2$; $C_2H_4/3O_2$; $C_2H_2/4O_2$. $L/d = 3$; $n = 3$. The dashed lines delimit the regions where a detonation and a deflagration are observed downstream of the grid section.

speed in the product gases is more or less the same, the limit criterion does not change a lot ($d/\lambda \simeq 2.2$-3.2). But for a $H_2/0.5O_2$ mixture where the sound speed is higher, it is more difficult to generate strong shocks, and the limit criterion is higher ($d/\lambda \simeq 4$-4.4). The higher the value of $(d/\lambda)_c$, the more efficient the damping. Figure 3 shows that for a given tube diameter ($d = 76.2$ mm) and a given number of grid layers, the most efficient damping is obtained with a $H_2/0.5O_2$ mixture, i.e., for the most explosive mixture among those studied presently.

The efficiency in damping the transverse waves depends on the number of grid layers n (i.e., the thickness of the damping material) once a particular grid is chosen. Figure 4 shows the variation of $(d/\lambda)_c$ with n for the different mixtures and the same tube ($d = 76.2$ mm) for a given grid characteristic and length. As expected, $(d/\lambda)_c$ for each mixture asymptotes to some constant value when the number of grid layers increases. That means that once the transverse waves are damped, a further increase in grid layers no longer influences the value of $(d/\lambda)_c$. It is of interest to note also that if one extrapolates

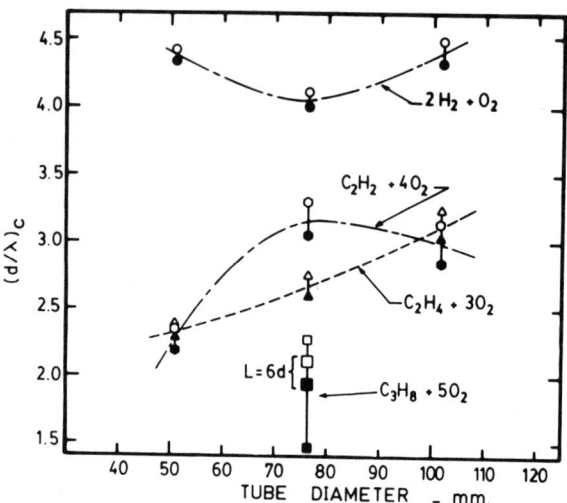

Fig. 3 Influence of tube diameter on the critical ratio $(d/\lambda)_C$ for four different mixtures: $H_2/0.5O_2$; $C_2H_4/3O_2$; $C_2H_2/4O_2$; $C_3H_8/5O_2$. $L/d = 3$; $n = 3$. For $C_3H_8/5O_2$, the results are given in a unique tube ($d = 76.2$ mm) with $L/d = 3$ (symbols □ ■) and $L/d = 6$ (symbols ▫ ▪).

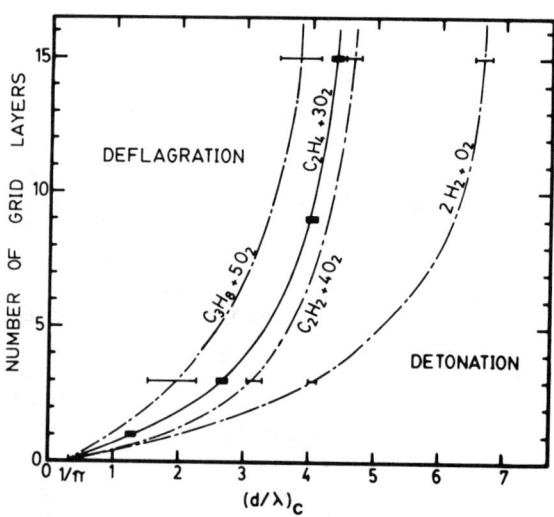

Fig. 4 Influence of number of grid layers on the critical ratio $(d/\lambda)_C$ for four different mixtures: $H_2/0.5O_2$; $C_2H_4/3O_2$; $C_2H_2/4O_2$; $C_3H_8/5O_2$. $L/d = 3$; $d = 76.2$ mm.

(as is reliable for the $C_2H_4/3O_2$ mixture) the results to the zero layer of grid (i.e., the original rigid wall of the tube), one reaches a value of $(d/\lambda)_c$ close to $1/\pi$. This corresponds to the detonation limit criterion ($\lambda \simeq \pi d$) proposed for a circular tube by Dupré et al. (1986).

The wave velocities V before and after the grid are shown in Figs. 5 - 7 for three different mixtures. Each plot represents the variation of V with initial pressure P_0 in the different tubes for the case of three layers of grid and a grid length of three diameters. For $H_2/0.5O_2$ in Fig. 5, for example, one notes that prior to entering the grid section, the velocity is closed to the normal C-J velocity of the mixture. At the critical pressure for a given tube diameter, the velocity suffers an abrupt decrease of about 57% of its initial near C-J value upon emerging from the grid section. Further decrease of the initial pressure far below the critical value only results in a small decrease in the velocity of the deflagration

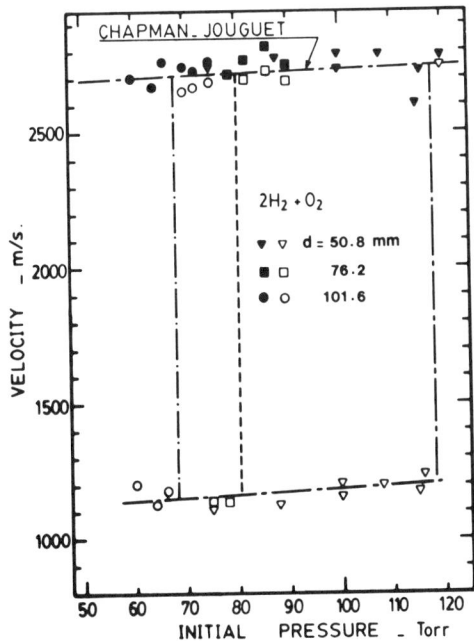

Fig. 5 Influence of initial pressure on the wave velocity upstream (black symbols) and downstream (open symbols) of the grid section, in the three tube diameters, for the $H_2/0.5O_2$ mixture. L/d = 3; n = 3. The vertical dashed lines (from right to left) correspond to the velocity drop at the critical pressures corresponding, respectively, to d = 50.8, 76.2, and 101.6 mm.

after the grid. The downstream deflagration velocity is about the same for the different tube diameters. Similar results are obtained for $C_2H_4/3O_2$ (Fig. 6) and $C_2H_2/4O_2$ (Fig. 7). This suggests that the velocity deficit of at least 40% is due to the damping of the transverse waves. Without its normal transverse waves, intense detonative combustion cannot be maintained across the wavefront downstream of the grid.

Further indication that the grid section serves to damp out the transverse waves only and plays little role in the propagation of the deflagration downstream is shown in Fig. 8. The wave velocity for the $C_2H_4/3O_2$ mixture is plotted vs initial pressure for different numbers n of grid layers in a given tube. The downstream deflagration velocity remains at about 50% of C-J detonation velocity for a range of numbers of grid layers from 1 to 15. Thus, heat and momentum losses during the propagation through the grid section have little influence on the subsequent propagation of the deflagration. The dominant mechanism

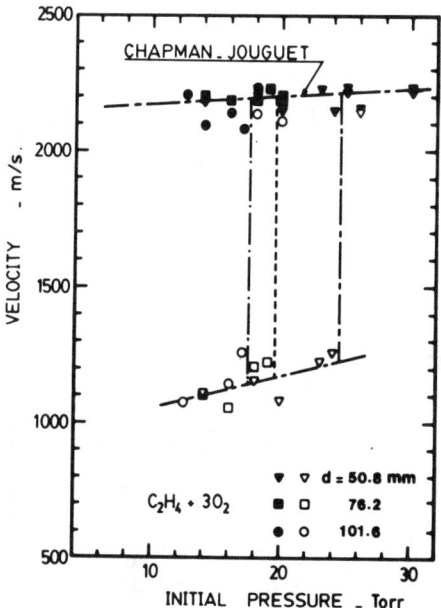

Fig. 6 Influence of initial pressure on the wave velocity upstream (black symbols) and downstream (open symbols) of the grid section, in the three tube diameters, for the $C_2H_4/3O_2$ mixture. L/d = 3; n = 3. The vertical dashed lines (from right to left) correspond to the velocity drop at the critical pressures corresponding, respectively, to d = 50.8, 76.2, and 101.6 mm.

DETONATION IN AN ABSORBING WALLED TUBE

for the velocity deficit is the destruction of the transverse waves across the detonation front itself.

This is better confirmed by the following experiment: The grid section was replaced by an approximately equivalent length of Schelkhin spiral made with a wire of 2 mm diameter and of pitch of 4 mm. Momentum and heat losses through a section with a Schelkhin spiral are known to be very large and result in substantial velocity deficits (Guénoche, 1949) of the same order of magnitude as in the present experiment (Knystautas et al., 1986). Figure 9 shows a streak record of the $C_2H_4/3O_2$ mixture at an initial pressure of 20 Torr, which is much below the critical value of 24.5 Torr, for d = 50.8 mm, L/d = 3, and n = 3. One notes that at this pressure, the detonation propagates through the Schelkhin spiral without significant attenuation and the velocity downstream is exactly the same as the upstream near C-J value. At such a pressure (20 Torr), the grid section would have made the detonation fail. Thus, the stack of grids is more efficient for detonation failure than the Schelkhin

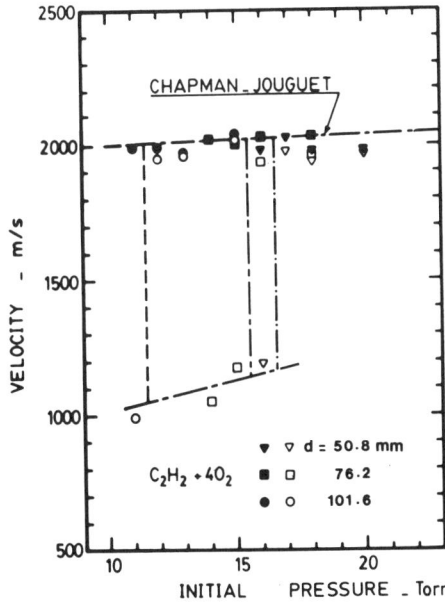

Fig. 7 Influence of initial pressure on the wave velocity upstream (black symbols) and downstream (open symbols) of the grid section, in the three tube diameters, for the $C_2H_2/4O_2$ mixture. L/d = 3; n = 3. The vertical dashed lines (from right to left) correspond to the velocity drop at the critical pressures corresponding, respectively, to d = 50.8, 76.2, and 101.6 mm.

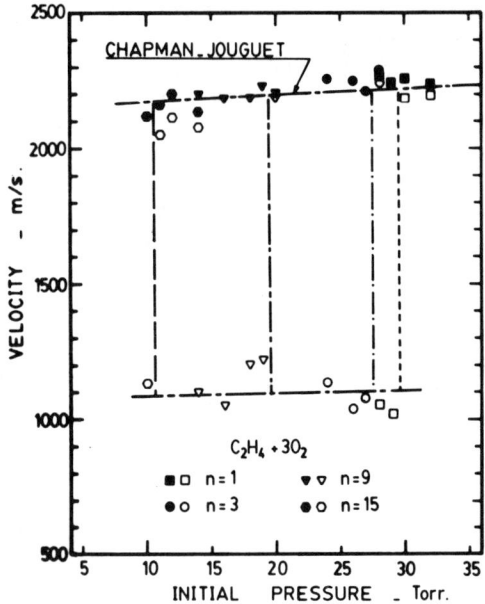

Fig. 8 Influence of initial pressure on the wave velocity upstream (black symbols) and downstream (open symbols) of the grid section for different numbers of grid layers (n = 1, 3, 9, and 15) and for the $C_2H_4/3O_2$ mixture. L/d = 3; d = 76.2 mm. The vertical dashed lines (from right to left) correspond to the velocity drop at the critical pressures corresponding, respectively, to n = 1, 3, 9, and 15.

spiral. Without totally rejecting the effect of the momentum and heat losses on the grid, the damping of transverse waves seems to be a more important mechanism.

Regarding the constancy of the deflagration velocity downstream of the grid, this is due to the fact that there exists no strong mechanism to generate the transverse waves again. Weak acoustic waves are always present in a turbulent flame zone; however, the amplification of these waves to amplitudes necessary to sustain a cellular detonation requires the traveling of many tube diameters, as shown previously by Donato (1982). Thus, within less than 1m travel downstream of the grid section in the present experiment, acceleration of the deflagration was rarely observed. Transverse waves can also regenerate via spontaneous transition, i.e., from an explosion center in the turbulent flame brush (Urtiew and Oppenheim, 1966) via the SWACER mechanism (Lee et al., 1976 ; Lee and Moen, 1980). However, the scale of turbulence in a smooth-walled tube is insufficient to cause immediate spontaneous

DETONATION IN AN ABSORBING WALLED TUBE

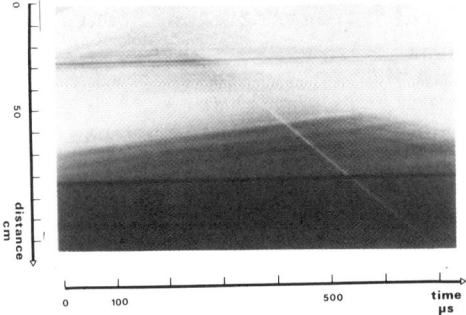

Fig. 9 Streak record obtained below the critical pressure in the case of a Schelkhin spiral replacing the grid section for the $C_2H_4/3O_2$ mixture. S = position of the Schelkhin spiral. P_0 = 20 Torr; d = 50.8 mm; L/d = 3; n = 3.

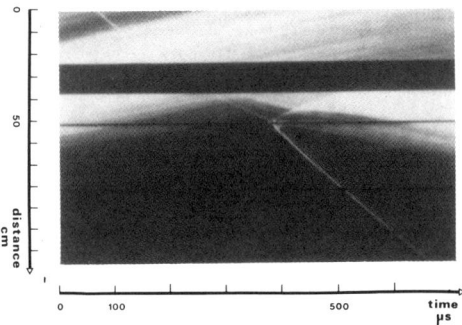

Fig. 10 Streak record obtained below the critical pressure in the case of a ring-shaped obstacle placed at the distance of 3d downstream of the grid section for the $C_2H_4/3O_2$ mixture. G and O = positions of the grid section and ring-shaped obstacle, respectively. P_0 = 20 Torr; d = 50.8 mm; L/d = 3; n = 3.

transition to detonation. An experiment was performed to cause transition downstream by placing a ring-shaped obstacle with a blocking ratio of 0.34 at a distance of three diameters after the grid. Figure 10 shows the streak record obtained with a $C_2H_4/3O_2$ mixture in the 50.8 mm tube at an initial pressure of 20 Torr (much below the critical pressure of 24.5 Torr). A spontaneous DDT is observed immediately after the obstacle. Whether the obstacle creates the necessary scale of a turbulent mixing zone or provides a strong reflected shock able to cause autoignition (or a combination of both mechanisms) is not certain. Nevertheless, the effect of an obstruction to create a large perturbation necessary for the deflagration to transit to detonation is clear.

In all of the foregoing experiments, the length of the grid section was kept constant at $L/d = 3$. This particular value was chosen since L is a little greater than the characteristic length, i.e., the distance traveled by a detonation wave during the characteristic time $t_r = d/a_{CJ}$ (a_{CJ} being the sound speed in the burnt gases). In the present experiments, t_r was mostly close to 1.85. Of course, in some cases, the transverse waves may need multiple reflections from the grid layers before being damped out. Figure 11 shows a streak record in the case of the $C_3H_8/5O_2$ mixture at 32 Torr initial pressure for $d = 76.2$ mm, $L/d = 3$, and $n = 3$. When the detonation

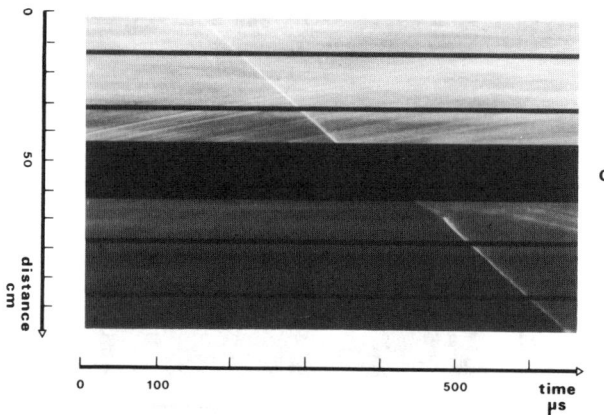

Fig. 11 Streak record obtained for the $C_3H_8/5O_2$ mixture. G = position of the grid section. $P_0 = 32$ Torr; $d = 76.2$ mm; $L/d = 3$; $n = 3$.

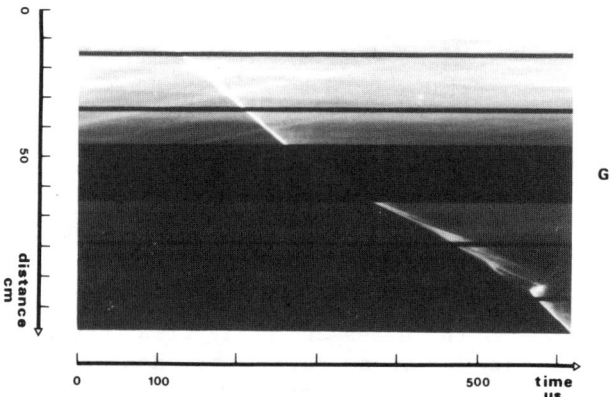

Fig. 12 Streak record obtained for the $C_3H_8/5O_2$ mixture. G = position of the grid section. $P_0 = 22$ Torr; $d = 76.2$ mm; $L/d = 3$; $n = 3$.

exits the grid, its velocity is significantly decreased. However, the presence of transverse waves results in DDT shortly downstream of the grid. As the initial pressure is lowered, the location of the transition point shifts downstream (Fig. 12) and hence, it is not possible to determine a unique value of the critical pressure for the case of $C_3H_8/5O_2$. With the present grid, even when up to 15 layers of grid are used, the length of 3d is not sufficient to provide a distinct value of the critical pressure. Thus, a longer grid section was used. The use of a very long grid section can produce complex results. The transverse waves may first be damped and the detonation decayed to a deflagration; however, long travel inside the grid section may generate sufficient turbulence in the flame zone to cause a transition back to a detonation inside the grid. If the length of the grid is increased to six diameters (with three layers of grid), similar phenomena as for the other mixtures are also observed: The critical pressure can now be determined within a few percent of the initial pressure. Nevertheless, for the case of $C_3H_8/5O_2$, a more effective grid should be used, with a length-to-diameter ratio equal to 3.

Left to themselves, the transverse waves reflect at the tube wall back toward the center of the tube. We have tried to estimate the degree of attenuation produced by the grid section by comparing the reflected shock off the grids with the normal reflection from the rigid, non-grid-covered tube wall. A series of runs were carried out using a detonation wave propagating in a tube, the end of which was covered by 15, 3, or 0 grid layers. The four different mixtures at three initial pressures were tested. The results indicate, as expected, that weaker reflections are obtained as the number of grid layers increases. If P_I and P_R, are, respectively, the incident and reflected pressures measured at the side wall slightly downstream off the grids, and P_R the reflected pressure measured at the tube end, it can be seen that in every case, P_R/P_I is much greater than $P_{R'}/P_I$ for n = 0. When the number of layers increases, the ratio P_R/P_I is decreasing quite exponentially. As an example, it varies from 7 without any grid to 0.2 when n = 15 in the case of the $H_2/0.5O_2$ mixture at 50 Torr initial pressure. Thus, the intensity of the transmitted shock through the grid layers decreases substantially with an increasing thickness of absorbing material, as previously noticed by Cloutier et al. (1971). As n increases, the ratio $P_{R'}/P_I$ has a tendency to decrease as well, indicating that the grids actually attenuate the shock and they are all the more efficient as the absorbing material is thick. However, the pressure records are not easy to interpret, perhaps because of multireflections coming off the different layers, and eventually because of the reflection from the backup wall. These last experiments have to be repeated with a non reactive shock wave instead of a detonation wave.

Conclusion

The present study conclusively demonstrates the essential role of transverse waves in the propagation of detonation. Transverse waves may, in fact, be the mechanism that renders the intense combustion of the detonation mode possible. These results offer an alternate mechanism (or perhaps the correct one) to account for the influence of confinement on the limits of detonation. The recent work of Murray (1985) on the transmission of a detonation from a rigid tube to a thin plastic-walled tube has shown very similar results. Depending on the wall thickness (i.e., the degree of confinement), it was found that detonation fails when d/λ is of a range similar to those of the present experiments. In Murray's study, the failure mechanism was credited to the curvature of the detonation front due to the divergence of the flow behind the detonation as the plastic tube yielded. In the present experiments, mass flow through the porous grid should also produce a large negative boundary-layer displacement thickness that would yield to a curved front. Observation of the detonation front in Murray's experiments generally showed a planar structure. Moreover, studies of normal shock reflection from a thin movable plastic diaphragm, carried out by Meyer (1957), show that the strength of the reflected waves depends on the mass of the plastic diaphragm. Effects such as the increase of heat and momentum losses to the porous wall seem to be small. Thus, rather than any other mechanism that can be involved to explain Murray's results, the failure of the detonation wave is more plausibly due to the damping of the transverse waves as they reflect off the thin plastic wall.

Clearly now, the present results show that the damping of transverse waves appears to be a consistent mechanism to account for detonability limits. Furthermore, they confirm the necessity of transverse waves (and hence of a cellular structure) in any detonative combustion.

Acknowledgments

This work was supported under the Natural Sciences and Engineering Research Council of Canada under NSERC grants A-7091 and A-3347.

References

Cloutier, M., Devereux, F., Doyon, P., Fifchett, A., Heckman, D., Moir, L., and Tardif, L. (1971) Reflections of weak shock waves from acoustic materials. J. Acoustic Soc. America, 50, 1393 - 1396.

Donato, M. (1982) The influence of confinement on the propagation of near limit detonation waves. Ph.D. Thesis, McGill Univ., Montreal, Canada.

Dupré, G., Knystautas, R., and Lee, J.H. (1986) Near-limit propagation of detonation in tubes. Progress in Astronautics and Aeronautics: Dynamics of Shock Waves, Explosions, and Detonations, Vol. 106. Edited by J.R. Bowen, J.-C. Leyer, and R.I. Soloukhin, AIAA, New York, 244 - 259.

Evans, M.W., Given, F.I., and Richeson, W.E. (1955) Effects of attenuating materials on detonation induction distances in gases. J. Appl. Phys., 26, 1111 - 1113.

Guénoche, H. (1949) Recherche sur la détonation et la déflagration dans les mélanges gazeux. Revue de l'Institut Français du Pétrole, 4, 48 - 69.

Knystautas, R., Lee, J.H., and Guirao, C.M. (1982) The critical tube diameter for detonation failure in hydrocarbon-air mixtures. Combust. Flame, 48, 63 - 83.

Knystautas, R., Lee, J.H., Peraldi, O., and Chan, C.K. (1986) Transmission of a flame from a rough to a smooth-walled tube. Progress in Astronautics and Aeronautics: Dynamics of Shock Waves, Explosions, and Detonations, Vol. 106. Edited by J.R. Bowen, J.-C. Leyer, and R.I. Soloukhin, AIAA, New York, 37 - 52.

Lee, J.H., Knystautas, R., and Yoshikawa, N. (1976) Photochemical initiation of gaseous detonation. Acta Astronautica, 5, 971 - 982.

Lee, J.H.S., and Moen, I.O. (1980) The mechanism of transition from deflagration to detonation in vapour cloud explosions. Prog. Energy Combust. Sci., 6, 359 - 389.

Meyer, R.F. (1957) The impact of a shock wave on a movable wall. J. Fluid Mech., 3, 309 - 323.

Murray, S.B. (1985) The influence of initial and boundary conditions on gaseous detonations. Ph.D. Thesis, McGill Univ., Montreal, Canada.

Urtiew, P.A., and Oppenheim, A.K. (1966) Experimental observations of the transition to detonation in an explosive gas. Proc. Roy. Soc. London, Series A: Math. and Phys. Sci., 295, 13 - 28.

Lateral Interaction of Detonating and Detonable Gaseous Mixtures

J. C. Liu,* C. W. Kauffman,† and M. Sichel‡
University of Michigan, Ann Arbor, Michigan

Abstract

Two adjacent 1.6 cm square detonation tubes, separated at the test section by a very thin cellulose film, were used to obtain Schlieren framing pictures of the interaction that occurs when a normal detonation in the primary explosive comes into contact with bounding explosive layers of various compositions. In earlier studies, it was observed that in some cases a microexplosion led to the rapid direct initiation of an oblique detonation; otherwise, an inert oblique shock that is not strong enough to initiate a detonation is induced in the bounding explosive. The formation of the initiating microexplosion was found to be very sensitive to the separating film, which was never more than 50 nm thick. The results of a detailed study of the effect of film thickness on the layered detonation interaction are reported in the present paper. Experiments were conducted with stoichiometric H_2/O_2 as the primary explosive and with various H_2/O_2 mixtures as the secondary explosive; some tests were conducted using methyl ether (C_2H_6O) as the primary explosive. Framing pictures and pressure traces for the interactions with different film thicknesses are presented and a map showing how the interaction mode varies with film thickness and the equivalence ratio of the bounding explosive has been constructed. As the film thickness is decreased from 50 to 15 nm, the range of secondary explosive equivalence ratios for which direct initiation occurs is broadened. These results suggest that direct initiation may involve the transmission of transverse waves across the boundary between the primary and secondary explosive.

Copyright © 1988 by the American Institute of Aeronautics and Astronautics, Inc. All rights reserved.

* Visiting Scholar, Department of Aerospace Engineering (presently with Xian Modern Chemistry Institute, Xian, Chi).
† Associate Professor, Department of Aerospace Engineering.
‡ Professor, Department of Aerospace Engineering.

Introduction

The diffraction and transmission of a detonation into a bounding explosive layer have been previously studied in this laboratory experimentally and analytically [see Sichel et al. 1984; Liu et al. 1985, 1986]. Four main modes of interaction have been observed as follows:

Mode 1 – an oblique shock with regular reflection from the lower test section wall of the detonation tube is induced in the bounding explosive mixture.
Mode 2 – a secondary detonation is directly initiated in the bounding mixture.
Mode 3 – a detonation in the bounding mixture is initiated by reflection of the induced incident oblique shock from the lower wall.
Mode 4 – Mach reflection occurs at the lower wall and, while a chemical reaction may be induced behind the Mach system, detonation is not observed.

Under what conditions and how the initiation of the secondary detonation can be induced is of special interest. Previous study suggested that the presence of the thin film separating the two explosive layers and the energy released from the primary detonation may play key roles. In fact, the thin film confines the primary explosive layer. The present paper describes the results of a systematic experimental study of the effects of different film thicknesses and different primary explosive mixtures.

The experimental arrangement is the same as presented in Liu et al. (1985, 1986). The essential features are briefly described as follows for convenience:
1) The special detonation tube used is shown in Fig. 1. The tube consists of two adjacent 1.6 × 1.6 cm square channels. In the test section, the primary explosive (in the top channel) and the bounding explosive (in the bottom channel) are separated by a 1.6 by 15 cm nitrocellulose membrane (thin film) made using a technique developed by Dabora et al. (1965).

2) The primary explosive is ignited at the end of the detonation tube using a glow plug. The wave propagation distance is more than 3 m

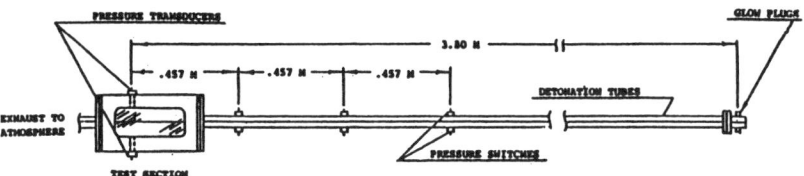

Fig. 1 Schematic diagram of the two-channel detonation tube.

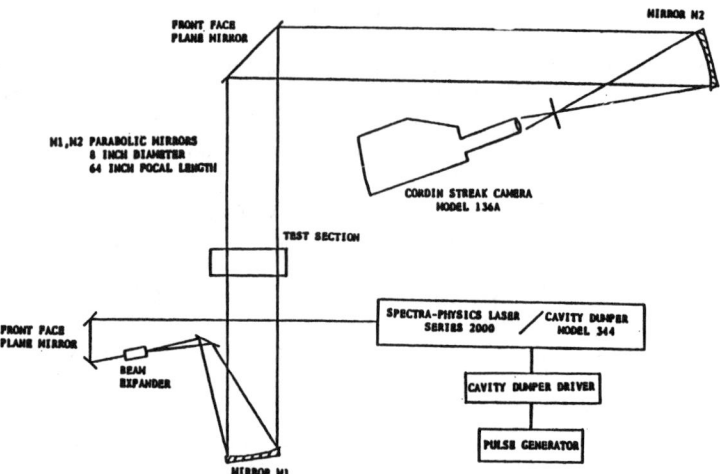

Fig. 2 Schematic diagram of the pulsed laser/highspeed camera system.

which, for the primary explosive mixture used, is long enough to provide a stable detonation, and the primary detonation velocity is monitored using pressure switches to insure that it is fully developed.

3) Two pressure transducers mounted on the top and bottom walls of the test section, respectively, are used to sense the pressure histories in both channels.

4) A pulsed argon-ion laser/high-speed camera system is used to photograph the interaction phenomena using the technique developed by Lu et al. (1982) and is shown schematically in Fig. 2. The laser pulse width, which plays the role of a shutter, is about 12 ns so that very sharp and distinct framing pictures can be obtained. The interframe delay, which depends on the repetition rate of the laser pulses, is chosen to be 2 μs for most of the runs and this gives a framing rate of 0.5×10^6/s.

5) A Spectra-Physics 2000 laser together with a model 344 Cavity Dumper is used. The high speed camera used is a Cordin streak camera model 136A using Kodak PAN film with an ASA rating of 1000.

6) The time interval, during which the dynamic event passes through the glass window of the test section, is about 100 μs, thus allowing 50 frames to be taken for a single run.

Influence of Membrane Thickness on Flow Patterns in the Bounding Explosive

The bounding explosive may be the same as the primary explosive, but more often is different from the primary one and this is why a separating membrane is necessary. The membrane should have sufficient strength and thickness to separate two mixtures and to prevent diffusion between them. On the other hand, the membrane should be thin enough to avoid interference with the interaction between the two explosive layers. Lu (1968) estimated, based on the acceleration of the membrane material due to sudden pressure increase, that the effect of the membrane in confining the detonation would be negligible when its thickness υ_f is

$$\upsilon_f < 0.1 \, (\rho_e/\rho_f) b$$

where ρ_e is the density of the primary explosive mixture, ρ_f the density of the membrane material, and b the reaction zone length of the primary detonation.

For a membrane material such as nitrocellulose, having a density of 1.58 g/cm^3 and for an explosive (H$_2$/O$_2$ mixture with an equivalence ratio $\phi = 1.0$), with a density of 5.362×10^{-4} g/cm^3 at normal temperature and pressure, and a reaction length of about 0.28 cm (see Dabora et al. 1965), it follows that

$$\upsilon_f = 9.48 \times 10^{-6} \text{ cm} = 950 \text{ Å} = 95 \text{ nm}$$

According to this calculation, when the primary detonation is exposed to a bounding mixture separated from the primary mixture by a membrane with a thickness equal to or less than the above critical value, the detonation wave will expand laterally as if no separating materials were present. For all runs reported in Liu et al. (1985, 1986), a 50 nm thick nitrocellulose film was chosen to satisfy the above requirements.

The validity of Dabora's estimation of the critical film thickness has been verified by Dabora et al. (1965), Lu (1968), and Liu et al. (1986) from the standpoint of the effect of the film on the detonation velocity decrement of a primary gaseous detonation in the presence of a compressible boundary. However, it has been noted by Liu et al. (1986) that, in considering the flow patterns induced in the bounding explosive, it is necessary to reconsider the critical film thickness. For example, some preliminary experiments were made with stoichiometric H$_2$/O$_2$ as both the primary and secondary mixture and both with and without the 50 nm separating film. Without the film, the direct initiation of detonation in the secondary explosive was observed for all runs; however, with the film, direct initiation was never observed. Detonation in the secondary explosive was, in this case, initiated by Mach reflection from the wall of the lower shock tube.

In order to study the influence of membrane thickness on the initiation modes in the bounding mixture, a series of experiments was

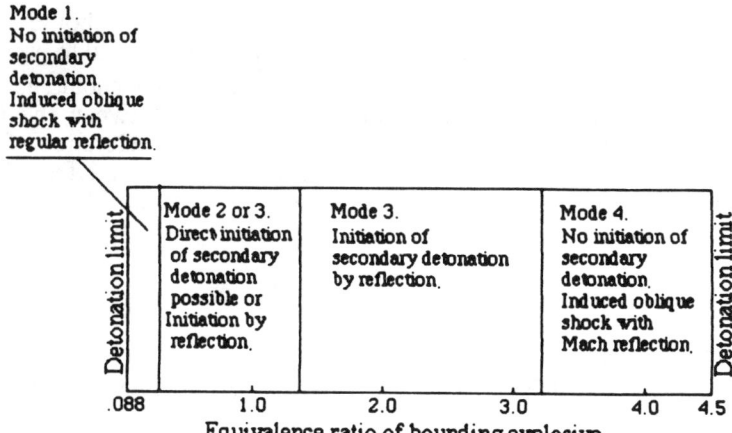

Fig. 3 Possible flow patterns in H_2/O_2 bounding explosives induced by the stoichiometric H_2/O_2 primary detonation.

therefore made using various thicknesses of the separating nitrocellulose membrane ranging from 15-70 nm, and using stoichiometric H_2/O_2 as the primary explosive with H_2/O_2 mixtures with a wide range of equivalence ratio ϕ as the bounding explosive. The 15 nm thick film was the thinnest one that could be made and, since the success of making the thinnest film was low, only two runs were usually made for each case using this minimum film thickness.

The nature of the flow patterns induced in the bounding mixture, based on all experimental results obtained so far, are summarized in Figs. 3 and 4 for stoichiometric H_2/O_2 as the primary explosive and variable equivalence ratio mixtures of H_2/O_2 as the bounding or secondary explosive. Figure 3 shows the general features of flow patterns induced in the bounding explosive. It can be seen that within the detonability limits of H_2/O_2, that is in the equivalence ratio range ϕ of 0.088-4.5 (Lewis and Elbe 1961), direct initiation of secondary detonation can be induced only in a range of ϕ of about 0.35-1.25, much narrower than the detonation limits.

Figure 4 presents a map showing how the direct initiation of detonation in the bounding mixture is influenced by the separating film thickness and the equivalence ratio of the bounding mixture. From this figure, it can be seen that direct initiation is very sensitive to the film thickness. In fact, within the "direct initiation possible region" (see Fig. 3, ϕ = 0.35 - 1.25), for all runs direct initiation can occur only when the thinnest film (15 nm) is used. When the film thickness is increased, direct initiation becomes uncertain. The frequency of direct initiation appears to be inversely proportional to the film thickness. If the film is thick enough, direct initiation is no longer observed.

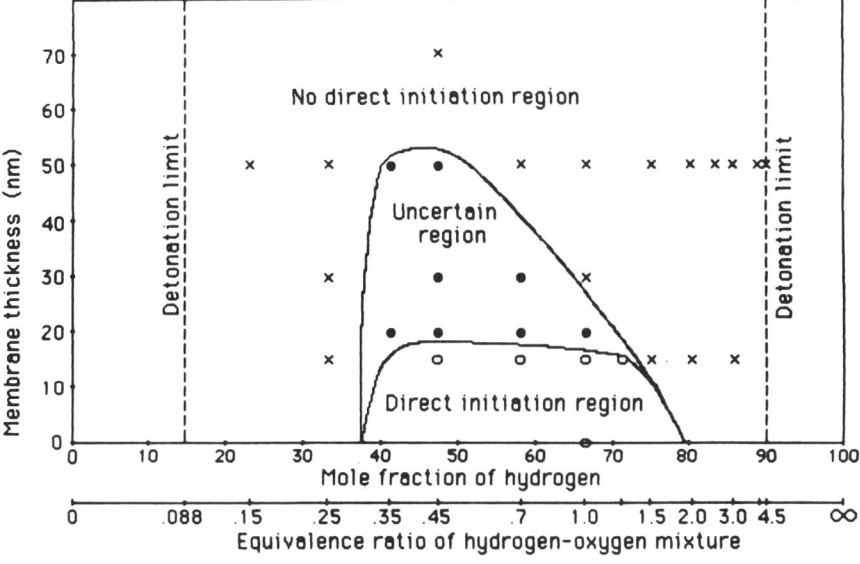

Fig. 4 Direct initiation of secondary detonation in H_2/O_2 bounding explosives as a function of separating membrane thickness (primary explosive H_2/O_2, $\phi = 1.0$).

Figures 5 and 6 show framing photographs and pressure traces illustrating the effect of film thicknesses. In these two runs, H_2/O_2 with $\phi = 1.0$ was used for both the primary and secondary mixtures. Figure 5 shows that when film is 15 nm thick direct initiation occurs. Figure 6 shows, on the other hand, that when the film is 30 nm thick direct initiation does not occur and the secondary detonation is initiated behind the Mach reflection from the lower wall.

For these two framing photographs and for all the photographs presented in this paper, the framing rate is 0.5×10^6/s, so that the time interval between adjacent frames is $2\mu s$. The images of four reference wires appear in every framing photograph. From right to left, the first wire marks the start of the separating membrane; the third indicates the location of the two pressure transducers mounted at the top and the bottom walls of the shock tube. The distances between the first and second and between the first and fourth wires are 5.08 and 10.16 cm, respectively.

The "uncertain region" in Fig. 4 refers to a combination of film thickness and equivalence ratio for which direct initiation occurs in some

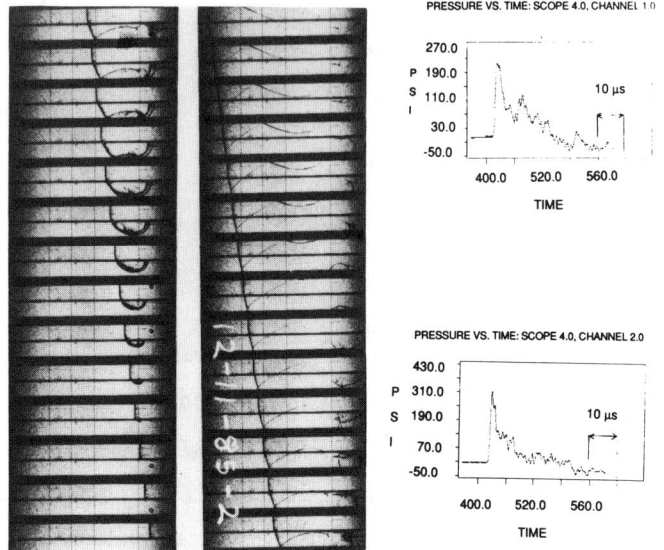

Fig. 5   Direct initiation of secondary detonation (primary explosive H_2/O_2, ϕ = 1.0; bounding explosive H_2/O_2, ϕ = 1.0; membrane thickness = 15 nm). There are four reference wires in this and all the framing photographs presented in this paper. The first wire shows the location where the membrane begins; the third one shows the location at which the two pressure transducers are mounted; the distances between the first and second wires and between the first and fourth are 5.08 and 10.16 cm, respectively.

runs, while in others, under the same conditions, initiation only occurs after Mach reflection. There are also cases in this region in which direct initiation occurs at first, but this detonation is later quenched. Figures 7-9 are typical framing photographs of interactions corresponding to the "uncertain region" in Fig. 4. The primary mixture in these figures is stoichiometric H_2/O_2, while the bounding mixture is H_2/O_2 with ϕ = 0.45, and the film thickness is 30 nm.

Figure 7 shows direct initiation, but under what might be called critical conditions in that there is a fairly long oblique shock connecting the primary planar detonation wave and the secondary oblique detonation. Figure 8 shows a case in which direct initiation does not occur, but an initiating Mach reflection can be seen in the last few frames. Figure 9 represents a marginal case between that shown in Figs. 7 and 8. An oblique detonation is induced by a localized explosion at the beginning, but the oblique detonation wave segment becomes shorter as the wave propagates and finally is replaced by a Mach stem that then develops into a detonation. For purposes of comparison, framing pictures of the direct initiation of a detonation in an H_2/O_2, ϕ = 0.45

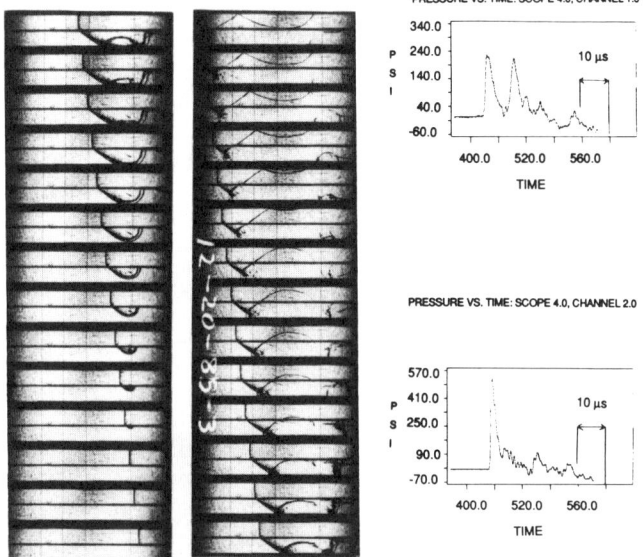

Fig. 6　No direct initiation: the secondary detonation is induced by reflection (primary explosive H_2/O_2, $\phi = 1.0$; bounding explosive H_2/O_2, $\phi = 1.0$; membrane thickness = 30 nm).

bounding mixture with a thinner 15 nm thick membrane as shown in Fig. 10.

It was already observed by Liu et al. (1985) that direct initiation starts from a localized explosion behind the shock propagating into the bounding mixture; a similar initiation mechanism was also observed by Bartlma and Schroder (1985) in the case of the diffraction of a detonation. According to Lee (1977), localized explosions are initiated when a hot spot resulting from gasdynamic fluctuations is induced in the shocked mixture, which is the bounding explosive in the present study. Evidently, even a relatively thin film can block the transmission of gasdynamic fluctuations due to turbulence or transverse waves from the primary to the secondary explosive; this may explain why direct initiation is so sensitive to the film thickness. From Fig. 4, it can also be seen that lean H_2/O_2 mixtures with an equivalence ratio $\phi = 0.45$ are most sensitive to initiation by a glancing wave, whereas, typically, stoichiometric or slightly rich H_2/O_2 mixtures are found to be most sensitive to direct initiation.

Table 1 presents some selected velocity and pressure data taken from framing photographs and pressure traces for three different bounding mixtures (H_2/O_2 with $\phi = 0.25$, 0.45, and 1.0) and with different membrane thicknesses. From these data, it can be seen that,

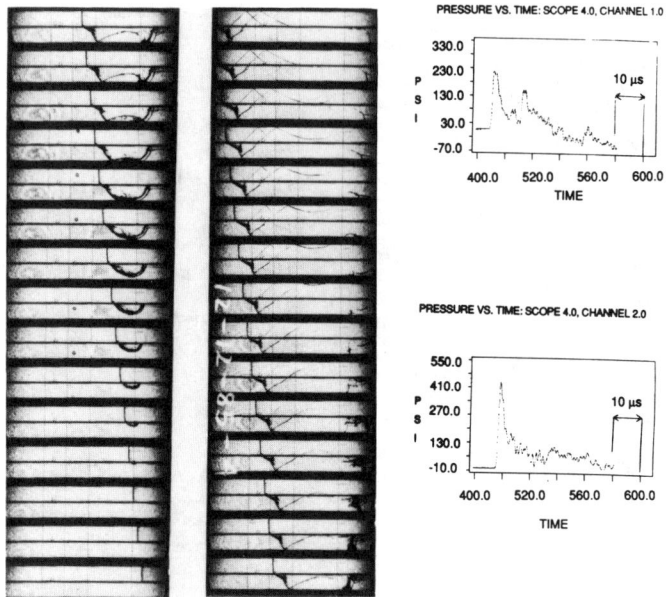

Fig. 7 Direct initiation of secondary detonation (primary explosive H_2/O_2, $\phi = 1.0$; bounding explosive H_2/O_2, $\phi = 0.45$; membrane thickness = 30 nm).

before meeting the separating film, the primary detonation propagates with the C-J velocity, but upon exposure to the bounding mixture the velocity decreases to a new steady-state velocity almost instantly. This behavior is more clearly shown by X-t graphs of which Fig. 11 is an example.

The data in Table 1 show that in the case of $\phi = 0.25$, the velocity deficit of the primary detonation increases slightly with decreasing membrane thickness. In the cases of $\phi = 0.45$ and 1.0, however, the velocity deficit is not affected by the film thickness. This difference may be related to the fact that the very lean bounding mixture with $\phi = 0.25$ is essentially unreactive, so that only an oblique shock with a regular reflection is observed.

It should be noted that in the case of H_2/O_2 with $\phi = 0.45$ as the bounding explosive layer, $\upsilon f=0$ means there was no separating film, so there was diffusion between two different mixtures. This may explain why the measured primary detonation velocity shown in Table 1 is, in this case, lower than that of the cases with a membrane present. But, the important result is that in the range 0-70 nm, the velocity deficit is independent of the film thickness, even though the film thickness strongly influences the interaction in the bounding explosive.

Table 1 Some selected velocity and pressure data from experiments with different membrane thickness

Primary explosive -- H_2/O_2, $\phi = 1.0$.
Bounding explosive -- H_2/O_2, $\phi = 0.25, 0.45$, and 1.0
For this mixture, $V_{CJ} = 2841.5$ m/s, $P_{CJ} = 270$ psi (Gordon-McBride, 1971)

Bounding explosive	ℓ_f, nm	0	15	20	30	50	70
$\phi=0.25$	V_1, m/s		2813±24	2834±7	2846	2825±12	
	V_2, m/s		2676±28	2732±22	2790	2737±64	
	$(V_1-V_2)/V_1$		4.9%	3.7%	2.6%	3.1%	
	$V_{o.s.}$, m/s		1462±8	1344±16	1286	1379±66	
	P_{top}, psi		193±1	199±4	213	212±3	
$\phi=0.45$	V_1 (m/s)	2769±2*	2084±17	2823±9	2830±22	2830±7	2870±7
	V_2 (m/s)	2680±9*	2712±9	2753±3	2769±16	2744±50	2796±8
	$(V_1-V_2)/V_1$%	3.2%*	3.3%	2.5%	2.2%	3.0%	2.6%
	$V_{o.s.}$(m/s)	---	---	1470±43	1392±12	1408±2	1419±17
	$V_{o.d.}$(m/s)	2318±3*	2239±13	2370	2270±135	2332±36	---
	$V_{m.s.}$(m/s)	---	---	2899	2783±86	2624±6	2850±16
	P_{top}(psi)	206±4*	199±3	214±5	220±4	222±8	217±3
$\phi=1.0$	V_1(m/s)		2874±4	2854±42	2852±8	2862±30	2875±31
	V_2(m/s)		2804±1	2790±15	2802±12	2784±14	2820±5
	$(V_1-V_2)/V_1$		2.5%	2.2%	1.8%	2.7%	1.9%
	$V_{o.s.}$(m/s)		---	---		1674±14	1655±33
	$V_{o.d.}$(m/s)		2718±7	2713±1	2712±12	---	---
	$V_{m.s.}$(m/s)		---	---		3262±66	3320±12
	P_{top}(psi)		229	217±2	221±4	223±3	231±2

Note: ℓ_f = membrane thickness, V_1 = Primary detonation velocity before exposure to bounding mixture, V_2 = Primary detonation velocity during exposure to bounding mixture, $V_{o.s.}$ = Velocity component normal to oblique shock wave front in bounding mixture, $V_{o.d.}$ = Velocity component normal to oblique detonation wave front in bounding mixture, $V_{m.s.}$ = Mach stem propagating velocity in bounding mixture, P_{top} = The measured peak pressure of primary detonation wave.

* No membrane to separate two different mixtures, diffusion between them existed.

The pressures behind the primary detonation as given in Table 1 are much lower than the computed C-J detonation pressures. This discrepancy arises because of the large sensing area of the pressure transducers used (5.5 mm diameter). However, the stoichiometric H_2/O_2 detonation wave has a 2.8 mm reaction length and a 2.8 mm/ms propagation velocity, making it impossible to sense the C-J pressure using the present transducer-oscilloscope system. However, the very consistent pressure data given in Table 1 also do support the conclusion

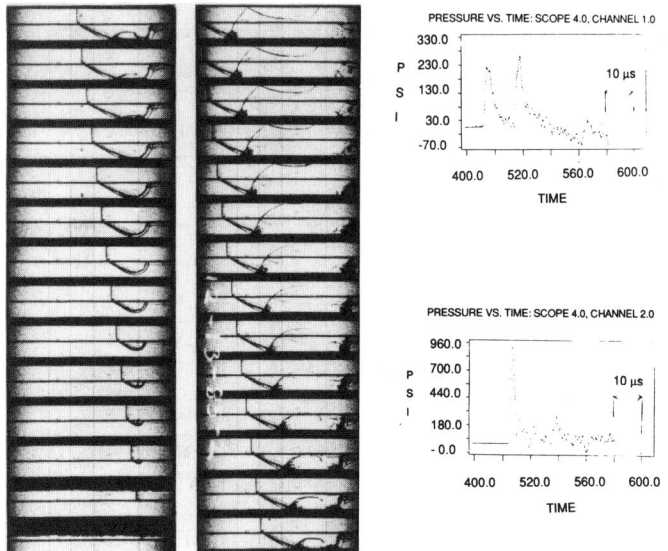

Fig. 8 No direct initiation: Mach reflection occurs and may lead to development of a secondary detonation (primary explosive H_2/O_2, ϕ = 1.0; bounding explosive H_2/O_2, ϕ = 0.45; membrane thickness = 30 nm).

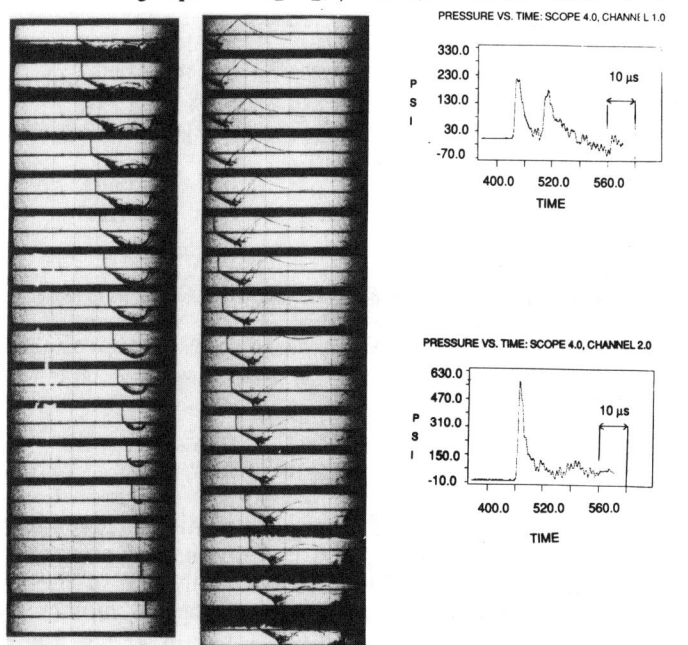

Fig. 9 A direct initiation occurs at the early period, but vanishes later – replaced by Mach reflection (primary explosive H_2/O_2, ϕ = 1.0; bounding explosive H_2/O_2, ϕ = 0.45; membrane thickness = 30 nm).

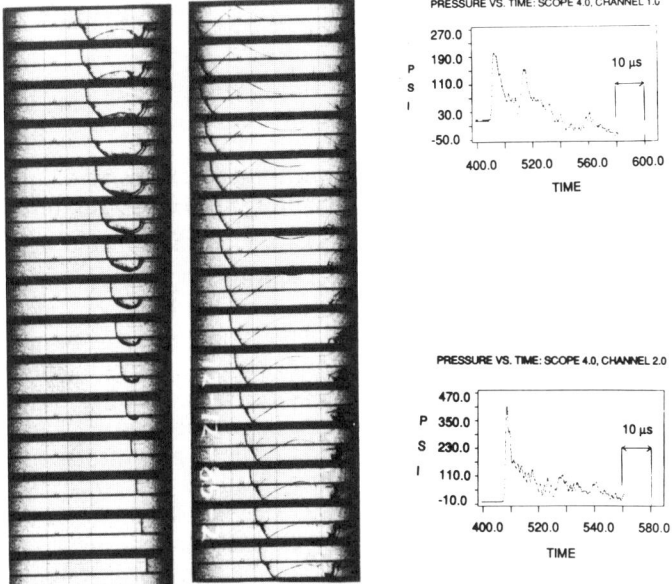

Fig. 10 Direct initiation of secondary detonation (primary explosive H_2/O_2, $\phi = 1.0$; bounding explosive H_2/O_2, $\phi = 0.45$; membrane thickness = 15 nm). Comparison with Fig. 7 shows the difference between them.

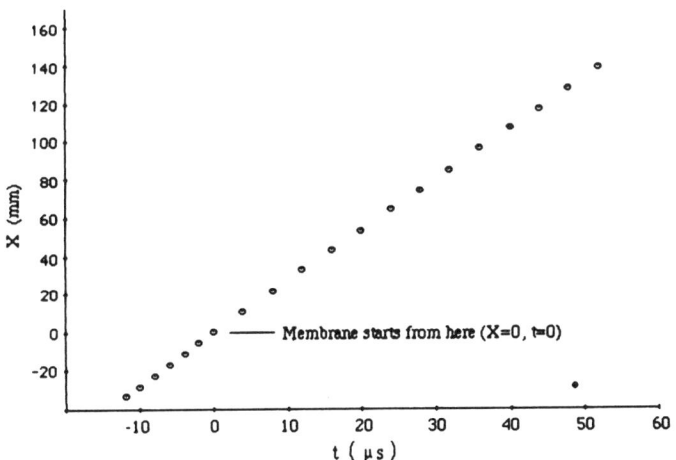

Fig. 11 X - t graph of primary detonation propagation (primary explosive H_2-O_2, $\phi = 1.0$; bounding explosive H_2/O_2, $\phi = 0.25$). The primary wave is exposed to the bounding mixture at X=0, t=0.

that when the film is thin enough, its influence on the primary detonation is negligible.

Influence of Primary Detonation Energy on Initiation Modes

In order to assess the effect of the primary explosive on initiation preliminary experiments were conducted using H_2/O_2 mixtures with equivalence ratios of $\phi = 0.45, 0.75, 0.85, 1.1, 1.5, 2.0$, and 2.5 as the primary explosive, while using the most sensitive H_2/O_2 mixtures with $\phi = 0.35$ and 0.45 as the bounding explosive. No basic changes were noted in the flow patterns induced in the bounding mixtures compared to those observed with stoichiometric H_2/O_2 as the primary mixture. This result suggested that a more energetic explosive was needed to study the effect of primary detonation energy on the interaction and an ether-oxygen mixture was therefore chosen. Table 2 provides detonation parameters of methyl ether $(C_2H_6O)/O_2$ mixtures computed using the Gordon-McBride computer code (1971). The corresponding parameters are also listed for stoichiometric H_2/O_2 for comparison. Because of the limited strength of the glass windows of the detonation tube, only a C_2H_6O/O_2 mixture with $\phi = 0.5$ was used as the primary explosive in the study described below.

A series of experiments was conducted using H_2/O_2 mixtures as the bounding explosive with equivalence ratios of 0.1-4.5. The membrane thickness was about 50 nm for all runs. Figure 12a shows the general features of induced flow patterns in the bounding mixtures. The interaction mode 1 was only observed when the H_2/O_2 mixture was very

Table 2 C-J detonation parameters of C_2H_6O-O_2 mixture with variable equivalence ratios computed using Gordon-McBride computer code

ϕ	Mol.wt. g	a_1 m/s	V_{CJ} m/s	M_{CJ}	γ_1	γ_{CJ}	P_{CJ}/P_1	T_{CJ}/T_1	ρ_{CJ}/ρ_1
				C_2H_6O/O_2					
0.25	32.969	317.25	1749.8	5.516	1.3390	1.1578	19.366	9.733	1.819
0.50	33.875	308.54	2016.7	6.536	1.3014	1.1308	26.552	11.848	1.851
0.75	34.722	301.57	2189.0	7.259	1.2743	1.1294	31.988	12.187	1.8577
1.0	35.516	295.78	2320.8	7.846	1.2539	1.1313	36.676	12.519	1.8598
1.75	36.263	282.38	2509.8	8.888	1.2107	1.1612	44.703	11.690	1.8419
				H_2/O_2					
1.0	12.010	537.8	2841.5	5.2833	1.4015	1.1291	18.844	12.352	1.8386

Note: a = sound speed, V = detonation velocity, M = Mach number, γ = ratio of specific heat, P = pressure, T = temperature, ρ = density. Subscripts: 1 = condition ahead of wave, CJ = condition at CJ plane.

DETONATING AND DETONABLE GASEOUS MIXTURES

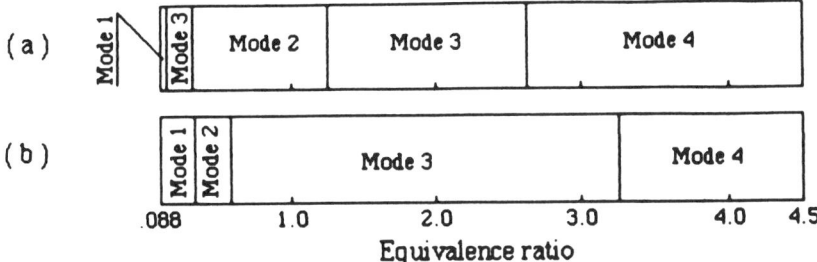

Fig. 12 Flow patterns in H_2/O_2 bounding explosives with 50 nm thick separating membrane.
a) Induced by the primary detonation of C_2H_6O/O_2, $\phi = 0.5$.
b) Induced by the primary detonation of H_2/O_2, $\phi = 1.0$.

lean, $\phi = 0.1$. Interaction mode 2 (direct initiation) was observed for all runs when the bounding mixture equivalence ratios were in the range of 0.35-1.0. Interaction mode 3 (initiation by reflection) was observed when ϕ ranged from 0.15-0.25 and 1.5-2.5. Interaction mode 4 (Mach stem without detonation) was observed when the bounding mixture was very rich, with equivalence ratios ϕ of 2.75-4.5. For comparison, Fig.12b repeats the interaction modes observed with H_2/O_2, $\phi = 1.0$ as the primary explosive layer with a 50 nm thick film. In comparison to Fig. 3, Fig. 12b shows only that range of mode 2 equivalence ratios for which direct detonation was actually observed.

From Figs. 12a and 12b, the difference caused by the change in the primary explosive can be seen clearly. When C_2H_6O/O_2, $\phi = 0.5$ is used direct initiation of the H_2/O_2 mixture can be induced for all runs over a wide range of equivalence ratio. Even though the bounding mixture is quite lean, $\phi = 0.15$ and 0.25, a secondary detonation can still be initiated by reflection in contrast to Fig. 12b. When H_2/O_2, $\phi = 1.0$ is used, direct initiation occurs only in a very narrow range of bounding mixtures, $\phi = 0.35$-0.45, and then, as also indicated by Fig. 4, only for some runs. The C-J pressure of C_2H_6O/O_2, $\phi = 0.5$ mixture is much higher than that of H_2/O_2, $\phi = 1.0$ mixture. Cook (1974) suggested that the pressure represents an energy density parameter – the amount of energy involved in unit volume of the explosive. The energy density or pressure of the primary detonation obviously plays a key role in its ability to induce a secondary detonation.

The dividing boundary between modes 3 and 4, on the other hand, moves to the left when the primary mixture is changed from H_2/O_2, $\phi = 1.0$ to C_2H_6O/O_2, $\phi = 0.5$, i.e., there is a very wide region with rich equivalence ratios where no secondary detonation can be induced under any circumstances. This effect can be explained by comparing the C-J velocities of the primary and secondary mixtures. For

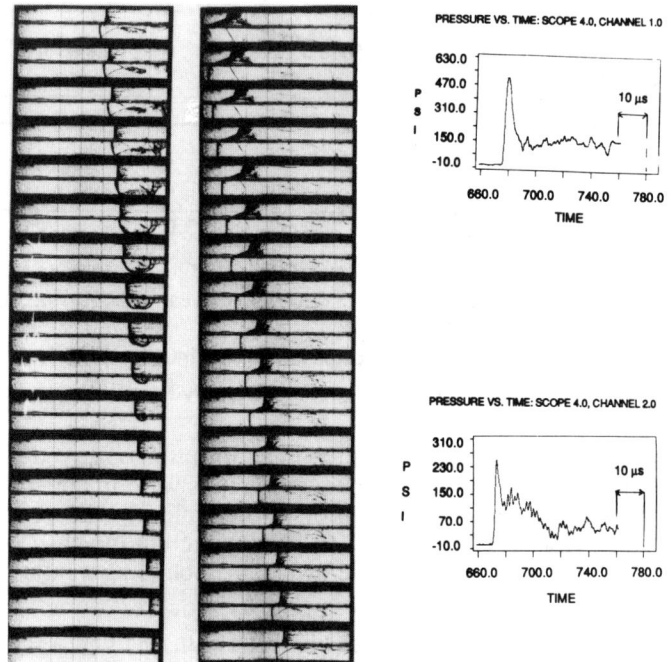

Fig. 13 Direct initiation of secondary detonation (primary explosive C_2H_6O/O_2, $\phi = 0.5$; bounding explosive H_2/O_2, $\phi = 1.0$; membrane thickness = 50 nm).

rich H_2/O_2 secondary mixtures (for example, $\phi > 2$), the C-J detonation velocity is much higher than that of the primary explosive (C_2H_6O/O_2, $\phi = 0.5$). Any detonation initially induced in the bounding mixture would thus be highly underdriven.

Figure 13 shows framing photographs of the direct initiation of a secondary detonation with C_2H_6O/O_2 as the primary explosive. The bounding mixture is stoichiometric H_2/O_2. The localized explosion that leads to the direct initiation can be seen clearly. The development from the appearance of the localized explosion to the formation of the secondary detonation is very rapid. Since the secondary detonation velocity is much higher than that of the primary one, the former overtakes the latter in a short time. After the secondary detonation overtakes the primary, the interaction becomes inverted so that the secondary detonation induces an oblique shock in the primary mixture which utimately engulfs the primary wave.

Figure 14 shows the situation when initiation of the secondary detonation occurs by shock reflection. The bounding explosive is H_2/O_2, $\phi = 2.0$. It can be seen clearly that initiation starts in the region behind the

DETONATING AND DETONABLE GASEOUS MIXTURES 279

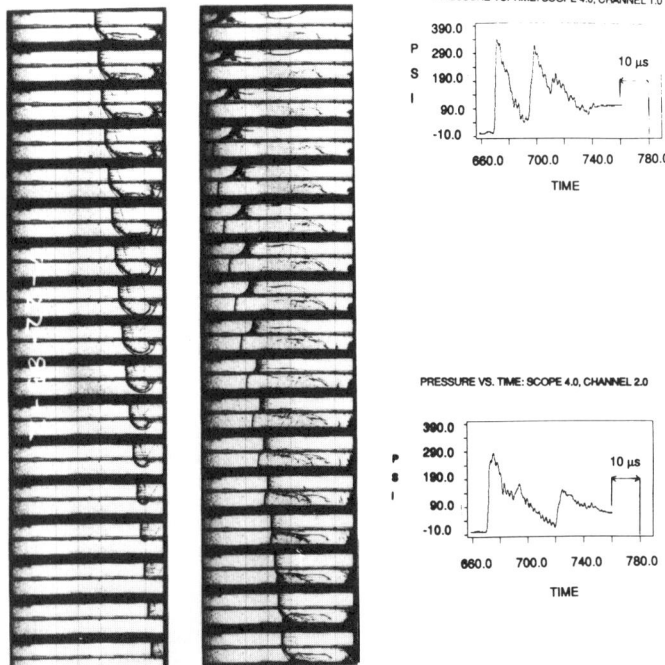

Fig. 14   Initiation of secondary detonation by reflection (primary explosive C_2H_6O/O_2, $\phi = 0.5$; bounding explosive H_2/O_2, $\phi = 2.0$; membrane thickness = 50 nm).

reflected wave and a Mach stem then forms instantly and rapidly engulfs the induced oblique shock. After the secondary wave overtakes the primary detonation, the flow pattern is identical to that shown in Fig. 13.

As shown in Fig. 15 for very rich bounding mixtures, a Mach reflection is induced in the bounding explosive and develops into a planar shock, but no detonation appears. The shock front is quite "thin" and the area behind the wave is quite "clear" that is, there are no traces of the transverse wave that characterizes the formation of detonation.

Figure 16 shows the variation with equivalence ratio of the velocities of the waves induced in the H_2/O_2 bounding mixture as determined from the framing photographs. For comparison, the theoretical C-J velocity is also shown. In region A, only interaction mode 1 was observed, so that only the oblique shock velocity (the component normal to the wave front) is presented. In region C, interaction mode 2 was observed and the planar secondary detonation velocities are shown.

In regions B and D, interaction mode 3 was observed and it is interesting that in region B where the C-J detonation velocities of the

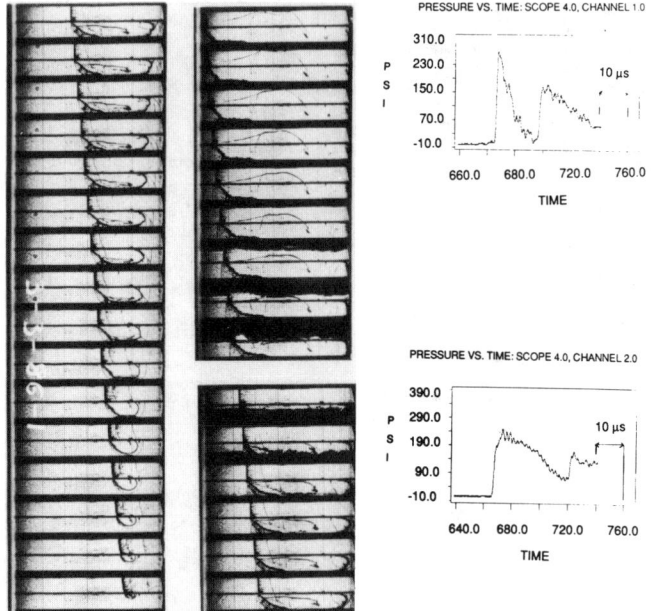

Fig. 15 Mach reflection is induced in bounding explosive and develops into a planar shock wave but no detonation appears (primary explosive C_2H_6O/O_2, $\phi = 0.5$; bounding explosive H_2/O_2, $\phi = 4.0$; membrane thickness = 50 nm).

bounding explosive are lower than that of the primary detonation, the induced Mach stem velocities are higher than the C-J velocities of the bounding explosives, i.e., overdriven waves are induced. On the other hand, in region D where the C-J detonation velocities of the bounding explosives are higher than that of the primary explosive, the induced Mach stem propagating velocities are lower than the C-J detonation velocities, i.e., underdriven waves are induced. However, transition to detonation is so rapid that planar secondary detonations are also observed. It should be noted that in both regions C and D, the measured planar detonation velocities in the bounding mixtures are below the C-J value. As mentioned above, this appears to be a result of side relief, since the secondary detonation now expands laterally into the primary explosive layer.

In region E, only a normal shock rather than a detonation was observed and from region D to region E there is a steep decrease in the induced wave velocity. This drop characterizes the change from a self-propagating detonation to an inert induced shock wave in the bounding explosive.

Finally, it should be noted that the influence of the film thickness on the interaction still exists, even when using ether-oxygen mixture as

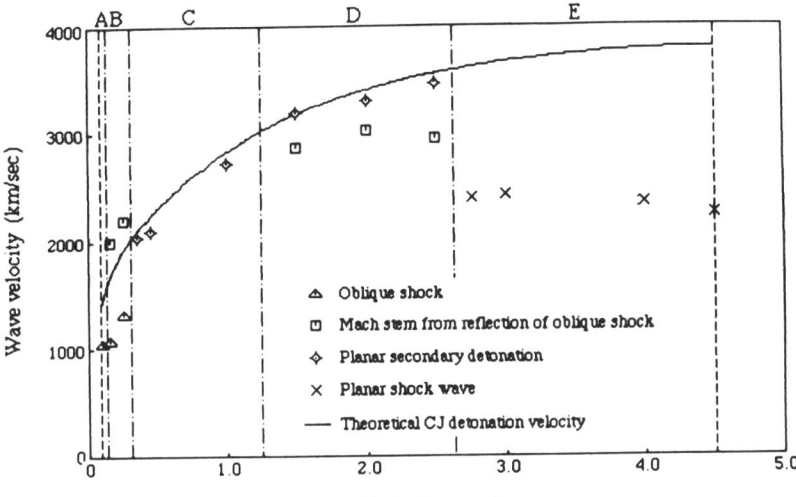

Fig. 16 Experimental secondary wave velocities induced by C_2H_6O/O_2, $\phi = 0.5$ primary detonation and theoretical C-J detonation velocities in H_2/O_2 bounding explosives.

the primary explosive. A couple of runs have been made by using C_2H_6O/O_2 with $\phi = 0.5$ as both the primary and secondary explosive and with and without the 50 nm membrane. In this case direct initiation was observed only when the film was absent.

Conclusions

1) The manner in which a detonation is induced in an explosive layer bounding a normal detonation is very sensitive to the thickness of any films used to separate the two mixtures.

2) In a certain range of membrane thickness, the propagation of the primary wave is not affected by the film even though the interaction and initiation in the bounding mixture is very sensitive to film thickness. The result suggests that the propagation of transverse waves or turbulent disturbances from within the primary detonation into the bounding mixture plays an important role and that under some conditions, even very thin films can block this effect.

3) In the case of direct initiation of detonation in the bounding explosive, there exists an "uncertain region" in which direct initiation occurs in only a fraction of the runs. The frequency of direct initiation in this regime varies inversely with the film thickness.

4) Detonations can be initiated in the bounding explosive for only a limited range of equivalence ratios. With H_2/O_2 as the bounding explosive, lean mixtures with an equivalence ratio of the order of 0.45 are found to be the most sensitive.

5) The direct initiation of the detonation in the secondary explosive appears to be enhanced by the use of primary explosives with a high C-J pressure.

Acknowledgment

This work was sponsored by the U.S. Army Research Office under Grant DAAG 29-83-K-009, Dr. David M. Mann project monitor. The authors are also grateful for the many helpful suggestions of Dr. P.L. Lu from The Army Research and Development Command. The assistance of Mr. Mike Denn in conducting the experiments and of Mr. C.J. Iott with the electronic and optical instruments was extremely helpful.

References

Bartlma, F. and Schroder, K. (1985) The diffraction of a detonation wave at a convex corner. 10th International Colloquium on Dynamics of Explosions and Reactive Systems, Berkeley CA, Aug. 4-9.

Cook, M. A. (1974) The science of industrial explosives, IRECO Chemicals, Salt Lake City, UT.

Dabora, E. K., Nicholls, J. A., and Morrison, R. B. (1965) The influence of a compressible boundary on the propagation of gaseous detonations, 10th Symposium (International) on Combustion, pp. 817-832, The Combustion Institute, Pittsburgh PA.

Gordon, S. and McBride, B. (1971) Computer Program for Calculation of Complex Chemical Equilibrium Compositions, Rocket Performance, Incident and Reflected Shocks, and Chapman-Jouguet Detonations. NASA Report SP-273.

Lee, J.H.S. (1977) Initiation of gaseous detonation. Ann. Rev. Phys. Chem. 28, 75.

Lee, J.H.S. (1984) Dynamic parameters of gaseous detonations. Ann. Rev. of Fluid Mech. 16, pp. 311-336.

Lewis, B. and von Elbe, G. (1961) Combustions, Flame and Explosions of Gases. 2nd ed. p. 530. Academic, New York.

Liu, J.C., Liou, J.J., Sichel, M., and Kauffman, C.W. (1985) Chemical and physical processes in combustion. Fall Technical Meeting of Eastern Section, p. 76-1. The Combustion Institute, Pittsburgh PA.

Liu, J.C., Liou, J.J., Sichel, M., Kauffman, C.W., and Nicholls, J.A. (1986) Diffraction and transmission of a detonation into a bounding explosive layer. 21st Symposium (International) on Combustion, pp.1639-1647. The Combustion Institute, Pittsburgh PA.

Lu, P.L., Landini, R. and Larson, H. (1982) 15th International Congress on High Speed Photography and Photonics. San Diego CA.

Lu, P.L. (1968) The structure and kinetics of the H_2-CO-O_2 detonations. Ph.D. dissertation. University of Michigan, Ann Arbor.

Sichel, M., Liou, J.J., Kauffman, C.W., and Tovey, J.L. (1984) The interaction phenomena between adjacent detonating layers. Fall Technical Meeting of Eastern Section. The Combustion Institute, Pittsburgh, PA.

Steady, Plane, Double-Front Detonations in Gaseous Detonable Mixtures Containing a Suspension of Aluminum Particles

B.A. Khasainov*
USSR Academy of Sciences, Moscow, USSR

and

B. Veyssière†
Laboratoire d'Energétique et de Détonique, ENSMA, Poitiers, France

Abstract

The quantitative numerical model developed previously for studying the double-front detonation (DFD) structure is used to analyze the influence of various parameters on the occurrence of DFD in two-phase mixtures of a detonable gas with aluminium particles in suspension. The role played by the gaseous composition, the losses to the tube walls, and the mechanism of aluminium burning are particularly examined. Results of computations confirm the experimental observations and the theoretical predictions.

Introduction

In a detonable mixture, when the heat release from chemical reactions occurs in an extended zone behind the leading front through a nonmonotonous process, the flow may exhibit a secondary discontinuity front behind the leading detonation front. This particular propagation regime has been called "double-front" detonation (DFD) and its existence in plane configuration has been displayed experimentally in gaseous detonable mixtures with a suspension of aluminium particles (Veyssière and Manson 1982; Veyssière 1986). Theory of heterogeneous media

---
Copyright © 1988 by the American Institute of Aeronautics and Astronautics, Inc. All rights reserved.
*Senior Researcher, Institute of Chemical Physics.
†Charge de Recherche, CNRS.

(Nigmatulin 1970) was used successfully to predict steady DFD parameters for a mixture of gaseous explosive with coal particles (Afanasieva et al. 1983) and for gas/droplet systems of liquid oxygen-hydrogen and liquid heptane-oxygen (Voronin 1984). In a previous work (Khasainov and Veyssière 1986), we have described a quantitative numerical model for studying the steady DFD structure in a plane configuration. In the present paper, this model is used to investigate quantitatively the influence of important parameters such as the size and concentration of particles and the diameter and roughness of the detonation tube on the characteristics of the DFD structure.

The DFD Model

A complete description of the numerical DFD model has been given in a previous work (Khasainov and Veyssière 1986). The two-phase flow behind the first front F_1 is treated as one-dimensional and plane. Momentum and convective energy losses to the tube walls are included. For describing the flow, we use the time-independent form of the set of conservation equations of the two-phase flow theory given by Nigmatulin (1970), with interaction terms for mass, momentum, and energy exchanges between particles and gases. For modeling the burning of aluminium particles, an empirical quasisteady burning law, with a correction term (see Gremiachkin et al. 1979) to take account of the particles and gas velocity differences, is used,

$$\frac{\dot{d}_p}{d_p} = - \frac{1 + 0.276 \sqrt{Re}}{t_b} \quad (1)$$

with
$$t_b = K \, d_{po}^n / \phi^{0.9} \quad (2)$$

where d_p is the diameter of a particle (index o corresponds to the initial diameter), $\dot{d}_p/2$ the regression rate of aluminium, t_b the burning time of the particle, ϕ the volume fraction of oxidizing species in the gaseous phase $\phi = X_{O2} + X_{H2O} + X_{CO2}$, and K and n empirical constants determined experimentally in quiescent oxidizing atmospheres.

To take account of the presence of different sizes of particles in the suspension, the particle size distribution is assumed to be the sum of fractions having discrete diameters, as

$$\sigma_o = \sum_{i=1}^{N} \sigma_{pi} \qquad (3)$$

where σ_{pi} is the concentration of particles of the i-th fractions and σ_o the total initial particle concentration.

Remarks about the Flow Pattern

Khasainov and Veyssière (1986) showed that the flow pattern behind the leading detonation front F_1 is determined by local differencies between the chemical heat release rate $\dot{q}_+$ and the effective rate of heat losses $\dot{q}_-$. For this purpose, the equation for the pressure gradient deduced from the governing system is used:

$$\frac{dp}{dz} = -\frac{\gamma_{CJ} - 1)M^2}{v_g} \frac{\dot{q}_+ - \dot{q}_-}{1 - M^2} \qquad (4)$$

where p is the pressure, v_g the velocity of the gases relative to the first front F_1, $M = v_g/a_g$ the Mach number, $z = D_{CJ}t - x$, $z = 0$ at F_1, and γ_{CJ} the isentropic coefficient. (Note: CJ = Chapman-Jouguet).

To compute the steady solution, it is necessary to satisfy the condition $M = 1$ behind the second front F_2 together with $\dot{q}_+ = \dot{q}_-$, which corresponds to the CJ rule for nonideal detonations.

It is noteworthy that $\dot{q}_+ = 0$ before ignition of aluminium particles. In the burning zone of aluminium particles, one can get a useful estimation of the chemical heat release rate from

$$\dot{q}_+ \cong \sigma_o Q_{A\ell} \phi_{CJ}^{0.9}/d_{po}^n \qquad (5)$$

On the opposite side, $\dot{q}_-$ tends to diminish with increasing distance from F_1.

Computed Results

A first series of computations has been done with the following parameters:
1) The friction coefficient to the walls λ_w (see Eqs. (33) and (34) of Khasainov and Veyssière (1986)) is (Guinzburg

1958)

$$\lambda_w = \begin{cases} 64/Re_w, & Re_w = D_w v_g \rho_g/\mu_g \leqslant 1200 \\ 0.316/Re_w^{0.25}, & 1200 < Re_w < 10^5 \\ 0.00332 + 0.221/Re_w^{0.237}, & Re_w \geqslant 10^5 \end{cases} \quad (6)$$

where D_w is the diameter of the detonation tube.
2) The ignition temperature of aluminium particles is taken $T_{ign} = 2310$ K, following the Friedman-Macek criterion (Friedman and Macek 1962) and the empirical parameters K and n in the burning law (Eq. (2)) are taken from Bouriannes (1971): n = 2 and $K = 4 \times 10^6 s/m^2$.
3) Following the experimental data from Veyssière (1983), an actual 5 modal aluminium particle size distribution has been used: particles of discrete diameters of 9, 11, 13, 15 and 17 μm with corresponding mass fraction 1, 15, 59, 22 and 3%.
4) The gaseous mixtures are the mixtures H1-H3, E1-E3, A1-A3 from Veyssière (1986). The gas parameters at F_1 have been computed earlier and are listed in Table 1 together with the volume fraction ϕ of the oxidizing species, the specific heat of reaction Q_{Al} and the stoechiometric coefficient α_{Al} (see Eqs. (6) of Khasainov and Veyssière 1986)) of the reaction of aluminium particles with a gaseous oxidizer.

The typical profiles for parameters obtained from a steady DFD computation, in the case of the mixture H2 with a mass particle concentration $\sigma_o = 40$ g/m³, are shown in Fig. 1. The shape of the calculated pressure profile is very similar to the experimental one. A more complete analysis has been provided previously in Khasainov and Veyssière (1986). Note that the point of the maximum Mach number coincides with the point of minimum pressure. Numerical investigation of the conditions for the existence of steady propagation of DFD have displayed (Khasainov and Veyssière 1986) that the steady DFD structure may occur only when the aluminium particle concentration is taken between two extreme values σ_{min} and σ_{max}, as

$$\sigma_{min} \leqslant \sigma_o \leqslant \sigma_{max} \quad (7)$$

The lower concentration limit of DFD corresponds to the minimum value σ_o for chocking to take place behind F_1 and the upper concentration limit of DFD corresponds to the maximum value of σ_o when the location of F_2 coincides with the point of maximum Mach number (or minimum pressure)

Table 1 Initial Composition of Gaseous Explosives, Chapman-Jouguet Parameters of Leading Detonation Wave and Parameters of Brutto Reaction between Aluminium Particles and Gaseous Oxidizing Components

Mixture	Initial Composition			C-J Parameters at Front F_1 [a]						Brutto Reaction Parameters		
	Fuel + a O_2 + b N_2			D_{CJ}	P	T	γ_{CJ}	W_{CJ}	Q_{gas}	ϕ_{CJ}	Q_{Al}	α_{Al}
	Fuel	a	b	m/s	bar	K		kg	MJ/kg		MJ/kg(Al)	
H1	H_2	0.64	2.41	1852	14.8	2739	1.188	0.02512	4.15	0.296	19.02	0.9881
H2	H_2	0.47	1.77	2000	16.0	2975	1.162	0.02349	5.32	0.303	17.80	0.9982
H3	H_2	0.38	1.45	2055	15.9	2959	1.176	0.02211	5.13	0.295	17.62	0.9997
E1	C_2H_4	3.85	14.46	1741	16.9	2727	1.183	0.02851	3.55	0.226	16.34	1.0692
E2	C_2H_4	2.61	9.81	1865	19.4	2993	1.168	0.02746	4.51	0.200	14.67	1.0781
E3	C_2H_4	1.76	6.27	1878	19.0	2741	1.239	0.02494	3.10	0.127	15.64	1.0477
A1	C_2H_2	4.39	16.49	1660	15.2	2554	1.209	0.02927	2.76	0.213	18.86	1.0512
A2	C_2H_2	2.34	8.79	1888	19.9	3154	1.158	0.02823	4.94	0.164	14.40	1.1013
A3	C_2H_2	2.00	7.52	1932	20.7	3231	1.163	0.02768	5.01	0.143	14.07	1.0995

[a] These parameters were computed by Veyssière (1986).

DFD'S IN MIXTURES WITH ALUMINUM PARTICLES

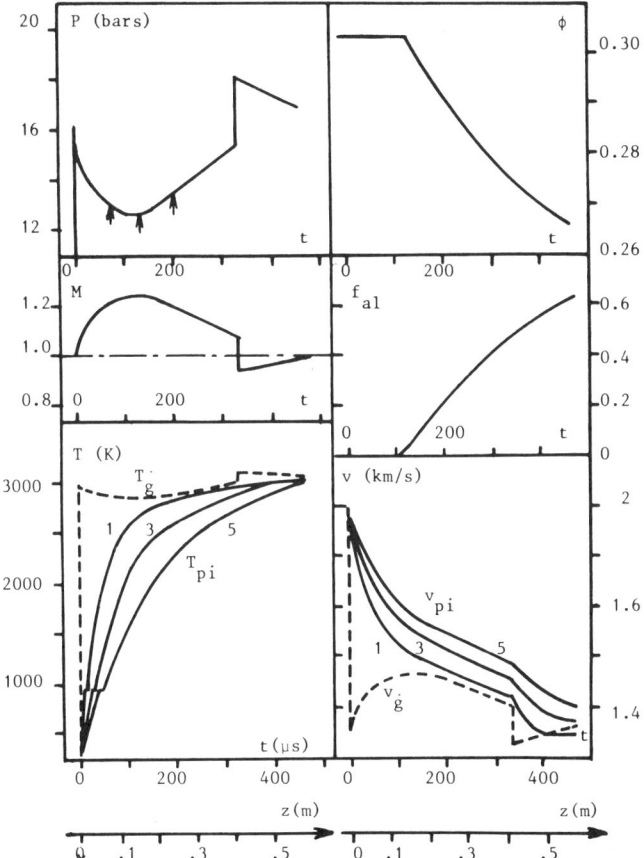

Fig. 1 DFD parameters profiles for a mixture H2 with a mass particle concentration σ_o = 40 g/m³ (i = 1, 3, 5 corresponds to 9, 13, and 17 μm aluminium particles).

behind F_1, i.e., σ_{max} corresponds to the threshold of DFD stationarity.

Computed values of σ_{min} and σ_{max} are summarized in Table 2 for the nine mixtures under consideration. Also listed are the values of the delay τ between the first (F_1) and the second (F_2) fronts, the pressure p_2 at the second front, and other details of the structure: delay and pressure at the ignition point (p_{ign}, τ_1) and at the point of minimum pressure (p_m, τ_m). It can be seen that the calculated values of the ignition delay τ_1 (based on an ignition temperature T_{ign} = 2310 K) are greater than the experimental ones. The influence of the ignition temperature will be discussed below.

In Fig. 2, the calculated dependences of τ and p_2 for mixtures H1-H3 (solid lines 1-3) are compared as a function of the particle mass concentration σ_o with the experimental results from Veyssière (1986). It is seen that the behavior of these parameters exhibited by the experiments are confirmed: when σ_o increases, τ decreases and p_2 increases. However, there remain some discrepancies on the quantitative results. Calculated values of τ (see Table 2 and Fig. 2) are higher and much more sensitive to the particle concentration variations than the experimental ones. The domain of existence of steady DFD is narrower than that corresponding to the conditions where a secondary front has been observed experimentally. However, the

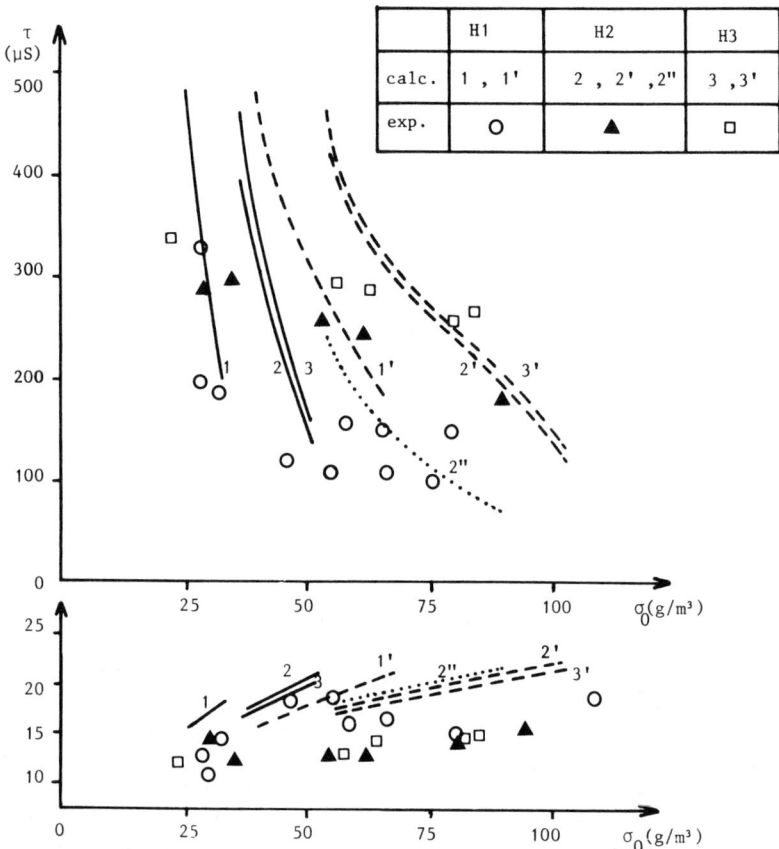

Fig. 2 Comparison of the experimental and calculated values of τ and p_2 as function of the particle concentration σ_o in mixtures H (———— "smooth" walls, ————— "rough" walls (h = 50 µm), ······ "rough" walls (h = 50 µm), and T_{ign} = 1700 K).

DFD'S IN MIXTURES WITH ALUMINUM PARTICLES 291

Table 2 Parameters of Computed Pressure Profiles
for Steady DFD in Various Gaseous Explosives
Laden with Aluminium Particles[a]

Mixtures	σ g/m^3	Ignition point		Point of minimum pressure		Front F2	
		p bar	τ_1 μs	p_m bar	t_m μs	p_2 bar	τ μs
Lower Concentration Limit of DFD							
H1	26	12.3	100	11.7	193	15.7	488
H2	38	13.3	70	12.6	137	17.5	393
H3	37	12.9	75	12.4	138	17.0	463
E1	40	13.7	101	13.2	192	18.1	549
E2	68	15.4	70	14.8	129	21.2	543
E3	70	14.3	120	13.8	240	21.1	725
A1	32	12.0	166	11.6	350	17.2	608
A2	102	15.4	59	14.5	112	21.9	642
A3	150	15.3	56	14.4	106	23.6	670
Upper Concentration Limit of DFD							
H1	33	11.9	105	11.5	200	18.3	202
H2	52	12.7	73	12.3	136	21.0	138
H3	51	12.5	77	12.0	146	20.2	156
E1	53	13.3	106	12.9	209	21.4	212
E2	101	14.7	73	14.1	142	25.6	145
E3	106	13.4	141	12.9	227	26.0	311
A1	38	11.9	175	11.5	370	19.1	392
A2	224	13.5	70	12.7	137	28.6	138
A3	250[b]	13.8	65	13.0	132	23.6	625

[a] Lower and upper concentration limits of DFD are shown for smooth shock tube with diameter 6.9 cm, T_{ign} = 2300 K, n and K from Bouriannes (1971).
[b] Upper concentration limit for the mixture A3 appeared to be higher than 250 gm^3; however, computations were made with σ_o < 250 g/m^3.

experimental results reported by Veyssière (1986) do not correspond entirely to a steady propagation. For low particle concentrations, unsteady propagation is observed with the delay τ between F_1 and F_2 either 1) increasing, due to an insufficient amount of heat release by aluminium burning for sustaining the second discontinuity, or 2) decreasing until a limiting value is reached. (However, the detonation tube is probably too short to reach the steady propagation. It must be noted that the computations display a DFD reaction zone length of about 0.5 m, which is quite high by comparison with the tube length used for the experiments - about 5 m. This aspect, however, needs to be clarified in the future by nonsteady modeling of DFD initiation and propagation.) The quasisteady propagation has been observed mainly for particle concentrations around the upper limit of the experimental concentrations. In some mixtures, the experimental concentrations were too low to reach a quasisteady propagation regime. Hence, one can conclude that with the parameter values used here, the computed particle concentration limit for steady DFD to occur is underestimated.

Calculated values of p_2 shown in Fig. 2 confirm the experimental observation that p_2 is far less sensitive than τ to particle concentration variations. This derives from the expression for the pressure jump at F_2

$$\frac{P_{2+}}{P_{2-}} = \frac{2\gamma_{CJ}M_-^2 + 1 - \gamma_{CJ}}{\gamma_{CJ} + 1} \qquad (8)$$

From Fig. 1, it follows that just before F_2, the Mach number remains quite close to unity.

Influence of Gaseous Composition

Further examination of the results of Table 2 reveals that, for mixtures having the same fuel component, the values of σ_{min} and σ_{max} increase when the equivalence ratio increases from a lean composition to a rich one. This may be explained by considering the estimation of $\dot{q}_+$ given by Eq. (5), which exhibits a proportionality of $\dot{q}_+$ to $(\sigma_o Q_{Al} \phi_{CJ}^{0.9})$. From Table 1, it can be deduced that the product $(Q_{Al} \phi_{CJ}^{0.9})$ substantially decreases when the equivalence ratio increases. Since, at the second C-J point, one must satisfy the condition $\dot{q}_+ = \dot{q}_-$ and, at this point, $\dot{q}_-$ does not practically depend on Q_{Al} and σ_o, the only way to satisfy the above condition with a decreasing value of $Q_{Al} \phi_{CJ}^{0.9}$ is to increase σ_o. That is why σ_{min} and σ_{max} increase with increasing equivalence ratio. This influence of the gaseous composition had been observed in experiments (Veyssière 1986). In the same way, such simple considerations can explain why mixtures H2 and H3 have very similar values of DFD parameters. (Table 1 indicates that the product $(Q_{Al} \phi_{CJ}^{0.9})$ is almost the same for these two mixtures.)

Influence of the Tube Walls

As reported previously (Khasainov and Veyssière (1986)), the existence of momentum and energy losses to the tube walls is a necessary condition for the steady DFD structure to exist. To provide additional confirmation of the strong dependence of the DFD structure on the wall losses, we have arbitrarily varied the diameter of the detonation tube with regard to the nominal diameter D_w (= 69 mm). Figure 3 shows the variations of τ vs the particle concentration σ_0 in the mixture H2 for the nominal diameter D_w of the tube and for diameters half ($D_w/2$) and twice ($2\ D_w$) the nominal diameter. As could be expected, the increase of $\dot{q}_-$ due to the smaller tube diameter results in a very important widening of the particle concentration domain where DFD may exist. Alternatively, decreasing $\dot{q}_-$ by increasing the tube diameter induces a strong narrowing of the concentration limits of DFD, with drop of σ_{min} and σ_{max}. In Fig. 4, concentration limits of σ_{min} and σ_{max} are plotted as function of the ratio of the diameter input value D to the nominal diameter D_w. It can be seen that, when the tube diameter becomes about 10 times greater than the nominal diameter, there is no longer steady DFD at even as small an aluminium particle concentration as 1 g/m³. Hence, in the vicinity of the second C-J point of DFD, wall losses play a much more important role than the losses due to particle velocity and temperature relaxation. This is confirmed in Fig. 1 on the velocity and temperature profiles showing that near F_2 differences in the temperatures and velocities between particles and gases are much less important than near the leading front F_1.

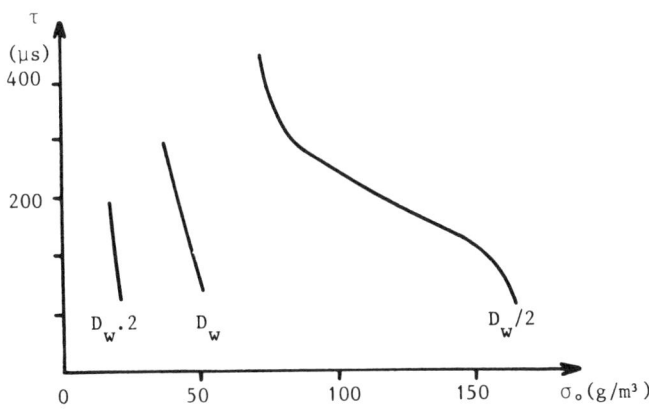

Fig. 3 Calculated values of τ as function of the particle concentration for different values of the tube diameter (mixture H2).

Computations made without momentum and energy losses to the tube walls gave no solution for steady DFD. The second C-J point could be achieved only with full burning of the aluminium particles, but this would require a higher velocity of the second discontinuity F_2 than that of the leading wave and the DFD would not be stable. This confirms the theoretical prediction of the impossibility of steady DFD in the ideal plane configuration (i.e., without wall losses).

Another way to vary the amount of wall losses is to introduce an effect of "roughness" on the tube walls. The form of the friction coefficient λ_w used in Eq. (6) corresponds to "smooth" walls. For simulating a roughness in the tube walls, we have also used the following expression (Guinzburg 1958) for λ_w:

$$\lambda_w = \frac{1}{\left(1.74 - 2 \, \text{Log}(2h/D_w)\right)^2} \qquad (9)$$

where h is the characteristic size of roughness.

Computations have been made for the mixtures H1-H3 with a characteristic size of roughness h = 50 µm, which

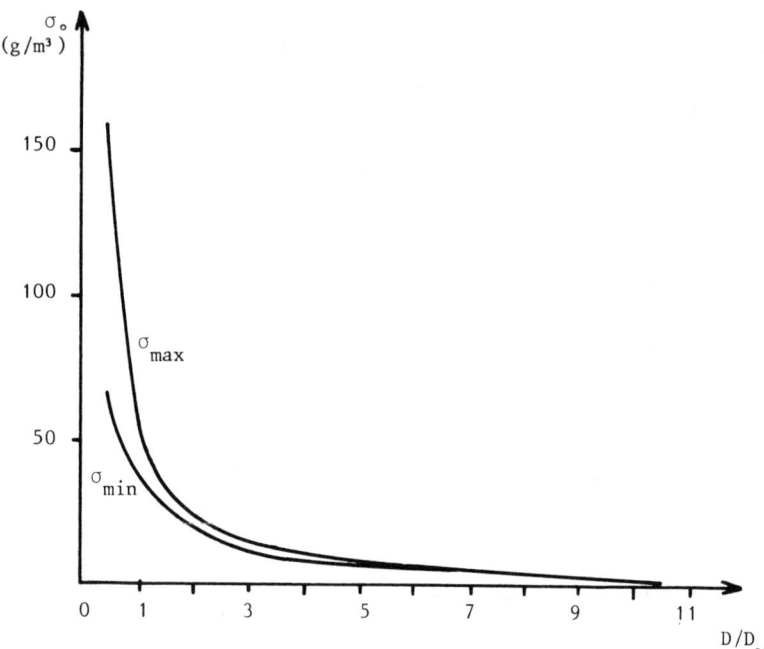

Fig. 4 Variation of the particle concentration range where steady DFD occurs as function of the tube diameter (mixture H2).

corresponds approximately to 5 times the size of the particles. The results are plotted in Fig. 2 (dashed curves 1'-3'). The increase in $\dot{q}_-$ due to the roughness widens the concentration limits of DFD propagation and increases the values of σ_{min} and σ_{max}. This effect can be explained as above: for a given gaseous mixture, an increase in $\dot{q}_-$ can be compensated only by an increase of the particle concentration σ_o. It is noteworthy that accounting for this "roughness" effect of the walls reduces the discrepancies between the calculated and experimental values of the DFD parameters (see Fig. 2). Particularly, the increase of roughness up to h = 50 μm results, in the case of mixtures A2, A3, and E3, in the increase of σ_{min} beyond 200 g/m³. This might be the explanation for why the DFD structure has not been observed until now in experiments with those gaseous compositions (the experimented particle concentrations were less than 200 g/m³). Anyway, this result agrees with the conclusions of Voronin (1984) about the strong effect of the wall losses on DFD parameters. However, there remains great uncertainty about the correct modeling of losses to the tube walls, as they have been computed on the basis of wall-friction coefficients deduced from measurements in steady duct flows; however, behind the leading detonation front, the interaction of the gas flow with the walls is not steady. The effects of boundary-layer development at the rear of the front may result in a transient increase of the friction coefficient by several orders of magnitude (Gel'fand et al. 1986). Such an effect will have to be taken into account in order to better describe the losses.

Influence of the Mechanism of Aluminium Burning

The results obtained above with the empirical parameters K and n in the burning law (Eq. (2)) taken from Bouriannes (1971) (n = 2 and K = 4 x 10⁶ in SI units) have been compared with those obtained with two other sets of parameters: n = 1.5 and K = 10,600 (Frolov et al. (1972)) and n = 1.75 and K = 230,850 (Price (1984)). This has been connected with the effect of particle diameter.

Typical results are shown in Fig. 5 in the case of the mixture H2 with "smooth" walls. Solid lines correspond to the nominal five-modal particle size distribution used above. Dashed lines correspond to coarser particles (the particle diameter of every fraction of the nominal size distribution was increased by a factor 1.5) and dotted lines correspond to finer particles (the particle diameter

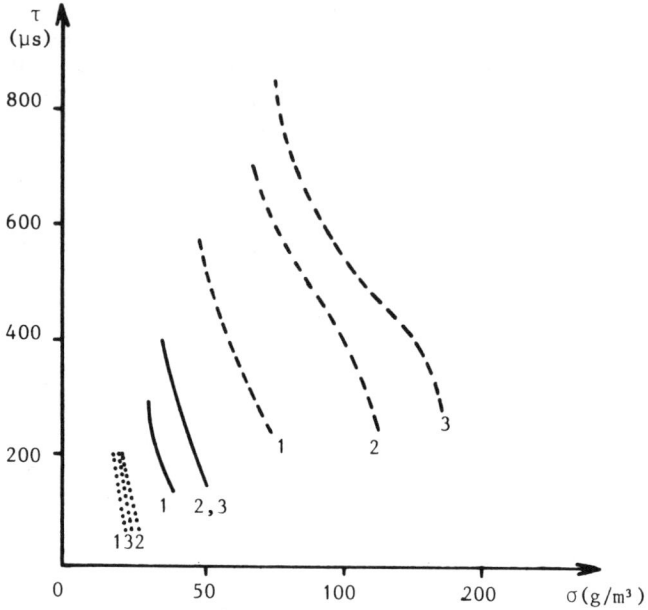

Fig. 5 Influence of the diameter of particles on the delay τ (..... smaller diameter, ——— nominal diameter, ----- larger diameter; 1 - n = 2, K = 4 x 10^6; 2 - n = 1.5, K = 10600; 3 - n = 1.75, K = 230850).

of every fraction was decreased by a factor 1.5). As could be expected on the basis of the estimation of $\dot{q}_+$ provided by Eq. (5), the increase of $\dot{q}_+$ due to the decrease of the diameter of particles results in diminishing the limiting concentrations σ_{min} and σ_{max}. Calculations using parameters K and n from Bouriannes (1971) and Price (1984) give very similar results in the studied range of particle diameter, while parameters from Frolov et al. (1972) give shorter burning times and higher values of $\dot{q}_+$ with a subsequent decrease of σ_{min} and σ_{max}. Moreover, the larger mean diameter of particles, the larger is the length of the steady DFD. For example, with K and n from Frolov et al. (1972), $Z_{DFD} \cong 0.25$ m for the smallest particles, $Z_{DFD} \cong 0.4$ m for the nominal ones, and $Z_{DFD} \cong 0.85$ m for the coarser ones. At last, if the nominal five-modal size distribution is replaced by a monodisperse suspension with a 13 μm mean diameter, the dependence of τ on the particle concentration σ_o is practically unchanged.

Taking account of the delay for melting of aluminium has relatively small effect on the dependence of τ on σ_{o_3}. For example, in the case of the mixture H2, at σ_o = 45 g/m and for smooth walls, τ = 233 µs; with melting ignored, this value only shifts to 217 µs.

Proper choice of the ignition criterion is also of great importance. As mentionned above, the computed τ_1 values (assuming an ignition temperature of particles T_{ign} = 2310K) appeared to be larger than experimental ones. Taking account of the remarks done in a previous discussion on this problem (Veyssière 1983), we have done some computations with T_{ign} = 1700 K. With this assumption, the computed values of τ_1 are decreased by a factor of about 1.6 and the agreement between calculated and experimental values of the ignition delay of particles becomes much better. Moreover, the decrease of T_{ign} from 2310 to 1700 K results in a decrease of the delay between F_1 and F_2. For example, in the mixture H2 at σ_o = 40 g/m³, τ decreases from 315 to 150 µs; at σ_o = 45 g/m³, τ decreases from 240 to 100 µs.

It is noticeable that the role of these different factors is increased when $\dot{q}_-$ is diminishing.

Conclusions

The present computed results, obtained from the proposed DFD model, confirm quite nicely the behavior of the structure of a detonation propagating in a fuel-air mixture containing a suspension of aluminium particles and provide a satisfying interpretation of some observed experimental features. The role played by the losses to the walls are shown to be of first importance for the steady propagation in plane configuration. DFD structure is also influenced by the modeling of aluminium burning. Proper adjustment of such important parameters provides a better agreement between the numerical results with the experimental observations. This is illustrated in Fig. 2 (curve 2") where calculated results for the mixture H2 are reported in the case of "rough" walls (h = 50 µm) and assuming an ignition temperature T_{ign} = 1700 K. As the DFD structure depends on the relative balance between the chemical energy release rate ($\dot{q}_+$) and all kinds of energy dissipation in the flow ($\dot{q}_-$), this last result displays the need to improve the modeling of kinetics parameters to reach a better agreement with experiments. This would permit, in the future, use of the present DFD model for making predictions in other kinds of mixtures.

References

Afanasieva, E.A., Levin, V.A., and Tunik, Yu.V. (1983) Multifront combustion of two-phase media. Progress in Astronautics and Aeronautics: Shock Waves, Explosions, and Detonations (edited by J.R. Bowen, N. Manson, A.K. Oppenheim, and R.I. Soloukhin), Vol. 87, pp. 394-413. AIAA, New York.

Bouriannes, R. (1971) Contribution à l'étude de la combustion de l'aluminium dans les mélanges oxygène - argon, dans l'azote et dans l'air. Thèse de Doctorat ès Sciences Physiques, University of Poitiers, France.

Friedman, R., and Macek, A. (1962) Ignition and combustion of aluminium particles in hot ambient gases. Combust. Flame 6(1), 9-19.

Frolov, Yu.V., Pokhil, P.F., and Logachev, V.S. (1972) Ignition and burning of powdered aluminium in a hot gaseous medium. Fiz. Goreniya Vzryva 8(2), 213-236.

Gel'fand, B.E., Zel'dovich, Ya.B., Kajdan, Ya.M., and Frolov, S.M. (1986) Effect of losses on the detonation propagation in tube. VIth USSR Congress on Theoretical and Applied Mechanics, Tashkent: abstracts of papers p. 188.

Gremiachkin, V.M., Istratov, A.G., and Leypunskiy, O.I. (1979) Influence of gas flow on a metal particle combustion. Fiz. Goreniya Vzryva 15(1).

Guinzburg, I.P. (1958) Prikladnaya Gazodynamika, Leningrad.

Khasainov, B.A. and Veyssière, B. (1986) Analysis of the steady double-front detonation structure for a detonable gas laden with aluminium particles. Presented at 2nd International Colloquium on Dust Explosions, Jadwisin, Poland.

Nigmatulin, R.I. (1970) Methods of mechanics of a continuous medium for the description of multiphase mixtures. Prik. Mat. Mekh. 34(6), 1097-1112.

Price, E.W. (1984) Combustion of metalized propellants. Progress in Astronautics and Aeronautics: Fundamentals of Solid-Propellant Combustion (edited by K.K. Kuo, and M. Summerfield) Vol. 90, Ch. 9, pp. 479-513. AIAA, New York.

Veyssière, B. (1983) Ignition of aluminium particles in a gaseous detonation. Progress in Astronautics and Aeronautics: Shock Waves, Explosions and Detonations (edited by J.R. Bowen, N. Manson, A.K. Oppenheim, and R.I. Soloukhin) Vol. 87, pp. 362-375. AIAA, New York.

Veyssière, B. (1986) Structure of the detonations in gaseous mixtures containing aluminium particles in suspension. Progress in Astronautics and Aeronautics: Dynamics of Explosions (edited by J.R. Bowen, J.C. Leyer, and R.I. Soloukhin), Vol. 106, pp. 522-544. AIAA, New York.

Veyssière, B. and Manson, N. (1982) Sur l'existence d'un second front de détonation des mélanges biphasiques hydrogène - oxygène - azote - particules d'aluminium. CRAS Paris 295(II), 335-338.

Voronin, D.V. (1984) On the existence of double-front detonations in gas-droplet systems. Dynamika Mnogophasnykh sred, Novosibirsk, No. 68, pp. 35-43.

Chapter IV. Condensed-Phase Detonations

Critical Conditions for Hot Spot Evolution in Porous Explosives

B. A. Khasainov,* A. V. Attetkov,† A. A. Borisov,‡ B. S. Ermolaev,§
and V. S. Soloviev¶
USSR Academy of Sciences, Moscow USSR

Abstract

An analysis of hot spot formation and evolution in porous explosives is performed for a viscoplastic model of pore deformation behind an initiating shock wave. Critical conditions for initiation of reaction in hot spots are estimated for the limiting cases when the chemical reactions occur only in the solid phase and when they occur only in the gas phase. Based on the comparison of mean pore size in the explosive material with the critical one that is sufficient for initiating and sustaining the chemical reaction, an explanation is given of the experimental data on nonmonotonic dependence of the shock sensitivity of porous explosives on their microstructure. The dependence of the critical pore size on the initiating shock wave amplitude is presented.

Introduction

During the last decade, substantial progress was made in studies of shock initiation of porous energetic materials. Special attention has been paid to the effects of the structure of porous high explosives on their shock sensitivity [Taylor et al. (1976); Setchell and Taylor (1984); Von Holle (1983)]. Several physical models of "hot spot" generation (i.e., origination of reaction centers in

---

Copyright © 1988 by the American Institute of Aeronautics and Astronautics, Inc. All rights reserved.
* Senior Researcher, Institute of Chemical Physics
† Senior Researcher, N. Bauman Moscow Technical University
‡ Head of Laboratory, Institute of Chemical Physics
§ Senior Researcher, Institute of Chemical Physics
¶ Professor, N. Bauman Moscow Technical University

a heterogeneous explosive), which precedes the spread of the reaction over the entire volume of shocked explosive, were developed. These models are based on the analysis of hydrodynamic and elastic-plastic interactions of an initiating shock wave (ISW) with: density discontinuities in the material, e.g., voids or solid microparticles [Mader (1979); Hayes (1983)], viscoplastic deformation of the pores caused by the action of an ISW on the material [Khasainov et al. (1981, 1983); Borisov et al. (1986); Attetkov (1986); Frey (1985); Kim and Sohn (1985); Maden (1987)] or friction heating in the shear bands [Frey (1981); Amosov (1982)].

Each of these mechanisms may lead to formation of hot spots in a suitable situation. However, theoretical studies of the stage of growth (or evolution) of the reaction centers are rather contradictory, as was demonstrated by Wackerle and Anderson (1983).

In the present work the formation and development of the reaction centers in porous explosives are analyzed based on the viscoplastic mechanism. Critical conditions for initiation of solid high explosives (HE) around a pore are calculated for the case where chemical reaction occurs in the solid phase. Within the framework of the gas-phase mechanism of chemical reaction initiation, extinguishment of certain reaction centers and transition to a self-sustained growth of the hot spot are described. The nonmonotonic dependence of the shock sensitivity of porous HE on the size of pores and HE grains (Setchell and Taylor 1984) is explained.

Dynamics of Pore Deformation

In the analysis of the dynamics of viscoplastic deformation of pores in solid materials, an effective spherical cell is usually considered (Carrol and Holt 1972). The initial radius of a void is assumed to be equal to the mean pore size a_o, and the outer radius of the cell b_o is defined in such a way that the cell porosity ϕ_o coincides with the HE porosity, i.e., $\phi_o = (a_o/b_o)^3 = 1-\rho/\rho_s$. Here ρ is the density of porous HE and ρ_s is the theoretical maximum density of the HE. Thus, in the model under consideration, the structure of the porous HE is characterized by a_o and ϕ_o.

Estimates made by Khasainov et al.(1981), indicate that, for Reynolds numbers $Re=a_o(\rho_s P_m/4\mu)^{\frac{1}{2}} < 1$ (P_m is characteristic ISW amplitude, and μ is the solid HE viscosity), the process of pore deformation behind the ISW front is viscosity-controlled and spherically symmetrical, that is, the cell together with its field of radical velocities moves as a whole at the particle velocity behind the ISW front. This hypothesis is supported by data (Hasegawa and Fujiwara 1982) which shows that, during the course of collapse of gaseous bubbles behind a shock wave in glycerol, the shape of the bubbles is nearly spherical, even for Re ≃ 10. For propagation in the HE of a compression wave with an extended pressure profile (ramp wave), the assumption of sphericity of pore deformation is substantiated by Khasainov et al. (1983), Frey (1985), and Attetkov (1986). In both cases (sharp and ramp waves), the bulk solid in the cell may be considered to be incompressible because specific volume changes in a shocked porous material are mainly due to collapse of voids. It should be noted, however, that the real pore shape is close to the spherical one solely in high-density explosives (Soloviev et al. 1981). Shock Hugoniots of porous materials calculated employing the model of spherical cells (Dunin and Surkov 1979) are nevertheless consistent with the experimental data, even for HE densities characteristic of loosely packed materials.

The solid material is assumed to obey the relationships relevant to a viscoplastic medium. As the wave amplitude, in a porous material, $P_m(t)$, exceeds the pore strength, the pore deformation occurs in the totally plastic regime (elastic behavior of the solid material is ignored). The time history of the macroscopic pressure P_m is assumed to be prescribed. In the model suggested, a response of a porous cell to a given load is considered alone, and the effect of the processes occurring in the cell as a result of chemical reactions on the macroscopic flow pattern is not considered here, in contrast to the closed model proposed by Khasainov et al. (1981).

Inasmuch as the radial flow of the solid material around a pore arising on passage of a shock or ramp wave through the cell results in heating the solid and gas phases, this may lead to initiation of the chemical reaction.

Mechanism of Local Chemical Reaction Initiation

The analysis of the viscoplastic mechanism of hot spot formation performed by Khasainov et al. (1981, 1983) has shown that this mechanism is very efficient and provides for a temperature rise at the surface of not too small pores up to T≥1000°K for P_m≥0.5 GPa. However, neither the melting of HE nor the detailed mechanism of chemical reactions is considered in these works.

Now we will formulate the main features of the mechanisms of chemical reactions that may occur during pore deformation. As the pore radius decreases, the temperature of the solid material at the pore surface may reach melting point. Melting points of solid explosives do not exceed 450°K at atmospheric conditions, a much lower temperature than that at which chemical decomposition becomes noticeable. Melting temperature increases nearly linearly with pressure by about 200°K per 1 GPa. With decreasing pore radius, a phase transition must occur at the pore surface. Since the pressures (≥ 1 GPa) and temperatures ($\geq 300°K$) under consideration are much higher than the critical point coordinates on the phase diagram of typical organic nitrocompounds, the solid material will not be converted on further heating first into liquid and then into gas. Instead, it will transfer to a new isotropic phase that for convenience is referred to below as a "gas," though under the conditions considered there is no difference between liquid and gas.

When the material is reactive, its heating during the course of pore deformation may initiate chemical reactions both in a layer of the solid adjacent to the pore surface and on the surface proper. For values of the kinetic parameters typical of solid HE pressures on the order of 1 GPa, and melting points of HE are below 700°K; at relatively low ISW amplitudes "gasification" of a solid at the surface of a pore being deformed will result from the phase transition starting when the pore surface temperature attains the melting point. At relatively high pressures $P_m \geq 3$ GPa and melting points of 1000°K, the major contribution to HE gasification results from homogeneous and heterogeneous solid-phase HE reactions occuring before the surface temperature reaches the melting point. The gaseous products evolved from the pore surface diffuse in both cases to the pore center, and their temperature finally turns out to be higher than that at the pore surface because of continuing gas compression (regardless of whether the pores have been preevacuated). As a result, an exothermic chemical reaction may also begin in the gas phase. The dynamics of the reaction center development (whether it will keep burning or fade out) is determined by the competing processes of mechanical pore deformation, heterogeneous and homogeneous solid-phase chemical reactions, and homogeneous chemical reaction and diffusion in the gaseous phase, and also by the processes of heat conduction in the gas and solid around the pore (Borisov et al. 1986).

Because of the complexity of the hot spot evolution process, it is expedient to analyze the solid-phase (mechanism A) and gas-phase (mechanism B) mechanisms of reaction center formation and growth separately. Within the framework of mechanism A we ignored the heterogeneous reaction on the pore surface and assumed that pores contain no gas at the initial time moment and that depletion of the solid HE during the course of the chemical reaction is insignificant. Thus, the problem is reduced to determining conditions of a thermal explosion of the material around a pore being deformed by ISW (Attetkov 1986). The critical ignition phenomena are associated in mechanism A with interactions of the processes of mechanical energy dissipation, chemical heat generation, and heat transfer in the material around the pore.

In analyzing mechanism B of chemical reaction initiation in hot spots, we ignored the contribution of solid-phase chemical reactions; i.e., we assumed that the ignition takes place in the gas phase. Unlike the current version of the solid-phase ignition model, mechanism B allows the entire sequence of the processes of reaction center origination and growth up to the onset of the self-sustaining regime of pore burnout to be followed.

Mathematical Formulation of the Problem

In the frame of reference fixed at the center of a pore cell, the conservation equations for an incompressible solid viscoplastic material read as

$$\frac{\partial}{\partial r}(r^2 v_s) = 0, \quad \rho_s = \text{const}$$

$$\rho_s \left[\frac{\partial v_s}{\partial t} + v_s \frac{\partial v_s}{\partial r}\right] = \frac{\partial \sigma_r}{\partial r} + \frac{2}{r}(\sigma_r - \sigma_\theta) \quad (1a)$$

$$\rho_s c_s \left[\frac{\partial T_s}{\partial t} + v_s \frac{\partial T_s}{\partial r}\right] = (\lambda_s/r^2)\left(\frac{\partial}{\partial r} r^2 \frac{\partial T_s}{\partial r}\right)$$

$$+ (2/3)(\sigma_r - \sigma_\theta)\left(\frac{\partial v_s}{\partial r} - \frac{v_s}{r}\right) + \rho_s Q_s z_s \exp(-E_s/RT_s) \quad (1b)$$

$$\sigma_r - \sigma_\theta = Y + 2\mu(\partial v_s/\partial r - v_s/r), \quad P_s = -(\sigma_r + 2\sigma_\theta)/3 \quad (1c)$$

for $a(t) \leq r \leq b(t)$

Here c_s, λ_s, and Y, are, respectively, the specific heat capacity, thermal conductivity, and yield strength, of solid HE, $P_s = P_s(r,t)$ is the solid-phase pressure, Q_s, z_s, and E_s are the heat, preexponential factor, and activation energy of the chemical reaction in the solid, respectively. The thermal and physical parameters of HE are assumed to be temperature-independent, since the solid-phase temperature T_s cannot exceed melting point T_m. When the gas-phase initiation mechanism is considered, the intensity of the chemical source in the solid is set at zero.

The conservation equations for the gas phase are written as

$$\frac{\partial \rho_g}{\partial t} + \frac{1}{r^2} \frac{\partial}{\partial r}(\rho_g v_g r^2) = 0 \tag{2a}$$

$$\rho_g \left(\frac{\partial v_g}{\partial t} + v_g \frac{\partial v_g}{\partial r}\right) = -\frac{\partial P_g}{\partial r} \tag{2b}$$

$$\frac{\partial \rho_g e_g}{\partial t} + \frac{1}{r^2} \frac{\partial}{\partial r}[v_g r^2(\rho_g e_g + P_g)]$$

$$= \frac{1}{r^2} \frac{\partial}{\partial r}\left(\lambda_g r^2 \cdot \frac{\partial T_g}{\partial r}\right) + Q_g z_g \rho_g^n A^n \exp(-E_g/RT_g) \tag{2c}$$

$$\frac{\partial \rho_g A}{\partial t} + \frac{1}{r^2} \frac{\partial}{\partial r}(\rho_g v_g r^2 A) = \frac{1}{r^2} \frac{\partial}{\partial r}\left(\rho_g D r^2 \frac{\partial A}{\partial r}\right)$$

$$- z_g \rho_g^n A^n \exp(-E_g/RT_g) \tag{2d}$$

for $0 \le r \le a(t)$. Here A is the concentration of the gasification products evolved into pores during melting or decomposition of solid HE, D is the gas ("vapor") diffusivity, Q_g, z_g, E_g, and n are the kinetic parameters of the gas-phase reaction. The Abel equation with a constant covolume b_g was used as an equation of state of the gas phase

$$P_g (1 - b_g \rho_g) = \rho_g R_g T_g, \quad e_g = R_g T_g/(k - 1) \tag{3}$$

$$k = (c_p/c_v)_g = \text{const}$$

It should be emphasized that these equations are employed in mechanism B alone.

The conservation equations for two phases are connected by the boundary conditions at the pore surface $r = a(t)$

$$P_{s+} = P_{g+} + 2Y/3 - 4\mu v_{s+}/a + j(v_{s+} - v_{g+})$$

$$j = \rho_s(\dot{a} - v_{s+}) = \rho_g(\dot{a} - v_{g+}), \quad (\dot{a} = da/dt) \tag{4}$$

$$T_{s+} = T_{g+}, \quad \lambda_g(\partial T_g/\partial r)_+ = \lambda_s (\partial T_s/\partial r)_+ - jQ_m$$

$$D\rho_g (\partial A/\partial r)_+ = j(1 - A_+)$$

Where j is the mass rate of gasification due to solid-phase chemical reactions or phase transition at $T_{s+} = T_m$, $j = 0$ before the reaction starts or for $T_{s+} < T_m$, and $T_m = T_{mo} + C_m P$ and Q_m are the melting temperature and heat.

Boundary conditions at the pore center ($r = 0$) and at the periphery of the spherical cell [$r = b(t)$ with $b(0) = a_o/\phi_o^{1/3}$] are

for $r = 0$, $v_g = 0$, $\partial T_g/\partial r = \partial P_g/\partial r = \partial A/\partial r = 0$

for $r = b(t)$, $\partial T_s/\partial r = 0$ \hfill (5)

The initial conditions for Eqs. (1-5) are

for $t = 0$, $v_s = v_g = 0$, $P_g = P_{go}$,

$T_s = T_g = T_o$, $j = 0$, $a = a_o$, $\phi = \phi_o$

$$\bar{P}_s(t=0) = 3[\int_{a_o}^{b_o} P_s r^2 dr/(b^3 - a^3)]_{t=0} \tag{6}$$

$$= [P_m(0) - P_{go}\phi_o]/(1 - \phi_o)$$

Equations (1) - (4) are analogous to those that describe the behavior of a gas bubble in a liquid (Nigmatulin 1978). In the present model, the response of the solid HE to compression and chemical reactions in the solid or gas phase are incorporated, and the phase transition kinetics are neglected. In particular, the temperature jump at the interface is ignored.

The pressure distribution in the solid material surrounding the pore is found by integration of the equations of solid-phase motion with respect to the radius from $a(t)$ to r, with the boundary conditions [Eq. (4)] taken into account. The mean pressure in the material around the pore $\bar{P}_s$ is found by averaging this pressure distribution over the solid phase volume. Since the mean pressure in a porous medium, also the ISW amplitude, P_m in the model considered is related to the mean pressure in the pore $\bar{P}_g$ and that in the solid material of the cell $\bar{P}_s$ by the expression

$$P_m = \bar{P}_s(1 - \phi) + \bar{P}_g \phi$$

we can establish the following equation for the motion of pore walls [from Eq. 4 it can be deduced that $v_{s+}=\dot{a} - j/\rho_s$]

$$P_m - \bar{P}_g = (2Y/3)ln(1/\phi) - (1-\phi)[4\mu v_{s+}/a - j(v_{s+} - v_{g+})]$$

$$- \rho_s\{(a\dot{v}_{s+} + 2jv_{s+}/\rho_s)[1 - \phi - 1.5(\phi^{1/3}-\phi)]$$

$$+ 1.5\ v^2_{s+}[1 - \phi - (\phi^{1/3}(2 +\phi) - 3\phi)]\} \quad (7)$$

where $\dot{v}_{s+}=dv_{s+}/dt$.

In the case of mechanism A, $\bar{P}_g=0$ and $\lambda_g=0$; hence, there is no need to use Eq. (2). For the sake of simplicity we also disregarded the effects associated with phase transitions and burning of HE before the ignition (j = 0 and $\dot{v}_{s+}=\dot{a}$ in this case) and assumed Y and μ to be constant. Thus, within the framework of mechanism A, the problem reduces to a solution of the heat conduction equation for the solid-phase jointly with Eq. (7). As this simplified model does not account for the generation of gaseous reaction products in the bulk of the solid material surrounding the pore, it is limited to the calculation of solely the critical ignition conditions in the vicinity of a pore being deformed. The dynamics of hot spot growth are beyond its scope.

In ignition mechanism B, the pressure in the pore is assumed to depend only on time, since the time of the pressure-wave journey in it (a_o/c_{sound}) is much less than the characteristic time of viscosity-controlled deformation of pores ($4\mu/P_m$). This assumption is equivalent to ignoring the gas motion in the pore; therefore, viscous heating was ignored in Eq. (2). Furthermore, in this model the temperature dependence of the viscosity and yield strength can also be disregarded, since the solid material temperature is bounded from above by the melting point, and Y and μ start decreasing with temperature rise only when T_s approaches T_m (Borisov et al. 1986). The heat of melting was assumed to be pressure-independent in calculations (Q_m is substantially less than heat of reaction), and the kinetic parameters for the gas-phase chemical reactions were taken equal to Q_s, Z_s, and E_s, respectively (n=1).

The governing equations were solved numerically for mechanism A and approximately by the method of integral relationships for mechanism B. A comparison of the approximate calculations of pore deformation and solid heating with numerical calculations showed reasonable agreement.

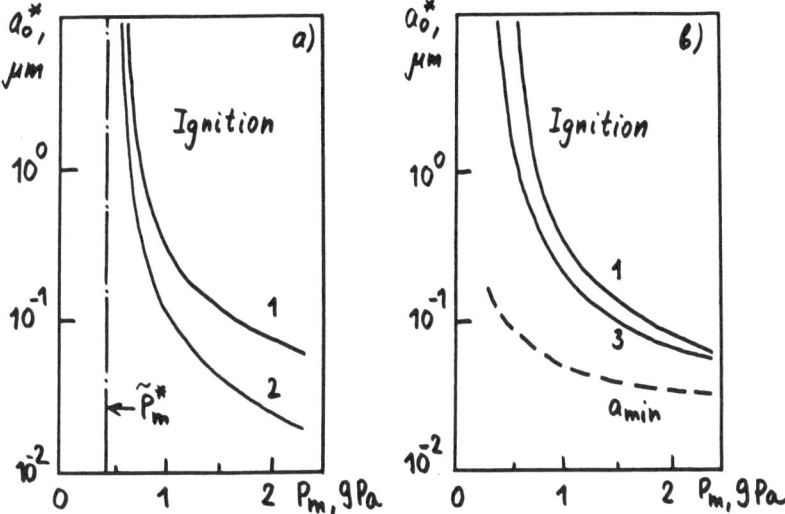

Fig. 1 The effect of viscosity (a) and yield strength (b) on the threshold of reaction initiation in TNT of initial porosity ϕ_0 = 0.1. Yield strength Y = 0.2 GPa for curves 1 and 2 and 0.1 GPa for curve 3. Viscosity μ = 10 Pa-s for curves 1 and 3 and 1 Pa-s for curve 2. The dashed line in b represents the threshold pore radius at which cooling of HE by heat conduction becomes important.

Results and Discussion

To illustrate the effect of the basic parameters on the critical conditions of solid HE ignition in the vicinity of pores being deformed behind the shock wave (mechanism A) and on the ignition and growth of hot spots in mechanism B, we performed calculations for a model solid HE similar in its properties to TNT: ρ_s=1.66 g/cm^3, Q_s=4.31 MJ/m^3, E_s=0.224 MJ/mole, z_s=10^{19} s^{-1}, λ_s=0.2 W/(m°K), and C_s=1205 J/(kg°K). The viscosity of the solid HE was varied over a wide range to demonstrate the limits within which the viscoplastic model can result in reaction center growth.

Critical Conditions of Hot Spot Initiation by Mechanism A

Fig. 1 shows the critical pore size a^*_o, at which the chemical reaction can still be initiated on the pore surface as a function of the ISW amplitude for various values of the viscosity and yield strength. It is assumed that $P_m(t) = P_{mo}$ = const. The explosive is ignited at a given $\tilde{P}_m$ when $a_o > a^*_o$. As seen from the results presented, the plastic properties of the solid HE dominate when the ISW amplitude is commensurate with the material strength ($P_m \approx Y$), but for relatively strong waves ($P_m > Y$) the effect of viscosity prevails. For TNT the material strength becomes unimportant when $P_m \geq 1.5$ GPa.

The influence of structural parameters of a porous HE (a_0 and ϕ_0) on the reaction initiation limit P_m^* is illustrated by the curves plotted in Fig. 2. The results show the following.

1) The structural parameters affect the initiation limit only for $P_m < 2$ GPa; at $P_m \geq 2$ GPa, practically all the pores ignite. Approximately the same pressure range ($P_m = 2.0$-2.2 GPa) was defined by Balinets and Karpukhin (1981) as the range in which the concentration of effective reaction centers reached its maximum value in TNT of the initial density $\rho = 1.66$ g/cm^3 ($\phi_0 \approx 0.06$). Taylor and Ervin (1976) considered the ISW amplitude $P_m \approx 1.7$ GPa to be the limit of detonation initiation in TNT with $\rho = 1.56$ g/cm^3. Recent experimental studies by Balinets and Gogulya (1986) employing light-emission measurements yielded nearly the same values of the detonation initiation limit, with Pm = 2.0 GPa for pressed TNT with $\rho = 1.6$ g/cm^3 ($\phi_0 \approx 0.036$).

2) Near the threshold wave amplitudes the effect of pore size inhomogeneity on the critical conditions for initiation of a chemical reaction in hot spots in TNT is most prominent at low porosities ($\phi_0 < 0.1$). As the initial porosity of HE increases, the dependence of P_m^* on the mean pore size becomes weaker.

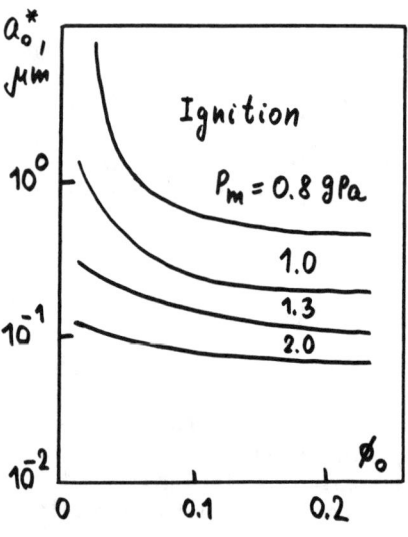

Fig. 2 Influence of structural parameters on initiation threshold of the solid-phase chemical reaction in TNT.

CRITICAL CONDITIONS FOR HOT SPOT EVOLUTION 313

3) The calculated dependence of a^*_o on ISW amplitude shows that a^*_o grows abruptly with decreasing shock-wave amplitude; i.e., a threshold pressure for reaction initiation exists. Earlier we derived the following expressions for estimating this pressure:

$$P^*_m = (2Y/3)\ln(1/\phi_o) + \rho_s c_s (T_{ign} - T_o)/3$$

(Khasainov et al. 1983) and

$$P^*_m = (2Y/3)\ln\{1 + [(1-\phi_o)/\phi_o]$$
$$\cdot \exp[(1.5\rho_s c_s (T_{ign} - T_o)/Y)^{\frac{1}{2}}]\}$$

(Attetkov 1986). Here $T_{ign} \simeq 1000\,°K$ is a typical temperature of pore surface ignition. The values of P^*_m in Fig. 1a are shown by the vertical bar. These two estimates agree with each other and with numerical calculations performed for mechanism A.

4) Since a decrease in the shock amplitude causes a^*_o to increase, the chemical reaction in a very porous HE (with a broad distribution of pore sizes) will be initiated by a relatively weak wave solely around sufficiently large pores. The number of such pores is greater in coarse HE. Therefore, at low shock amplitudes HE's with fine grains are less sensitive than the coarse ones having the same porosity. On the other hand, when an initiating wave is sufficiently strong (to the extent that a^*_o turns out to be much less than the mean pore size), a fine-grained HE will be more sensitive than the coarse one with the same porosity. Hence, the nonmonotonic behavior of the shock sensitivity of porous explosives as a function of their microstructure (Setchell and Taylor 1984) appears to be a natural consequence of the dependence of a^*_o on the ISW amplitude. This effect can also be accounted for based on an estimate of the threshold pore radius a_{min} at which conductive cooling of the hot spot during pore deformation becomes essential. According to Khasainov et al. (1981)

$$a_{min} = 2[m\lambda_s/(\rho_s c_s P_m)]^{\frac{1}{2}}$$

The dashed curve in Fig. 1b shows the a_{min} vs shock amplitude dependence for $\mu = 10$ Pa-s. The a^*_o and a_{min} dependences on the shock amplitude are seen to be qualitatively similar, and their quantitative discrepancy diminishes as the shock amplitude rises and the strength of HE becomes unimportant.

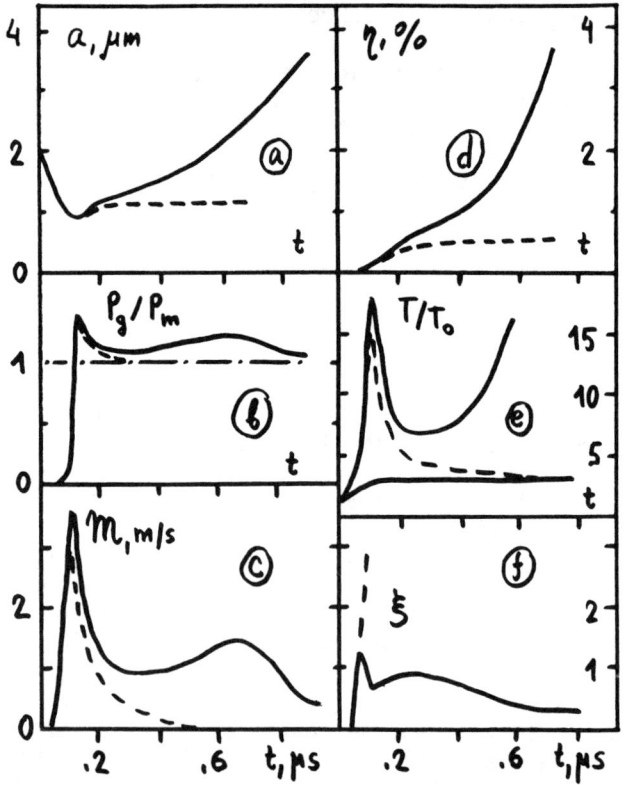

Fig. 3 Dynamics of formation and evolution of a reaction center in TNT. Results are shown for $P_m=1.75$ GPa, $\mu = 50$ Pa-s, $Y = 0$, $\phi_o = 0.05$, $a_o = 2\mu m$, $k = 2$, $b_g = 1$ cm^3/g, $W_g = 22.7$ g/mole, $D = \lambda_g/\rho_g c_g$, $\lambda_g = 0.08$ W/(m°K), the constants in dependence of melting point on pressure $T_m = T_{mo} + C_m P$: $T_{mo} = 478°K$, $C_m = 200°K$/GPa, and $Q_m = 1$ MJ/Kg, with (solid curve) and without (dashed lines) gasphase decomposition of TNT.

Development of Reaction Centers According to Mechanism B

In Fig. 3 the evolution of a reaction center is shown with and without allowance for the gas-phase exothermic reaction of TNT "vapor" in a hot spot formed in an HE charge loaded with a long-duration shock wave. The results diverge only after vapor ignition occurs. First, we discuss the case of inert vapors. At the initial stage of pore deformation a reduction of its radius (Fig. 3a) is accompanied by a rise of the pore surface temperature (the bottom curve in Fig. 3e) and the gas temperature [the medium-dashed curve in Fig. 3e represents a temperature of gas in the center of the pore, $T_g(r=0)$]. At $t=0.07\mu s$ the pore surface temperature reaches the melting point and the phase transition starts, as a result of which vapors of the solid material begin to flow into the pore and thus enhance the pressure rise in the pore (Fig. 3b). The linear regression rate for the solid undergoing the phase transition is shown in Fig. 3c by the dashed curve. The dimensionless parameter ξ, inversely proportional to the gradient of vapor concentration at the pore surface, is presented in Fig. 3f. In the absence of the chemical reaction in the gas, the vapors rapidly fill the pore volume, and their concentration becomes almost uniformly distributed (dashed curve). Characteristics of this stage are a fast rise of the gas pressure, regression rate, and gas temperature in the pore center. The mass fraction of gasified solid material η also grows (Fig. 3d). The pore pressure becomes even higher than the mean pressure P_m in the two-phase medium; therefore, the pore radius ceases to grow and the pore starts to expand (Fig. 3a) at the expense of both the phase transition and the pressure difference. This immediately causes a pressure decay in the pore to the ambient pressure, which in turn leads to a drop of the temperature in the pore and of the regression rate (Figs. 3e and 3c) and η to level off (fig. 3d). Finally the gas temperature drops to the melting point as the result of pore expansion and injection into the pore of a large amount of relatively cold gas at $T_{g+} = T_m$, and solid gasification stops. Starting with this instant (about 0.6µs), the surface layers of the pore are slowly cooled by heat conduction in the solid phase. Because of the phase transition of an inert material, about 0.5% of it converts into gas during pore deformation.

The situation is quite different for reactive vapors. Almost immediately after gasification begins, HE vapors diffuse to the pore center, where the high temperature ignites them. As a result, the vapor front appears as if it is pushed backward to the pore surface (Fig. 3f, solid line) because of the burning of the vapor. With exothermic reaction in the pore the pore pressure and temperature do not drop as rapidly as in the case of inert vapor. The gas temperature then grows to a high value and causes the regression rate to increase (Fig. 3c). After a relaxation period, from 0.1 to 0.8 µs, the regression rate attains a quasisteady value of about 0.4 m/s, and the reaction center grows further in a self-sustaining regime.

The burnt fraction of the HE as a function of time during pore deformation is shown in Fig. 4 for various shock-wave amplitudes. Although the characteristic reaction center burnout times may differ appreciably,

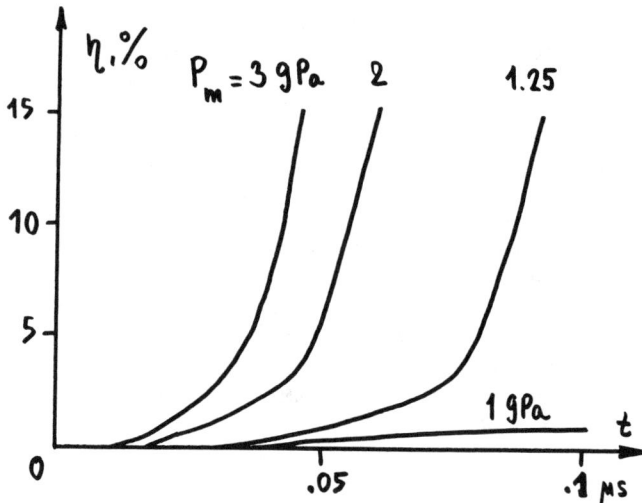

Fig. 4 Burnt fractions of TNT vs time for different shock-wave amplitudes. Results are for: μ = 10 Pa-s, a_o = 1 μm, and k = 2.

depending on values of the main parameters, the time histories of the burnt fractions possess some common features. A critical shock amplitude exists below which no self-sustaining chemical reaction is initiated in the hot spot. In the examples considered this critical pressure is close to 1 GPa, in reasonable agreement with the experimental data [Taylor and Ervin (1976); Balinets and Karpukhin (1981); Balinets and Gogulya (1986)]. The results (Fig. 4) suggest that decomposition of HE around the pore occurs in two stages corresponding to the clearly distinguished portions of the η (t) curves with fundamentally different sensitivities to the shock amplitude. In the initial stage (up to burnt fractions of the order of a few percentage points) the hot-spot burnout dynamics is strongly dependent on the shock amplitude. After the quasisteady self-sustaining regime is attained, the burning rate is only slightly affected by the shock strength. This result agrees qualitatively with the experimental data (Von Holle and Tarver 1981) and can be explained by the fact that, at the pressures considered, the gas density grows only slightly with the growth of pressure, and the burning velocity increases with the rising gas density. At the reaction initiation limit the burnt fraction of the solid HE ($\eta \approx$ 1%) may be sufficient to sustain propagation regimes of low velocity detonations (Khasainov et al. 1977). Calculations also show an interesting effect; namely, in some cases an increase of P_m enhanced HE burning up to $\eta \approx$ 15-30% only. After this the burning rate vs P_m dependence for a growing hot spot is reversed. This result may indicate that the solid-phase reaction should be incorporated into the model.

Figure 5 demonstrates η (t) dependence for various initial pore radii a_0. The smaller the mean pore size, i.e., the higher the HE-specific surface for a fixed value of porosity (ϕ_0 = 0.05 in the case of issue), the faster the reaction center grows. This trend is observed only for mean pore sizes exceeding the critical value $a^*_o \approx a_{min}$. For example, at a_o = 0.3µm the chemical reaction cannot be initiated with the other parameters being equal. Thus, the gas-phase ignition mechanism also may explain the nonmonotonic dependence of shock sensitivity of porous explosives on their microstructure.

The time histories of the burnt fraction and gas pressure in the pore, for the case when the porous HE is pressurized by a shock wave and ramp compression wave with an extended front of the same amplitude (2.5 GPa), are

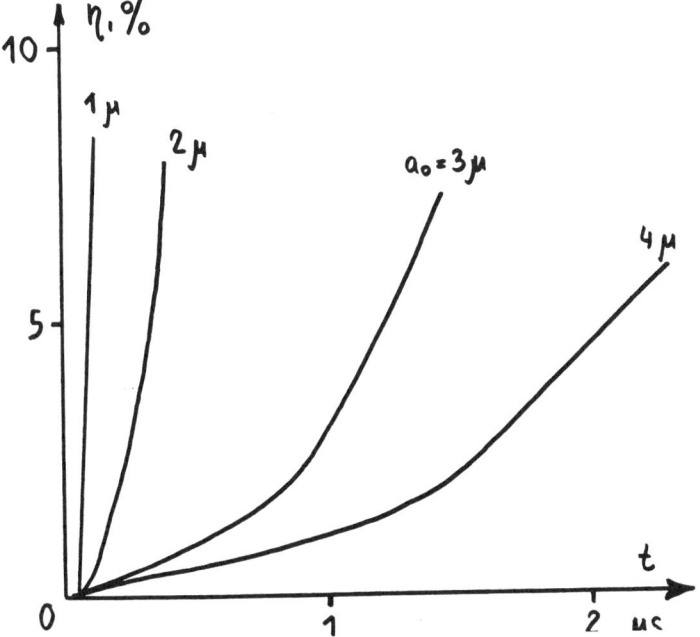

Fig. 5 Burnt fractions of TNT vs time for shock wave initiation at different values of the mean pore size. Results are for P_m = 2.5 GPa and μ = 100 Pa-s

presented in Fig. 6. The calculations demonstrate that viscosity increases result in longer hot spot ignition delays, since a_{min} rises as the viscosity grows and may become commensurate with a_o. The pressure in the pore after ignition also increases, since the forces resisting the pore expansion increase. In a wave with a longer rise time (2 μs instead of 1 μs), the reaction centers form much

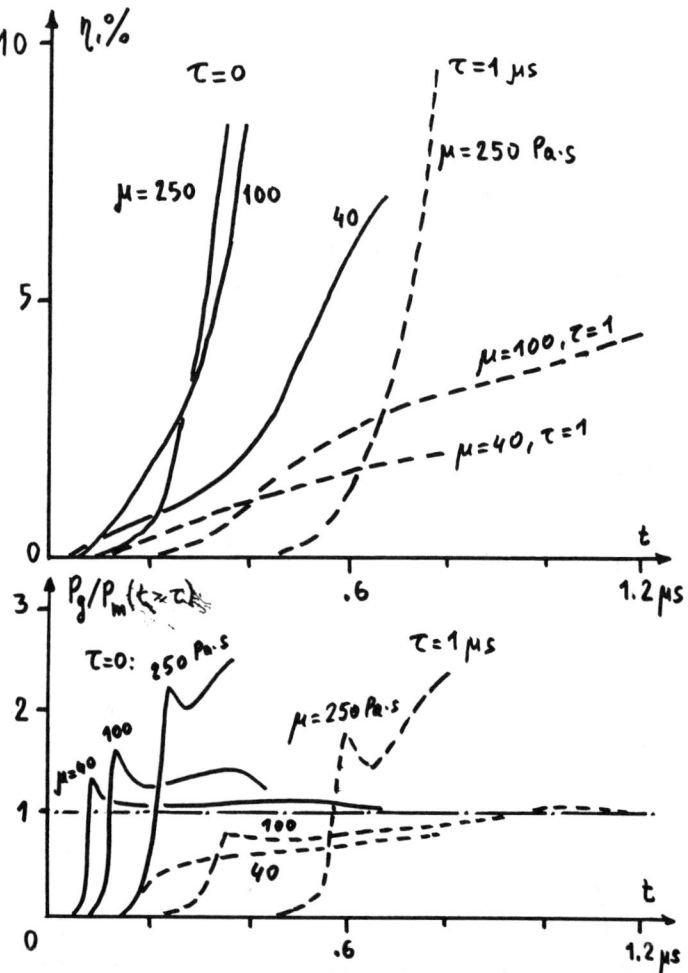

Fig. 6 Comparison of burnt fractions vs time curves for porous TNT (ϕ_o = 0.05) of various viscosity loaded by a shock and compression waves of the same amplitude (2.5 GPa). The pressure in the ramp wave changes stepwise to 0.5 GPa at time zero [to provide $P_m(t=0) > (2Y/3)\ln(1/\phi_o)$] and then to grow linearly with time to 2.5 GPa for τ = 1 and 2 μs.

more slowly, and, when $\tau = 2\mu s$ and $\mu = 40$ Pa-s, no reaction is initiated in the hot spot.
Additional calculations with gas heat conductivity set at zero show that at the initial stage of pore deformation thermal conduction from the gas phase contributes little to solid phase heating. During this time the gas pressure is much lower than the solid phase one (i.e., the HE is heated at the expense of the viscous energy dissipation and the work of plastic deformation). After the gas and solid pressure become equal, the energy is transferred to the fresh solid around the pore solely by gas-phase heat conduction rather than by viscous energy dissipation. If the HE strength is taken into account, the gas pressure at the stage of self-sustaining hot spot growth exceeds the solid-phase pressure by $(2Y/3)\ln(1/\phi) \approx 0.1$ GPa; v_{s+} in this case is nearly zero. The proposed model ignores the compressibility of solid HE and hence the possible role of hydrodynamic flow in the energy transfer mechanism. After postignition pressure relaxation in the gas phase, heat conduction becomes the major energy transfer mechanism from the gas to the solid in the model considered.

Summary

The main results of the computations are as follows:
1) The gas-phase ignition mechanism may describe both the development of a reaction center (from ignition to self-sustaining burning) and its extinction.
2) The two ignition mechanisms considered explain the experimentally observed nonmonotonic dependence of shock sensitivity of porous high explosives on grain and pore size through a comparison of the mean pore size with the critical one at which the self-sustaining chemical reaction can still be initiated in the hot spot. This critical pore size increases when the amplitudes of constant or ramped strength shock waves decline.
3) According to the gas-phase initiation mechanism, the rate of reaction center growth is determined by the initiating wave amplitude at the initial stage of the reaction but is practically independent of wave amplitude when the burnt fraction of the solid explosive around the pore exceeds several percentage points. This prediction is also consistent with the experimental data.
4) The model suggested may be employed in modeling detonation buildup (particularly low-velocity detonations) and estimating critical pressures of local chemical reaction initiation in porous high explosives behind shock and compression waves.

References

Amosov, A. P. (1982) Heating and ignition of solid reactive systems due to high velocity friction accompanied by formation of plastic and liquid layers. Khim. Fiz. 1, 1401.

Attetkov, A. V. (1986) Critical conditions of chemical reaction initiation in solid heterogeneous substances. Sixth Soviet Congress on Theoretical and Applied Mechanics. Abstracts, Tashkent, USSR, Vol. 57.

Balinets, Y. M. and Karapukhin, I. A. (1981) On initial stage of detonation initiation process in pressed TNT. Fiz. Goreniya Vzryva, 17, 103.

Balinets, Y. M. and Gogulya, M. F. (1986) Emissivity of shocked charges of pressed TNT. Khim Fiz. 5, 263.

Borisov, A. A., Ermolaev, B. S., and Khasainov, B. A. (1986) The model of hot spot growth during visco-plastic pore deformation. Sixth Soviet Congress on Theoretical and Applied Mechanics. Abstracts, Tashkent, USSR, Vol. 129.

Carrol, M. M. and Holt, A. C. (1972) Static and dynamic pore-collapse relations for ductile porous materials. J. Appl. Physiol., 43, 1626.

Dunin, S. Z. and Surkov, V. V. (1979) Structure of a shock wave front in a porous solid. Zh. Prikl. Mekh. Tekh. Fiz., 5, 106.

Frey, R. B. (1981) The initiation of explosive charges by rapid shear. Seventh Symposium (International) on Detonation, NSWC MP 82-334, p. 36.

Frey, R. B. (1985) Cavity collapse in energetic materials. Eighth Symposium (International) on Detonation, Preprints of papers, CONF-850706, Vol. 1, p. 385.

Hasegawa, T. and Fujiwara, T. (1982) Detonation in oxyhydrogen bubbled liquids. 19th Symposium (International) on Combustion, The Combustion Institute, Pittsburgh, PA, p. 675.

Hayes, D. B. (1983) Shock induced hot-spot formation and subsequent decomposition in granular, porous HNS explosive. Progress in Astronautics and Aeronautics: Shock Waves, Explosions, and Detonations, Vol. 87, edited by J. R. Bowen, N. Manson, A. K. Oppenheim, and R. I. Soloukhin, 445-467.

Khasainov, B. A., Ermolaev, B. S., Borisov, A. A., and Korotkov, A. I. (1977) Low velocity detonations in high density high explosives. Khimicheskaya Fizika Protsessov Goreniya i Vzryva Detonatsya, Chernogolovka, USSR, p. 79.

Khasainov, B. A., Borisov, A. A., Ermolaev, B. S. and Korotkov, A. I. (1981) Two-phase viscoplastic model of shock initiation of detonation in high density pressed explosives. Seventh Symposium (International) on Detonation, NSWC MP 82-334, p. 435.

Khasainov, B. A., Borisov, A. A. and Ermolaev, B. S. (1983) Shock wave predetonation processes in porous high explosives. AIAA Progress in Astronautics and Aeronautics: Shock Waves, Explosions, and Detonations, Vol. 87, edited by J. R. Bowen, N. Manson, A. K. Oppenheim, and R. I. Soloukhin, AIAA, New York, pp 492-504.

Kim K. and Sohn, C. H. (1985) Modeling of reaction build-up processes in shocked porous explosives. Eighth Symposium (International) on Detonation, Preprints of papers, CONF-850706, Vol. 2, p. 641.

Maiden, D. E. (1987) A model for calculating the threshold for shock initiation of pyrotechnics and explosives, 12th International Pyrotechnic Seminar, France, p. 17.

Mader, C. L. (1979) Numerical Modeling of Detonations. Univ. of California Press, Berkeley, CA.

Nigmatullin, R. I. (1978) Fundamentals of Heterogeneous Media Mechanics, Nauka, Moscow.

Setchell, R. E. and Taylor, P. A. (1984) The effects of grain size on shock initiation mechanisms in Hexanitrostilbene (HMX) explosive. Progress in Astronautics and Aeronautics: Shock Waves, Explosives, and Detonations, Vol. 94, edited by J. R. Bowen, N. Manson, A. K. Oppenheim, and R. I. Soloukhin, AIAA, New York, 350-368.

Soloviev, V. S., Attetkov, A. V., and Pyriev, V. A. (1981) Investigation of cast explosive microstructure. Detonatsya. Materialy Vses. Soveshchanya po Deonatsii. Chernogolovka, USSR, Vol. 2, p. 61.

Taylor, B. C. and Ervin, L. H. (1976) Separation of ignition and buildup to detonation in pressed TNT. Sixth Symposium (International) on Detonation, Office of Naval Research, ACR-221, p. 3.

Von Holle, W. G. and Tarver, C. M. (1981) Temperature measurements of shocked explosives by time-resolved infrared radiometry - a new technique to measure shock induced reaction. Seventh Symposium (International) on Detonation, NSWC MP 82-334, p. 993.

Von Holle, W. G. (1983) Shock wave diagnostics by time resolved infrared radiometry and nonlinear Raman spectroscopy. Shock Waves in Condensed Matter. Proceedings of the American Physical Society Topical Conference, p. 283.

Wackerle, J. and Anderson, A. B. (1983) Burning topology in the shock induced reaction of heterogeneous explosives. Shock Waves in Condensed Matter. Proceedings of the American Physical Society Topical Conference, p. 601.

Mechanism of Deflagration-to-Detonation Transition in High-Porosity Explosives

A. A. Sulimov,* B. S. Ermolaev,† and V. E. Khrapovskii‡
USSR Academy of Sciences, Moscow, USSR

Abstract

This work represents the key results of a complex investigation of deflagration-to-low-velocity detonation transition, carried out for granular nitrocellulose propellants confined in steel tubes. The transition occurs according to the mechanism in which the detonation wave rises upstream of the convective flame front in the unburnt explosive column. Experiments were performed employing a combined technique that comprises simultaneous optical and piezometric monitoring of the process. The details of the flame trajectory and the space pressure profile are discussed. Precompaction of the explosive material ahead of the flame front leads to stabilization of the flame propagation, explosive particle movement, and a decrease in the pressure rise rate. Before the onset of detonation, the explosive density ahead of the flame front approaches the theoretical maximum value, and the particle velocity increases up to 200 m/s. The distance to low-velocity-detonation onset is proportional to the initial particle diameter. Experimental results are compared with the numerical modeling. Fairly good agreement is obtained for explosive performance. Limits of applicability of modern theory of convective burning are determined. Essential deviations of the theory from the experimental data are found to exist for fine explosives [i.e., in the case of picric acid and pentaerythritol tetranitrate (PETN)]. Deflagration-to-detonation transition (DDT) in the explosives occurs via formation of a strong secondary compresssion wave in the combustion zone downstream of the flame front. The detonation occurs when this secondary wave overtakes the leading flame front. Criteria are derived to distinguish between the two mechanisms of DDT considered.

Copyright © 1988 by USSR Academy of Sciences. Published by the American Institute of Aeronautics and Astronautics, Inc. with permission.
*Head of Laboratory, Institute of Chemical Physics.
†Senior Researcher, Institute of Chemical Physics.
‡Senior Researcher, Institute of Chemical Physics.

TRANSITION TO DETONATION IN POROUS EXPLOSIVES

Introduction

The results of the recent investigation by the authors on the deflagration-to-detonation transition (DDT) in porous explosives are systematized. Particular attention is given to the mechanism and regularities of the propagation of an unsteady-state convective burning and its transition to a low-velocity detonation (LVD) [Sulimov and Ermolaev (1986); Sulimov et al. (1987); Belyaev et al. (1973)].

The experimental data on convective burning development are compared with the results of the numerical modeling. The quantitative criteria are established, making it possible to distinguish between the conditions under which two types of DDT are realized.

Experiments were made with nitrocellulose grain charges contained in strong confinements. The initial grain size varied from 0.6-3.3 mm. Porosity of charges was near 0.45. The complex experimental technique enabled the simultaneous optical and piezometric recording of the process up to the initiation of a low-velocity detonation (Fig. 1). Nitrocellulose grains were placed into the cylindrical channel of a thick-wall confinement of 15 mm i.d. and 200-800 mm long. On one end the confinement was closed with a plug in which an igniter was mounted; the other end carried a plug with a membrane. Pressure measurements (up to 1 GPa) were provided by high-frequency piezoelectric pressure gages described elsewhere (Belyaev et al. 1973). Such pressure gages (up to 8 pieces) were located along the confinement, one of them being near the igniter. The process was optically recorded either

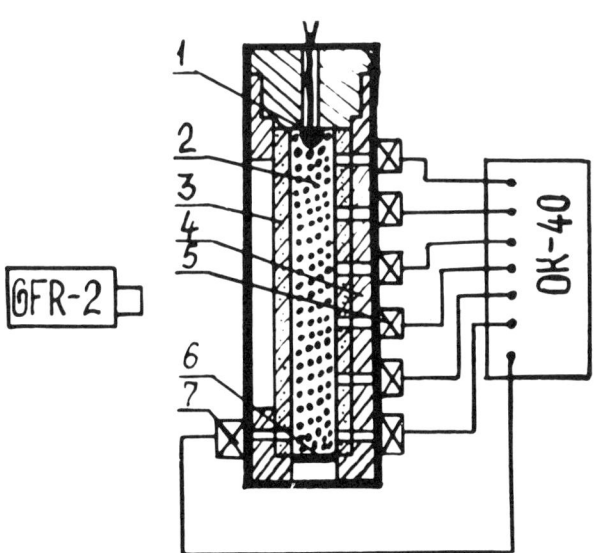

Fig. 1 DDT transparent pipe test configuration. 1: Igniter; 2: explosive; 3: Plexiglas pipe; 4: steel tube with slit; 5: pressure gage; 6: membrane; 7: photodiode.

through a transparent unit consisting of a Plexiglas cylindrical tube inserted into the ground channel of the steel confinement with a longitudinal slit or through a number of small-diameter radial holes drilled in the encasement. Synchronization of piezometric and optical measurements was provided by a photodiode positioned in the same section as a piezoelectric pressure gage. Signals from pressure gages were displayed on the screen of an eight-ray oscilloscope (OK-40); the process was photographed with a streak camera (GFR-2).

The following characteristics were determined: the distance traveled by the flame front (glow front) and its velocity, the movement velocity of burning explosive grains near the flame front, spatial-temporal pressure profiles, and the distance in which the transition from convective burning to low-velocity detonation occurred.

The results obtained can be summarized as follows [Sulimov and Ermolaev (1986); Sulimov et al. (1987)]:

1) There are two stages of convective burning: an accelerated process and a stabilized process. In the latter stage, despite an exponential rise in pressure in the combustion zone, the flame velocity is approximately constant, being about 400-500 m/s (Fig. 2). Increasing the grain size or decreasing the conductive combustion velocity with an inhibitor leads to a decrease in the pressure gradients and an increase in the length and duration of both stages. However, the flame velocity at the stage of steady-state convective burning remains almost constant.

2) The spatial pressure profiles in the convective burning wave represent flat steps with steeply rising regions in which the flame front is localized. As time passes, the step height and the steepness of the rising

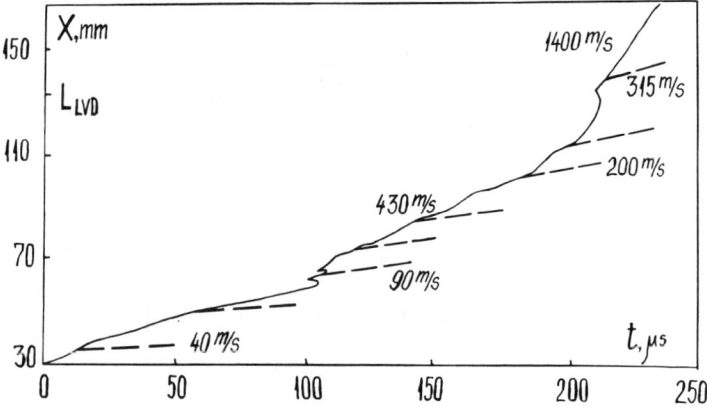

Fig. 2 Sketch of streak camera data of DDT (type 1) for nitrocellulose, grain size 0.8 mm. Solid line, luminous front, (numerical value is velocity of the front); dashed line, tracks of igniting grains near the flame front (numerical value is the grain movement velocity).

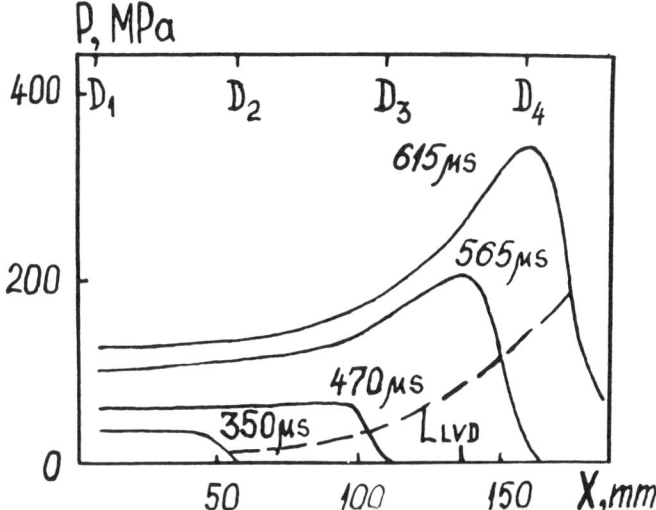

Fig. 3 Evolution of experimental spatial pressure profiles, nitrocellulose, grain size 0.8 mm. D_i, place of the pressure gage (numerical value is time of process). Dashed line, trace of luminous front.

region become large, and a "hump" appears near the flame front. This hump increases and then passes into a triangular peak when transition to a low-velocity detonation occurs (Fig. 3).

3) The pressure in the combustion zone near the igniter rises nearly exponentially, with a characteristic time proportional to the time constant calculated from the law of pyrostatics. For charges of gravimetric density, the coefficient of proportionality is equal to 1.5-2, regardless of grain size.

4) Compaction of the explosive material in the layers immediately before the flame front leads to flame velocity stabilization resulting from a decrease in the gas permeability of these layers. It also leads to a lower rate of pressure rise and to grain movement with increased velocities that amount to about 200 m/s immediately before the transition to a low-velocity detonation.

5) The distance within which the DDT occurs varies in direct proportion to the initial grain size (Fig. 4). The transition mechanism belongs to the first type in which the detonation wave is formed before the flame front in the "nonburning" explosive.

Convective burning was numerically modeled (Ermolaev et al. 1985) to better understand the mechanics of two-phase reacting media underlying the model. The following assumptions were made: 1) a detailed chemical kinetics of ignition and combustion reactions has no appreciable effect on convective burning, and the evolution of the process is mainly determined by the physiomechanical regularities of

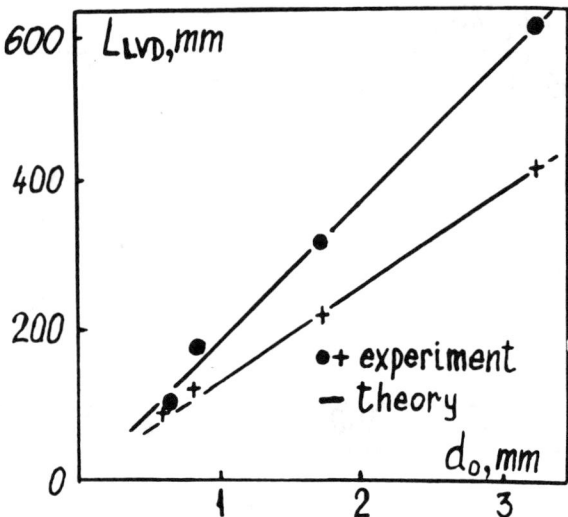

Fig. 4 Distance of the low-velocity detonation onset as a function of the initial grain size for nitrocellulose of two different levels of conductive combustion velocity. +, Experiment; Dashed line, theory.

Table 1 Parameters used in numerical modeling of convective combustion

Properties		
Physical and geometrical for porous medium	Thermophysical and thermodynamic	Physiochemical
Effective grain size, porosity Charge length Constants in the filtration law Compressibility	Corresponding constants of explosive and combustion products Combustion heat Constants in the law of interphase heat exchange	Ignition temperature Constants in the law of conductive combustion velocity

the motion of a reacting two-phase mixture; and 2) after ignition, combustion proceeds all over the internal surface of the porous explosive (which depends on the degree of dispersion), with a regression rate equal to the velocity of steady-state conductive combustion.

The preceding assumptions enable the number of parameters used for numerical modeling to be restricted (Table 1).

The numerical modeling results were compared with the experimental data for nitrocellulose. As distinct from

TRANSITION TO DETONATION IN POROUS EXPLOSIVES   327

earlier works in which comparison was fragmentary (Devis and Kuo 1979), this comparison (Sulimov et al. 1987) was made with respect to a set of characteristics that include the x-t diagram of the flame-front travel, the velocity of travel of burning grains near the flame front, pressure change at the ignition zone, and the pre-low-velocity--detonation distance, which was determined by the instant collapse of pores in the compression zone before the flame front. Good quantitative agreement between the theory of convective burning and the experimental observations was found (see Fig. 4). The mechanism of energy transfer in the wave was refined, and the dynamics of substance compression was investigated in detail. The "traditional" mechanism of convective heating based on the transfer of the energy released in the combustion zone by the filtration stream of hot gases (the "leading" filtration mechanism in which the velocity of gases in pores is higher than the flame velocity) is found to play a decisive role, but not at very high velocities and pressures. At flame velocities above 150-200 m/s and a grain size of about 1 mm, another mechanism of the convective heating of the pore surface is realized in which the leading role belongs to gasdynamic dissipative processes that accompany a high-velocity friction of gases on the pore walls.

Earlier investigations [Belyaev et al (1973); Korotkov et al (1969)] and flash radiography [Sandusky and Bernecker (1985)] demonstrated that the explosive density in the compacting zone (ahead of the flame front) approaches a maximum (porosity tends to zero) immediately before detonation. The calculated values of the pre-low-velocity-detonation distance determined by the instant of the complete collapse of pores coincide with the experimental values for

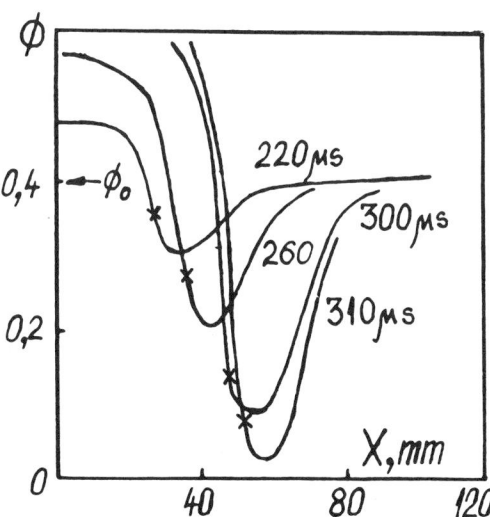

Fig. 5 Evolution of porosity profiles for nitrocellulose, grain size 0.4 mm, theory (Belyaev et al. 1973). x, Flame-front location.

nitrocellulose charges with different grain sizes. Taking
into account the evolution of pressure profiles (Fig. 3) and
of porosity profiles (Fig. 5) and the data on shock-wave
sensitivity of porous explosives (Soloviev 1977), we can
conclude that it is the compression of a porous body accompanied by energy accumulation on the pore surface that is
the main source of formation of hot spots during DDT. To
make the DDT model closed, it is necessary for a detailed
mechanism of formation of reaction sites to be involved,
e.g., the mechanism of viscoplastic heating during deformation of pores (Khasainov et al. 1981).

The comparison of the theoretical conclusions with the
experimental findings made it possible to determine the
applicability limits of the convective burning model.
Essential deviations from theoretical predictions were found
in the case of explosives whose grains were less than
50-100 μm in size because of the incompleteness of ignition
of the grain surface in the flame front. In this case DDT
proceeds by the second mechanism: an intense secondary pressure wave (Ermolaev et al. 1985) is formed in the combustion
zone, and when it overtakes the flame front (Sulimov and
Ermolaev 1986), a detonation wave (Belyaev et al. 1973)
arises (Fig. 6).

The concept of two types of DDT was first advanced in
works devoted to a study of DDT in fine-grained PETN
[Belyaev et al. (1973); Korotkov et al. (1969)] and was then
considered by Bernecker and Price (1974). Using the
literature data [Sokolov and Aksenov (1963); Bernecker et
al. (1976)] and findings from our laboratory for fine-
grained PETN, tetryl, and picric acid [Sulimov and Ermolaev
(1986); Belyaev et al. (1973); Korotkov et al. 1969)], we
have found the quantitative regularities that permit us to
distinguish between the two types of DDT.

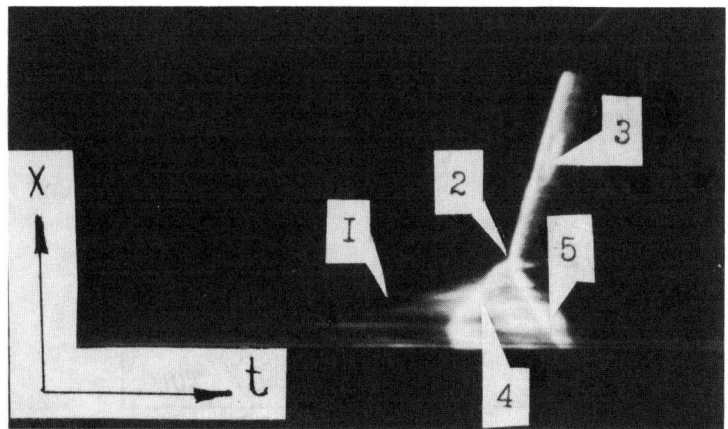

Fig. 6 Streak camera data of DDT (type 2) in fine PETN (Korotkov
et al. 1969). 1: Convective burning front; 2: detonation onset;
3: detonation wave; 4: secondary wave; 5: detonation wave.

The first type of DDT studied in detail in the previously mentioned experiments on coarse-grain nitrocellulose (grain size at least 0.6 mm) is characterized by the following features: 1) a pronounced and regularly reproducible flame front corresponding to the convective burning can be observed on the photographs obtained with the streak camera; 2) the pressure in the combustion zone rises at a rate close to that predicted by the mathematical modeling, assuming the complete surface combustion of the explosive grains behind the flame front; and 3) the pre-low-velocity-detonation distance varies approximately in proportion to the grain size. The characteristic features of the second tytpe of DDT are as follows: 1) no pronounced regular flame front can be observed in the photographs obtained during convective burning, and the glow represents a set of separate nearly horizontal bright streaks; 2) the rate of pressure rise, up to the moment of the appearance of the secondary wave, is considerably lower (1-2 orders) than the calculated one that corresponds to the ignition of the complete grain surface; and 3) the fineness of the explosive makes the explosion development difficult and leads to an increase in the predetonation distance.

These features make it possible to explain the difference in the character of explosion development by the fact that during convective burning under certain conditions the flame does not penetrate into the small pores, and most of the explosive surface is not ignited. As a result, the flame propagates only through large pores (Belyaev et al. 1973). As the pressure in the combustion zone rises, the favorable conditions are created for convective ignition of small pores when the critical pressure is reached, the combustion surface increases sharply, and the secondary wave with a high pressure rise rate appears. Thus, the type of DDT is determined by the physical state of an explosive rather than by its nature. Depending on the grain size and charge density, both types of transition can be realized with the same explosive (Korotkov et al. 1969). This concept differs from that of American investigators (Bernecker et al. 1982), who connect the difference in the DDT mechanism only with the nature of explosives.

References

Belyaev, A. F., Bobolev, V. K., Korotkov, A. I., Sulimov, A. A., and Chuiko, S.. (1973) The Transition of Burning to Explosion in Condensed Media. Nauka, Moscow, USSR, (In Russian).

Bernecker, R. R. and Price, D. (1974) The study of DDT in porous explosives. Combust. Flame, 22, 111, 129, 161.

Bernecker, R. R., Price, D., Erkman, I. O., and Clairmont-IR, A.R. (1976). DDT Behavior of Tetryl. VI Symposium (International) on Detonation Proceedings. ACR-21, ONR, p. 426.

Bernecker, R. R., Sandusky, N. W., and Clairmont-IR, A. R. (1982) DDT studies of porous explosive charges in plastic tubes. 7th Symposium (International) on Detonation, MP 82-334, NSWS, pp. 119-138.

Devis, T. K. and Kuo, K. K. (1979) Experimental study of the combustion process in granular propellant beds. J. Spacecraft Rockets, 16, 203-209.

Ermolaev, B. S., Novozhilov, B. V., Rosviansky, V. S., Sulimov A. A. (1985) The results of numerical modeling of convective burning in explosive materials at the rising pressure. Fiz. Goreniya Vzryva 21, 3-12. (In Russian).

Khasainov, B. A., Borisov, A. A., Ermolaev, B. S., and Korotkov, A. I. (1981) Two-phase viscoplastic model of shock initiation of detonation in high density pressed explosives. 7th Symposium (International) on Detonation. NSWCMP 82-334, p. 435.

Korotkov, A. I., Sulimov, A. A., Obmenin, A. I., Dubovitsky, V. F., and Kurkin, A.I. (1969) Deflagration to detonation transition in porous explosives. Fiz. Goreniya Vzryva 5, 315.

Sandusky, H. W. and Bernecker, R. R. (1985) Compressive reaction in porous beds of energetic materials. 8th Symposium (International) on Detonation, Preprint, ACC. Vol. 2, pp. 631-640.

Sokolov, A. V. and Aksenov, Y. N. (1963). Initiation and development of detonation in RDX. Explosion Work, 5219, GGTI, Moscow, USSR, p. 201.

Soloviev, V. S. (1977) The shock-wave initration in condensive explosives. Chemical Physics of combustion and explosion. Detonation. Chernogolovka, Moscow, USSR, pp. 12-20.

Sulimov, A. A. and Ermolaev, B. S. (1986) Deflagration to detonation transition in solid explosives. Detonation and Shock Waves, 8th Symposium (USSR) on Combustion and Explosives, Chernogolovka, Moscow, USSR, p. 134.

Sulimov, A. A., Ermolaev, B. S., Korotkov, A. I., Okunev, V. A., Dosviansky, V. S., and Foteenkov, V. A. (1987) The properties of convective burning wave in confinement. Fiz. Goreniya Vzryva 23.

Effect of Graphite and Diamond Crystal Form and Size on Carbon-Phase Equilibrium and Detonation Properties of Explosives

S. A. Gubin,* V. V. Odintsov,† S. S. Sergeev ‡
Moscow Physical Engineering Institute, Moscow, USSR
and
V. I. Pepekin§
USSR Academy of Sciences, Moscow, USSR

Abstract

The effect of a condensed phase (graphite and diamond) fineness on the detonation properties of carbon-rich explosives is considered. The contribution of the surface area to the energy of a fine crystalline substance depends on the size and form of a crystal and can be taken into account by changing the corresponding standard enthalpy of formation of the substance. The changes of pressure of coexistence of the carbon solid phases as functions of a form of a crystal and number of atoms in it are given. The coexistence of the two carbon solid phases in detonation products results in a specific transition portion in a graph of calculated dependence of detonation velocity on initial density. The parameters of the carbon solid phases pertaining to the beginning of the transition portion of the detonation velocity curve of TNT at the initial density of 1550 kg/m³ are given. Taking into account the diamond formation in the detonation products and the effect of solid carbon particle size allows one to explain, at least qualitatively, the peculiar behavior of the TNT's detonation velocity vs initial density curve that had been previously noted.

Copyright © 1988 by the American Institute of Aeronautics and Astronautics, Inc. All rights reserved.
*Reader.
†Senior Research Fellow.
‡Engineer.
§Professor.

Introduction

Presently, the phase diagram of carbon attracts the attention of researchers who deal with shock-wave and detonation processes. On the one hand, the interest is caused by a possibility of explosive diamond synthesis [Gustov et al. (1982); Trefilov et al. (1978); Staver et al. (1984)], and on the other hand, correct information on the thermodynamic state of condensed carbon is needed to improve the accuracy and reliability of computations of explosive processes [Gubin et al. (1984); Baute et al. (1985); Gubin et al. (1986a, b)].

Formation of the diamond phase of carbon in the detonation products has been unequivocally established. It has been confirmed by both computations and experiments. At the same time, the experiments on explosive diamond synthesis show that the condensed phase is generated in the form of a fine powder. The question about the effect of a condensed-phase fineness on the detonation properties of carbon-rich explosives arises in this connection.

Effect of Crystal Form and Size on the Thermodynamic Properties of Diamond and Graphite

The formation of a unit surface area of a condensed substance requires some work, which is usually referred to as a surface energy density:

$$\sigma = (\partial U/\partial A)_{S,V,N} = (\partial H/\partial A)_{S,p,N} = (\partial F/\partial A)_{T,V,N}$$
$$= (\partial G/\partial A)_{T,p,N}$$

The values of temperature T and pressure P being constant, the change in the surface area A of a given amount of substance is accompanied by changes in enthalpy H and entropy G:

$$\Delta H_{T,p} = \int_A [\sigma - T(\partial \sigma/\partial T)_p] dA, \quad \Delta S_{T,p} = -\int_A (\partial \sigma/\partial T)_p dA$$

The surface tension of liquids, which is equal to the surface energy density, depends linearly on temperature up to the critical point temperature. On the other hand, entropy of a substance tends toward zero when the temperature approaches zero, regardless of the dispersion of the substance. This means that for liquids the linear dependence of the surface energy density on temperature becomes weaker as the temperature decreases; and in the vicinity of absolute zero, the surface energy density of crystals no longer depends on the temperature.

It can be assumed that the surface energy density of crystals is independent of temperature up to the melting point. Then, the equation

$$(\partial \sigma / \partial T)_p = 0$$

holds in the entire region of existence of the crystalline phase of a substance, and the entropy change is equal to zero when the surface area changes.

A change in the enthalpy of a crystalline substance associated with variation of the surface area is independent of temperature under the previous assumption. Thus, the effect of surface area (i.e., of the particle size) on the thermodynamic functions of a crystalline substance reduces to temperature-independent changes of the energy parameters and can be taken into account by correcting the standard enthalpy of formation of the substance.

The contribution of the surface energy of a fine crystalline substance can be assessed by direct counting of broken bonds on the surface of the crystals. This number depends not only on the size but also on the form of the crystals.

Based on the physical reasoning and for the sake of simplicity and convenience, two cases have been chosen for analysis: the equilibrium of graphite and diamond crystals in their 1) thermodynamic and 2) compact forms.

An octahedron is one of the thermodynamic forms of a diamond crystal, and a film in the plane of a stratified graphite structure is the thermodynamic form of a graphite crystal. A rectilinear hexagonal prism with the eccentricity $\epsilon = D/(2B) = \sqrt{2}/2$ (where D is the prism height and B is the length of the prism base) has been assumed as the compact form of diamond and graphite crystals. The compactness of a prism with such a ratio of height and base-edge length is demonstrated by the fact that, with the number of atoms being fixed, the most distant points of the prism (i.e., the vertices) are located at the minimum distance from the prism center. It should be noted that the minimum surface energy of a graphite crystal in prism form is reached at an eccentricity value of less than 10^{-2}, which allows one to consider such a prism to be a film. Since the definition of a direction in a crystal is of great importance, the orientation of crystal form with respect to the crystal lattice should be specified. The bases of the graphite prism coincide with the layers of carbon atoms, and the bases of a cubic diamond prism are parallel to the crystallographic plane (1, 1, 1).

Compact forms of crystals need to be considered because the form of carbon crystals produced in detonation products can be determined not only by thermodynamic characteristics but by kinetic ones as well. Under these conditions, the compact form may turn out to be closer to reality than the thermodynamic one.

Without counting the broken bonds and detailing the geometrical properties of crystals, we present some results of calculations that have taken into account the effects of carbon particle size.

Table 1 shows the least-squares coefficients of the equation

$$\chi = \alpha \cdot N_{at}^{\beta}$$

approximating the dependence of a typical dimension of crystals (the length of octahedron edge or the prism base

Table 1 Coefficients of approximation of characteristics of the dispersed carbon solid phases

Approximated parameter	Form of a crystal	Carbon phase	α	β
Dimension of a crystal, nm	Thermodynamic	Graphite	0.4464	0.3533
		Diamond	0.2064	0.3397
	Compact	Graphite	0.1238	0.3384
		Diamond	0.1110	0.3357
A_M, 10^{-5} m^2/mole	Thermodynamic	Graphite	6.065	-0.2906
		Diamond	0.8887	-0.3206
	Compact	Graphite	1.245	-0.3222
		Diamond	0.9614	-0.3253
$\Delta(\Delta_f H°_{298})$, kJ/mole	Thermodynamic	Graphite	185.1	-0.3386
		Diamond	586.5	-0.3334
	Compact	Graphite	526.6	-0.3351
		Diamond	670.5	-0.3333

A_M, molar surface area; $\Delta(\Delta_f H°_{298})$, change of standard enthalpy of formation. Dimensions of a crystal pertain to the edge length of the octahedron of prism base.

edge), a molar surface area A_M, and a change in the standard enthalpy of formation $\Delta(\Delta_f H°_{298})$ of graphite and diamond for the thermodynamic and compact forms on an amount of atoms in a crystal in the range $N_{at}=10^4 - 10^7$.
The equation describes the geometric parameters of dispersed diamond and graphite, except for the thermodynamic form of graphite (i.e., for a film). In the latter case the discreteness of a film thickness (i.e., of the prism height) associated with the appearance of new carbon atom layers when the prism base length attains certain critical values strongly affects the geometric parameters of the crystals. Such a strong effect by the structure discreteness is due to the small number of layers along the prism height in the case of the thermodynamic form of graphite crystals. The discreteness of graphite film layers imposes far less effect on the change of the standard enthalpy of formation. Consequently, the previously mentioned equation gives a good approximation of a change of the standard enthalpy of formation for graphite and diamond as a function of a number of atoms in a crystal for both the thermodynamic and compact forms.

Table 2 shows that the form and number of atoms in a crystal (defining its dimensions) considerably affect the phase diagram of carbon. The changes of pressure of coexistence of the solid carbon phases as functions of the form of a crystal and the number of atoms in it are shown here. The equilibrium computations for dispersed carbon phases were carried out by using approximate values of changes of the standard enthalpy of formation of graphite and diamond. It was assumed that the grain size did not affect the thermal equation of state written in the form of Cowen's equation proposed originally for graphite [Cowen et al. (1956)]. Coefficients of that equation for diamond were borrowed from Baute et al. (1985). The assumption of incompressibility of diamond results in some increase of the change in the phase-equilibrium pressure.

It follows from Table 2 that a decrease in the number of atoms in a crystal for the considered forms leads to a rise in graphite-diamond equilibrium pressure. This increase is much less for the compact form than for the thermodynamic one. It results from the significantly stronger effect a crystal form has on the formation enthalpy of graphite than it has on that of diamond. As a consequence, there is even a possibility that graphite-diamond equilibrium pressure decreases for some forms of crystals. Thus, for example, for crystals in the form of a rectilinear hexagonal prism with the eccentricity exceeding approximately 2.56, the pressure of carbon solid-phase

Table 2 Changes of pressure of dispersed graphite and diamond equilibrium (in GPa)

T, °K	N_{at}	Thermodynamic form				Compact form			
	10^4	10^5	10^6	10^7	10^4	10^5	10^6	10^7
500	17.92	6.270	2.630	1.146	4.806	2.087	0.9491	0.4397
1000	19.46	6.742	2.817	1.212	5.162	2.235	1.015	0.4698
1500	21.27	7.282	3.033	1.284	5.569	2.404	1.090	0.5042
2000	-	7.899	3.275	1.362	6.283	2.595	1.175	0.5431
2500	-	8.601	3.551	1.448	6.896	2.812	1.271	0.5872
3000	-	9.407	3.865	1.541	7.633	3.058	1.380	0.6374
3500	-	10.34	4.223	1.641	8.538	3.339	1.504	0.6941
4000	-	11.42	4.634	1.749	9.698	3.660	1.646	0.7589

equilibrium decreases with increasing grain size. The
larger the eccentricity, the stronger the effect.

Effect of Phase State of Carbon on the Detonation

Parameters

It was mentioned earlier (Gubin et al. 1986b) that,
when a possibility of formation of the two carbon solid
phases in the detonation products was considered in
computations of detonation parameters of carbon-rich
explosives, a specific transition portion in a graph of
calculated dependence of detonation velocity on initial
density appeared. This change is caused by a transition
from one phase state of condensed carbon to another in the
detonation products. At low initial densities of
explosive, the temperature and pressure of detonation
products in the Chapman-Jouquet (CJ) point lie in the
region of graphite stability in the equilibrium phase
diagram of carbon. As the initial density of explosive
increases, the detonation product pressure rises sharply;
the point corresponding to the CJ state on the carbon phase
diagram moves upward, crosses the phase equilibrium line,
and falls into the region of diamond stability. In
accordance with this, diamond rather than graphite is
formed in the detonation products when the initial density
of the explosive increases.
It is the coexistence of both graphite and diamond in
the detonation products when crossing the phase equilibrium
line that leads to the transition portion of the detonation
velocity-initial density curve. Inclusion of the particle
size effect for carbon in the thermodynamic and compact
forms of crystals results in a shift of the transition
portion of the curve toward greater values of initial
density. This shift in compact form is considerably less
than in the thermodynamic form. Along with the shift of
the transition portion of the curve toward higher explosive
initial density, a reduction of the carbon particle size
leads to an increase of the slope of the transition portion
up to positive values. The positive slope of the
transition portion appears when crystals of graphite and
diamond are sufficiently fine.
In his monograph, C. Mader (1977) discusses the
experimentally observed sharp change in the slope of the
TNT's detonation velocity on initial density dependence
(Urizar et al. 1961). To obtain such a change in the slope
it is necessary that, in the density range between 1550 and
1640 kg/m^3, either a change in the carbon condensation

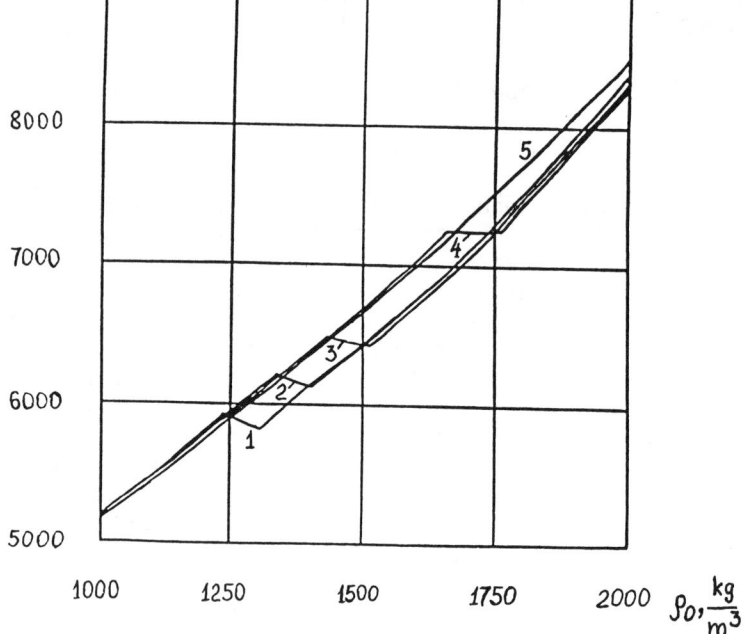

Fig. 1 Dependence of detonation velocity on initial density of TNT computed with the thermodynamic form of graphite and diamond crystals. Line 1: nondispersed solid carbon; lines 2, 3, 4, and 5: $N_{at} = 10^7$, 10^6, 10^5, and 10^4, respectively.

mechanism (which results in an increase of the heat of formation by 6 kcal/mole of carbon) or a change in the detonation product composition (which decreases their energy) should occur. Furthermore, Mader details the effect of the condensation mechanism on the detonation parameters for various degrees of carbon condensation in the CJ point.

We believe that taking into account the diamond formation in the detonation products and the effect of solid carbon particle size allows one to explain, at least qualitatively, the peculiar behavior of the detonation velocity vs initial density curve, as noted by Mader. He was right in associating this peculiarity with condensed carbon. However, among its possible causes he did not mention the chief one, namely, the change of the condensed phase density due to the phase conversion.

The dependence of detonation velocity on initial density for TNT is presented in Figures 1 and 2. The results of computations with the thermodynamic form of

EFFECT OF GRAPHITE AND DIAMOND CRYSTALS

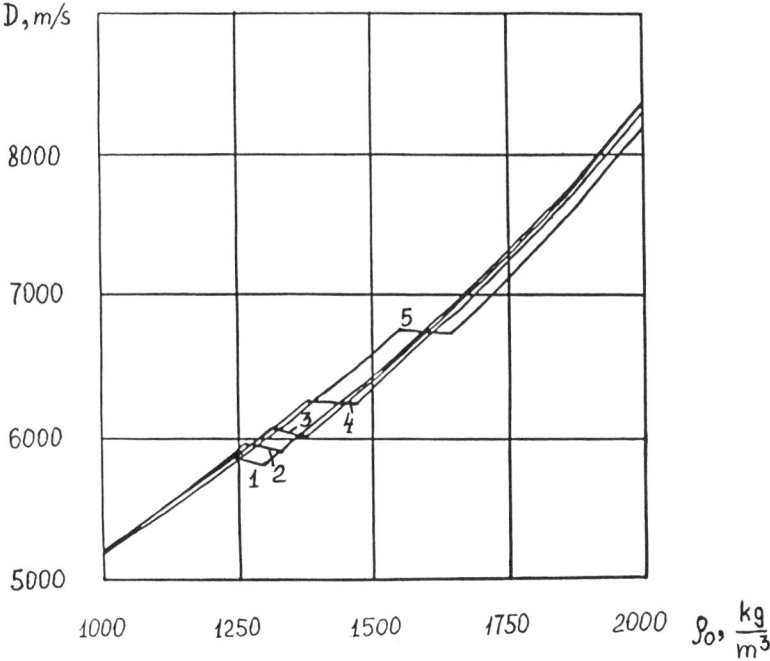

Fig. 2 Dependence of detonation velocity on initial density of TNT computed with the compact form of graphite and diamond crystals. Definitions of lines as in Fig. 1.

graphite and diamond crystals are shown in Fig. 1, and those with compact form are shown in Fig. 2.

In conclusion, we present the parameters of the carbon solid phases pertaining to the beginning of the transition portion of the TNT's detonation velocity curve at an initial density of 1550 kg/m^3.

For the thermodynamic form, the number of carbon atoms in a crystal is 2.5×10^5, the length of the diamond octahedron edge is 14 nm, the length of a base edge of the graphite hexagonal film is 36 nm, and the increase in the enthalpy of formation is 2.2 kcal/mole for diamond and 0.66 kcal/mole for graphite.

For the compact form, the number of carbon atoms in a crystal is 10^4, the length of the prism base edge is 2.5 nm for diamond and 2.8 nm for graphite, and the increase in the enthalpy of formation is 7.4 kcal/mole for diamond and 5.8 kcal/mole for graphite.

References

Baute, J. and Chirat, R. (1985) An extensive application of WCA equation of state for explosives. Prepr. 8th Symp. (Int.) Deton. USA, 94-102.

Cowen, R. D. and Fickett, W. (1956) Calculation of the detonation parameters of solid explosives with the Kistiakowsky-Wilson Equation of State. J. Chem. Phys. 24, 932.

Gustov, V. V. and Pikaev, A. K. (1982) The works of Soviet scientists on high energy chemistry. Khim. Vys. Energ., 16, 483.

Gubin, S. A., Odintsov, V. V., and Pepekin, V. I. (1984) On the role of a carbon phase state when evaluating detonation parameters of explosives. Khim. Fiz., 3, 754-759.

Gubin, S. A., Odintsov, V. V., and Pepekin, V. I. (1986a) The phase diagram of carbon and its counting in detonation parameter computations. Khim. Fiz., 5, 111-120.

Gubin, S. A., Odintsov, V. V., and Pepekin, V. I. (1986b) Thermodynamic computations of detonation of condensed substances. Prepr. Inst. Chem. Phys. Sci. Acad. USSR, Chernogolovka, USSR.

Mader, C. L. (1977) Numerical Modeling of Detonation. Univ. of California Press, Berkeley, CA.

Staver, A. M., Gubareva, N. V., Liamkin, A. I., et al. (1984) Ultradispersed diamond powders produced by using explosion energy. Fiz. Goreniya i Vzryva, 20, 100-104.

Trefilov, V. I., Savvakin, G. I., Skorohod, V. V., et al. (1978) Structure of ultradispersed diamonds produced under conditions of explosion. Dokl. Akad. Nauk SSSR, 239, 838.

Urizar, M. J., James, E., Jr., and Smith, L. C. (1961) Detonation velocity of pressed TNT. Phys. Fluids, 4, 262.

Two-Phase Steady Detonation Analysis

J. M. Powers,[*] D. S. Stewart,[†] and H. Krier[‡]
University of Illinois at Urbana-Champaign, Urbana, Illinois

Abstract

Steady solutions to a set of two-phase reactive flow model equations are studied to test the hypothesis that observed deviations from Chapman-Jouguet (C-J) detonation states in porous solid propellants are manifestations of the two-phase nature of the flow. Shock jump relations are presented. A simple expression for a minimum detonation wave speed analogous to a C-J detonation for a single phase is given. The analogous C-J point is a sonic point. In the appropriate limit, the minimum detonation velocity varies linearly with initial bulk density and the corresponding detonation pressure varies with the square of initial bulk density. Nonideal gas effects play an important role in determining C-J conditions. The effect of reaction zone structure is studied to determine its effect on detonation end states.

I. Introduction

Detonation, a shock-induced chemical reaction, has been described by the theory of Chapman and Jouguet (C-J). The simple C-J theory analyzes a detonation as a jump discontinuity and solves algebraic equations that represent the conservation of mass, momentum, and energy for the detonation end states (pressure, density, etc.) as functions of the detonation velocity. C-J theory identifies a unique wave speed and sonic end state for an unsupported detonation. In addition to end state analysis, the structure of the detonation (i.e., the distribution of pressure, density, etc., in the reaction zone behind the wave head) must be considered when classifying possible steady solutions. In some cases, certain end states can be ruled out because the end state cannot be attained from the initial state at the wave head. (The discussion of detonation structure is the topic of ZND theory.)

Copyright © 1988 by J. M. Powers. Published by the American Institute of Aeronautics and Astronautics, Inc. with permission.
[*] Visiting Assistant Professor, Dept. of Mech. and Ind. Eng.
[†] Associate Professor, Dept. of Theo. and Appl. Mech.
[‡] Professor, Dept. of Mech. and Ind. Eng.

States and wave speeds other than those predicted by C-J theory may be realized. Under appropriate circumstances, it can be shown for one-phase models that a so-called weak detonation (which travels supersonically relative to the detonation products) is possible. As discussed by Fickett and Davis (1979), weak detonations are observed in experiments. Also, the numerical work of Butler, et al. (1982) suggests that both C-J and weak detonations may be admitted by equations of two-phase reactive flow used to model solid propellants and granulated explosives.

The evaluation of numerical solutions to the two-phase reactive flow equations [the work of Butler, et al. (1982) is an example] requires a systematic study of the steady solutions. Knowledge of the one-dimensional steady solutions is necessary in order to develop the stability theory for two-phase detonation models. Steady solutions are also necessary to verify that numerical predictions of apparently steady solutions by unsteady model equations are not numerical artifices.

Our work analyzes the steady form of a two-phase, unsteady detonation model similar to Butler and Krier's (1986) and Baer and Nunziato's (1986). The first step in a complete structure analysis is the identification of detonation shock states and detonation end states. We first discuss shock jumps admitted by the equations. We then consider end states admitted by the governing equations and, in particular, identify the minimum detonation wave speed admitted by our two-phase model. In this process, we identify a C-J wave speed analogous to the C-J wave speed of the simple one-phase theory. These steps require the analysis of algebraic conditions. We next pose the steady equations in a reduced form in which the structure can be studied. The additional constraints that a physically admissible detonation structure places on this problem and the identification of other admissible detonation states is given by Powers (1988).

II. The Unsteady Two-Phase Model

Here a two-phase model is proposed. The model is similar to models used by Butler and Krier (1986) and Baer and Nunziato (1986). The proposed model utilizes simpler drag and heat transfer relations and a simpler solid equation of state in order to facilitate the analysis. As in the Butler-Krier model, but not in the Baer-Nunziato model, the proposed model allows momentum changes in response to a gradient in the porosity of the gas phase. Also, as Butler-Krier did, but Baer-Nunziato did not, we do not consider compaction work in our energy equations. Details about why these choices concerning the momentum equation and compaction work are given by Powers (1988). Intraphase diffusive thermal and momentum transport is ignored. Details about the rationale for the model in the context of continuum mixture theory can be found in Wallis (1969), Baer and Nunziato (1983), and Krier and Gokhale (1978). For the proposed model, it is assumed that each phase is a continuum; consequently, partial differential equations resembling one-phase equations are written to describe the evolution of mass, momentum, and energy in each constituent. In addition, each phase is described by a thermal state relation and a corresponding caloric state relation. One phase is assumed to be a solid, the other, gas. In order to close the system, a compaction equation similar to that of Baer and Nunziato (1986) is adopted. As discussed there, the compaction law is suggested by an assumed form of the second law of thermodynamics. The compaction law states that the solid volume fraction changes in response to pressure differences between phases and to combustion. Choosing such a compaction equation also insures that the characteristics are real; thus, the initial value problem is well-posed. We emphasize that the choices we have adopted for the closure problem place a premium on simplicity so that explicit analytic calculations can be made whenever possible. At the same time, the model adopted here is thought to be representative of a wider class of two-phase detonation models currently in use.

In order to describe the transfer of mass, momentum, and energy from one phase to another, phase interaction terms must be specified. The solid phase is assumed to be composed of spherical particles so that certain empirical correlations can be invoked [Butler and Krier (1986), Kuo and Nydegger (1978), Denton (1951)]. These correlations describe the burning of spherical particles, particle/gas drag, and particle/gas heat transfer. In contrast to Butler and Krier (1986) and Baer and Nunziato (1986), we explicitly enforce the conservation of number of particles; that is, we assume the particles do not agglomerate or disintegrate.

The unsteady equations are

$$\frac{\partial}{\partial t}\left[\rho_1\phi_1\right] + \frac{\partial}{\partial x}\left[\rho_1\phi_1 u_1\right] = \left(\frac{3}{r}\right)\rho_2\phi_2 aP_1^m \tag{1}$$

$$\frac{\partial}{\partial t}\left[\rho_2\phi_2\right] + \frac{\partial}{\partial x}\left[\rho_2\phi_2 u_2\right] = -\left(\frac{3}{r}\right)\rho_2\phi_2 aP_1^m \tag{2}$$

$$\frac{\partial}{\partial t}\left[\rho_1\phi_1 u_1\right] + \frac{\partial}{\partial x}\left[P_1\phi_1 + \rho_1\phi_1 u_1^2\right]$$
$$= u_2\left(\frac{3}{r}\right)\rho_2\phi_2 aP_1^m - \beta\frac{\phi_2\phi_1}{r}(u_1-u_2) \tag{3}$$

$$\frac{\partial}{\partial t}\left[\rho_2\phi_2 u_2\right] + \frac{\partial}{\partial x}\left[P_2\phi_2 + \rho_2\phi_2 u_2^2\right]$$
$$= -u_2\left(\frac{3}{r}\right)\rho_2\phi_2 aP_1^m + \beta\frac{\phi_2\phi_1}{r}(u_1-u_2) \tag{4}$$

$$\frac{\partial}{\partial t}\left[\rho_1\phi_1\left(e_1+u_1^2/2\right)\right]+\frac{\partial}{\partial x}\left[\rho_1\phi_1 u_1\left(e_1+u_1^2/2+P_1/\rho_1\right)\right]$$

$$=\left(e_2+u_2^2/2\right)\left(\frac{3}{r}\right)\rho_2\phi_2 aP_1^m-\beta\frac{\phi_1\phi_2}{r}u_2\left(u_1-u_2\right)-h\frac{\phi_1\phi_2}{r^{1/3}}\left(T_1-T_2\right) \quad (5)$$

$$\frac{\partial}{\partial t}\left[\rho_2\phi_2\left(e_2+u_2^2/2\right)\right]+\frac{\partial}{\partial x}\left[\rho_2\phi_2 u_2\left(e_2+u_2^2/2+P_2/\rho_2\right)\right]$$

$$=-\left(e_2+u_2^2/2\right)\left(\frac{3}{r}\right)\rho_2\phi_2 aP_1^m+\beta\frac{\phi_1\phi_2}{r}u_2\left(u_1-u_2\right)+h\frac{\phi_1\phi_2}{r^{1/3}}\left(T_1-T_2\right) \quad (6)$$

$$\frac{\partial}{\partial t}\left[\phi_2/r^3\right]+\frac{\partial}{\partial x}\left[u_2\phi_2/r^3\right]=0 \quad (7)$$

$$\frac{\partial \phi_2}{\partial t}+u_2\frac{\partial \phi_2}{\partial x}=\frac{\phi_1\phi_2}{\mu_c}\left[P_2-P_1-\frac{P_{20}-P_{10}}{\phi_{20}}\phi_2\right]-\left(\frac{3}{r}\right)\phi_2 aP_1^m \quad (8)$$

$$P_1=\rho_1 RT_1\left(1+b\rho_1\right) \quad (9)$$

$$e_1=c_{v1}T_1 \quad (10)$$

$$P_2=\left(\gamma_2-1\right)c_{v2}\rho_2 T_2-\frac{\rho_{20}s}{\gamma_2} \quad (11)$$

$$e_2=\frac{P_2+\rho_{20}s}{(\gamma_2-1)\rho_2}+q \quad (12)$$

$$\phi_1+\phi_2=1 \quad (13)$$

where the subscript "0" denotes the undisturbed condition, 1 the gas phase, 2 the solid phase, ρ the density, ϕ the volume fraction, u the velocity, r the solid particle radius, P the pressure, m the burn index, a the burn constant, β the drag parameter, e the internal energy, h the heat-transfer coefficient, R the gas constant, b the covolume correction, c_v the constant volume specific heat, s the nonideal solid sound parameter, μ_c the compaction viscosity, γ_2 the Tait parameter, and q the heat of reaction. Undisturbed conditions are specified as

$$\rho_1 = \rho_{10}, \quad \rho_2 = \rho_{20}, \quad \phi_1 = \phi_{10}, \quad u_1 = 0$$

$$u_2 = 0, \quad r = r_0, \quad T_1 = T_0, \quad T_2 = T_0$$

Equations (1) and (2) describe the evolution of each phase's mass, Eqs. (3) and (4) the momentum evolution, and Eqs. (5) and (6) the energy evolution. Here, a principle of two-phase modeling has been employed; the motion of an isolated constituent is determined by the action of other constituents on it. Homogeneous mixture equations are formed by adding Eqs. (1) and (2), Eqs. (3) and (4), and Eqs. (5) and (6), respectively. Thus, for the mixture, conservation of mass, momentum, and energy is maintained.

Inhomogeneities in Eqs. (1-6) model interphase momentum, energy, and mass transfer. The model incorporates momentum transfer in the form of Stokes drag and energy transfer as constant-coefficient convective heat transfer. For mass transfer, we utilize an empirical relation for the regression of particle radius. It is observed that the rate of change of particle radius is proportional to the surrounding pressure raised to some power. As we consider a constant number of compressible particles, our model must also allow the particle radius to change in response to a change in density of the particle phase. By combining the particle mass evolution equation (2) with Eq. (7) for number conservation, it is seen that the proposed model has the desired features, as

$$\frac{\partial r}{\partial t} + u_2 \frac{\partial r}{\partial x} = -aP_1^m - \frac{r}{3\rho_2}\left(\frac{\partial \rho_2}{\partial t} + u_2 \frac{\partial \rho_2}{\partial x}\right) \quad (14)$$

that is, as the motion of a particle is followed, its radius may change in response to both burning and compression.

The compaction law is expressed in Eq. (8). Constituent 1 is a gas described by a virial equation of state (9). Likewise, constituent 2 is a solid described by a Tait equation of state (11). Assumption of a constant specific heat at constant volume for each phase allows caloric equations (10) and (12) consistent with the assumptions of classical thermodynamics to be written for each phase. The variable ϕ is defined as a volume fraction,

$\phi \equiv$ constituent volume/total volume. All volume is occupied by constituent 1 or 2; no voids are permitted. This is enforced by Eq. (13). By using standard techniques [Whitham (1974)], it can be shown that the characteristic wave speeds are u_1, u_2, $u_1 \pm c_1$ and $u_2 \pm c_2$, where c_i represents the sound speed of each phase. The uncoupling of the characteristic wave speeds is a consequence of the assumed form of the compaction law. Other closure relations will, in general, couple the characteristic wave speeds.

III. Steady State Analysis

Steady state analysis of two-phase equations has been performed by Kuo and Summerfield (1974a, 1974b), Krier and Mozaffarian (1978), Taylor (1984), and Drew (1986). The studies of both Kuo and Summerfield and Drew considered only the deflagration of incompressible particles. Krier and Mozaffarian numerically studied two-phase detonation reaction zones with the assumption of incompressible particles. As discussed by Macek (1959), for typical condensed phase detonation pressures, this assumption is unrealistic. Others such as Taylor (1984) studied shock jumps in two-phase reactive flow equations. Here, we consider a steady two-phase detonation model with compressible particles. The discussion we present of two-phase detonation end states is new to the best of our knowledge.

Equations (1-13) can be recast in a more tractable form. First, we write the equations in dimensionless form. For a right-running steady wave, we make the Galilean transformation $\xi = x - Dt$, $v = u - D$ under which Eqs. (1-8) collapse to eight ordinary differential equations. Here, D is a constant defined as the steady wave speed. Next, Eqs. (1), (3), and (5) may be eliminated in favor of homogeneous mixture equations formed by the addition of the steady form of Eqs. (1) and (2), (3) and (4), and (5) and (6). The resultant mixture equations and the steady form of Eq. (7) may be integrated to form algebraic equations. Thus, the steady two-phase model is described by four ordinary differential equations and nine algebraic equations.

Dimensionless Model Equations

Define dimensionless variables where "*" indicate a dimensionless quantity, as follows:

$$\xi_* = \xi / L \quad v_{*i} = v_i / D \quad P_{*i} = P_i / (\rho_{i0} D^2)$$

$$\rho_{*i} = \rho_i / \rho_{i0} \quad e_{*i} = e_i / D^2 \quad T_{*i} = c_{vi} T_i / D^2 \quad r_* = r / L$$

$$i = 1, 2$$

where D is the wave speed, L an undefined length scale to be associated with the reaction zone length, ρ_{i0} the initial density of phase i, and c_{vi} the specific heat at constant volume of phase i. Define the following independent dimensionless parameters as

$$\pi_1 = 3a\rho_{10}^m D^{2m-1} \qquad \pi_2 = \beta/(\rho_{20} D) \qquad \pi_3 = hL^{2/3}/\rho_{20} c_{v1} D$$

$$\pi_4 = m \qquad \pi_5 = \rho_{10}/\rho_{20} \qquad \pi_6 = c_{v1}/c_{v2} \qquad \pi_7 = R/c_{v1} + 1$$

$$\pi_8 = s/(\gamma_2 D^2) \qquad \pi_9 = \rho_{20} DL/\mu_c \qquad \pi_{10} = q/D^2 \qquad \pi_{11} = \phi_{10}$$

$$\pi_{12} = r_0/L \qquad \pi_{13} = b\rho_{10} \qquad \pi_{14} = c_{v2} T_0/D^2$$

$$\pi_{17} = \gamma_2$$

and the following dependent dimensionless parameters as:

$$\pi_{18} = \pi_{11} + \frac{1 - \pi_{11}}{\pi_5}$$

$$\pi_{19} = \left[\pi_7 - 1\right] \pi_6 \pi_{14} \left[1 + \pi_{13}\right]$$

$$\pi_{20} = 1 - \pi_{11}$$

$$\pi_{21} = \left[\pi_{17} - 1\right] \pi_{14} - \pi_8$$

$$\pi_{22} = \pi_{11} \left[\pi_6 \pi_{14} + \frac{1}{2} + \pi_{19}\right] + \frac{1 - \pi_{11}}{\pi_5} \left[\pi_{14} + \pi_{10} + \frac{1}{2} + \pi_{21}\right]$$

$$\pi_{23} = \pi_{11} \pi_{19} + \frac{1 - \pi_{11}}{\pi_5} \pi_{21}$$

$$\pi_{15} = \frac{\pi_{21} - \pi_5 \pi_{19}}{1 - \pi_{11}}$$

then the dimensionless model equations (for compact notation the stars are dropped) can be written economically as

$$\frac{d}{d\xi}\left[\rho_2 \phi_2 v_2\right] = -\pi_1 \frac{\rho_2 \phi_2 P_1^{\pi_4}}{r} \tag{15}$$

$$\rho_2 \phi_2 v_2 \frac{dv_2}{d\xi} + \frac{d}{d\xi}\left[P_2 \phi_2\right] = -\pi_2 \left[v_2 - v_1\right]\frac{\phi_1 \phi_2}{r} \tag{16}$$

$$\rho_2 v_2 \frac{de_2}{d\xi} + P_2 \frac{dv_2}{d\xi} = -\pi_3 \left[\pi_6 T_2 - T_1\right]\frac{\phi_1}{r^{1/3}} \tag{17}$$

$$v_2 \frac{d\phi_2}{d\xi} = \pi_9 \phi_1 \phi_2 \left[P_2 - \pi_5 P_1 - \pi_{15}\phi_2\right] - \pi_1 \frac{\phi_2 P_1^{\pi_4}}{r} \tag{18}$$

$$\rho_1 \phi_1 v_1 + \frac{1}{\pi_5}\rho_2 \phi_2 v_2 = -\pi_{18} \tag{19}$$

$$\rho_1 \phi_1 v_1^2 + P_1 \phi_1 + \frac{1}{\pi_5}\left[\rho_2 \phi_2 v_2^2 + P_2 \phi_2\right] = \pi_{18} + \pi_{23} \tag{20}$$

$$\rho_1 \phi_1 v_1 \left[e_1 + \frac{v_1^2}{2} + \frac{P_1}{\rho_1}\right] + \frac{1}{\pi_5}\rho_2 \phi_2 v_2 \left[e_2 + \frac{v_2^2}{2} + \frac{P_2}{\rho_2}\right] = -\pi_{22} \tag{21}$$

$$r = \pi_{12} \sqrt[3]{\frac{-v_2 \phi_2}{1 - \pi_{11}}} \tag{22}$$

$$P_1 = \left[\pi_7 - 1\right] \rho_1 T_1 \left[1 + \pi_{13} \rho_1\right] \tag{23}$$

$$e_1 = T_1 \tag{24}$$

$$P_2 = \left[\pi_{17} - 1\right] \rho_2 T_2 - \pi_8 \tag{25}$$

$$e_2 = \frac{P_2 + \pi_{17} \pi_8}{(\pi_{17} - 1) \rho_2} + \pi_{10} \tag{26}$$

$$\phi_1 + \phi_2 = 1 \tag{27}$$

Undisturbed conditions are

$$\rho_2 = 1 \quad v_2 = -1 \quad \phi_2 = \pi_{20} \quad T_2 = \pi_{14}$$

Inert Shock Jump Conditions

Equations (15-27) are valid throughout the steady flowfield. If a detonation solution with a leading shock is assumed, it is necessary to know shock jump conditions. In order to form two-phase shock jump conditions, it is assumed that a thin shock wave exists in which no appreciable reaction, drag, heat transfer, or compaction occurs. Thus, Eqs. (15-18) may be replaced by the shock jump conditions,

$$\left[\rho_2 \phi_2 v_2\right]_0^s = 0 \tag{28}$$

$$\left[P_2 \phi_2 + \rho_2 \phi_2 v_2^2\right]_0^s = 0 \tag{29}$$

$$\left[\rho_2 \phi_2 v_2 \left(e_2 + v_2^2/2 + P_2/\rho_2\right)\right]_0^s = 0 \tag{30}$$

$$\left[\phi_2\right]_0^s = 0 \tag{31}$$

where "s" denotes the shocked state and "0" the undisturbed state. Equations (28-31) and state relations [Eqs. (25) and (26)] are sufficient to calculate the shock jumps for phase 2. The shock jumps for phase 1 are implied by the mixture Eqs. (19-21) and state relations [Eqs. (23) and (24)]. Since porosity does not change through the shock, the shock jumps for the two phases are calculated independently.

Three solutions to the gas shock jump relations exist: the inert solution, a nonphysical solution, and a shock solution. In the limit as π_{13} (or b) approaches zero, the nonphysical prediction of gas density approaches $-1/\pi_{13}$ and is therefore rejected. In the same limit, the shock solution can be thought of as a perturbation to the ideal-gas shock solution. The asymptotic gas shock solution is

$$P_{1s} \cong \left[\frac{2 - \pi_{14}\pi_6(\pi_7 - 1)^2}{\pi_7 + 1}\right] - \left[\frac{2 + \pi_{14}\pi_6(\pi_7 - 1)^2}{\pi_7 + 1}\right]\pi_{13} \tag{32}$$

For $\pi_{13} > 0$, the nonideal shock relations predict lower shock pressures than the ideal relation. Figure 1 shows a plot of the dimensional gas phase shock pressure vs the shock wave speed. Unless otherwise noted, dimensional input parameters for this and all other cases are listed in Table 1. These input parameters are representative of the granular solid

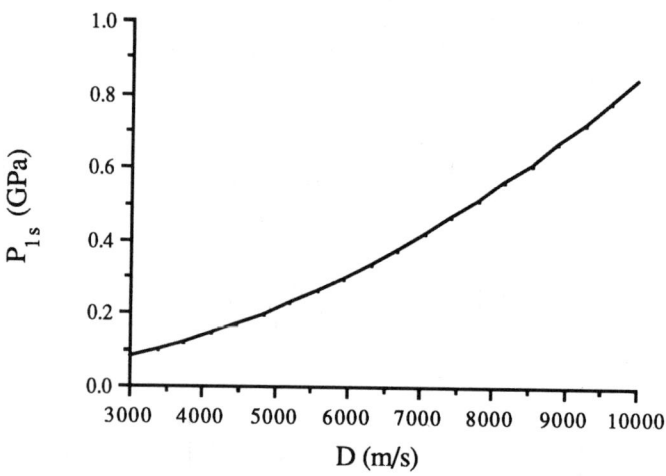

Fig. 1 Gas-phase shock pressure vs shock wave speed.

TWO-PHASE STEADY DETONATION ANALYSIS

Table 1 Dimensional input parameters

a, m / (s Pa)§	2.90×10^{-9}	R, J / (kg K)	8.50×10^{2}
ρ_{10}, kg / m^3	1.00×10^{1}	s, (m / s)$^{2\#}$	8.98×10^{6}
$m^\S$	1.00×10^{0}	q, J / kg§	5.84×10^{6}
β, kg / (s m^2)	1.00×10^{4}	r_0, m$^{\S,\yen}$	1.00×10^{-4}
ρ_{20}, kg / m$^{3\ \S,\yen}$	1.90×10^{3}	b, m^3 / kg	1.10×10^{-3}
h, J / (s K m^2)	1.00×10^{7}	$\gamma_2^\#$	5.00×10^{0}
c_{v1}, J / (kg K)$^\yen$	2.40×10^{3}	μ_c, kg / (m s)	1.00×10^{6}
c_{v2}, J / (kg K)$^{\S,\yen}$	1.50×10^{3}	T_0, K	3.00×10^{2}

propellant HMX. Table 1 notes references for some of these parameters. Those parameters without reference have been estimated to match data or are able to be arbitrarily specified. We allow the initial gas density and temperature to be arbitrarily specified. The heat transfer and drag parameters, not important in this work, have been estimated to roughly match the trends predicted in experiments. The Tait equation parameters have been estimated to match experimental shock and compaction data [Powers, et al. (1988)]. The gas state equation parameters have been chosen to match detonation state predictions of the thermochemistry code TIGER [Coperthwaite and Zwisler (1974)] as reported by Butler and Krier (1986).

For the solid phase, two solutions are admitted: the inert solution and the shock solution. The exact expression for solid phase shock pressure is

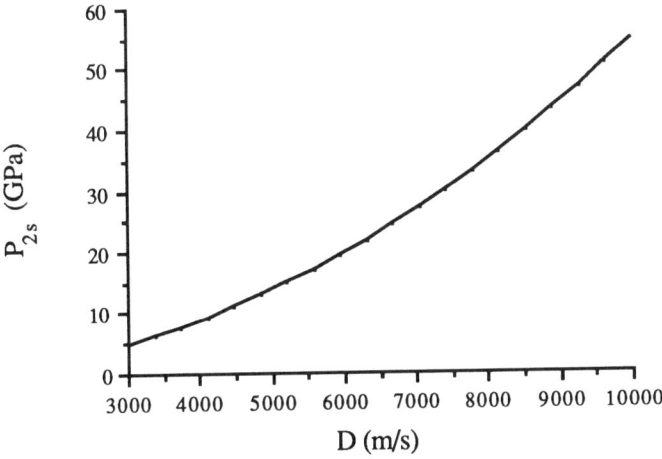

Fig. 2 Solid-phase shock pressure vs shock wave speed.

$$P_{2s} = \frac{2 - \pi_{14}(\pi_{17} - 1)^2}{\pi_{17} + 1} - \pi_8 \qquad (33)$$

In Fig. 2, the dimensional solid phase shock pressure is plotted vs the shock wave speed.

Similar relations are available for shock density, temperature, and energy. For the case of two ideal gases ($\pi_{13} \equiv \pi_8 \equiv 0$), these equations reduce to familiar shock relations in terms of the shock Mach number. For the ideal gases, shock Mach numbers may be expressed as

$$M_{1s}^2 = \frac{1}{\pi_7(\pi_7 - 1)\pi_{14}\pi_6}, \quad M_{2s}^2 = \frac{1}{\pi_{17}(\pi_{17} - 1)\pi_{14}} \qquad (34\text{-}35)$$

One criterion for a solid state equation is that the state equation in conjunction with the Rankine-Hugoniot jump conditions be able to match experimental shock loading data. By solving the shock jump equations for the final velocity, an expression for shock wave speed D vs piston velocity u_2, can be developed. The relation in terms of dimensional parameters is

$$D = \frac{1 + \gamma_2}{4} u_2 + \sqrt{\left(\frac{1 + \gamma_2}{4} u_2\right)^2 + \gamma_2(\gamma_2 - 1) c_{v2} T_0} \qquad (36)$$

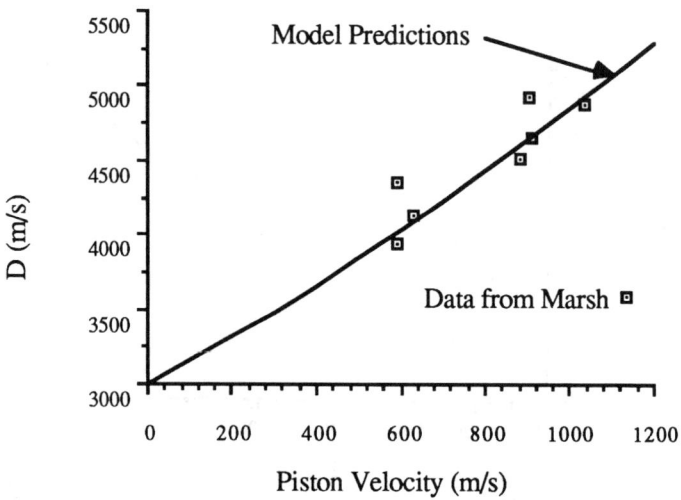

Fig. 3 Experimental and predicted shock wave speed vs pistion velocity for solid HMX.

As shown in Fig. 3 by choosing $\gamma_2 = 5$ the experimental shock loading data of Marsh (1980) is accurately matched.

End State Analysis

An end state has been reached when, at a given location ξ, the inhomogeneities of Eqs. (15-18) are identically zero. Equations (15-18) suggest that an end state has been reached when $\phi_2 = 0$ and $T_1 = \pi_6 T_2$, corresponding to complete reaction and temperature equilibrium. As discussed more fully in Sec. IV, it is not established if other end states exist or even if the complete reaction/temperature equilibrium state is a legitimate end state. By assuming that the complete reaction/temperature equilibrium state is an end state, we may draw conclusions about the steady wave speeds admitted by the two-phase model and show that in many ways the two-phase model is analogous to one-phase models.

In the complete reaction state, the mixture equations (19-21) allow for the gas phase properties to be determined. For $\phi_2 = 0$ ($\phi_1 = 1$), Eqs. (19-21) can be combined to form an equivalent two-phase Rayleigh line [Eq. (37)] and two-phase Hugoniot [Eq. (38)], as,

$$P_1 - \pi_{23} = \pi_{18}^2 \left(1/\pi_{18} - 1/\rho_1 \right) \quad (37)$$

$$\left[e_1 - \frac{\pi_{11} \pi_6 \pi_{14}}{\pi_{18}} \right] + \frac{1}{2} \left[P_1 + \pi_{23} \right] \cdot \left[1/\rho_1 - 1/\pi_{18} \right]$$

$$= \frac{\left(1 - \pi_{11} \right) \left(\pi_{14} + \pi_{10} \right)}{\pi_5 \pi_{18}} \quad (38)$$

From the state relations [Eqs. (23) and (24)], we can write $e_1 = e_1(P_1, \rho_1)$, which is substituted into the Hugoniot equation (38). The Rayleigh line equation (37) allows ρ_1 to be eliminated in favor of P_1. Substituting this in Eq. (38) results in a cubic equation for P_1. Three cases are possible: 1) three distinct real solutions, 2) two equal real solutions and a third real solution, and 3) a real solution and a complex conjugate pair of solutions. When three distinct real solutions exist, two are analogous to the weak and strong solutions predicted by the simple one-phase theory. The third solution has no such counterpart and often is a nonphysical solution with $\rho_1 < 0$. The third solution has not been thoroughly studied.

By imposing the condition that two real roots are degenerate (which forces the Rayleigh line and Hugoniot to be tangent), a minimum detonation velocity can be found. We will call this condition the C-J condition. Because the detonation velocity D is embedded in the dimensionless parameters, it is convenient to return to dimensional variables to express C-J conditions. Define the bulk density, bulk pressure, and bulk internal energy as

$$\rho_a = \rho_{10} \phi_{10} + \rho_{20} \phi_{20} \ , \quad P_a = P_{10} \phi_{10} + P_{20} \phi_{20}$$

$$e_a = \frac{\rho_{10}\phi_{10}e_{10} + \rho_{20}\phi_{20}e_{20}}{\rho_{10}\phi_{10} + \rho_{20}\phi_{20}}$$

By applying the tangency condition, it can be shown that in the limit as $b\rho_a$ and $P_a/(\rho_a e_a)$ approach zero that the Taylor series approximation for the C-J wave speed and pressure are given by

$$D_{CJ} \cong \frac{\sqrt{2e_a R\left(R + 2c_{v1}\right)}}{c_{v1}} \left(1 + b\rho_a - \frac{c_{v1}^2}{2R\left(R + 2c_{v1}\right)} \frac{P_a}{\rho_a e_a}\right) \quad (39)$$

$$P_{CJ} \cong \frac{2e_a R}{c_{v1}} \rho_a \left(1 + b\rho_a - \frac{c_{v1}^2}{2R\left(R + 2c_{v1}\right)} \frac{P_a}{\rho_a e_a}\right) \quad (40)$$

Similar expressions can be obtained for C-J density, temperature, and energy.

For $P_a = b = 0$, our formulas show that it is appropriate to treat the two-phase C-J condition as a one-phase C-J condition using ρ_a and e_a as effective one phase properties. Fickett and Davis (1979) give equations for one-phase C-J properties. In these equations, one can simply substitute the bulk density for the initial density and the bulk internal energy for the chemical energy to obtain the two-phase C-J equations.

From Eqs. (39) and (40), the C-J velocity depends linearly on ρ_a, the bulk density, and the C-J pressure varies with the square of bulk density in agreement with the trends discussed by Johansson and Persson (1970). This dependence is demonstrated in Figs. 4 and 5. These plots were obtained by solving the full set of nonlinear equations to determine the C-J state. The bulk density was varied by varying the initial porosity. Also plotted in Figs. 4 and 5 are predictions of the approximations (39) and (40) and predictions of the thermochemistry code TIGER reported by Butler and Krier (1986).

Equations (39) and (40) indicate that the C-J state is quite sensitive to the nonideal parameter b. In particular, it suggests that, when the dimensionless group $b\rho_a$ is of order 1, that nonideal effects become quite important. This is demonstrated in Fig. 6, which for constant bulk density, plots C-J wave speed vs the nonideal parameter b. The figure was obtained by solving the full nonlinear equations. In other works (Butler and Krier 1986), the parameter b has been treated as adjustable in order to match C-J data.

By numerically studying exact two-phase C-J conditions, it can be inferred that the C-J point is a sonic point -- that is the gas velocity is equal to the local gas phase sonic velocity. In addition, numerical studies

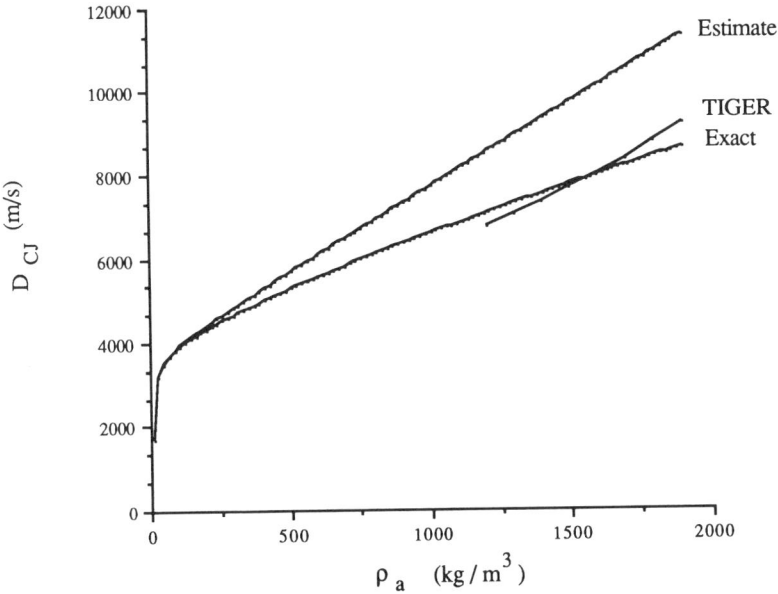

Fig. 4 Minimum detonation velocity vs bulk density for a nonideal gas.

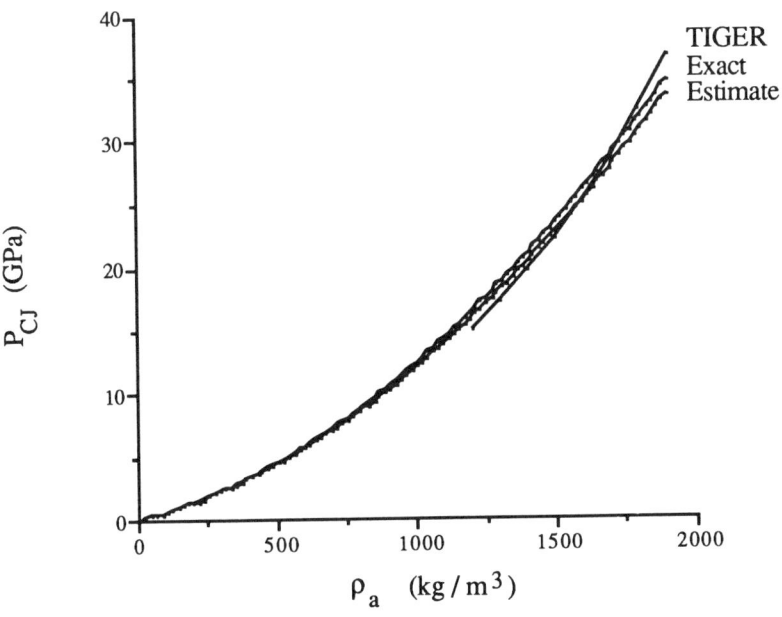

Fig. 5 C-J pressure vs bulk density for a nonideal gas.

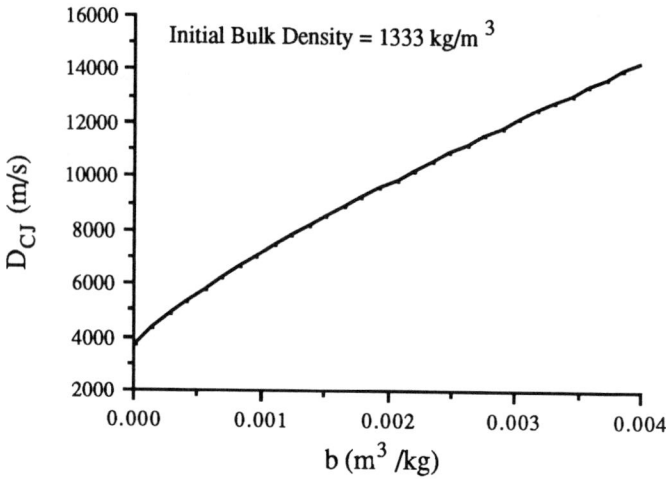

Fig. 6 Minimum detonation wave speed vs nonideal gas parameter b with constant bulk density.

indicate that for $D > D_{C-J}$ the nonideal strong point is subsonic, while the nonideal weak point is supersonic. This corresponds to the results of the simple one-phase theory.

IV. Reaction Zone Structure

It is important to note that the two-phase C-J solution is not necessarily admitted by our model equations when detonation structure is considered. A global phase space analysis to the full set of differential-algebraic equations will determine which wave speeds correspond to physically admissible solutions. In this section, we outline the problem of two-phase detonation structure and describe solution techniques.

It is possible to reduce Eqs. (15-27) to a set of four uncoupled differential equations in four unknowns ρ_2, v_2, P_2, and ϕ_2. For $\pi_8 = 0$ (an ideal solid equation of state), Eqs. (25) and (26) can be used to write e_2 as a function of P_2 and ρ_2. The expression for e_2 can be substituted into Eq. (17). The resulting set of ordinary differential equations can be uncoupled using the techniques of linear algebra to solve explicitly for $d\rho_2/d\xi$, $dv_2/d\xi$, $dP_2/d\xi$, and $d\phi_2/d\xi$. These equations are written below.

$$\frac{d\rho_2}{d\xi} = \rho_2 \frac{A\left[\rho_2 v_2 - \frac{P_2(\pi_{17}-1)}{v_2}\right] - B\rho_2 + C\frac{\rho_2}{v_2(\pi_{17}-1)}}{r\left(v_2^2 - \pi_{17} P_2/\rho_2\right)} \quad (41)$$

$$\frac{dv_2}{d\xi} = \frac{-AP_2 + B\rho_2 v_2 - C\rho_2(\pi_{17} - 1)}{r(v_2^2 - \pi_{17} P_2/\rho_2)} \qquad (42)$$

$$\frac{dP_2}{d\xi} = \rho_2 \frac{AP_2 v_2 - B\pi_{17} P_2 + C\rho_2 v_2(\pi_{17} - 1)}{r(v_2^2 - \pi_{17} P_2/\rho_2)} \qquad (43)$$

$$\frac{d\phi_2}{d\xi} = \pi_9 \frac{\phi_1 \phi_2}{v_2}\left[P_2 - \pi_5 P_1 - \pi_{15}\phi_2\right] - \pi_1 \frac{\phi_2 P_1^{\pi_4}}{r v_2} \qquad (44)$$

$$\rho_1 = \Theta\left[\Omega \pm \sqrt{\Omega^2 - \Lambda}\right] \qquad (45)$$

$$P_1 = \Delta - \frac{\Theta}{\Omega \pm \sqrt{\Omega^2 - \Lambda}} \qquad (46)$$

$$v_1 = \frac{-1}{\Omega \pm \sqrt{\Omega^2 - \Lambda}} \qquad (47)$$

$$r = \pi_{12} \sqrt[3]{\frac{-v_2 \phi_2}{1 - \pi_{11}}} \qquad (48)$$

$$T_1 = \frac{1}{\pi_7 - 1}\frac{P_1}{\rho_1} \qquad (49)$$

$$e_1 = \frac{1}{\pi_7 - 1}\frac{P_1}{\rho_1} \qquad (50)$$

$$T_2 = \frac{1}{\pi_{17} - 1}\frac{P_2}{\rho_2} \qquad (51)$$

$$e_2 = \frac{1}{\pi_{17} - 1} \frac{P_2}{\rho_2} + \pi_{10} \tag{52}$$

$$\phi_1 = 1 - \phi_2 \tag{53}$$

$$A = -\pi_9 \frac{r \phi_1 \phi_2}{v_2} \left[P_2 - \pi_5 P_1 - \pi_{15} \phi_2 \right] \tag{54}$$

$$B = -\pi_2 (v_2 - v_1) \phi_1 - \frac{P_2}{v_2} \left[\pi_9 r \phi_1 \left(P_2 - \pi_5 P_1 - \pi_{15} \phi_2 \right) - \pi_1 P_1^{\pi_4} \right] \tag{55}$$

$$C = -\pi_3 \left(\pi_6 T_2 - T_1 \right) \phi_1 r^{2/3} \tag{56}$$

$$\Omega = \frac{\pi_7}{\pi_7 - 1} \frac{\pi_{18} + \pi_{23} - \frac{1}{\pi_5} \left(\rho_2 \phi_2 v_2^2 + P_2 \phi_2 \right)}{2 \left[\pi_{22} + \frac{1}{\pi_5} \rho_2 \phi_2 v_2 \left(\frac{P_2}{\rho_2} \frac{\pi_{17}}{\pi_{17} - 1} + \pi_{10} + \frac{v_2^2}{2} \right) \right]} \tag{57}$$

$$\Lambda = \frac{\pi_7 + 1}{\pi_7 - 1} \frac{\pi_{18} + \frac{1}{\pi_5} \rho_2 \phi_2 v_2}{2 \left[\pi_{22} + \frac{1}{\pi_5} \rho_2 \phi_2 v_2 \left(\frac{P_2}{\rho_2} \frac{\pi_{17}}{\pi_{17} - 1} + \pi_{10} + \frac{v_2^2}{2} \right) \right]} \tag{58}$$

$$\Delta = \frac{\pi_{18} + \pi_{23} - \frac{1}{\pi_5} \left(\rho_2 \phi_2 v_2^2 + P_2 \phi_2 \right)}{1 - \phi_2} \tag{59}$$

$$\Theta = \frac{\pi_{18} + \frac{1}{\pi_5}\rho_2\phi_2 v_2}{1 - \phi_2} \qquad (60)$$

In these equations, A, B, C, D, Ω, Λ, Δ, and Θ are introduced as intermediate variables. From these equations, it is straightforward to see how all variables may be written as functions of the variables ρ_2, v_2, P_2, and ϕ_2. In these equations, we take $\pi_{13} = \pi_8 = 0$ corresponding to two ideal gases. For a description of the uncoupled equations which account for nonideal effects, see Powers (1988).

It is not established whether this two-phase reactive flow model admits steady solutions. With a similar model Butler and Krier (1986) have noted difficulties in numerically integrating the time-dependent equations near the complete reaction state. These difficulties could be associated with the singular nature of the two-phase flow equations near the complete reaction state. The complicated algebraic problem of determining the end states of Eqs. (41-44) has not been solved analytically in closed form. While a complete reaction condition is a likely end state, incomplete reaction end states may be admitted by Eqs. (41-44). It is generally thought that low-velocity detonations are associated with incomplete reaction, which suggests that two-phase reactive flow equations have the potential for describing a low-velocity detonation.

It is noted that Eqs. (41-43) are singular when $v_2^2 = \pi_{17} P_2/\rho_2$ or $r = 0$, assuming that there is no common factor in the numerator to remove the singularity. That is, if either the solid velocity is sonic or the reaction is complete, then there is a zero in the denominator of the right-hand sides of Eqs. (41-43). Although not apparent, these equations must also imply a singularity when the gas velocity is sonic, for if differential equations for the gas are written instead of the solid, a similar gas-phase sonic condition would appear in the denominator of uncoupled gas phase equations. At these singular points, a necessary condition for finite gradients is that the numerators of the right hand sides of Eqs. (41-44) identically vanish. That the numerators vanish is also a necessary condition for an end state, a more general condition than the hypothesized condition described in Sec. III.

Equations (45-47) describe gas-phase variables in terms of solid-phase variables. These equations are obtained by using the mixture Eqs. (19-21) to solve for the gas-phase variables. These equations have two branches. The branch associated with the positive sign corresponds to a strong solution and that with the negative sign corresponds to a weak solution. It can be shown that following a shock jump the solution curve is on the strong branch.

By making appropriate substitutions from Eqs. (45-60) into Eqs. (41-44), it is possible to write expressions for $d\rho_2/d\xi$, $dv_2/d\xi$, $dP_2/d\xi$, and $d\phi_2/d\xi$ in terms of the variables ρ_2, v_2, P_2, and ϕ_2. Then, assuming an end state can be identified, it is possible to linearize the differential equations about the end state. By examining the linear system of differential equations, the nature of the end state will be revealed as either a source, sink, or saddle. End states that are saddle points exist in some one-phase detonation models and are associated with eigenvalue detonations. The eigenvalue detonation wave speed is the only steady wave speed admitted by the model equations. Such eigenvalue detonations may also be associated with a two-phase detonation model. If so, the wave speed associated with the eigenvalue detonation could be used in a numerical integration of the full nonlinear set of differential equations to describe the detonation structure. A numerical integration at any other wave speed would necessarily diverge from the end state, should that end state be a saddle. It is for this reason that great care must be exercised in undertaking a numerical integration of Eqs. (41-60).

Acknowledgments

This work was supported by the U. S. Air Force Office of Scientific Research under Grant AFOSR-85-0311 and the U. S. Department of Energy, Los Alamos National Laboratories under Contract DOE LANL 9XR6-5128C1.

References

Baer, M. R., and Nunziato, J. W. (1983) A theory for deflagration-to-detonation transition (DDT) in granular explosives. SAND82-0293, Sandia National Laboratories, Albuquerque.

Baer, M. R., and Nunziato, J. W. (1986) A two-phase mixture theory for the deflagration-to-detonation transition (DDT) in reactive granular materials. *Int. J. Multiphase Flow* **12**, 861-889.

Butler, P. B., and Krier, H. (1986) Analysis of deflagration to detonation transition in high-energy solid propellants. *Combust. Flame* **63**, 31-48.

Butler, P. B., Lembeck, M. F., and Krier, H. (1982) Modeling of shock development and transition to detonation initiated by burning in porous propellant beds. *Combust. Flame* **46**, 75-93.

Coperthwaite, M., and Zwisler, W. H. (1974) "TIGER" Computer code documentation. Report PYV-1281, Stanford Research Institute.

Denton, W. H. (1951) The heat transfer and flow resistance for fluid flow through randomly packed spheres. Proc. General Discussion on Heat Transfer, IME and ASME, London.

Drew, D. A. (1986) One-dimensional burning wave in a bed of monopropellant particles. *Combust. Sci. Technol.* **47**, 139-164.

Fickett, W., and Davis, W. C. (1979) *Detonation*, University of California Press, Berkeley.

Johansson, C. H., and Persson, P. A. (1970) *Detonics of High Explosives*. Academic, London.

Krier, H., and Gokhale, S. S. (1978) Modeling of convective mode combustion through granulated propellant to predict detonation transition. *AIAA J.* **16**, 177-183.

Krier, H., and Mozaffarian, A. (1978) Two-phase reactive particle flow through normal shock waves. *Int. J. of Multiphase Flow* **4**, 65-79.

Kuo, K.K., and Nydegger, C.C. (1978) Flow resistance measurements and correlation in a packed bed of WC 870 spherical propellants. *J. Ballistics* **2**, 1-25.

Kuo, K. K., and Summerfield, M. (1974a) Theory of steady-state burning of gas-permeable particles. *AIAA J.* **12**, 49-56.

Kuo, K. K., and Summerfield, M. (1974b) High speed combustion of mobile granular solid propellants: wave structure and the equivalent Rankine-Hugoniot relation. *Fifteenth (International) Symposium on Combustion*, The Combustion Institute, Pittsburgh, PA.

Macek, A. (1959) Transition from deflagration to detonation in cast explosives. *J. Chem. Phys.* **31**, 162-167.

Marsh, S. P., ed. (1980) *LASL Shock Hugonoit Data*, University of California Press, Berkeley.

Powers, J. M. (1988) Ph.D. Thesis, *Theory of detonation structure for two-phase materials*. Department of Mechanical and Industrial Engineering, University of Illinois at Urbana-Champaign.

Powers, J. M., Stewart, D. S., and Krier, H. (1988) Analysis of steady compaction waves in porous materials. To appear in *J. Appl. Mech.*

Taylor, P. A. (1984) Growth and decay of one-dimensional shock waves in multiphase mixtures. *Acta Mech.* **52**, 239-267.

Wallis, G. B. (1969) *One-Dimensional Two-Phase Flow*. McGraw-Hill, New York.

Whitham, G. B. (1974) *Linear and Nonlinear Waves*. Wiley, New York.

Heterogenous Detonation Along a Wick

B. Plewinsky,* W. Wegener,† and K.-P. Herrmann‡
Bundesanstalt für Materialforschung und -prüfung
Berlin, Federal Republic of Germany

Abstract

The combustion behavior of tetramethyl-dihydrogen-disiloxane (TMDS) in pure oxygen [$p(O_2) \leq 40$ bar] was investigated. The reaction vessel used for this purpose contained a PMMA tube (inside diameter 3 cm, length 100 cm) in the center of which a cotton wick - soaked with the TMDS - was fixed. The combustion processes in the reaction vessel were observed with a framing camera, an image-converter camera, and a rotating drum camera. The investigation showed that, from oxygen pressures of 10 bar upward, a detonation takes place along the wick soaked with the TMDS. This new type of a heterogeneous detonation is named wick detonation. The velocity of the heterogeneous detonation varies 400 - 1480 $m \cdot s^{-1}$. During a single experiment considerable changes in the detonation velocity have been observed frequently. The observations can be interpreted as follows: The combustion is confined initially to a narrow region along the wick. This first reaction zone is caused by a leading shock front that spreads the fuel into the gas phase in a somewhat irregular manner. A following reaction zone, which is characterized by intensive burning, is caused by a secondary shock wave.

Introduction

In contrast to gaseous detonations, which have been studied for almost a century, detonations in two-phase systems have been known for only about 30 years. In 1952, Loisson was the first to report on film detonations. Burgoyne and Cohen reported in 1954 on detonations or deto-

---
Copyright © 1988 by the American Institute of Aeronautics and Astronautics, Inc. All rights reserved.
*Research Chemist
†Mechanical Engineer
‡Mechanical Engineer

nation-like phenomena when they investigated the combustion behavior of liquid aerosols in an oxidizing gaseous atmosphere. Heterogeneous detonations have been observed in numerous solid/gas and liquid/gas systems (Dabora and Weinberger 1974) and in powders, sprays, single droplet columns, fogs, monopropellant sprays, fuel films and quartz sand coated with combustible liquids (Lyamin 1984). While either the oxidizer or fuel can be in the condensed phase, in most practical systems the oxidizer is gaseous.

This paper deals with a new type of two-phase detonation: a heterogeneous detonation along a wick. As fuel tetramethyl-dihydrogen-disiloxane (TMDS) was used. This substance was chosen because of its particularly reactive H-Si-bonds (Plewinsky et al. 1985).

It was also of some interest to investigate the burning behavior of TMDS in quartz sand whose surface was wetted with this liquid fuel. In all the experiments, pure oxygen was used as the oxidizer.

Experimental Details

Heterogeneous Detonation along a Wick

To establish heterogeneous detonation waves tubes made of PMMA (inside diameter 3 cm, length 100 cm) were used. The apparatus (Fig. 1) consists of a tube of this type in the center of which a cotton wick was fixed, a feed mechanism to fill the tube with pure oxygen [$p(O_2) \leq 40$ bar], and a storage vessel containing 2.5 ml of the fuel to impregnate the wick. The combustion was started with the aid of an incandescent wire (Cu/Ni/Zn alloy nickelin, diameter 0.1 mm). As fuel tetramethyl-dihydrogen-disiloxane [$(CH_3)_2HSi-O-SiH(CH_3)_2$, TMDS, purity 97% min, boiling point 71°C] was used. The combustion processes in the reaction tubes were observed with a rotating drum camera, rotating prism camera (HYCAM), and image converter camera (IMACON 790).

Heterogeneous Detonation in Quartz Sand

These experiments were carried out with clean quartz sand. The sand grains were 0.5 - 1 mm in size. As fuels TMDS and dodecane $C_{12}H_{26}$ (purity 99% min, boiling point 216°C) were used. The sand (40.7 g) was wetted with 2.5 ml of the fuel. As oxidizer pure oxygen was used (TMDS: $5 \leq p(O_2)/\text{bar} \leq 30$, $C_{12}H_{26}$: $p(O_2)/\text{bar} = 20$ and 40). The reaction tubes were made from PMMA (inside diameter 0.6 cm, length 100 cm). The combustion process was also initiated

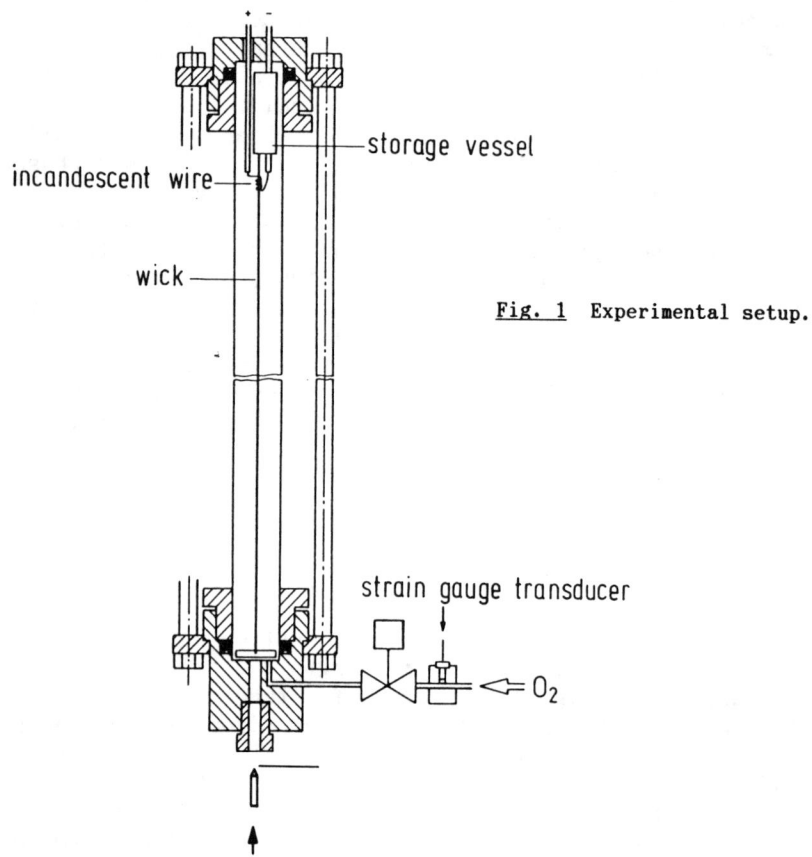

Fig. 1 Experimental setup.

by means of an incandescent wire. The propagation of the reaction zone in these tubes was observed with a rotating drum camera.

All experiments were carried out at a temperature of 18±1°C.

Experimental Observations and Discussion

Heterogeneous Detonation along a Wick

Framing camera (picture framing speed 8,000 s^{-1}) and image converter photographs (picture framing speed 50,000 s^{-1}, exposure time 4 μs) of light emission from a wick detonation of TMDS are shown in Figs. 2 and 3, respectively. These photographs represent measurements at oxygen pressures of 15 bar. Although the shock front is not visible, it is apparent that the combustion is confined ini-

WICK DETONATION 365

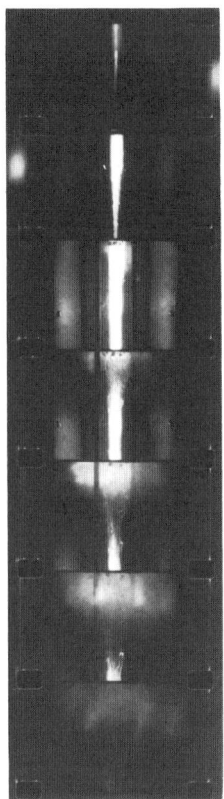

Fig. 2 Framing camera photograph of a wick detonation, $p(O_2)$=15 bar, picture framing speed 8,000 s^{-1}.

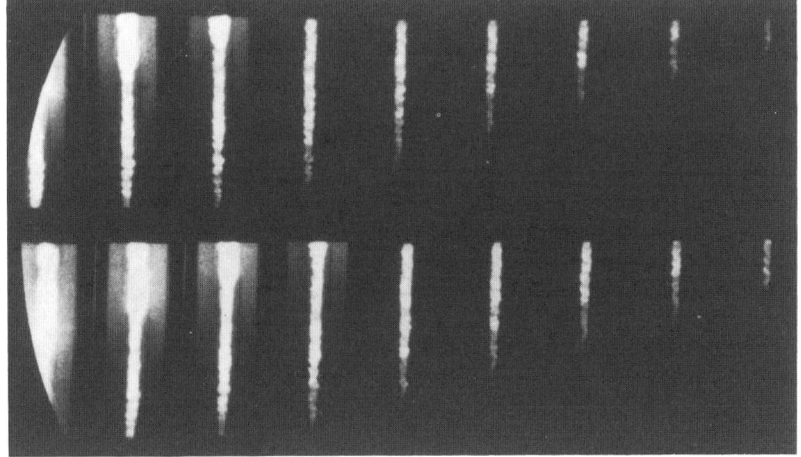

Fig. 3 Image converter photograph of a wick detonation, $p(O_2)$=15 bar, picture framing speed 50,000 s^{-1}, exposure time 4 μs.

tially to a narrow region along the wick and spreads the
fuel into the gas phase in a somewhat irregular manner at
some distance behind the shock front. What is remarkable is
the light dot that can be seen in front of the combustion
zone on the image converter photographs (Fig. 3). The leading shock front is probably near this dot.

A streak photograph for the case of a detonation along
a wick is presented in Fig. 4. The oxygen pressure is again
15 bar. This photo shows a transition from deflagration to
detonation, which is also typical for gas reactions. The
detonation along the cotton wick is represented in this
photograph by the line of minor intensity. The TMDS, which
cannot burn due to the lack of sufficient oxygen, is - as
already mentioned - spread by the leading shock front into
the gas phase (Ragland and Nicholls 1969, Borisov et al.
1981). In addition to this line, the streak photopraph
(Fig. 4) shows an extended luminous area. This is caused by
a zone of intensive burning of the TMDS. Photographs from
both the framing camera and image converter camera (Figs. 2
and 3) demonstrate that in this region not only the surrounding of the wick, but also the entire interior of the
tube is burning. As Ragland and Nicholls (1969) pointed
out, this reaction zone is caused by a secondary shock wave
that is probably generated by the combustion process.

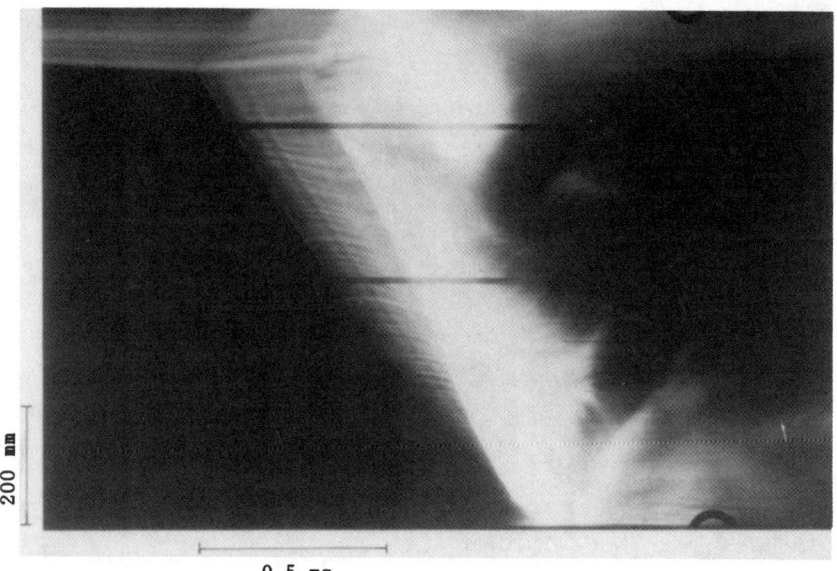

Fig. 4 Streak photograph of a wick detonation, $p(O_2)=15$ bar.

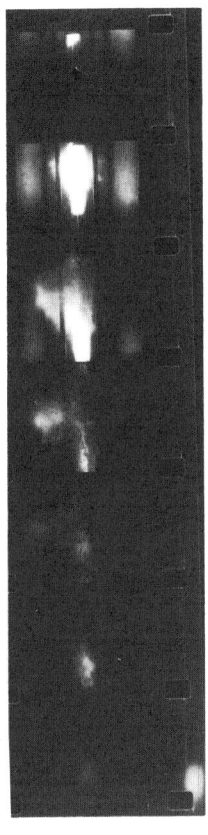

Fig. 5 Framing camera photograph of a wick detonation, p(O$_2$)=25 bar, picture framing speed 8,000 s^{-1}.

At 18°C, detonations in the TMDS-saturated gas phase cannot take place at an oxygen pressure as low as 10 bar due to the fact that this concentration is below the lower explosion limit.

When the oxygen pressure is increased, the combustion behavior of TMDS alters its characteristics. The burning zone around the wick nearly vanishes. This can be seen by comparing the framing camera and the image converter photographs of a wick detonation at 25 bar (Figs. 5 and 6) with those recorded at an oxygen pressure of 15 bar. On the other side, it should be mentioned that the streak photographs at 25 bar (Fig. 7) are very similar to those taken at 15 bar. But it should be kept in mind that the deflagration/detonation transition distance decreases with increasing oxygen pressure. At 25 bar, the value of this quantity becomes nearly zero (Fig. 7).

Fig. 8 shows the velocity of the wick detonation as a function of the oxygen pressure. The data were obtained from streak photographs. The detonation velocity varies

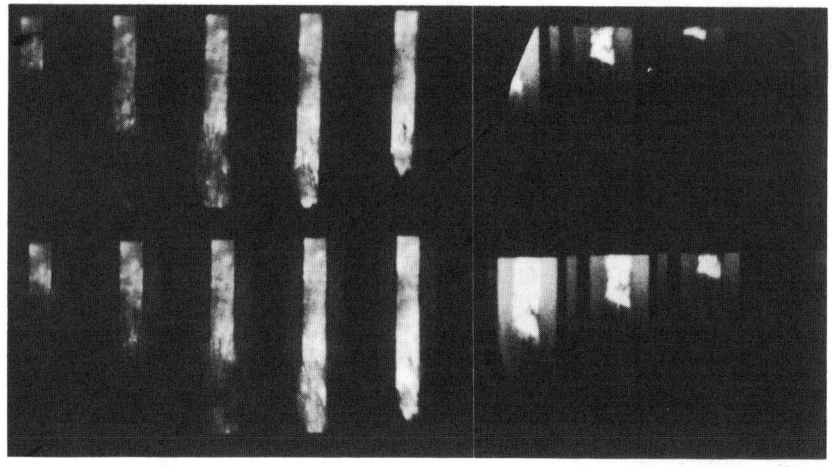

Fig. 6 Image converter photograph of a wick detonation, $p(O_2)=25$ bar, picture framing speed 50,000 s^{-1}, exposure time 4 µs.

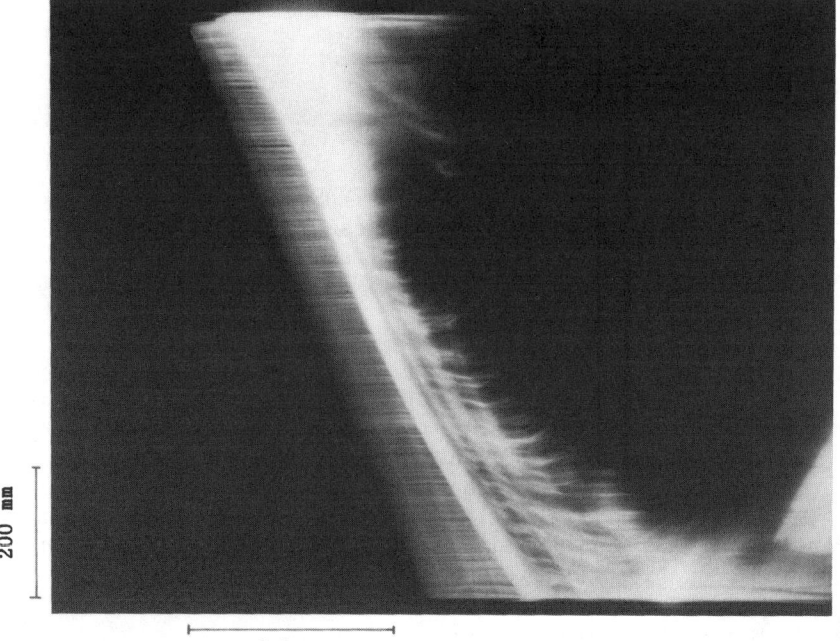

Fig. 7 Streak photograph of a wick detonation, $p(O_2)=25$ bar.

WICK DETONATION 369

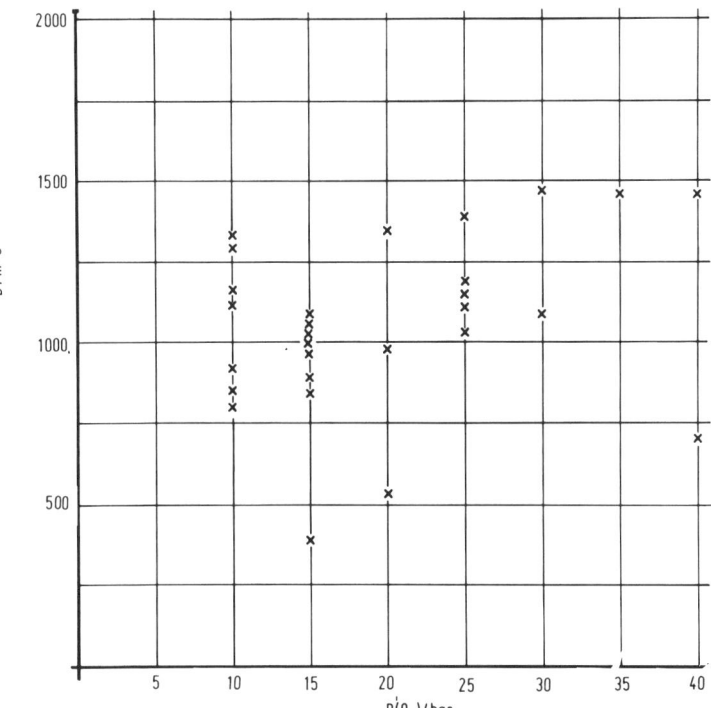

Fig. 8 Velocity of wick detonation D vs oxygen pressure, p(O_2).

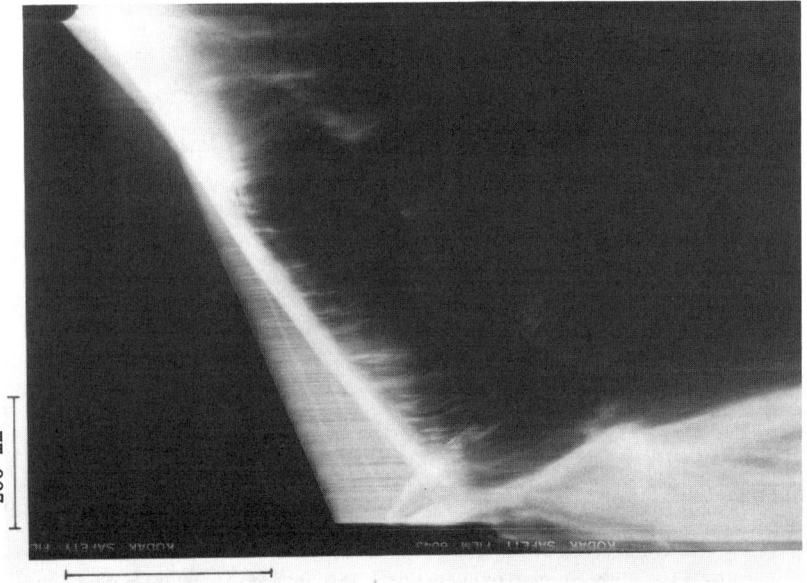

Fig. 9 Streak photograph of a wick detonation, p(O_2)=40 bar.

from about 400 to about 1500 m/s. At the same oxygen pressure different values of the detonation velocity can be observed. Even in one experiment, the detonation velocity sometimes changes its value. This effect can be seen in Fig. 9. Here, the streak photograph of a wick detonation of TMDS at an oxygen pressure of 40 bar is shown. As a result of the variations of the measured values, no correlation of the detonation velocity and the oxygen pressure can be established. For this and for other reasons, no attempt has been made to calculate detonation velocity and other data from initial conditions.

Heterogeneous Detonation in Quartz Sand

The experiments showed that in the heterogenous quartz sand system containing liquid TMDS as fuel a detonation is established if the oxygen pressure $p(O_2)$ is greater than or equal to 10 bar. The velocity of the heterogenous detonation D increases with increasing oxygen pressure [$p(O_2)=10$ bar, D=948 m·s^{-1}; $p(O_2)=15$ bar, D=1035 m·s^{-1}; $p(O_2)=20$

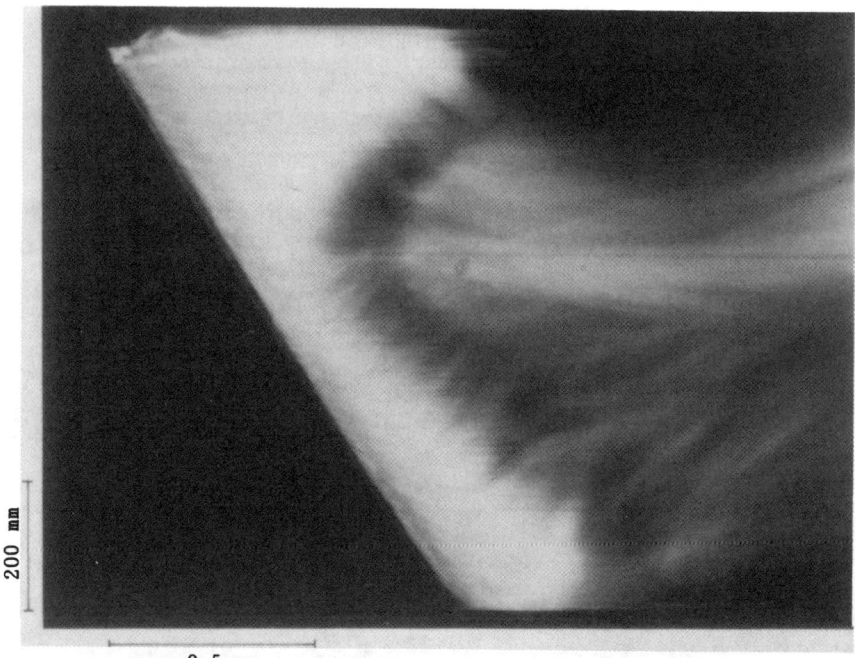

Fig. 10 Streak photograph of a detonation in quartz sand, $p(O_2)=15$ bar.

bar, D=1110 m·s⁻¹; p(O₂)=30 bar, D=1180 m·s⁻¹]. Fig. 10 shows a typical streak photograph for a detonation in quartz sand [p(O₂)=15 bar]. As can be seen in this photograph, the transition distance is nearly zero. The same small detonation induction distance can be observed at the other oxygen pressures.

In the heterogeneous quartz sand system containing dodecane $C_{12}H_{26}$ as fuel, we did not observe detonations. The applied oxygen pressures were 20 and 40 bar. In contrast to our results, Lyamin (1984) found in a similar system containing hexadecane $C_{16}H_{34}$ as fuel heterogeneous detonation at oxygen pressures from above 5 bar.

Acknowledgment

We are very grateful to Messrs. H.-J. Seeger and R. Kowall for their technical assistance.

References

Borisov, A. A., Gel'fand, B. E., Sherpanev, S. M., and Timofeev, E. I.(1981) Mechanism for mixture formation behind a shock sliding over a fluid surface. Fiz. Goreniya Vzryva 17(5), 86-93.

Burgoyne, J. H. and Cohen, L. (1954) Effect of drop size on flame propagation in liquid aerosols. Proc. R. Soc. London A225, 375-392.

Dabora, E. K. and Weinberger, L. P. (1974) Present status of detonations in two-phase systems. Acta Astronautica 1, 361-372.

Loison, R. (1952) The propagation of deflagration in a tube covered with an oil film. C. R. Acad. Sci. 234, 512-513.

Lyamin, G. A. (1984) Heterogeneous detonation in a rigid porous medium. Fiz. Goreniya Vzryva 20(6), 134-138.

Plewinsky, B., Wegener, W., and Herrmann, K.-P. (1985) Measurements of explosion pressures in calorimetric bombs. Thermochim. Acta 94, 33-42.

Ragland, K. W. and Nicholls, J. A. (1969) Two-phase detonation of a liquid layer. AIAA J. 7, 859-863.

Photographically Observed Waves in Detonation of Liquid Nitric Oxide

Garry L. Schott* and Kenneth M. Chick[†]
*Los Alamos National Laboratory, University of California
Los Alamos, New Mexico*

Abstract

Homogeneous liquid nitric oxide (empirical formula: NO; boiling point: 121 K) functioning as a detonating, prototype high explosive has recently been investigated systematically at Los Alamos National Laboratory. The present extension of these studies addresses experimentally the delayed formation of detonation in a layer of this cryogenic material situated between a plastic window and a metal plate when an externally driven, flat shock wave is transmitted normally from the plate and impulsively accelerates the interface to a speed near 0.90 km/s. Four such experiments, done with variants of slit-plus-rotating-mirror streak photography through the window, show the following. Beyond a 3 mm depth, the wave is recognized as a detonation, by its 5.5 km/s frontal velocity and, more interestingly, by its finely segmented transverse structure. The input shock flow, however, is smooth; its front advances steadily until detonation abruptly overtakes it. The sequence and character of events correspond closely with those obtained in other homogeneous liquids such as nitromethane and also in reflected shock waves in typical detonable gases. The techniques, however, are novel and appropriate to condensed explosives, where macroscopic compression and ultimate destruction of the surroundings are dominant characteristics and viewing from the side is disadvantageous.

Introduction

Nitric oxide (NO) is quasistable and moderately energetic with respect to its constituent elements, O_2 and N_2 (nitric oxide is termed "nitrogen oxide" in IUPAC nomenclature and "nitrogen (II) oxide" in some literature; ΔH^o_{fo} (NO) = +89.8 kJ/mole), and it detonates

---

Copyright ©1988 by the American Institute of Aeronautics and Astronautics, Inc. All rights reserved.
* Staff Member, Reaction Sciences, Dynamic Testing Division.
† Graduate Research Assistant (1984-86), Dynamic Testing Division; Present affiliation: University of Michigan.

reproducibly at small diameters in its homogeneous, cryogenic liquid state [Miller (1968); Ramsay and Chiles (1976); and Ribovich et al. (1977)]. Various molecular, optical, thermodynamic, and fluid dynamic aspects of the detonation of this explosive material have recently been studied in a concerted program of fundamental research on explosives at Los Alamos National Laboratory [Rivera (1985)].

The present extension of these studies addresses the extreme "reaction zone" situation found in the initiation of detonation. A weak shock wave transforms quiescent explosive to a compressed state in which chemical reaction begins at a resolvable rate. Unsteady fluid-dynamic consequences ensue and unsupported detonation subsequently is established. The variety of unsteady wave behavior we might anticipate from existing experience is much less than when detonation develops through transition from deflagration; and this is especially so for fluid explosives, either liquid or gaseous. The generally observed shock-initiation behavior in these mechanically homogeneous media is conventionally associated with thermal explosion in an adiabatic layer near the impact interface. Detonation develops first in the hotter, more dense preshocked fluid and later overtakes the input shock. Selected accounts, with references to relevant earlier experiments and theory, are: with liquids, [Campbell et al. (1961a) and Hardesty (1976)]; with gases, [Strehlow and Dyner (1963)]. Even this prototype mechanism exhibits recognizable variations as consequences of significant differences of reac-

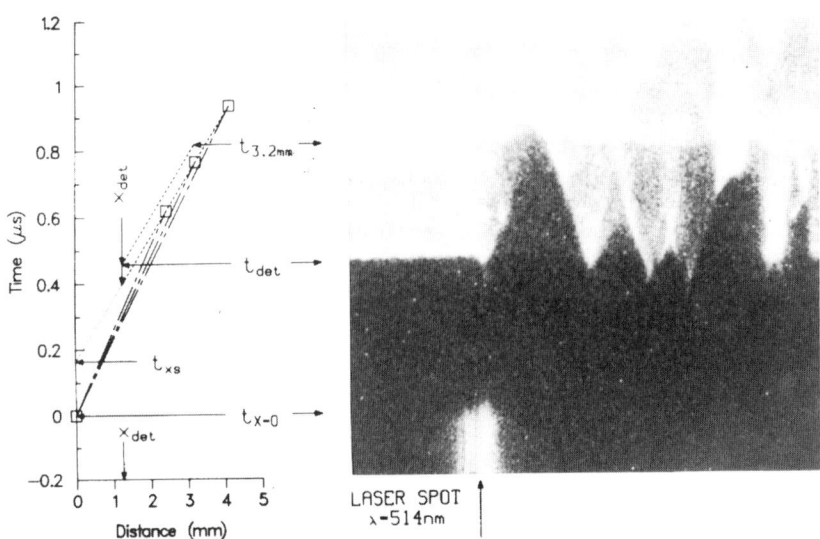

Fig. 1 Composite results of two instances of delayed detonation in shocked liquid nitric oxide, measured by wave arrivals at sample boundaries in one shot and direct, image-intensified streak photography of visible wavelength light in another. See text and Fig. 2 for details.

tion kinetics and Mach number within the initially shocked layer; an example is the phenomenological distinction between "mild" and "strong" ignition of detonable gaseous fuel-oxygen mixtures [Saytzev and Soloukhin (1958; 1962); Meyer and Oppenheim (1971)].

Liquid nitric oxide is described from a safety standpoint as "sensitive" [Ramsay and Chiles (1976)], but controlled initiation of its detonation has been only minimally investigated heretofore. Hence, we have adopted the general objective of exploring the fluid-dynamical situation in which chemical decomposition of weakly shocked nitric oxide precedes detonation. By means of streak photographic experiments, a strength of shock wave that leads to delayed detonation has been determined first and then the character of waves during the development of detonation has been elucidated. The light recorded is either emitted internally or produced externally and reflected from selected surfaces that confine the explosive.

The point of departure for our experiments to explore the shock initiation of detonation in liquid nitric oxide was a completed study of overdriven detonation waves done to produce Hugoniot equation-of-state data on the hot, dense product mixture. Manufactured solid high-explosive systems uniformly shocked an aluminum plate in which sealed, slot-shaped pockets of liquefied NO were embedded. Transit times of the wave through paths containing each sample and alongside it were registered by flashes of photographable light from covered argon-filled gaps at the terminal face of the plate. The pertinent explosive, cryogenic, and optical details have been reported [Schott et al. (1985)].

Experiments and Results

Wave Transit Times

High-explosive-driven piston systems were adapted from the product state study [Schott et al. (1985)] by substituting weaker driving explosives, interposing mismatched barrier layers between the explosive and the plate, and eliminating free-run spaces between the assembled layers. Krypton gas was used in the flash gaps for recording of arrival times by the rotating-mirror streak camera. When detonation is unsupported but begins promptly, its front traverses the sample in a transit time t_J equal to the ratio of the depth d to the Chapman-Jouguet (C-J) wave speed D_J. Less strongly driven pistons produce transit times t that measurably exceed t_J, so that there is an excess transit time, $t_{xs} = t - t_J$. Ideally, t_{xs} is independent of the choice of d, indicating that transient initiation processes are completed in a lesser length of run. This is the regime to be addressed here.

In the transit-time experiment that guided the remainder of this study, three sample depths were present, having $d = 2.4, 3.2$, and 4.1 mm. From measurement of the aluminum free surface motion, it was determined that the well-supported shock in the piston had pressure $p = 10.4$ GPa and mass velocity $u_p = 0.60$ km/s as it approached the liquid nitric oxide sample. For the input shock wave in the nitric oxide, the unmeasured u_p is estimated to be 0.89 km/s.

Transit-time results from this shot are shown graphically in the left side of Fig. 1. The dashed line linking the points for the three depths has a slope of 5.36 km/s, an acceptably close match to the known [Ramsay and Chiles (1976); Ribovich et al. (1977)] D_J for the initial sample conditions ($T_o = 123$ K, $\rho_o = 1.26$ g/cm^3). Hypothetical, dotted extrapolation of this inferred detonation trajectory to the ordinate axis indicates t_{xs} to be 0.17 μs.

Photography of Explosive Interior

The next step to interrogate the events leading to an excess transit time was made by photographing the full depth of a single, window-covered, disk-shaped nitric oxide sample viewed from downstream and resolved by a diametrical slit. Figure 2 illustrates this experiment. The shock source was closely matched to the previous one. The depths of the sample and of the aluminum layer beneath it were 3.2 and 4.8 mm, respectively. The window aperture exposed only the central 19 mm of the sample disk, so that two-dimensional effects at the periphery were not viewed. A channel-plate image intensifier tube was used to amplify the faint emitted light from the nitric oxide products, which are less luminous than those from many condensed explosives. The intensifier occupied a 29 mm diam area

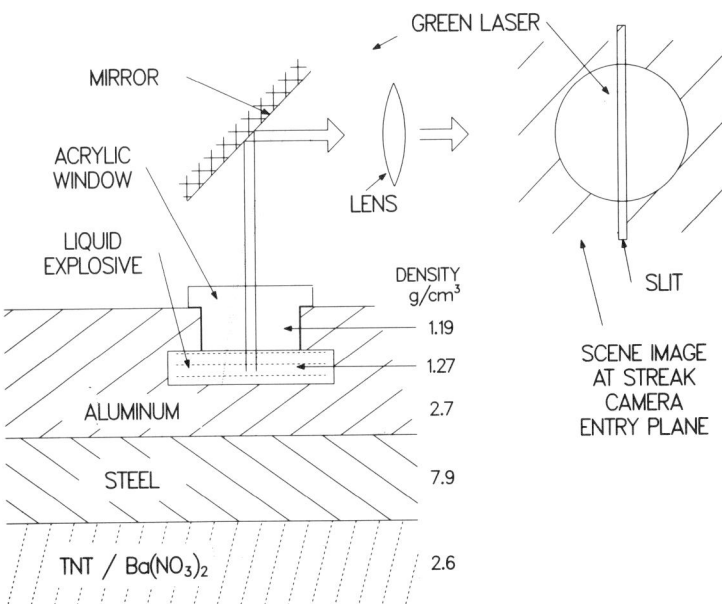

Fig. 2 Schematic optics and explosive-shock driver for direct streak photography of delayed nitric oxide detonation.

at the image surface of the rotating-mirror streak camera; the slit image swept across it in 2.29 μs within a 10 μs electronically shuttered span synchronized with the detonator and the mirror rotation phase.

Besides the light emitted by the exploding sample, the camera system also viewed a small laser-illuminated spot on the diffusely reflecting aluminum surface at the base of the sample, for the purpose of detecting the perturbation expected when the input shock entered the sample. The spot was made by crudely focusing a continuous-wave, 514 nm argon-ion laser beam routed through the window and static nitric oxide sample on a different optical axis from that used by the camera. The resulting photograph forms the right side of Fig. 1 and its scale matches the ordinate from the transit-time plot. The laser spot recorded before the input shock arrived was extinguished quickly afterward and this event is taken as the zero of time, precise to perhaps 0.05 μs. The photographed self light, which is not spatially uniform, was distinctly delayed from the input wave arrival, by 0.4 μs and more.

At about 0.8 μs, the self-light intensity was temporarily and nearly synchronously perturbed across the window diameter. As Fig. 1 shows, this time acceptably matches the transit time for 3.2 mm sample depth found in the three-sample shot. Moreover, the uncoated acrylic plastic window remained transparent and light emitted from the reaction products continued to be recorded until the slit image exited the circular photographic frame. This happened after an additional interval of 0.7–1.1 μs, and during this interval no other indication of arrival of a wave front at the window occurred. We thus identify the event at $\sim$ 0.8 μs as the already luminous detonation wave reaching the inner window surface.

At the time near 0.45 μs when the observable luminosity began, the wave front is logically deduced to have been partway along its 3.2 mm path through the sample. Immediate, steady detonation would traverse the 3.2 mm depth in $t_J = 0.58$ μs. As a provisional conclusion, we associate the change from nonluminous to luminous with the transition to detonation. For illustration in the left side of Fig. 1, the time of this transition and the inferred detonation front position at that time are denoted t_{det} and x_{det}.

Auxiliary Experiments and Assessment

The two experiments combined in Fig. 1 appear to be in agreement that detonation was established after a delay and subsequently propagated at D_J, within the indicated times and distances of run. In the absence of confirmatory evidence, however, the assignments of events responsible for the three optical changes exhibited in the Fig. 1 streak photograph would be left as somewhat conjectural. Thus, it is appropriate to introduce briefly the results of three auxiliary experiments done with similar diagnostic methods and other strengths of the input shock.

First, a three-depth transit-time experiment like that in Fig. 1 (left) was done with a weaker input shock source, for which the respective pressure and mass velocity in the aluminum piston approaching the liquid nitric oxide were $p = 8.7$ GPa and $u_p = 0.52$ km/s. The transit times exceeded those in Fig. 1 and t_{xs} was not constant, as in Fig. 1, but increased progressively. The ratios t/d diminished with increasing d, indicating that the wave front accelerated, but establishment of detonation within the 4 mm range of d was not confirmed.

This same weak input shock system was used for another shot in which we attempted to perform laser Doppler velocimetry of the aluminum-to-nitric oxide interface, as was done by Hardesty (1976) to observe delayed reaction and subsequent detonation in liquid nitromethane. At $d = 3.2$ mm, the initial liquid nitric oxide was transparent enough that static interferometer fringes were recorded until the time the shock emerged from the aluminum. However, the earliest event we sensed was extinguishment of the reflected light; no interpretable dynamic fringes were recorded thereafter. Evidently, even a weak shock promptly renders liquid nitric oxide highly opaque at $\lambda = 514$ nm. This result guided the use of the green laser beam and the interpretation of the outcome in the photographic experiment of Figs. 1 and 2.

Finally, a cylinder of liquid nitric oxide with $d = 41$ mm was detonated by prompt initiation from a stronger shock source and its self-light was photographed end-on as illustrated in Fig. 2. The streak record was exposed sufficiently across the full 20 mm charge diameter throughout the 2.3 µs recording interval. The detonation front reached the acrylic window during this interval and, when it did, the light intensity increased suddenly and then subsided more slowly, much as occurred at $t \simeq 0.8$ µs in Fig. 1.

Even with the measure of confirmation provided by these auxiliary experiments, the results in Fig. 1 leave us without data on the speed and steadiness of the input shock during the time before detonation takes over. Likewise, onset of early reaction in the initially shocked layer or commencement of laterally contiguous detonation within that compressed layer is not detected by either the wave transit times at the selected discrete depths or by the streak-photographed self-light. Photographic sensitivity to reaction or detonation behind the input shock is strongly diminished for us by the prompt opacity that the shock imparts to the liquid nitric oxide. The situation in Fig. 1 contrasts with the anologous one in liquid nitromethane, where end-on streak photography has been particularly informative [Campbell et al. (1961a); Hardesty (1976)], and corresponds more closely to conditions in melted TNT and certain other liquid explosives examined by Campbell et al. (1961a).

Nevertheless, the photographed onset of self-light in Fig. 1 is highly informative, even without identifying a detonation that later overtakes the input shock. On the left side, the onset was coherent and sensibly smooth, while on the right, sporadic and initially localized luminous sites spread and interacted in a very coarse transverse pattern that persisted in the photograph until the explosion front overran the window, and even afterward. In the nitromethane experience [Campbell et al. (1961a)], similar variations from coherent to sporadic onset of luminosity were systematically associated with mechanically uniform or locally defective impact interfaces. However, the range of structures present in our Fig. 1 was unintentional and the lesson from it is that with liquid nitric oxide, as with nitromethane, the transition from weak shock to detonation greatly amplifies slight mechanical irregularities.

Wedge-Shaped Sample

Some criteria for our next experiments are apparent from the foregoing assessment of the results thus far. First and foremost, a method was needed that continuously and directly tracked the input-shock wave. Beyond this, we required precisely controlled amplitude of the weak input-shock wave and extreme mechanical homogeneity of the piston interior and surface that delivered it. Our most informative tests of the wave processes during establishment of detonation in liquid nitric oxide observed these considerations. The wedge technique so commonly applied for plane-shock initiation studies of solid explosives [Campbell et al. (1961a); Majowicz and Jacobs (1958); and Wackerle et al. (1976)] was selected to monitor the input shock through a depth of only a few millimeters. This technique was implemented with liquid explosive by using a ramped interior window surface as the downstream bounding plane of the wedge and as the mirror substrate. Continuous, precise selection of piston velocity in the needed range was obtained from a light-gas gun. High-purity copper [oxygen free, high conductivity (OFHC) grade] was used for the piston plate and impactor, on the basis of its excellent mechanical homogeneity and low elastic limit. Even though our input shocks were not quite strong enough to eliminate the elastic precursor wave as in Hardesty's (1976) nitromethane work, this potential complication was made much less than with aluminum at the same piston speeds. The copper piston interface with the explosive sample was diamond-tool machined and the resulting mirror-finished surface was determined by microscopic examination to be free of extraneous pits in the important interior area. Figure 3 illustrates schematically the orthodox wedge-shaped configuration of the explosive in our gas-gun target [Wackerle et al. (1976)], including the illumination but omitting most of the cryogenic and other accessories.

The dynamic measurement occurred at the downstream boundary of the explosive sample, formed by a transparent plastic window aluminized on its interior surface. This was illuminated by a synchronized flash source and photographed during the impact event by a continuously recording streak camera. Being inclined at 10 deg to the piston surface and set back only 0.2 mm from it at the wedge

DETONATION OF LIQUID NITRIC OXIDE

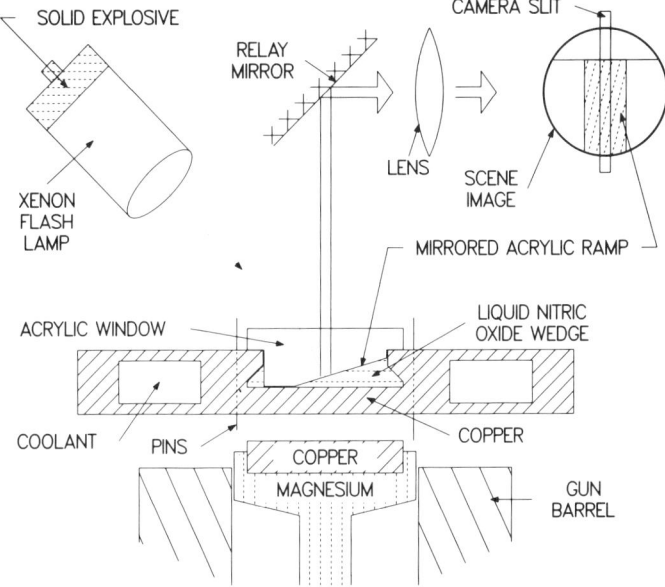

Fig. 3 Schematic arrangement of wedge-shaped liquid nitric oxide volume in dynamically photographed gas gun target.

toe, the mirrored window spread along the slit length the continuous progress of the "planar" explosion-wave system advancing through the depth of the sample. Alteration of intensity of the light reflected to the camera recorded advancing waves in the flow as continuous tracks with positive distance-time slope. Avoiding a large wedge angle or a window material with high sound speed assured that the flat input-shock front in the liquid progressed up the ramp faster than any other disturbance could move over the window surface.

Figure 4 displays the record from such a wedge experiment in liquid nitric oxide. Early in the record, where the explosive wedge was thinnest, the input shock made a smooth, straight track as it reached the inclined mirror. Farther along, detonation occurred and advanced faster along the mirror, making a second track. Transition occurred at a depth of 3.4 mm, where the input-wave track in the photograph ends and the detonation track abruptly changes slope. To the left of the transition, the detonation track traverses the same domain of space and time as the final portion of the input-shock track; the sign and magnitude of its slope identify the beginning of detonation as a separate, fast wave in the compressed explosive behind the input shock. The separation diminishes linearly to zero, showing that the input shock was overtaken supersonically from the rear.

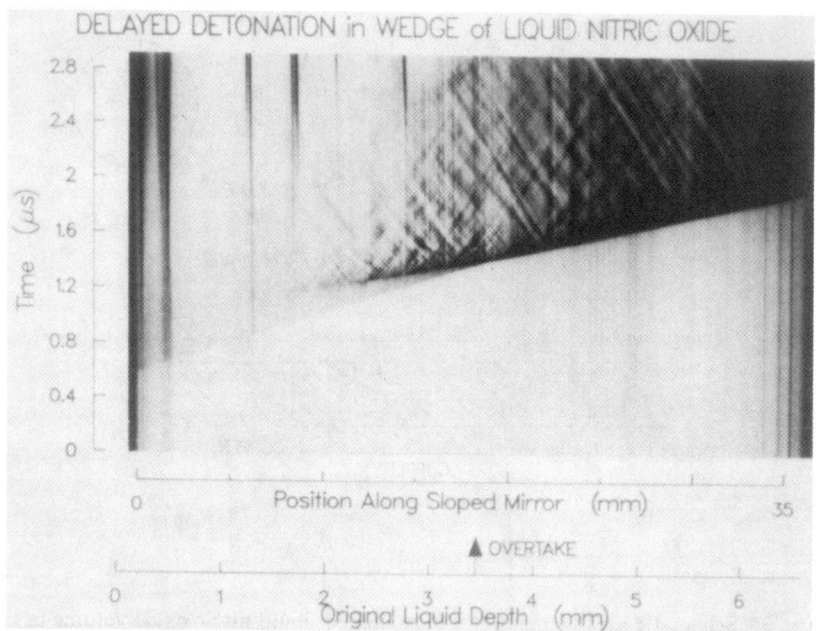

Fig. 4 Streak-camera record from gas gun target shown in Fig. 3. Time origin is arbitrary.

Figure 4 shows that both of the detonation track segments, but not the input-shock track, are followed by a contiguous pattern of fine-grained undulations of intensity in the photographic image of the mirrored interface between the detonation products and the adjacent window. The pattern persists after consumption of the explosive for a time that increases in approximate proportion to the depth of detonated explosive beneath it. Beyond the field reproduced in Fig. 4, the record was ultimately terminated by waves converging from the periphery.

Chemical Contamination Test

The experiment illustrated in Figs. 3 and 4 was done twice, to test reproducibility and to investigate whether the shock sensitivity of liquid nitric oxide is noticeably affected by its chemical purity. This matter is significant because nitric oxide gas becomes progressively contaminated with its "low-temperature" decomposition products, N_2O and NO_2, during storage and handling. Mole fractions of these contaminants near 10^{-5} are difficult for us to control or eliminate and NO_2 forms blue-colored N_2O_3 when the gas is liquefied.

The two wedge shots used the same materials and dimensions for all their solid components and the impactor speeds were the same $(1.009 \pm 0.001 \text{ km/s})$. The sample temperature and density were 120 K and 1.28 g/cm^3 in both shots. The nitric oxide supplies were deliberately different, however. In the first experiment, the NO gas was recently distilled and neither the N$_2$O band near 2240 cm^{-1} nor the NO$_2$ band near 1620 cm^{-1} was detected in its infrared absorption spectrum, taken at 83 kPa pressure and 10 cm path. The contaminant levels [Fateley et al. (1959)] thus indicated are <20 ppm of N$_2$O and <40 ppm of NO$_2$. The NO gas used in the second shot, which is the one shown in Fig. 3, had been stored for one year. Its infrared analysis showed 145 ppm ($\pm 20\%$) of N$_2$O and 325 ppm ($\pm 25\%$) of NO$_2$. The qualitative features of the two experimental records were indistinguishable and the depth at which detonation overtook the input shock was 3.0 mm with the purer sample and 3.4 mm with the deliberately impure material. This difference is small enough that much more extensive statistics and examination of other variables would be needed to determine its significance.

Discussion

The primary result of our gas gun, wedge-initiation experiments is photographic demonstration that detonation begins in liquid nitric oxide, as in other shocked fluid explosives, in the precompressed layer that accumulates behind the input-shock front. The initiation mechanism is completed somewhat later, after the input shock has been overrun, eliminating dense, unreacted material in the preshocked layer. Detonation begun by this mechanism is overdriven at the transition when it first enters ordinary, quiescent explosive, but this is temporary and the wave front decays continuously into "permanent" unsupported detonation. As the early detonation moves forward in the preshocked layer, it has unusually high pressure and wave speed associated with Chapman-Jouguet behavior for the local state of compression and motion and is termed a "superdetonation."

A sufficient basis for recognizing this shock-to-detonation transition of homogeneous explosive and distinguishing it from the usual behavior of polygranular solid explosives [Campbell et al. (1961b)] is that the input shock exhibits no protracted acceleration before it is overtaken. By its initial arrival tracks, Fig. 4 shows first this straight-line characteristic of the input wave and then the briefly overdriven and subsequently stable trajectory of the faster, permanent detonation. Evidently, our control of piston and interface uniformity was sufficient that smoothness of the arrival tracks was not distorted. However, tilt of the input shock was enough to interfere with direct wave speed measurement [see Schott (1988)] at our 10 deg wedge angle, which is unfavorably small.

The foregoing is as much information as wedge initiation experiments customarily yield, when the terminal surface of the solid explosive, or a thin plastic film affixed to it, serves as the mirror substrate and is promptly destroyed in registering the earliest wave arrival. In the present setup, however, the mirror remained reflective and photographically visible at the moving interface between its acrylic plastic substrate and the shocked liquid nitric oxide and/or subsequent detonation products. As a result, the trajectory of the superdetonation wave was recorded in as direct and clear a manner as has yet been accomplished in any but gaseous systems [Strehlow and Dyner (1963)]. (The ingenious demonstration of nitromethane's superdetonation trajectory presented in Figs. 9–11 of Campbell et al. (1961a) can be understood to have used a right-angle wedge, illuminated through adequately transparent preshocked explosive.)

The time-swept photographic field recorded from the interface between detonation products and the shocked acrylic window after arrival of the detonation front shows evidence of the detonation's transverse structure. Once the superdetonation is coherently formed, some 10 mm along the sloped-mirror scale in Fig. 4, the arrival track itself is not resolvably rough under our photographic resolution. Sensitive demonstration of the transverse structure depends on having the continuing record. The most prominent feature of the transverse structure field is a cross-hatching of light and dark alternations that form two nearly parallel, symmetric sets at positively and negatively oblique angles to the time axis. Finer examination shows that close behind the detonation these alternations are curved and approach a condition of tangency with their arrival track envelope.

High-resolution streak photographs of self-light from many detonations in homogeneous liquids viewed as in our Fig. 2 have exhibited similar families of obliquely oriented alternations of intensity [e.g., Persson and Persson (1976)]. The apparent motions of contiguous features back and forth along the slit direction are known not to be simply interpretable [Fickett and Davis (1979)]. Nevertheless, the presence of transverse fluid-dynamic effects in detonating homogeneous liquids and qualitative effects on their scale have come to be recognized by the cross-hatching that appears in streak photographs.

Photography of mirrored acrylic windows at boundaries of detonating liquids has also received some prior application that may be compared with the present wedge-face results [Mallory (1966; 1967); and Mallory and Greene (1969)]. In streak photography of transversely structured liquid detonations incident normally on such an "impedance mirror," whose illumination was also normally incident, the irregularities became stationary in the image soon after the wave entered the mirror substrate, and cross-hatching was absent. In contrast, a similarly illuminated framing picture of such a mirror on a flat sidewall of a detonating slab of liquid did register cross-hatching. Lacking a full optical treatment of these situations, we observe that a feature of the sidewall configuration in common with our wedge application, but absent in normal wave-front impact, is a temporally growing triangular prism of shocked acrylic plastic. The outer prism face in the photographic path also may carry transverse structure transmitted from the passing detonation.

Two position-dependent fluid-dynamic features of the detonation wave arrival at the inclined face of the wedge shown in Fig. 4 are noteworthy. First, the transverse structure was present from the beginning of recognizable, delayed events in the flow following the input shock wave. In fact, the superdetonation that was demonstrably present and moving steadily before it overtook the input shock cannot be traced as a contiguous entity back to the impact surface. Fluid-mechanical reverberations through the thin toe of liquid explosive between the piston and the window may have biased the recording of the earliest part of the superdetonation. Nevertheless, some minimum thickness of decomposed explosive is expected to be involved in launching any macroscopic detonation front [Hardesty (1976)] and it is only approximately accurate that superdetonation originates "at" the impact interface. The short run of the input shock before it was overtaken in our experiments, between 3 and 3.5 mm, makes the finite formation zone relatively conspicuous.

Second, the transverse structural manifestations in Fig. 4 are not noticeably different between the areas of the mirror reached before the overtake, by the superdetonation, and early afterward. The record does extend beyond the zone in which overdriving of the detonation produced an unsteady wave speed. Cross-hatchings pass through the overtake point without leaving a demarcation line. Rather, transverse waves and cells spontaneously established in the superdetonation appear to have been transmitted without prompt change through the transition into lower-density explosive. Smoked-wall imprints of cellular structure during "strong" reflected-shock initiation of gaseous detonation [Strehlow (1964); Strehlow et al. (1967)] also have shown this behavior; recording of the adjustment of the transverse wave spacing was achieved in those cases by virtue of a sufficiently long recording time and distance for interaction with the lateral boundary of the explosive gases.

Acknowledgment

This work is supported by Institutional Supporting Research and Development funds provided by the Los Alamos National Laboratory of the University of California under the auspices of the U. S. Department of Energy, Contract W-7405-ENG-36. Key scientific advice was provided by W. C. Davis, J. J. Dick, W. Fickett, and J. Wackerle; indispensable technical contributions to the results in Fig. 1 were made by D. Harkleroad, S. Salazar, and H. Stacy and to those in Figs. 3 and 4 by R. Alcon, W. Chiles, T. Elder, and W. Spencer.

References

Campbell, A. W., Davis, W. C., and Travis, J. R. (1961a) Shock initiation in liquid explosives. *Phys. Fluids* 4, 498–510.

Campbell, A. W., Davis, W. C., Ramsay, J. B., and Travis, J. R. (1961b) Shock initiation of solid explosives. *Phys. Fluids* 4, 511–521.

Fateley, W. G., Bent, H. A., and Crawford, Jr., B. L. (1959) Infrared spectra of the frozen oxides of nitrogen. *J. Chem. Phys.* **31**, 204–217.

Fickett, W. and Davis, W. C. (1979) *Detonation*, Appendix 7A. University of California Press, Berkeley.

Hardesty, D. R. (1976) An investigation of the shock initiation of liquid nitromethane. *Combust. Flame* **27**, 229–251.

Majowicz, J. M. and Jacobs, S. J. (1958) Initiation to detonation of high explosives by shocks. *Bull. Am. Phys. Soc., Ser. II* **3**, 293 (abstract).

Mallory, H. D. (1966) Evidence of turbulence in the reaction zone of detonating liquid explosives. *J. Appl. Phys.* **37**, 4798–4803.

Mallory, H. D. (1967) Turbulent effects in detonation flow: diluted nitromethane. *J. Appl. Phys.* **38**, 5302–5306.

Mallory, H. D. and Greene, G. A. (1969) Luminosity and pressure aberrations in detonating nitromethane solutions. *J. Appl. Phys.* **40**, 4933–4938.

Meyer, J. W. and Oppenheim, A. K. (1971) On the shock-induced ignition of explosive gases. *Thirteenth Symposium (International) on Combustion*, pp. 1153–1164. The Combustion Institute, Pittsburgh, PA.

Miller, R. O. (1968) Explosions in condensed-phase nitric oxide. *Ind. Eng. Chem. Process Des. Dev.* **7**, 590–593.

Persson, P. A. and Persson, G. (1976) High resolution photography of transverse wave effects in the detonation of condensed explosives. *Sixth Symposium (International) on Detonation*, pp. 414–425. ACR-221, U. S. Office of Naval Research.

Ramsay, J. B. and Chiles, W. C. (1976) Detonation characteristics of liquid nitric oxide. *Sixth Symposium (International) on Detonation*, pp. 723–728. ACR-221, U. S. Office of Naval Research.

Ribovich, J., Murphy, J., and Watson, R. (1977) Detonation studies with nitric oxide, nitrous oxide, nitrogen tetroxide, carbon monoxide, and ethylene. *J. Hazard. Mater.* **1**, 275–287.

Rivera, T. (1985) Fundamental research on explosives program. *Ind. Eng. Chem. Prod. Res. Dev.* **24**, 440–442.

Saytyev, S. G. and Soloukhin, R. I. (1958) Combustion in an adiabatically heated gaseous mixture, *Dokl. Akad. Nauk SSSR* **122**, 1039–1041; English translation: *Proc. Acad. Sci. USSR, Phys. Chem. Sect.* **118–123**, 745–747.

Saytyev, S. G. and Soloukhin, R. I. (1962) Study of combustion in an adiabatically-heated gas mixture. *Tenth Symposium (International) on Combustion*, pp. 344–347. The Combustion Institute, Pittsburgh, PA.

Schott, G. L. (1988) Initiation of detonation from a shocked state of liquid nitric oxide. Bull Am. Phys. Soc., Ser. II **33**, 536 (abstract).

Schott, G. L., Shaw, M. S., and Johnson, J. D. (1985) Shocked states from initially liquid oxygen – nitrogen systems. *J. Chem. Phys.* **82**, 4264–4275.

Strehlow, R. A. (1964) Detonation initiation. *AIAA J.* **2**, 783–784.

Strehlow, R. A. and Dyner, H. B. (1963) One-dimensional detonation initiation. *AIAA J.* **1**, 591–595.

Strehlow, R. A., Liaugminas, R., Watson, R. H., and Eyman, J. R. (1967) Transverse wave structure in detonations. *Eleventh Symposium (International) on Combustion*, pp. 683–692. The Combustion Institute, Pittsburgh, PA.

Wackerle, J., Johnson, J. O., and Halleck, P. M. (1976) Shock initiation of high-density PETN. *Sixth Symposium (International) on Detonation*, pp. 20–28. ACR-221, U. S. Office of Naval Research

Chapter V. Explosions

Overpressures Imposed by a Blast Wave

J. Brossard,* P. Bailly,† C. Desrosier,‡ and J. Renard§
University of Orléans, Bourges, France

Abstract

The prediction of damage caused by blast waves from accidental gaseous explosions requires a knowledge of the dynamic loads imposed. Generally these dynamics loads are deduced from the incident pressure signal or from TNT curves, taking into account an equivalency coefficient. From numerous small scale experiments (two experimental setups) we propose several curves that correlate the characteristics ΔP_r^+, t_r^+, i_r^+, ΔP_r^-, t_r^- and i_r^- of both the positive and negative phases of the reflected pressure signal on a plane surface as functions of the reduced radial distance $R/E^{1/3}$. The parameters of the study are the perpendicular front distance of the plane surface from the center of explosion and the chemical energy E confined in the hemispherical soap bubbles. Comparisons are presented with the incident characteristics and with TNT curves when they are available. The least-squares polynomials are deduced, and one model of the reflected pressure signal is proposed. The experimental data appear particularly useful for prediction by structural designers.

Introduction

The prediction of damage caused by blast waves from accidental gaseous explosions is frequently deduced from curves that relate damage to the overpressure imposed by the detonation of an equivalent charge of TNT. Recently (Brossard et al. 1985), an important effort was made to collect numerous experimental data on the pressure waves generated by gaseous detonations. In the case of spherical

---
Copyright © 1988 by the American Institute of Aeronautics and Astronautics, Inc. All rights reserved.
* Professor, Department of Mechanical Engineering
† Assistant, Department of Civil Engineering

symmetry, the characteristics of the positive and negative phases of the pressure signal of the incident wave are clearly correlated by means of the reduced distance ($\lambda = R/E^{1/3}$) in which R and E are, respectively, the radial distance from the center of explosion and the chemical energy released. A set of practical curves similar to the well-known TNT curves, are defined independently of equivalency consideration.

In the damage zone, the incident pressure wave interacts with mechanical structures, and the dynamic loads imposed by the blast wave are strongly dependent on the location of the structure and on the characteristics of the explosion. Frequently, the overpressure interactions concern large surfaces, and the dynamic loads are conveniently derived from the reflected overpressures at various times throughout the decay period. The theoretical prediction of the characteristics of that reflected wave are clearly defined for the peak overpressure only. Some experimental results are available in the well-known technical manual (TM5.1300, 1969), but they concern high explosives and thus require the knowledge of an equivalency coefficient for gaseous explosives.

When the incident pressure wave is well defined (Brossard et al. 1985), the properties of the oblique reflected wave on a plane surface are theoretically deduced by means of practical diagrams that take into account the regular reflection and the transition to the nonregular reflection. Through this technique the calculated properties are restricted to the amplitude of the shock front, and no information is available on the decay period, in particular the negative pressure amplitude and the negative phase duration.

The purpose of this paper is to supply several useful curves, that are similar to those established for TNT but that concern the detonation of gaseous charges and take into account both the positive and the negative phases of the pressure signal of the reflected wave on a plane surface. This pressure signal characterizes the dynamic load imposed by the blast wave.

In a previous work (Brossard et al. 1985), it has been demonstrated that the results obtained from the incident pressure waves generated by small-scale detonation of gaseous mixtures are very similar to large-scale experimental data when compared to a single parameter, the reduced distance ($\lambda = R/E^{1/3}$). This observation is then considered valid for the properties of the reflected pressure signal observed on a large plane surface. Consequently, the results of numerous measurements have been collected from two different experimental setups. With very satisfactory accuracy, the experimental results are correlated through the reduced distance ($\lambda = R/E^{1/3}$) independently of the type of reflection. The results concern the overpressure of the reflected shock front, the positive phase duration, the positive impulse, the negative overpressure, the negative phase duration, and the negative impulse. Then a model of the reflected pressure signal is proposed as a practical too for the structural designer.

Experimental

Two different experimental setups are schematized in Fig. 1. The incident pressure wave is generated by a hemispherical charge (radius $2.5 \leqslant R_o \leqslant 12$ cm) of stoichiometric propane-oxygen mixture confined in a soap bubble. The specific energy is 15.25 MJ/m^3. The chemical energy E contained in the spherical bubble with the same radius is the first parameter of the study. The detonation is triggered by an exploding wire, which delivers the electrical pulse. The pressure gages (Kistler 603 B) are distributed (radial distances R, R_1 and R_2 from the center of explosion) on the plane' surface (wood or concrete) in front of the soap bubble (first device) or at a right angle (second device). The electric signal (charge amplifier SEDEME 5007) is captured by means of numerical oscilloscopes and transferred to the computer memory for systematic calculations. The measured characteristics of the incident wave (pressure gage G in Fig. 1) are in good agreement with the previous results (Brossard et al. 1985). The perpendicular frontal distance R ($35 \leqslant R \leqslant 102$ cm) is the second experimental parameter. The parallel plane surface and the perpendicular one are large enough to assure that the observed signal at the maximum value R is slightly modified by the refracted wave on the edge of the surface. The minimum and maximum radial distances of the pressure gages from the center of explosion are, respectively, 36.8 and 110.6 cm. The angle of incidence varies from zero on the axis of symmetry to 49 deg on the farther pressure gage at the minimum radial distance R. The minimum and maximum values of the reduced distances are $\lambda = 0.75$ and $\lambda = 11.5$, respectively, and correspond to the extreme values (2.0 bars and 0.045 bar) of the incident peak pressure. From a preliminary calculus we know that the two modes of reflection are investigated according to the initial conditions defined by the normal distance R and the chemical energy E.

392 J. BROSSARD ET AL.

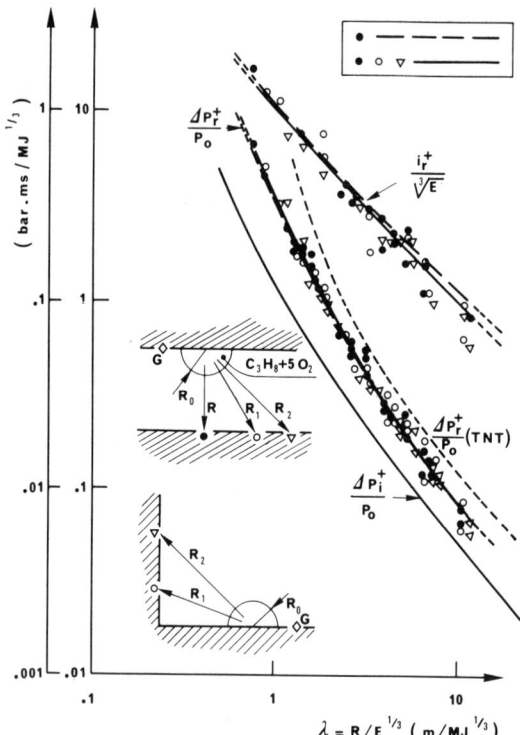

Fig. 1 Sketches of experimental configurations, peak overpressure and positive impulse of reflected wave as a function of the deduced radial distance.

Results

One typical pressure record is presented in Fig. 2. The six characteristics of the reflected pressure are defined and automatically calculated through the numerical technique. These characteristics, as reduced, are:

$\Delta p_r^+/p_0$ = peak overpressure of positive phase
$t_r^+/E^{1/3}$ = duration of positive phase
$i_r^+/E^{1/3}$ = impulse of positive phase
$\Delta p_r^-/p_0$ = maximum negative pressure
$t_r^-/E^{1/3}$ = duration of negative phase
$i_r^-/E^{1/3}$ = impulse of negative phase

p_0 is the ambient initial pressure, and the units are bar, millisecond (ms), and megajoules (MJ). The experimental data are collected in Figs 1, and 3-5 as functions of the scaled distance ($\lambda = R/E^{1/3}$) (m/MJ$^{1/3}$). The solid symbols are specific to the normal reflection, and the open ones

OVERPRESSURES IMPOSED BY A BLAST WAVE

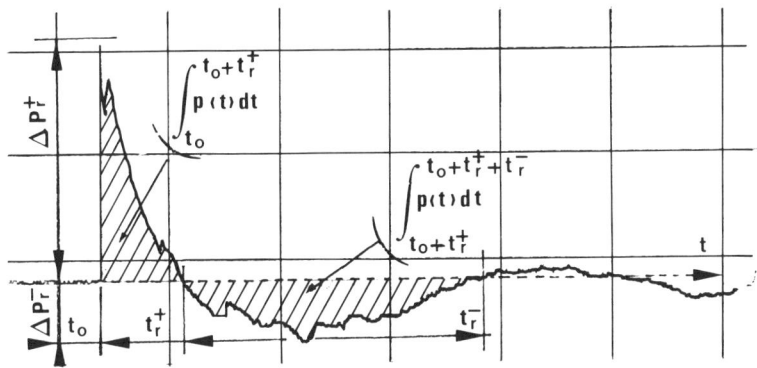

Fig. 2 Typical pressure record and characteristics of the signal.

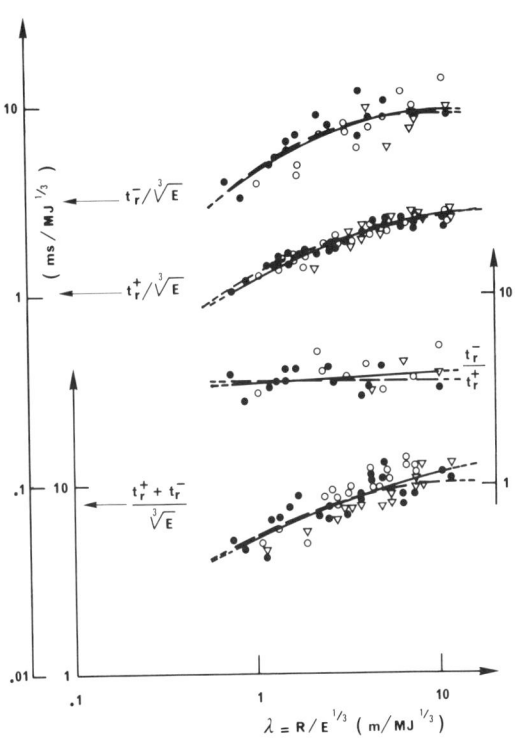

Fig. 3 Positive and negative phase durations as a function of reduced radial distance.

characterize the oblique reflection (see Fig. 1). The observed nonnegligible scattering of the data is the consequence of the bubble radius R_o fluctuations, the double-peak pressure records (see Fig. 2, for example) frequently observed in the case of oblique reflection, and the effects of additional pressure waves near the point where the decay curve crosses the time axis. In particular the scattering is more sensitive to the characteristics of the negative phase especially the duration t_r^- (e.g. the far-field pressure gages which are sometimes perturbed by the refracted wave from the edge of the plane surface). The scattering is not so sensitive (Fig. 3 and 5) when we plot the sums $(\Delta p_r^+ + /\Delta p_r^-/)$, $(t_r^+ + t_r^-)$, and $(i_r^+ + /i_r^-/)$: this is evident from the comparison of $(\Delta p_r^+ + /\Delta p_r^-/)$ in Fig. 5 with $/\Delta p_r^-/$ in Fig. 4. We have observed too, that the scattering is more sensitive from the measurements on the first device (parallel plates) than from the second one (perpendicular plates) because the symmetry and the stability of the reversed suspended bubble are not as perfect as the lying ones on the horizontal surface.

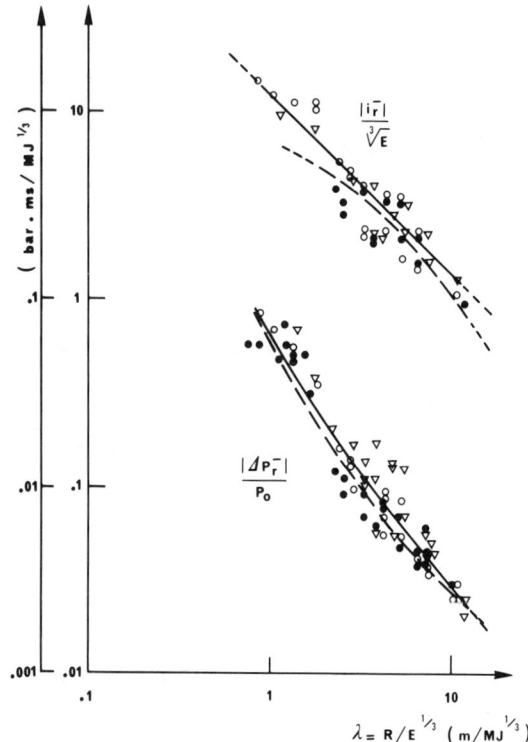

Fig. 4 Overpressure and impulse of the negative phase as a function of reduced radial distance.

OVERPRESSURES IMPOSED BY A BLAST WAVE

Discussion

At first, independent of the experiment device and the mode of reflection (which depends on both variables R and E), all the experimental data are well correlated as functions of the single parameter, the reduced distance ($\lambda = R/E^{1/3}$). Then the least-squares second-order polynomials are deduced for each series of data: - the first one (noted "o"), which includes all the results deduced from both normal and oblique reflections (the continuous curves in the figures); - the second one (noted

Table 1 Characteristics of the reflected pressure wave: least-squares polynomials

$\ln(\Delta p_r^+/p_o)$	=	o	$1.293 - 2.116 \ln\lambda + 0.204 \, (\ln\lambda)^2$
		n	$1.217 - 2.021 \ln\lambda + 0.182 \, (\ln\lambda)^2$
$\ln(\Delta p_r^-/p_o)$	=	o	$-0.476 - 1.472 \ln\lambda + 0.067 \, (\ln\lambda)^2$
		n	$-0.524 - 1.711 \ln\lambda + 0.167 \, (\ln\lambda)^2$
$\ln\left[(\Delta p_r^+ + /\Delta p_r^-/)/p_o \right]$	=	o	$1.466 - 2.102 \ln\lambda + 0.221 \, (\ln\lambda)^2$
		n	$1.378 - 2.012 \ln\lambda + 0.204 \, (\ln\lambda)^2$
$\ln(i_r^+/E^{1/3})$	=	o	$0.011 - 1.092 \ln\lambda + 0.027 \, (\ln\lambda)^2$
		n	$0.051 - 1.089 \ln\lambda + 0.038 \, (\ln\lambda)^2$
$\ln(/i_r^-//E^{1/3})$	=	o	$0.230 - 1.002 \ln\lambda + 0.011 \, (\ln\lambda)^2$
		n	$-0.446 - 0.163 \ln\lambda - 0.269 \, (\ln\lambda)^2$
$\ln\left[(i_r^+ + /i_r^-/)/E^{1/3} \right]$	=	o	$0.714 - 0.954 \ln\lambda + 0.001 \, (\ln\lambda)^2$
		n	$0.841 - 0.805 \ln\lambda - 0.032 \, (\ln\lambda)^2$
$\ln(t_r^+/E^{1/3})$	=	o	$0.182 + 0.519 \ln\lambda - 0.074 \, (\ln\lambda)^2$
		n	$0.248 + 0.481 \ln\lambda - 0.075 \, (\ln\lambda)^2$
$\ln(t_r^-/E^{1/3})$	=	o	$1.455 + 0.599 \ln\lambda - 0.113 \, (\ln\lambda)^2$
		n	$1.488 + 0.666 \ln\lambda - 0.161 \, (\ln\lambda)^2$
$\ln\left[(t_r^+ + t_r^-)/E^{1/3} \right]$	=	o	$1.662 + 0.461 \ln\lambda - 0.049 \, (\ln\lambda)^2$
		n	$1.695 + 0.471 \ln\lambda - 0.082 \, (\ln\lambda)^2$

$t_r/E^{1/3} = ms/MJ^{1/3}$ $i_r/E^{1/3} = bar.ms/MJ^{1/3}$ $\lambda = R/E^{1/3} = m/MJ^{1/3}$

"n"), which concerns only the normal reflection (the dotted curves in the figures).
The numerical coefficients are listed in Table 1.

An interesting observation is based on Fig. 3 in which the ratio (t_r^+/t_r^-) is plotted as a function on λ. We observe the quasi-constant value (~ 3.6) of this ratio.

The data points can be fitted accurately by means of the two straight lines:

$$\ln(t_r^-/t_r^+) = \begin{cases} 1.259 + 0.058 \ln \lambda & \text{(o)} \\ 1.274 + 5.77 \times 10^{-5} \ln \lambda & \text{(n)} \end{cases}$$

respectivly from all the data (o) and from normal reflection (n) only.

Generally speaking, in the case of a large number of numerical data, the two curves differ slightly; this generality confirms the possible mixing of the results for reduced distances λ ranging from 0.7 to 15 independently of the mode of reflection.

When it is feasible the experimental data deduced from the normal reflections are compared (see Fig. 1) with the similar TNT curves (TM5-1300 1969; Baker et al. 1983). The correlation seems clear for (Δp_r^+), $(\Delta p_r^+ / \Delta p_r^-/)$, and (i_r^+), and the mean energy equivalency is approximately 2. The positive phases of the reflected pressure on a large surface are equivalent for TNT (energy E_{TNT}) and gaseous charges (energy E_{GAS}) if the energies are related by $E_{GAS} = 2.E_{TNT}$. This correlation is not valid for (Δp_r^-) and (i_r^-), which lead to the same value of the energy equivalency coefficient equal to 1. It is important to note the different values of the energy equivalency coefficients deduced from the incident pressure measurements (Brossard et al. 1983) and the reflected pressure measurements, respectively, 5 and 2.

It is interesting to compare the characteristics of reflected pressure signal with the incident ones. First, we observe that the positive phase duration $t_r^+/E^{1/3}$ of the reflected signal is higher than the positive phase duration $t_r^+/E^{1/3}$ of the incident signal (Brossard et al. 1985) by a factor of approximately 1.4. Second, we compare the peak overpressure Δp_r^+ of the normal reflected shock front with the peak overpressure Δp_i^+ of the incident shock. For a given shot and the fixed value of the reduced distance, the reflection coefficient defined by the ratio $(\Delta p_r^+ / \Delta p_i^+)$, is generally lower than the theoretical value by about 5-10 % and, sometimes, in the far field and for large values of λ, is lower than the theoretical limiting value (=2). More generally, comparison of the two curves $(\Delta p_r^+)n$ (see Table 1) and Δp_i^+ (Brossard et al. 1983) in Fig. 1 leads to the polynomial relation:

$$\ln(\Delta p_r^+/\Delta p_i^+) = 0.864 - 0.203\ln\lambda - 0.059(\ln\lambda)^2 + 0.032(\ln\lambda)^3$$

of the overpressure ratio, and the numerical values are 20% lower than the theoretical ones (Baker et al. 1983). The comparison $(\Delta p_r^+)_0$ with Δp_i^+ in Fig. 1 must be considered carefully because the reflected overpressure is a function of both parameters, the angle of incidence and the amplitude of the side on overpressure, which are hidden through a single parameter, the reduced radial distance. In fact, the two least-squares polynomials $(\Delta p_r^+)_n$ and $(\Delta p_r^+)_0$ are not very different. A similar conclusion about the values of the reflected overpressure ratio is observed by Kogarko (1966) from small-scale experiments.

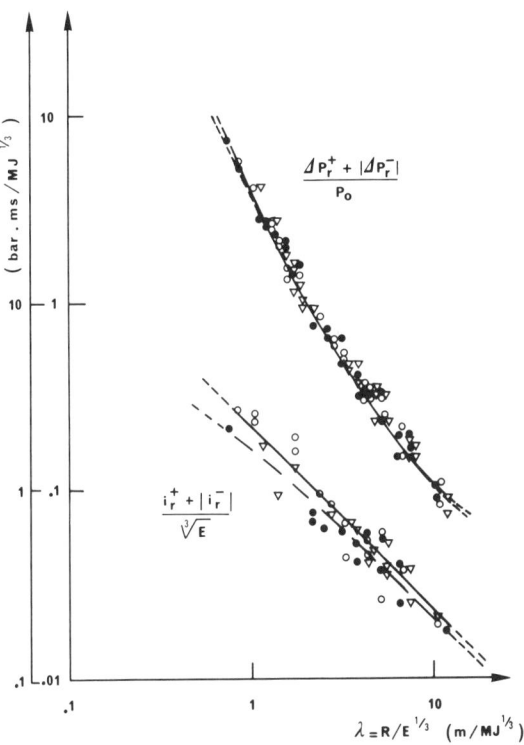

Fig. 5 Pressure amplitude and total impulse as a function of reduced radial distance.

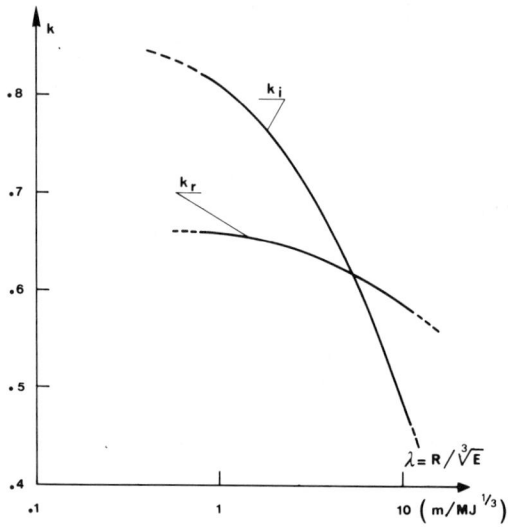

Fig. 6 Damping coefficients for incident (i) and reflected (r) pressure signal model as a function of reduced radial distance.

Modeling

Frequently, the first approach to the dynamic interactions of blast waves with mechanical structures requires the knowledge of imposed dynamic loads. In the case where the blast wave can be considered as generated by a gaseous detonation, the purpose of this experimental investigation is to provide a simple and easily usable model of the reflected pressure for numerical calculations.

The selected model is the one proposed by Lannoy (1984), based on the damped sinusoidal function of both parameters, the time t, and the reduced radial distance $\lambda = R/E^{1/3}$:

$$\Delta p(t,\lambda) = \Delta p^+ \frac{\sin\left[\pi(t-t^+)/t^-\right]}{\sin\left[-\pi t^+/t^-\right]} e^{-kt/t^+}$$

defined for t ranging from 0 to $(t^+ + t^-)$. Taking into account the two conditions,

$$i^+(\lambda) = \int_0^{t^+} \Delta p(t,\lambda) dt \qquad i^-(\lambda) = \int_{t^+}^{t^+ + t^-} \Delta p(t,\lambda) dt$$

OVERPRESSURES IMPOSED BY A BLAST WAVE

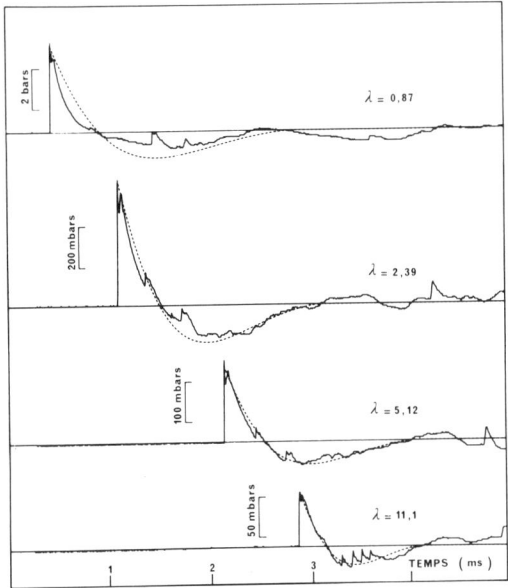

Fig. 7 Comparisons of reflected pressure records with model signals.

the coefficient k is clearly deduced from t^+, t^-, i^+, and i^-. Such an approach has been applied to the incident wave and to the reflected signal introducing the incident values (index i) and reflected values (index r) dependent on the single parameter λ (Lannoy (1984) and Table 1). The coefficients k_i and k_r are plotted in Fig. 6 as a function of λ. The least squares polynomials are, respectively,

$$k_i = 0.811 - 0.055 \ln\lambda - 0.035 (\ln\lambda)^2$$

$$k_r = 0.657 - 0.006 \ln\lambda - 0.011 (\ln\lambda)^2$$

In Fig. 7 are presented four typical reflected pressure records (solid lines) and the modeled signals (dotted lines).

The larger the value of λ (far field of the explosion), the better the coincidence. In fact, even if an underestimation of the reflected peak pressure is conceivable because of the low values of the reflection coefficient, the impact on the positive impulse is negligible. The most important characteristics of the mechanical response of a structure are the impulses and the phase durations.

Conclusions

Experimental data were successfully acquired on the reflection of an air blast on a large surface from spherical gaseous detonations. Substantially different reflected pressure signal parameters were measured and correlated from different small scale setups. Scaling has proved successful for variations of the explosible charge and the radial distance. Several curves similar to those relative to TNT are available now as a function of the single parameter ($\lambda = R/E^{1/3}$) in the range 0.5-20. They provide a good approach to both the important positive and negative phases of the reflected pressure signal. When it was possible, these curves were compared with those deduced from TNT. From the numerous experimental results, a simple model of the reflected pressure signal is proposed, easily usable for numerical calculations.

In spite of the disagreement on the observed value of the coefficient of reflection compared to the theoretical one, the coherence of the results is preserved and leads to a set of practical curves useful for prediction by structural designers.

References

Baker W.E., Cox P.A., Westine P.S., Kulesz J.J., and Strehlow R.A. (1983) Explosion hazards and evaluation,(Elsevier, New York).
Brossard J., Leyer J.C., Desbordes D., Garnier J.L., Hendrickx S., Lannoy A., Perrot J. and Saint Cloud J.P.,(1983) Experimental analysis of unconfined explosions of air-hydrocarbon mixtures. Characterization of the pressure field. 4th Symp. Int. on Loss Prevention and Safety Promotion in the Process Industries. European Fed. of Chem. Eng. Event no. 290, D 10-19.
Brossard J., Leyer J.C., Desbordes D., Saint Cloud J.P., Hendrickx S., Garnier J.L., Lannoy A., and Perrot J. (1985) Air blast unconfined gaseous detonations. Progress in Astro. and Aero. vol. 94, 556-566.
Kinney G.F. (1962) Explosive shocks in air. Macmillan, London.
Kogarko S.M., Adushkin V.V. and Lyamin A.G. (1966) An investigation of physical detonations of gas mixtures. Int. Chem. Eng. 6(3), 393-401.
Lannoy A. (1984) Analyse des explosions air-hydrocarbures en milieu libre: Etudes déterministe et probabiliste du scénario d'accident. Prévision des effets de surpression. Bulletin Direct. Etudes et Recherches EDF. A4.
TM5.1300 (1969) Departments of the Army, the Navy, and the Air Force. Structures to resist the effects of accidental explosions. Technical Manual, NAFVAC-P397/AFM88.

A Model of Point Explosions with Multistep Kinetics

H. Salem,* M. A. Fouad,† and M. M. Kamel ‡
Cairo University, Cairo, Egypt
and
M. A. El Kady§
Al-Azhar University, Cairo, Egypt

Abstract

This paper presents the development of a model for blast waves propagating into premixed propane-air mixtures of different fuel-to-air ratios. The chemistry of fuel combustion has been taken into account through the use of a multistep overall kinetic mechanism. The blast wave model, on the other hand, employs the quasi self-similar structure. The combined model provides an accurate prediction of the energy distribution as well as the flame location within the flowfield. The results are obtained for different values of the wave front Mach number and for equivalence ratios of 0.5 - 1.8. The results show that varying the equivalence ratio affects both the delay time and the energy release function which, in turn, affects the gasdynamic parameters within the flowfield. This effect is more pronounced when the wave propagates near its Chapman- Jouguet limit.

Introduction

The determination of the gasdynamic effects of detonation waves requires the accurate modeling of the structure of blast waves propagating into

---

Copyright © 1988 by the American Institute of Aeronautics and Astronautics, Inc. All rights reserved.
* Associate Professor, Mech. Power Eng. Dept.
† Associate Professor, Mech. Power Eng. Dept.
‡ Professor, Mech. Power Eng. Dept.
§ Associate Professor, Mech. Eng. Dept.

detonable fuel-air clouds. Several theoretical models for detonating blast waves have been developed incorporating the interaction between the gasdynamic and chemical kinetic processes [Sichel ,(1977), Bull et al. (1979)]. Most of the kinetic submodels in the literture do not take into account the delay and oxidation times of the mixture since, unfortunately, the simple single-step global model of fuel oxidation used by most authors fails to predict the wide variations in the detonation characteristics of different fuels since it cannot predict the delay time [kuhl et al. (1973), Strehlow and Baker (1975), Lee et al.(1977)]

On the other hand, while detailed kinetic models can accurately predict both delay and oxidation times, they are well established for a limited number of simple hydrocarbons. Besides, if detailed kinetic calculations are to be coupled with complicated flow models, extensive computer programming would be needed.

In a previous paper [Fouad et al. (1986)], the point explosion problem in a methane-air cloud was analyzed using an empirical formula for calculating the delay time associated with a single step global model for fuel oxidation.

For higher hydrocarbons, however, the initial fuel pyrolyses into intermediate hydrocarbon fragments whose combustion strongly affects both the delay time and the heat release rate. Consequently, the formation and oxidation of these intermediate hydrocarbons have to be taken into account when modeling the kinetics of higher hydrocarbons.

In the present study, the modeling of point explosions in homogeneous detonable mixtures of propane and air is carried out. A multistep overall kinetic model is used in conjunction with a quasiself-similar gasdynamic model to solve the flowfield for different wave front Mach numbers and equivalence ratios of the cloud. The decay coefficient of the wave front is also followed from its extreme overdriven value corresponding to a selfsimilar point explosion to its assymptotic value of zero at the Chapman-Jouguet (C-J) limit.

Kinetic Model

Kinetic information is generally incorporated by either an overall or a detailed mechanism. The assumption invoked when using an overall kinetic model is that a macroscopic chemical event, which involves a number of chemical reactions, can be represented by a single overall reaction that is assumed to have the same form as an elementary step. This approach considers the oxidation process to proceed directly to carbon dioxide and water, namely,

$$C_nH_m + (n + m/4) O_2 = n CO_2 + 1/2 m H_2O$$

The rate at which this overall reaction progresses is defined in terms of a semiemperical expression that is functionally similar to one resulting from the law of mass action.

The advantage of using such a scheme is that only four chemical species are involved in the formulation, while the heat release calculation is quite simple. This mechanism, however, does not account for the characteristics of hydrocarbon oxidation. Also, the formation of intermediate hydrocarbons and carbon monoxide is not taken into account.

The next stage of complexity is represented by the two-step kinetic mechanism that separates the highly exothermic oxidation of carbon monoxide to carbon dioxide from the less exothermic oxidation of the hydrocarbon to carbon monoxide, namely,

$$C_nH_m + (n/2 + m/4) O_2 = n CO + 1/2 m H_2O$$

and

$$CO + 1/2 O_2 = CO_2$$

Since no prediction is made as to the formation of intermediate species, this mechanism does not account for the time delay manifested the initial release of a significant amount of energy.

On the other hand, if a detailed kinetic mechanism is considered, the law of mass action is used to formulate ordinary differential equations for each species from the set of elementary reactions assumed to describe the reaction mechanism. These equations are usually stiff and require special integration techniques. Besides, the specific rate constants of the elementary reactions, let alone those of the elementary reactions themselves, are not necessarily well established and can be a large source of error [Kiehne et al,(1984)].

Recently, multistage overall kinetic mechanisms have been developed that are capable of modeling the main processes involved in hydrocarbon combustion, namely, the endothermic cracking of the parent hydrocarbon into intermediate hydrocarbon species, the formation of carbon monoxide and hydrogen, and the highly exothermic carbon monoxide and hydrogen oxidation to final products. This is carried out through the use of a few lumped reactions that greatly reduce the computation requirements as compared with detailed kinetic schemes.

Hautman et al. (1981) developed a multistage kinetic mechanism based on experimentally obtained species profiles for the oxidation of various hydrocarbons. In all cases investigated by Hautman et al., the oxidation is described in terms of three sequential and overlapping macroscopic events. First, the hydrocarbon is transformed into smaller intermediate species, primarily olefins. This process is iso-ergic, either slightly exothermic or slightly endothermic, and depends on the stoichiometry of the process that in turn, determines how much H_2O is formed in comparison to H_2. The intermediate hydrocarbons are then oxidized to CO, which is subsequently oxidized to CO_2. These two last steps are the exothermic ones responsible for the release of energy during hydrocarbon oxidation.

Another overall reaction may be added to represent the oxidation of H_2. In the present investigation, the following four-step overall kinetic mechanism developed by Hautman et al. has been used to describe the kinetics of combustion of alphatic hydrocarbons:

$$C_nH_{2n+2} = 1/2\ n\ C_2H_4 + H_2$$
$$C_2H_4 + O_2 = 2CO + 2H_2$$
$$CO + 1/2\ O_2 = CO_2$$
$$H_2 + 1/2\ O_2 = H_2O$$

Ethylene (C_2H_4) was chosen as the intermediate species because it is found experimentally [Hautman et al. (1981)] to be the dominant intermediate one in higher hydrocarbon oxidations. The reaction rates proposed here are as follows:

1) The parent alphatic hydrocarbon oxidation takes the form

$$d[\ C_nH_{2n+2}\ ]\ /\ dt = -\ 10^x\ \exp\ (-E_a/RT)$$
$$[C_nH_{2n+2}]^a\ [O_2]^b\ [C_2H_4]^c \quad \text{mole/ccs}$$

2) The overall intermediate oxidation step may be depicted as

$$d[C_2H_4]\ /\ dt = -\ 10^x\ \exp\ (-E_a/RT)$$
$$[C_2H_4]^a\ [O_2]^b\ [C_nH_{2n2}]^c \quad \text{mole/ccs}$$

3) The rate expression for CO oxidation yields

$$d[CO]\ /\ dt = -\ 10^x\ \exp\ (-E_a/RT)$$
$$S\ [CO]^a\ [O_2]^b\ [H_2O]^c \quad \text{mole/ccs}$$

4) The hydrogen oxidation step in this overall scheme has been represented as

$$d[H_2]\ /\ dt = -\ 10^x\ \exp\ (-E_a/RT)$$
$$[H_2]^a\ [O_2]^b\ [C_2H_4]^c$$

where

for C_nH_{2n+2}

$x = 17.32 \quad 0.88$, $E_a = 49,600 \pm 2400$

$a = 0.5 \pm 0.02$, $b = 1.07 \pm 0.05$

$c = 0.4 \pm 0.03$

for C_2H_4

$x = 14.70 \pm 2.00$, $E_a = 50,000 \pm 5000$

$a = 0.9 \pm 0.08$, $b = 1.18 \pm 0.10$

$c = -0.37 \pm 0.04$

for CO

$x = 14.53 \pm 0.25$, $E_a = 40,000 \pm 1200$

$a = 1.0 \pm 0.00$, $b = 0.25 \pm 0.00$

$c = 0.5 \pm 0.00$

for H_2

$x = 13.52 \pm 2.20$, $E_a = 41,000 \pm 1200$

$a = 0.85 \pm 0.16$, $b = 1.42 \pm 0.11$

$c = -.56 \pm 0.20$

and $S = [7.93 \exp(-2.48 F_R)] \leqslant 1.0$

where F_R is the equivalence ratio. It should be noted that the supression factor S has been added to account for the reduction in the rate of CO for fuel rich mixtures.

Blast Wave Model

The model considered treats the blast wave as a one dimensional unsteady phenomenon. The most general form of the governing conservation equations, while neglecting viscosity and transport terms, is given by Oppenheim et al. (1971). Only the final version of these equations in the non dimensional form is given here.

1) physical space coordinates

$$x = r / r_n \quad , \quad ẋ = r_n / r_o$$

where x is the field coordinate, ẋ is the front non dimensional radius, r the radius, with subscript n referring to conditions at the front and o to an arbitrary position in the flowfield.

2) front parameters

$$y = [a_a / w_n]^2$$

$$L = -2 [d \ln w_n / d \ln r_n]$$

$$= [d \ln y / d \ln ẋ] - 2 [d \ln a_a / d \ln ẋ]$$

where w_n is the front velocity, a the speed of sound, M the front Mach number, t the time, L the decay coefficient, and subscript a refers to the undisturbed medium.

3) gasdynamaic parameters of the flowfield
The nondimensional gasdynamic parameters are

$$f = u / w_n \quad , \quad h = R / R_a \quad , \quad g = p / [R_a w_n^2]$$

where f, h, and g are the nondimensional particle velocity, density, and pressure, respectively, while u, R, and p are the particle velocity, density, and pressure, respectively.

The nondimensional energy source term is given by

$$W_E = [r_n / w_n^3] O_E$$

where O_E is the source of energy per unit mass of the flowing substance.

The quasisimilar solution of the governing equations of the blast waves assumes that the rate of change of variables with respect to the front coordinate y is constant throughout the flowfield and equals to its value at the front [Oshima (1964)]. The application of this assumption to the equations transforms them into ordinary differential equations, which yield after suitable modifications

$$df/dx = [2kL/(2k -y(k-1)) -jkf/x -(1-y)(f-x)Lhf/2g(1-y) +E_E] / [k -(f-x)^2 h/g]$$

$$dg/dx = -h(f-x) [df/dx -Lf(1+y)/2(1-y)(f-x)]$$

$$dh/dx = -[h/(f-x)] [df/dx +jf/x -2Ly/(k-1+2y)]$$

where E_E is the nondimensional energy deposition term defined as

$$E_E = (k-1)O_E h/g$$

where k is the specific heat ratio of the mixture, with j being a geometric index = 0, 1, and 2 for planar, cylindrical, and spherical waves, respectively.

The above three equations are solved, starting at the wave front by considering the Rankine-Hugoniot conditions at the shock front, with the energy source appearing in the first equation being supplied from the kinetics model.

As ignition occures, a high speed blast wave is gene rated, which quickly decays to a limiting Macm number. This limiting Mach number for detonation is the Chapman-Jouguet Mach number M_{CJ} (or alternateively y_{CJ}). As the present work treats different mixtures of gases, M_{CJ} differs according to the fuel-air mixtures involved and their equivalence ratios.

The value of y_{CJ} is given by Kamel and Abdel Raouf (1984) by the following relationship

$$y_{CJ} = [(k^2-1) Q +1] - [((k^2-1) Q +1)^2 -1]^{1/2}$$

where k is the specific heat ratio of the gas mixture and Q the nondimensional chemical energy parameter defined as

$$Q = q/a_a^2$$

with q being the energy deposited at the front per unit mass of the medium.

Results and Discussions

The computations carried out in the present investigation are for planar, cylindrical, and spherical blast waves, propagating in homogeneous reacting propane-air mixtures. Three values of the blast wave Mach number are considered in the calculations, namely 5.77, 7.07, and 10 (y = 0.03, 0.02, and 0.01, respectively). The equivalence ratio F_R for the propane-air mixture is 0.5 - 1.8. For the sake of bravity, the results presented here are those for the planar case only. The cylindrical and spherical cases, on the other hand, are qualitatively similar except that, due to geometry, the effects due to chemistry are shifted closer to the front.

Distributions of the species concentrations and heat release rate, as well as the different gasdynamic parameters, throughout the flowfield have been predicted for various Mach numbers and equivalence ratios. The results are plotted in Figs. 1-10.

Figure 1 shows the species concentration profiles vs the time elapsed after the fuel air mixture has been swept by the shock for an equivalence ratio of 0.8 and a Mach number of 5.77 (y = 0.03). From Fig. 1, it can be noted that the concentration of the parent hydrocarbon remains nearly constant for a short time and then decreases at a relatively slow rate, followed by a higher rate of fuel depletion that results in a sharp increase in the C_2H_4 intermediate species concentration. However, the conversion to C_2H_4 is an endothermic process and, hence, the heat release during this period is negligible.

The inclusion of the formation of an intermediate hydrocarbon in the kinetics mechanism seems to have an effect on delaying the start of

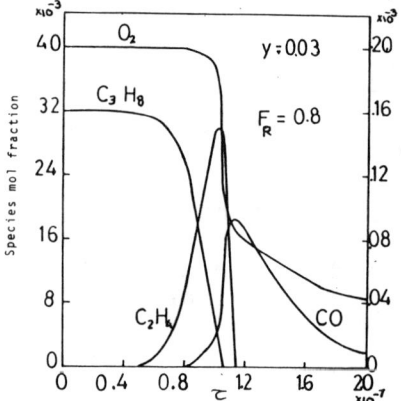

Fig. 1 Variation of species concentration with time for lean mixture.

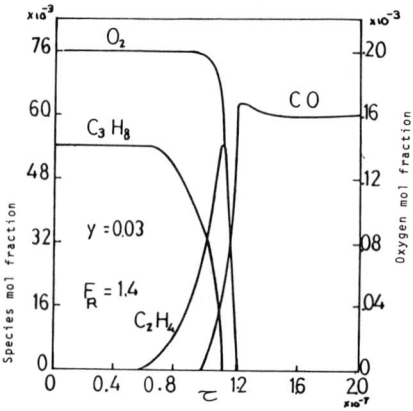

Fig. 2 Variation of species concentration with time for rich mixture.

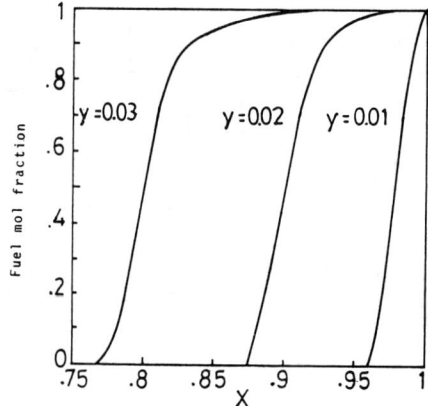

Fig. 3 Variation of fuel mole fraction along nondimensional coordinates of the flowfield for correct mixture.

the heat release and, hence, governs the duration of the delay period. Following the delay period, the intermediate hydrocarbon formed oxidizes rapidly to carbon monoxide, which, in turn, oxidizes to carbon dioxide. Also, hydrogen formed during the delay period or that resulting from C_2H_4 oxidation transforms into H_2O. This will substantially increase the rate of oxygen consumption.

Since all these oxidation reactions occurring in the postinduction region are exothermic, significant heat release rates are obtained to characterize the main reaction zone. In the following postreaction zone, excessive CO concentrations formed within the main reaction zone start to oxidize slowly in association with a slow drop in the oxygen concentration.

Similar trends are experienced for rich mixtures, as shown in Fig. 2, except that all the oxygen is consumed within the main reaction zone and that high CO levels, formed during the early stages of combustion, persist in the postreaction zone. However, for the temperature range encountered in the present study, the kinetic model predicts the delay time and, hence, the start of heat release. Nevertheless, its ability to predict the rate of heat release cannot be tested in the present complicated flow problem and it seems to need further investigation [Kiehne et al, (1984)].

Figure 3 shows the variation of nondimensional fuel concentration, referred to its value at the shock front, along the nondimensional flowfield coordinate x for a correct mixture at a different wave wave front velocity. As expected, at high wave velocity, the fuel is rapidly consumed closer to the front, due to the high postshock temperatures associated with these velocities. That, in turn, strongly reduces the delay time. As the wave velocity decreases, the delay time increases and decoupling between the shock and flame fronts may occur.

When varying the equivalence ratio at constant wave speeds, Fig. 4 illustrates that the

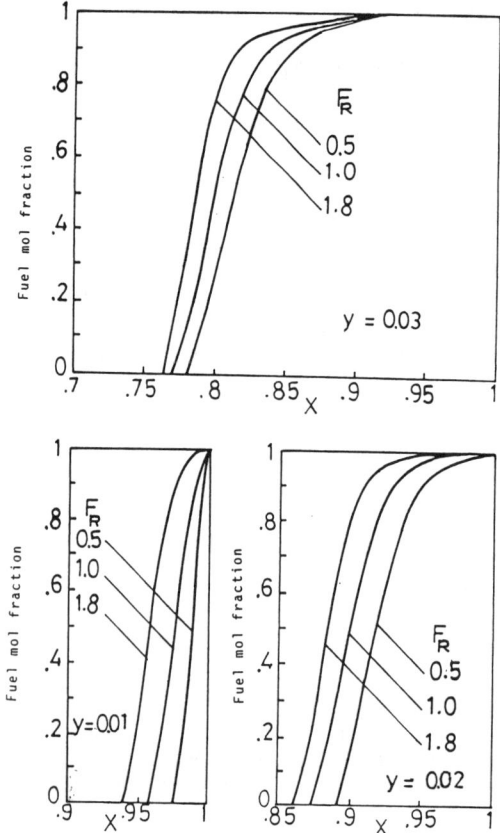

Fig. 4 Variation of fuel mole fraction along nondimensional coordinates of the flowfield for different equivalence ratios.

flame front gets closer to the wave front as the mixture becomes leaner. This may give the impression that a lean mixture has lower delay time than the chemically correct mixture. However, this is not the case, since at constant wave speeds different postshock temperatures are obtained for different equivalence ratios. Higher temperatures are associated with lean mixtures because of their smaller heat capacity, which results in shorter delay times than expected.

At the limiting detonation conditions (Chapman-Jouguet), the blast wave will propagate at C-J Mach number that varies with the mixture

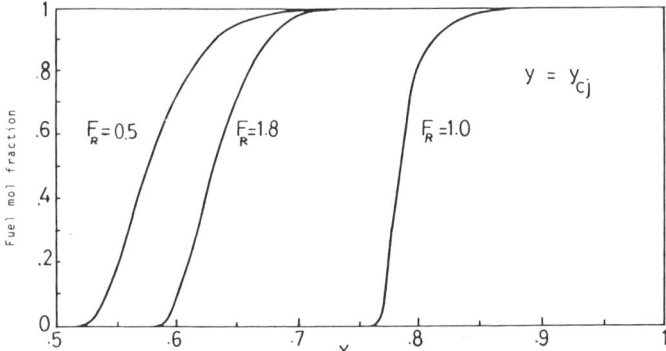

Fig. 5 Variation of fuel mole fraction along nondimensional coordinates of the flowfield at the limiting Chapman-Jouguet Mach number for different equivalence ratios.

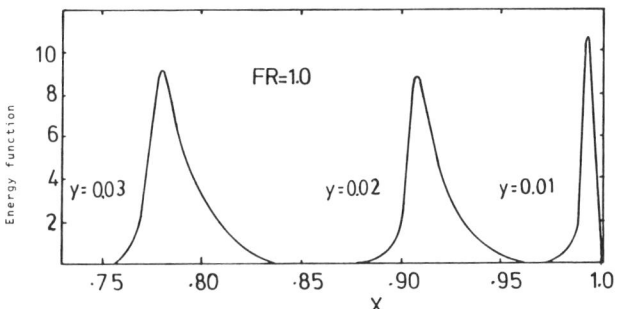

Fig. 6 Variation of energy function along nondimensional coordinates of the flowfield for correct mixture.

equivalence ratio. At this condition, Fig. 5 shows that both very lean and very rich mixtures possess longer delay times when compared with the chemically correct mixture. This comparison eliminates the misleading effect of temperatures that could lead to the wrong conclusions which might be deduced from Fig. 4.

Figure 6 shows the variation of nondimensional energy deposition function Q along the flowfield for the correct mixture at different wave speeds. In this figure, the peak point may be taken to indicate the position of the flame front. As expected, the flame front lags behind the shock front as the wave speed decreases. The noticed

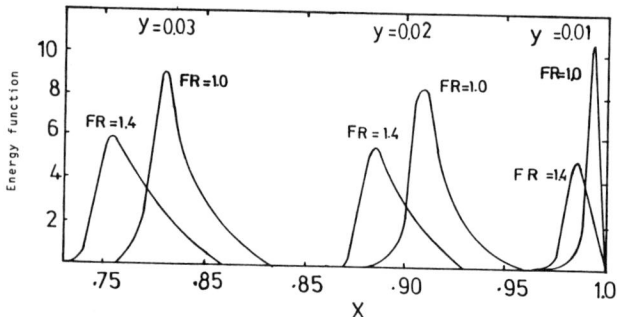

Fig. 7 Variation of energy function along nondimensional coordinates of the flowfield for different equivalence ratios.

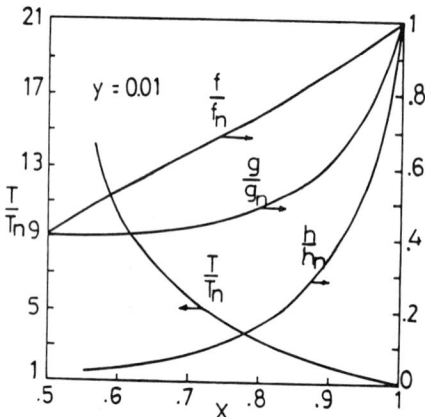

Fig. 8 Flowfield gasdynamic parameters for correct mixture.

rapid drop in energy behind the flame front may be attributed to the deposition of most of the energy within the main reaction zone, which is a characteristic of the kinetic medel used. The energy deposition function is greatly affected by the mixture strength as shown in Fig. 7. In addition, it should be pointed out that, for high wave speeds, the combustion energy is released very close to the shock front, which suggests that detonation models assuming instantaneous deposition of energy at the front may be applicable under these conditions. It as thus evident, from Figs. 6 and 7, that one may assume, for the self-similar case, that the chemical energy is deposited right at the front.

Figure 8 demonstrates the familiar pattern of distribution of gasdynamic parameters within the flowfield. The mass is accumulated behind the shock front, as can be seen from the density distribution h/h_n, while the temperature T/T_n increases toward the center of the explosion. The pressure g/g_n drops gradually from its value at the shock front, reaching an asymptotic value near the center of the flowfield. The particle velocity f/f_n drops from its frontal value, reaching zero velocity at the center of the explosion.

It was noted that for the same wave speed, the flowfield parameters are insensitive to variations in the equivalence ratio. Figures 9 and 10 show that higher particle velocities are

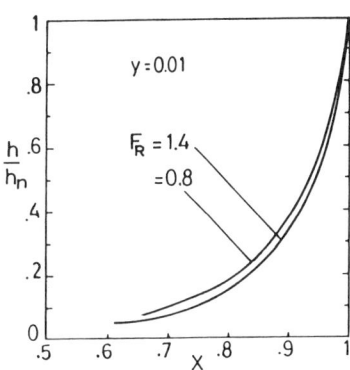

Fig. 9 Variation of nondimensional density along the flow field for different equivalence ratios

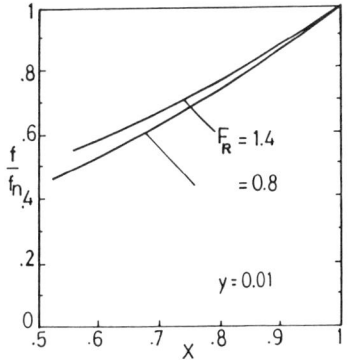

Fig. 10 Variation of nondimensional particle velocity along the flow field for different equivalence ratios

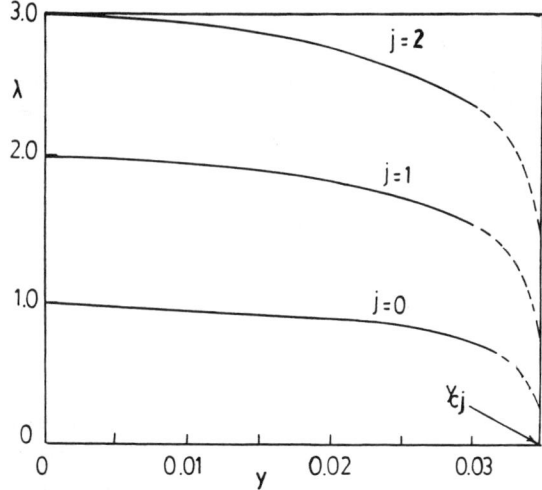

Fig. 11 Variation of decay coefficient with wave speed.

associated with rich mixtures due to higher energy deposition while the mass is accumulated behind the front for this rich mixture due to the higher average molecular weight of the mixture. No appreciable effect is predicted on the pressure distribution within the flow field.

The variation in the decay coefficient for different geometries with the front parameter y is depicted in Fig. 11. As expected, the decay coefficient starts at y=1 with the self-similar value of j+1. As the wave decays, however, the dacay coefficient reaches a value of zero asymptotically at a value of y corresponding to the C-J detonation velocity. This indicates that the velocities remain constant at the C-J value. It is also apparent from this figure that the spherical point explosion in a detonating gas decays to the C-J velocity faster than in the cylindrical and plane cases due, of course, to geometry considerations.

Conclusions

A mathematical model has been developed for a reacting blast wave propagating into propane-air mixtures at high Mach numbers. The model incorporates a quasisimilar flow model with a multistep global fuel oxidation model.

The present model can predict the position of flame front and the start of heat release that is important to the study of the sustenance of detonation.

At the same wave speed, the effect of equivalence ratio on the delay time is small with lean mixture burning nearer to the shock front.

At the C-J conditions the effect of equivalence ratio is substantial with growing possibility of decoupling between shock and flame fronts as the mixture becomes leaner or richer.

The distributions of nondimensional flow parameters are not seriously affected by the equivalence ratio at a fixed wave speed, while the spherical overdriven detonation waves decay to their C-J levels faster than the cylindrical and planer ones.

Acknowledgment

This work has been supported in part by the Foreign Relations Coordination Unit of the Supreme Council of Egyptian Universities through Grant FRCU-81015.

References

Bull, D.C., Elsworth, J.A., and Hooper, G.(1979) Concentration limits to unconfined detonation of ethane- air. Combust. & Flame 23, 27-40.

Burcat, A., Lifshitz, N., Scheller, K., and Skinner, G.B. (1978) Shock tube investigation of ignition in propane-oxygen-argon mixtures. Seventeenth Symposium (International) on Combustion, pp. 745-755. The Combustion Institute, Pittsburgh, PA.

Fouad, M.A., Salem, H. and Kamel, M.M.(1986) Blast waves propagating into methane-air clouds with chemical kinetics considered. Proc. Sixth ICOMPE, Vol.4,pp 45-58 Menoufia University, Egypt.

Hautman, D.J., Dryer, F.L., Schug, K.P., and Glassman, I.(1981) A multi-step overall kinetic mechanism for oxidation of hydrocarbons. Combust. Sci. Technol. 25, 219-235.

Kamel, M.M. and Abdel Raouf, A.M.(1984) Current status of blast wave theory and computations,VII: The quasi-similar solutions. CEFR Rept. 2, European Research Office, U.S. Army.

Kiehne, T.M., Mathews, R.D., Wilson, D. (1984) Computational evaluation of a four Step overall kinetic mechanism for the oxidation of propane in laminar premixed freely propagating and wall quench flame environments. Paper WSS/CI 84-5 presented at Spring Meeting of the Western States Section of The Combustion Institute, University of Colorado, Boulder, Colorado.

Kuhl, A.L., Kamel, M.M., and Oppenheim, A.K. (1973) Pressure waves generated by steady flames. Fourteenth Symposium (International) on Combustion, pp. 1201-1215. The Combustion Institute, Pittsburgh, PA.

Lee, J.H., Guirao, G.M., Chiu, K.W., and Bach, G.G. (1977) Loss prevention. CEP Technical Manual, Vol.II pp.59, American Institute of Chemical Engineers, New York.

Oppenheim, A.K., Lundstrom, E.A., Kuhl, A.L., and Kamel, M.M. (1971) A systematic exposition of the conservation equations for blast waves. J. Appl. Mech. 38, 783-794.

Oshima, K. (1964) Quasi-similar solutions of blast waves. Rept. 385, Aero-Research Institute, University of Tokyo.

Sichel, M. (1977) A simple analysis of blast initiation of detonation. Acta Astron. 4, 409-424.

Strehlow, R.A. and Baker, W.E. (1975) The characterization and evaluation of accidental explosions. NACA Baker, W.E. (1975) The characterization and evaluation of accidental explosions. NACA CR-134779.

Air-Blast Cumulation in Gaseous Detonating Systems

D. Desbordes*
Laboratoire d'Energétique et de Détonique, Poitiers, France
and
A. L. Kuhl†
R&D Associates, Marina del Rey, California

Abstract

A technique for experimentally simulating the interaction effects of multiple explosions on a small scale is described. The problem considered is that of the simultaneous detonation of multiple (N = 3, 6, 8, 12, and 16) spherical charges symmetrically distributed around a central point. The charge consisted of a 5 cm radius hemispherical bubble of C_2H_4/O_2 located on a charge radial 50 cm from the central point. The single-burst peak overpressure at the center was 0.3 bar. Rigid, smooth walls on the charge radial and charge bisector were used to simulate the effects of nearby charges. By varying the half-angle α between the confined planes, the effects of N simultaneous bursts can be simulated in the wedge-shaped channel, with N and α being related by N = 180°/α. Static pressure waveforms were measured on the charge radial and bisector by means of Kistler 603B pressure gages. Shock structure and wave patterns were recorded by high-speed schlieren photography. The experiments demonstrated that a shock cumulation process occured as the blast wave approached the central point. For N $\geq$ 6 the incident blast wave reflected off the charge bisector plane as a Mach reflection; the Mach-stem width grew until it reflected off the charge radial, then it reflected again on the charge bisector. In this way the peak pressure was found to increase discontinuously due to Mach reflection processes as the wave approached the center of symmetry. This process was confirmed by the schlieren photography and by

---

Copyright © 1988 by the American Institute of Aeronautics and Astronautics, Inc. All Rights reserved.
* Assistant Professor.
† Senior Staff Scientist, Nuclear Effects Department.

reflection factor analysis. Large increases (e.g., a factor between 5 and 16 times the single-burst value, depending on α) were measured in both the peak pressures and impulses near the center of symmetry.

Introduction

Laboratory-scale experiments performed at Laboratoire d'Energétique et de Détonique of Poitiers, France (Desbordes et al. 1978) on air-blast waves, produced by the detonation of spherical charges of gaseous mixtures (confined in soap bubble of 10 cm diam) are a very good simulation of large-scale fuel/air or oxygen explosions. The similarity of the blast waves, when the explosion regime is detonation, is clearly demonstrated in the paper of Brossard et al. (1984). Atmospheric gaseous detonations produce shock waves with a maximum overpressure Δp of about 20 bar. The TNT equivalent of this type of explosion is strictly valid only for the far field ($\Delta p < 1$ bar) because of the nonideality of the gaseous explosion source (lower energy density and power than TNT - see Baker et al. 1983).

Here, we deal with the problem of air-blast cumulation in correlation with the focusing effect of explosions. Particularly, these experiments are concerned with the air-blast overpressure enhancement as it approaches the center of symmetry of a regular polygon, when identical explosive charges located at the vertex of the polygon detonate simultaneously (Fig. 1). In this configuration, shock cumulation will occur in discontinuous jumps,

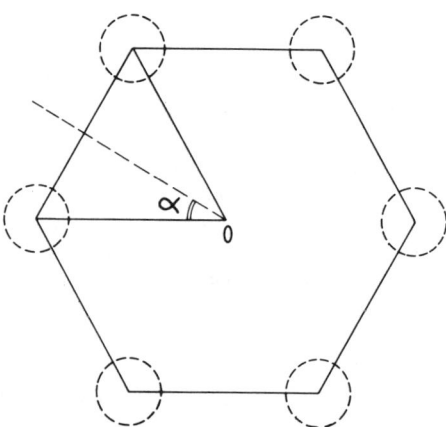

Fig. 1 Scheme of identical detonating charges equally distant from the center 0 and definition of the α angle.

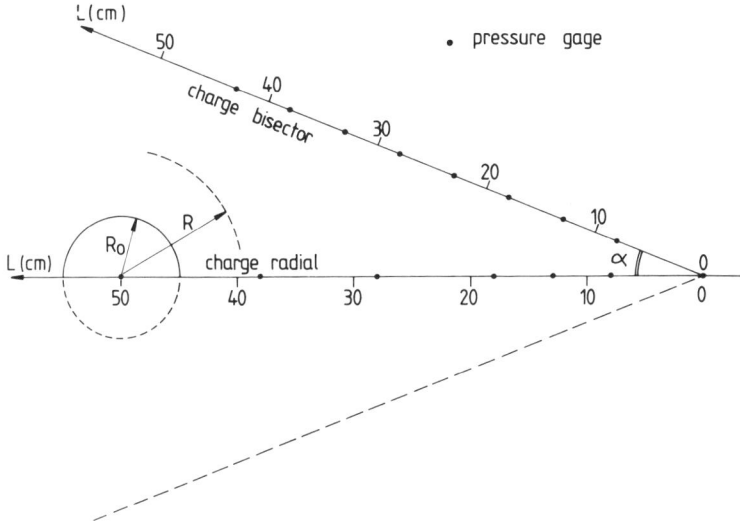

Fig. 2 Small-scale apparatus for studying air-blast cumulation.

produced by oblique interaction of shock waves (see Kuhl 1979).

Experiment Protocol

To simulate this type of wave focusing, two rigid smooth walls constituting a wedge are used (Fig. 2). The wedge angle is marked α.
Detonation is produced at the center of a hemispherical soap bubble of radius R_o = 5 cm, containing a C_2H_4/O_2 stoichiometric mixture at room temperature. The center of explosion is located at a normal distance of 50 cm from the edge 0.
For these experiments α takes values of 180, 60, 30, 22.5, 15, and 11.25 deg, corresponding, respectively, to free-space detonations of a single charge, and simultaneous detonations of 3, 6, 8, 12, and 16 charges.
One-microsecond rise time pressure gages (Kistler 603B) are mounted on an explosion radius perpendicular to the edge on the charge radial and on the charge bisector in different positions indicated on Fig. 2. A pressure gage is specially located near the edge at a distance of about 2.5 mm because of the 5 mm diameter of the gage. Observations of air-blast focusing are carried out by a schlieren photographic technique (cf. Coupez et al. 1985).

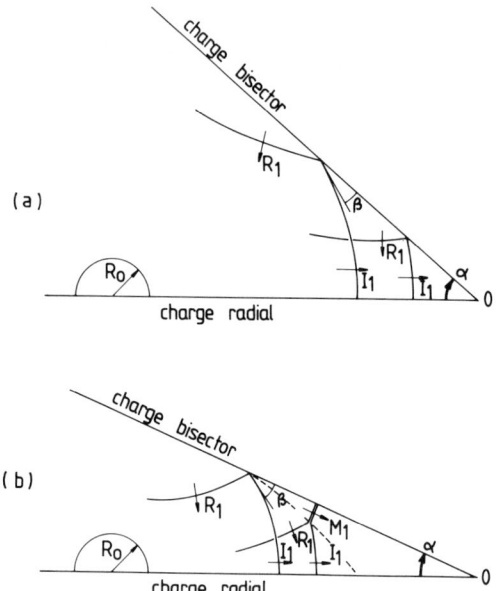

Fig. 3 Illustration of the two-type of oblique reflection on the charge bisector depending on α angle: (a) regular reflection, (b) Mach reflection.

Results

As a reference, air blast-wave characteristics (peak overpressure Δp and positive impulse I^+) of free-space detonation are given (Brossard et al., 1984) at each pressure measurement location on the charge radial (determined by reduced distance R/R_o from the center of the explosion, as shown in Fig. 2). These characteristics are those of the incident shock wave I_1.

In the range of air-blast overpressures produced by gaseous detonation, as used here, two mechanisms of strengthening the wave occur during its propagation toward the edge. Considering local strength and reflection angle β (Fig. 3) between I_1 and the charge bisector, reflection occurs: thus, it is possible to observe either 1) the regular reflection, consisting of I_1 and reflected wave R_1, both able to satisfy the local geometric flow constraints or 2) the Mach reflection, consisting of a third wave M_1 (the Mach stem) by adjusting flow to local geometric constraints. When Mach reflection appears, the three-shock system $I_1 - R_1 - M_1$ (triple-point) expands, and the width of the Mach-stem increases.

In our experiments we can distinguish between these two cases: 1) the regular reflection for $\alpha > 30°$, and 2) the Mach reflection for $\alpha \leqslant 22.5°$.

In both cases we consider centripetal and centrifugal propagation of waves (toward and backward from the edge) on the charge radial and bisector.

Centripetal propagation is concerned with I_1 and its associated R_1 and its modification by successive Mach-stem collisions on the charge bisector and radial.

Centrifugal propagation is concerned only with the main reflected wave MRW, which is the result of the implosion centered at the edge.

Examples of pressure histories at various stations along the charge radial and bisector are given in Figs 4-7 for two cases (α = 30 and 11.25 deg.). Time $t = 0$ corresponds to the initiation of spherical detonation at R = 0. Are deduced from pressure histories the trajectories (R/R$_o$ or L vs t) of the different successive waves and quantitative measurements of peak overpressure Δp and positive impulse I^+ function of R/R$_o$ or the distance from the edge L (see Fig. 2).

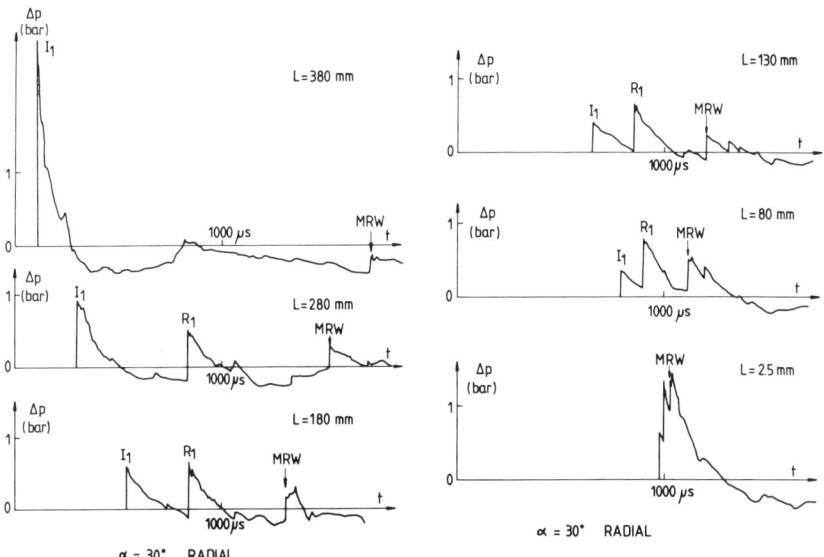

Fig. 4 Representative pressure histories at various stations along the charge radial for α = 30 deg.

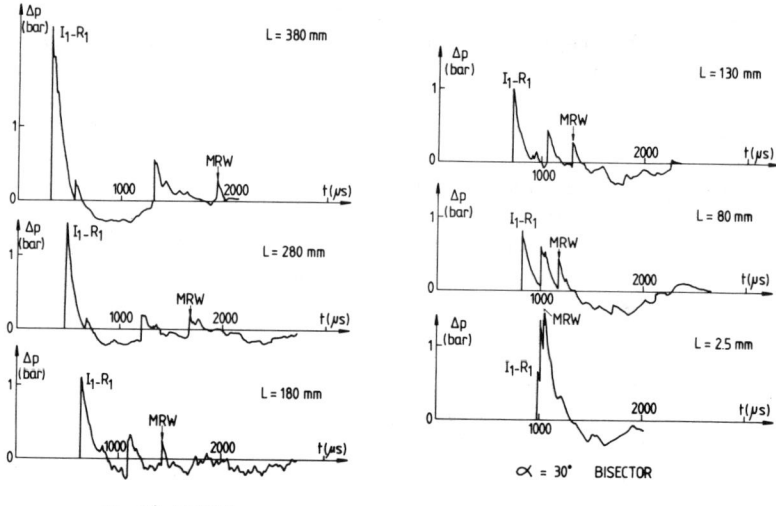

Fig. 5 Representative pressure histories at various stations along the charge bisector for α = 30 deg.

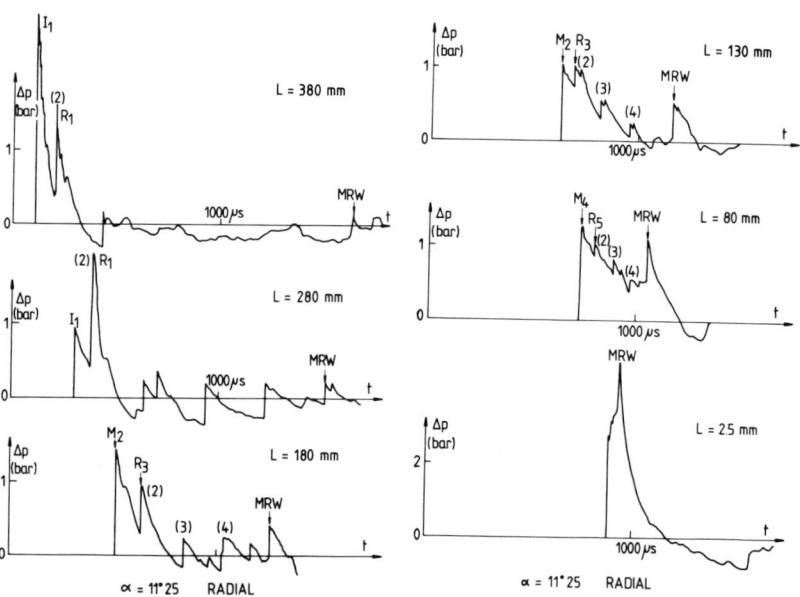

Fig. 6 Representative pressure histories at various stations along the charge radial for α = 11.25 deg.

AIR-BLAST CUMULATION

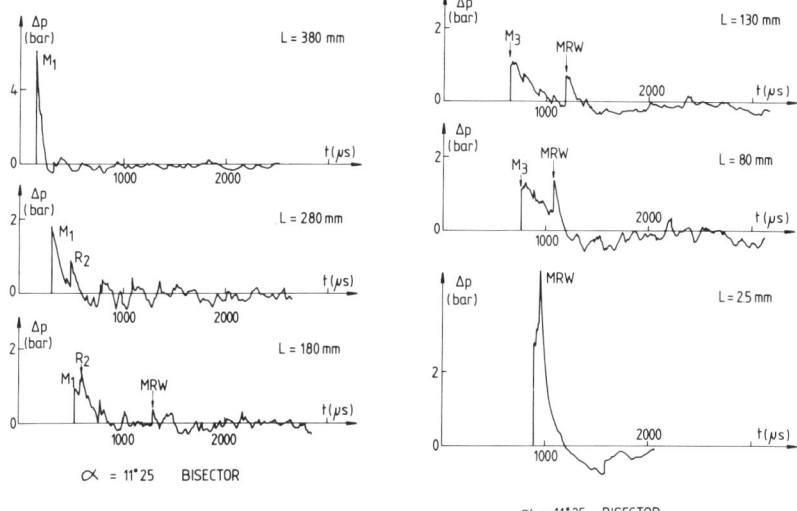

Fig. 7 Representative pressure histories at various stations along the charge bisector for α = 11.25 deg.

α ⩾ 30°

Maximum overpressure on the charge radial of I_1, R_1, and MRW waves are reported as a function of R/R_0 or L for α = 60 deg and α = 30 deg in Fig. 8 and 9, respectively.

Figures 10 and 11 summarize results obtained (Δp vs L and t vs L) on the charge bisector for α = 30°.

As can be seen on Fig. 12 and 13, when α becomes greater than 45 deg, only I_1 and MRW are visible on the pressure signal from the charge radial.

α ⩽ 22.50°

First M_1 appears on the charge bisector. Growth of the width of the Mach-stem causes triple-point collision with the charge radial and creates a new triple point that consists of the classical triple wave $I_2 - R_2 - M_2$, where $I_2 \equiv M_1$ is the new incident wave. The number of triple-point collisions with the charge radial and charge bisector depends on the value of α.

Figures 14-15 summarize shock overpressure data on the charge radial and on the charge bisector for the different successive waves, recorded at different locations L for α =

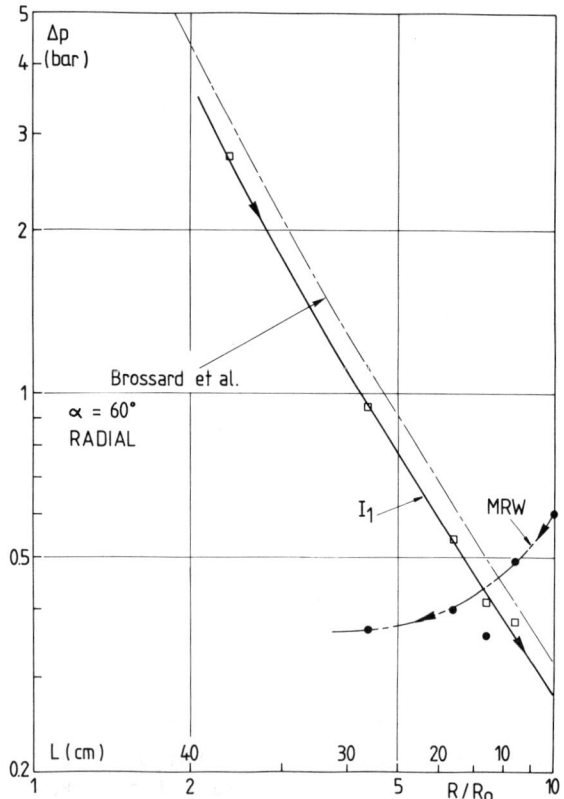

Fig. 8 Incident and reflected shock overpressure Δp as a function of L on the charge radial for α = 60 deg.

11.25 deg. Corresponding distance-time diagrams of the different successive waves are given in Figs 16-17.

A sample sequence of schlieren photographic records is given in Fig. 18 for α = 11.25 deg. Triple-point trajectories and collisions with the charge radial and bisector are displayed in Fig. 19. Near the edge, triple-point trajectory is not clearly observed; thus, the number of triple-point collisions with the charge radial and bisector is not really known.

Figure 20 represents a typical pressure record near the center of symmetry. A discontinuity in pressure jump is observed systematically (Kuhl 1979), perhaps because the pressure gage is not really located at the center of symmetry.

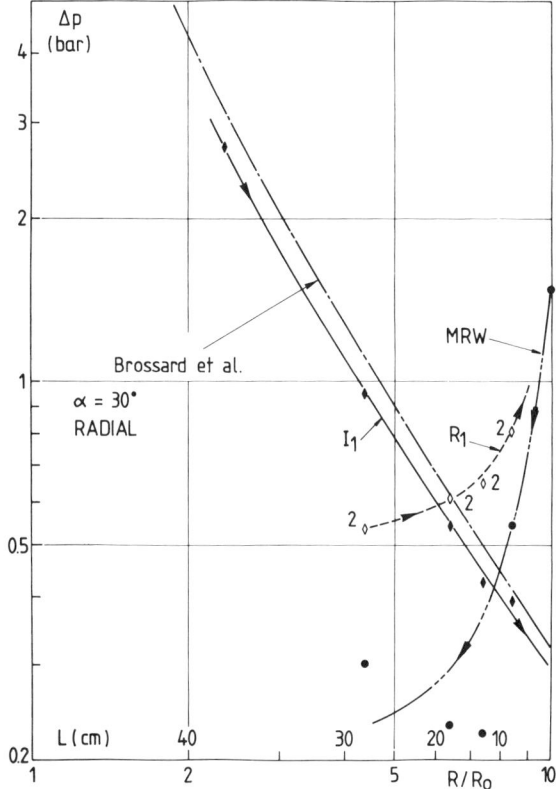

Fig. 9 Incident and reflected shock overpressure Δp as a function of L on the charge radial for α = 30 deg.

Table 1 Measured Overpressure Δp and Positive Impulse I^+ at the Edge for Different Wedge Angle α or Number of Identical Detonating Charge N

α, deg.	N	Δp, bar	I^+, 10^{-4} bar·s
180	1	.3	.3
60	3	.6	1.0
30	6	1.5	2.1
22.5	8	2.15	2.5
15	12	3.25	3.7
11.25	16	4.85	5.0

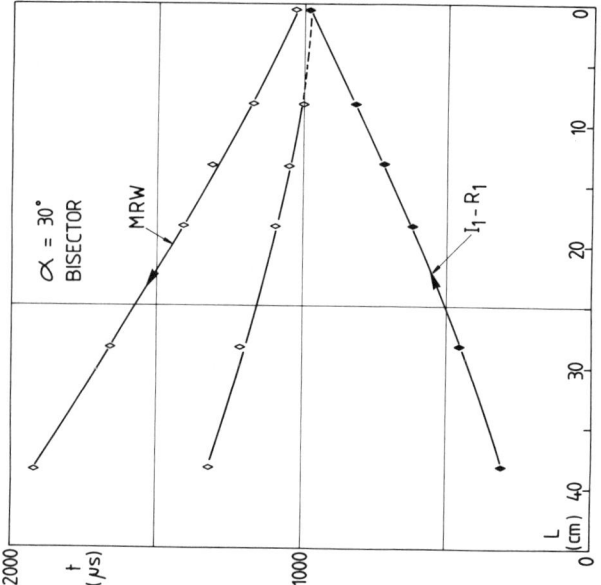

Fig. 11 Shock front trajectories along the charge bisector for α = 30 deg.

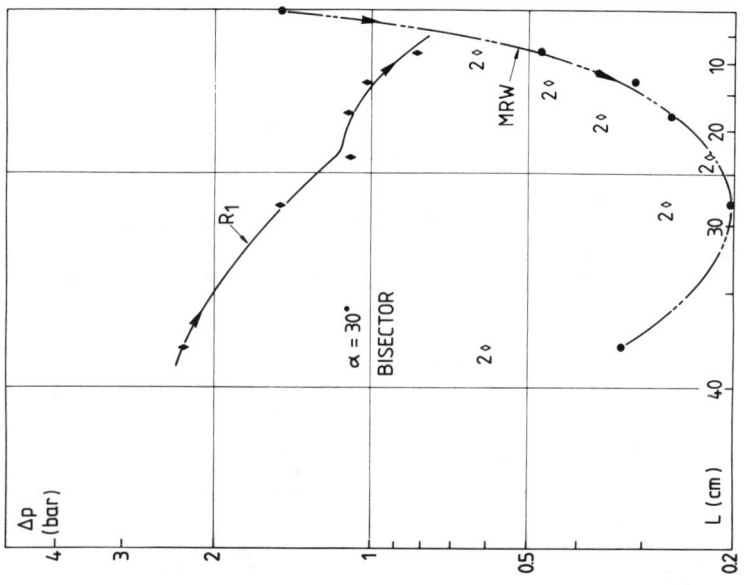

Fig. 10 Incident and reflected shock overpressure Δp as a function of L on the charge bisector for α = 30 deg.

AIR-BLAST CUMULATION 429

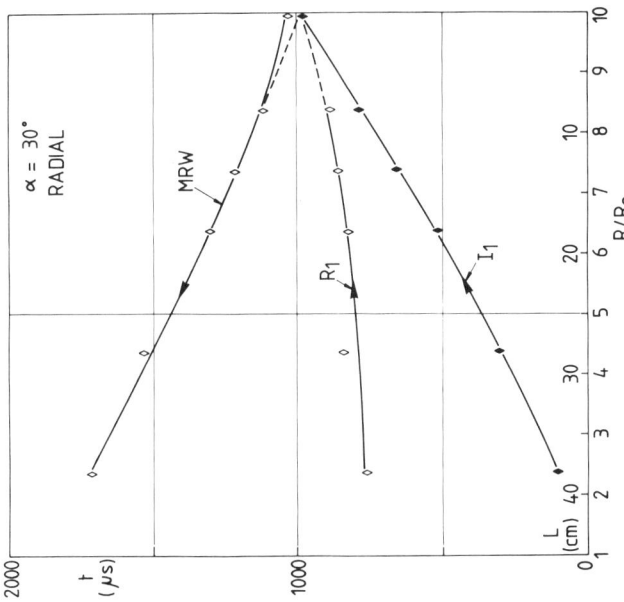

Fig. 13 Shock front trajectories along the charge radial for α = 30 deg.

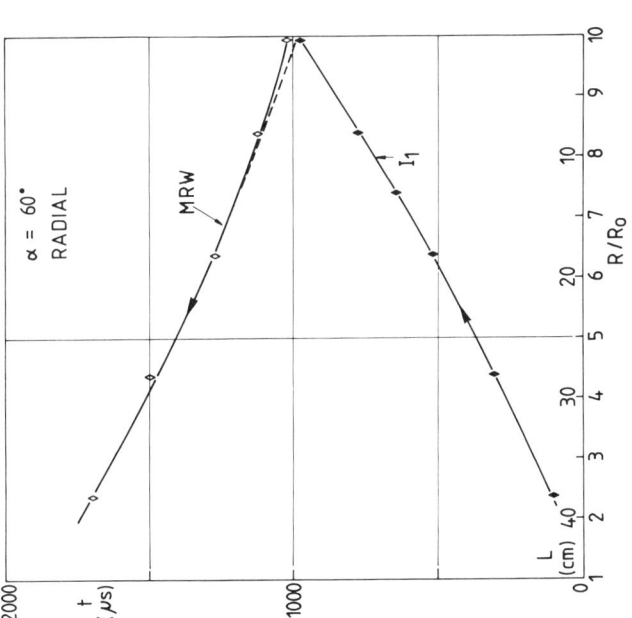

Fig. 12 Shock front trajectories along the charge radial for α = 60 deg.

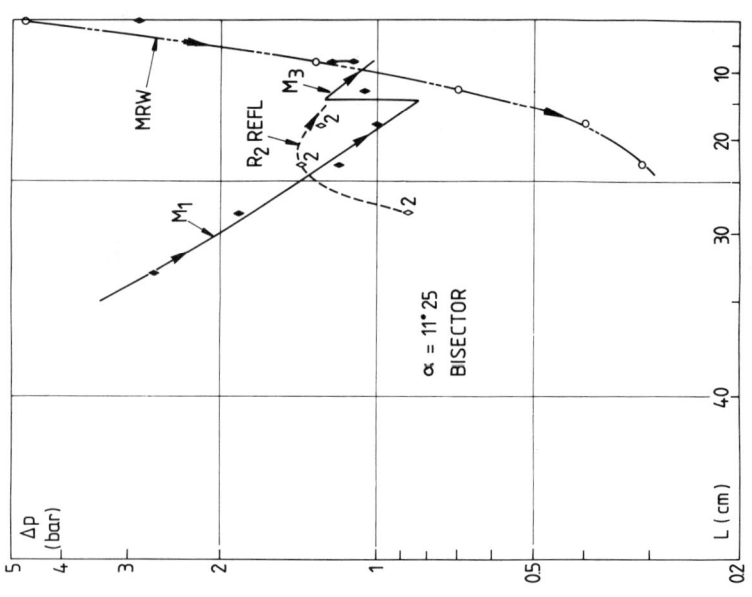

Fig. 15 Incident and reflected shock overpressure Δp as a function of L on the charge bisector for α = 11.25 deg.

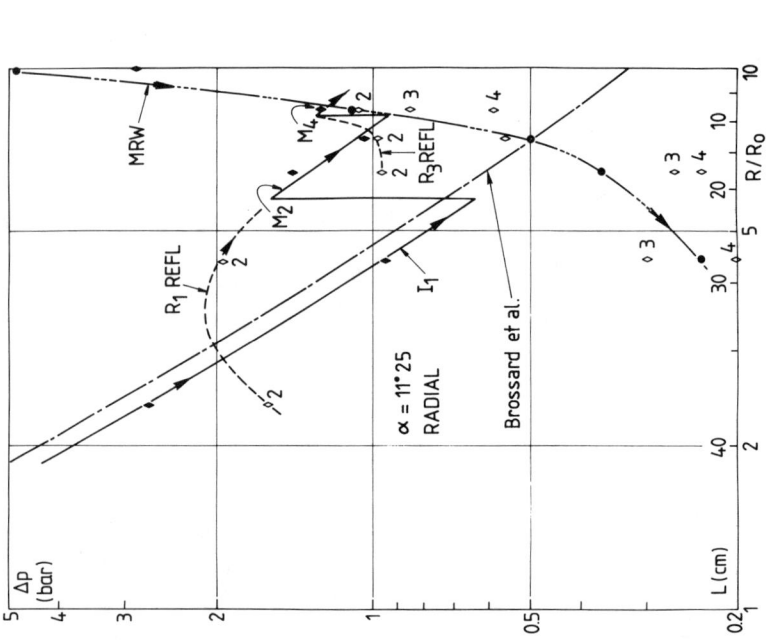

Fig. 14 Incident and reflected shock overpressure Δp as a function of L on the charge radial for α = 11.25 deg.

AIR-BLAST CUMULATION 431

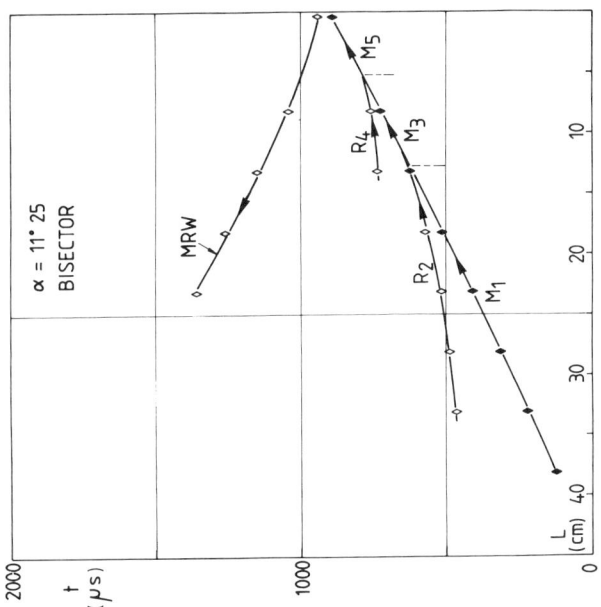

Fig. 17 Shock front trajectories along the charge bisector for $\alpha = 11.25$ deg.

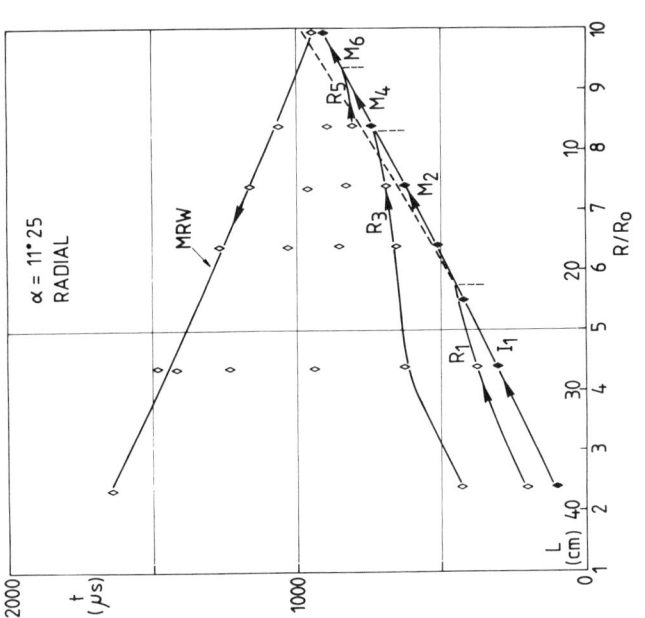

Fig. 16 Shock front trajectories along the charge radial for $\alpha = 11.25$ deg.

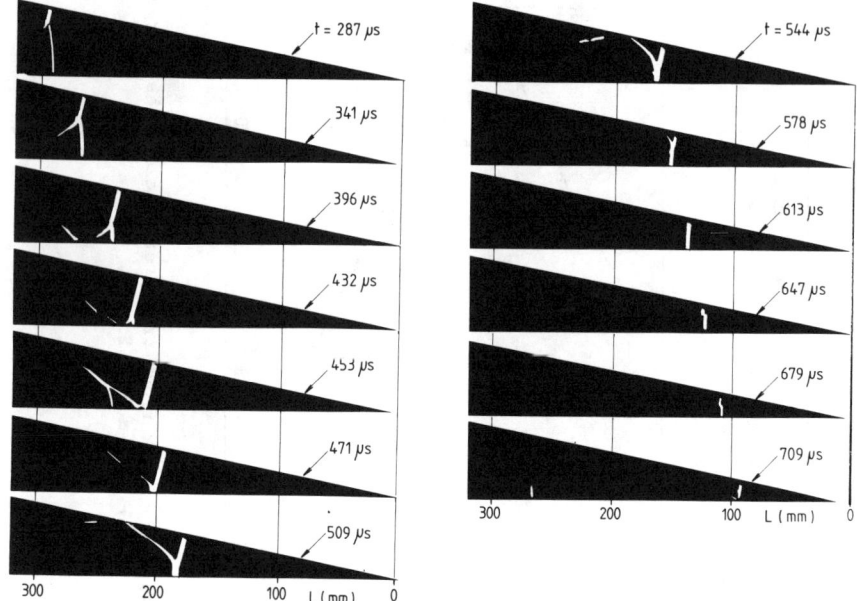

Fig. 18 Sequence of schlieren photographic records of incident shock wave propagating toward the center for α = 11.25 deg.

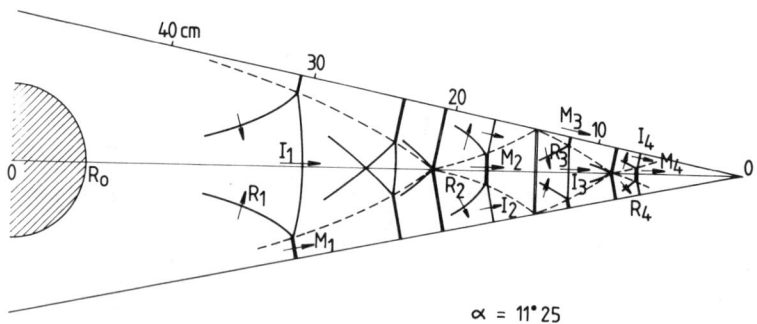

Fig. 19 Scheme of wave interactions for α = 11.25 deg.

Finally, smoothed curves of positive impulse I^+ vs L measured on the charge radial is given in Fig. 21 for different values of α.

In Table 1 we have reported characteristics of the wave: peak overpressure and positive impulse obtained experimentally at the edge for different values of α or N. Overpressure and impulse grow proportionally to N. In fact,

AIR-BLAST CUMULATION

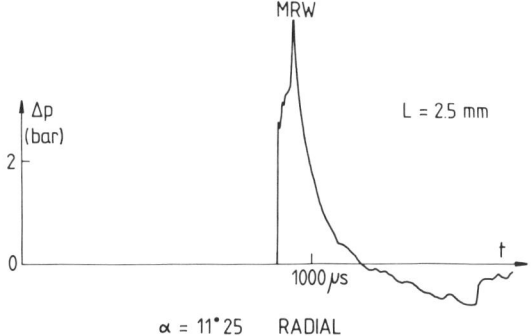

Fig. 20 Pressure history at the cumulation center for $\alpha = 11.25$ deg.

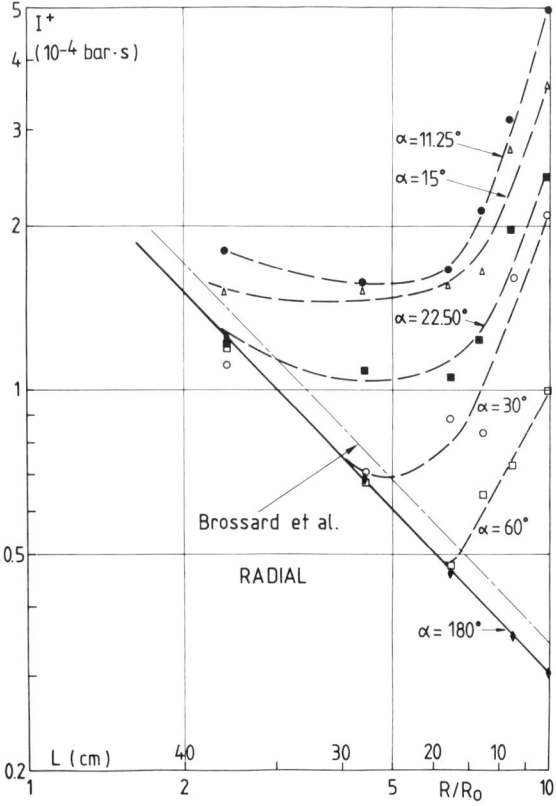

Fig. 21 Positive impulse as a function of L for different α.

the actual peak overpressure at the center could have been much higher. The measured value was no doubt limited by the gage sensing size. For positive impulse measured values are closer to reality.

In free-space explosion conditions (α = 180 deg), characteristics of the incident wave at the edge (R/R$_o$ = 10) are $\Delta p \cong 0.30$ bar and $I^+ \cong 0.3 \times 10^{-4}$ bar·s. For $\alpha = 11.25$ deg, the new characteristics are $\Delta p \cong 4.85$ bar and $I^+ \cong 5.0 \times 10^{-4}$ bar·s.

Summary and Conclusion

Experiments of air-blast cumulation have been performed in the case of detonation of fuel/air (or oxygen) mixtures. In this study, the cumulation center is located at distance R, 10 times R$_o$ of gaseous explosive spherical charges. Using the Sachs scale of explosion $\bar{R}$ ($\bar{R} = R/(E/p_o)^{1/3}$, where E is the chemical energy contained in a sphere and p_o the surrounding air pressure), $\bar{R}_o$ is equal to 0.12 (Brossard et al. 1984). $\bar{R}$ varies from $\bar{R}_o$ to 10 $\bar{R}_o$, corresponding (in reference to the TNT explosion source) to intermediate and far-field air blast.

Air blast strengthening proceeded by a series of discontinuous pressure jumps. These jumps are the result of regular or Mach reflections along planes of symmetry (the charge radial and bisector).

Maximum overpressure and positive impulse are observed at the center of symmetry and depend on α or the number of identical detonating charges. If it is possible with the characteristics (Δp vs R/R$_o$) of I_1 to determine locally after shock interaction analysis R_1, R_2, ... or M_1, M_2... overpressures, the trajectory of the triple-point cannot be determined a priori. Since the frequency of triple-point reflection grows as α decreases and as the wave approaches the center of symmetry, a prediction of implosion overpressure at the center is not feasible.

Nevertheless, measurements show that Δp and I^+ at the edge vary proportionally to N. So, in the particular value of α = 11.25 deg, the multiplying factor is of about 16.

Finally, considering the similarity of the incident blast-waves produced by gaseous detonating charge in air, such small-scale experiments can be used, as the first approach, to predict the results of large-scale events.

Acknowledgments

The authors would like to gratefully acknowledge the assistance of J.L. Brugier and C. Guerraud in performing some of the experimental work.

References

Baker, W.E., Cox, P.A., Westine, P.S., Kulesz, J.J., and Strehlow, R.A. (1983) Explosion hazards and evaluation. *Fundamental Studies in Engineering*, 5, Elsevier, Amsterdam.

Brossard, J., Leyer, J.C., Desbordes, D., Saint-Cloud, J.P., Henrickx, S., Garnier, J.L., Lannoy, A., and Perrot, J.L. (1984) Air blast from unconfined gaseous detonations, *Progress in Astronautics and Aeronautics: Dynamics of Shock Waves, Explosions and Detonations*, edited by J.R. Bowen, N. Manson, A.K. Oppenheim, and R.I. Soloukhin, Vol. 94, pp. 556-566. AIAA, New York.

Coupez, D., Percheron, T., Driessens, O., and Rhalam, B. (1985) Experiments on focusing of the blast-wave produced by the detonation of an hemispherical bubble of C_2H_4/O_2 mixtures. LED Report. University of Poitiers, Poitiers, France.

Desbordes, D., Manson, N., and Brossard, J. (1978) Explosion dans l'air de charges sphériques non confinées de mélanges réactifs gazeux. *Acta Astronaut.* 5, 1009-1026.

Kuhl, A.L. (1979) Air blast cumulation near the center of symmetry for six ANFO charges. RDA Report, TR-110006-003.

Steam Explosions: Major Problems and Current Status

John H. Lee* and David L. Frost†
McGill University, Montreal, Quebec, Canada

Abstract

The present paper discusses the fundamental mechanisms of vapor explosions together with comments on the current understanding of the phenomenon. The thermodynamic requirements for a rapid phase transition are discussed first. The energy required for generating a coarse mixture of fuel and coolant is estimated. A brief survey of triggering, fragmentation and propagation mechanisms is given. Although much qualitative information is available about vapor explosions, a number of difficult problems remain. In particular, a quantitative description of the influence of initial and boundary conditions (e.g., mixture geometry, trigger amplitude, degree of confinement) on the propagation and yield of a vapor explosion has yet to be established.

Introduction

Sudden contact between a hot and a cold liquid can result in a violent explosion. The explosion is due to the rapid vaporization of the cold liquid from the heat transfer from the hot liquid. Such explosions are referred to as vapor, steam, physical, or thermal explosions; rapid phase transitions; and molten fuel-coolant interactions. Accidental vapor explosions have occurred in the metallurgical, cryogenic and pulp and paper industries. After the Three Mile Island incident and the more recent nuclear reactor accident at Chernobyl, the nuclear industries have also shown great interest in the phenomenon. Extensive summaries of the occurrence of accidental vapor explosions in a variety of situations have been given by Buxton and

---

Copyright © 1985 by the American Institute of Aeronautics and Astronautics, Inc. All rights reserved.
*Professor, Mechanical Engineering Department.
†Assistant Professor, Mechanical Engineering Department.

Nelson (1975) and Reid (1983). General reviews of the various aspects of vapor explosions can be found in Cronenberg and Benz (1980) and Corradini et al. (1987). The purpose of this paper is to discuss the fundamental mechanisms of vapor explosions and comment on the current understanding of the phenomenon.

Thermodynamic Considerations

A vapor explosion from the rapid mixing of two liquids involves complex hydrodynamic processes. It is of interest to consider the thermodynamic requirements for the occurrence of a rapid phase transition for the ideal case in which a volume of cold liquid is heated uniformly by some external means. If the volume of liquid can be brought to a metastable superheated state, a rapid phase transition takes place when the system is disturbed. For a given pressure, there exists a maximum degree of superheat, at which point homogeneous nucleation occurs from the infinitesimal perturbations due to molecular fluctuations. The superheat limit (or the homogeneous nucleation) temperature can be determined experimentally. The apparatus consists of a vertical column of differentially heated host liquid. A small droplet of the test liquid is introduced at the bottom of the heated column. The host liquid is chosen so that it is immiscible with the test liquid and with a density slightly higher than the test liquid to permit the test droplet to rise slowly from buoyancy. The temperature of the test droplet then corresponds to the local temperature of the host liquid as it rises slowly. The homogeneous nucleation temperature T_N can be determined from the position in the column at which the rapid phase transition occurs. The value of T_N can also be obtained approximately from thermodynamic correlations, using the expression

$$T_N = T_c (0.11\ P_r + 0.89)$$

where $P_r = P/P_c$ and P_c and T_c denote the critical pressure and temperature of the liquid, respectively. For water, where $T_c = 647.3$ K and $P_c = 22.09$ MPa, the value of $T_N \approx 576.4$ K (or 303°C) at atmospheric pressure. The locus of the superheat limit states on the thermodynamic phase diagram is referred to as the spinodal curve. A measure of the violence of a vapor explosion can be obtained from the ratio of the enthalpy difference of the liquid between its superheat limit and its saturation state to its latent heat of vaporization at the given pressure. This ratio is referred to as the Jacob number, i.e.,

$$Ja = C_P(T_N - T_S)/h_{fg}$$

where T_N and T_S denote the superheat limit and the saturation temperatures, respectively, and h_{fg} is the latent heat of vaporization. For water at atmospheric pressure, where T_N = 576.4 K, T_S = 373 K, C_P = 4.2 kJ/kg, and h_{fg} = 2257 kJ/kg, Ja ≈ 0.38. In other words, about 38% of the superheated water can be converted into steam in the rapid phase transition.

Energy Requirements

A violent explosion requires a very fast vaporization rate for the cold liquid. Thus, the total heat-transfer rate from the hot to the cold liquid must be high. This requires a large surface area of contact. To achieve this, the volume of hot liquid must break up into fragments and distribute themselves rapidly in the cold liquid prior to the onset of rapid vaporization. Fragmentation and rapid dispersion of the fragments require energy. It is of interest to estimate this energy requirement to establish the vapor explosion itself. The total energy required consists of the surface energy, kinetic energy, and the work done against hydrodynamic drag on the fragments. The energy requirements for mixing have been discussed by Cho et al. (1976).

To form a liquid droplet of radius r requires a surface energy $E_S = 4\pi r^2 \sigma$, where σ is the surface tension of the liquid. The droplet acquires a kinetic energy

$$E_K = (4\pi r^3/3)\rho u^2/2$$

where ρ and u denote the density and the average velocity, respectively, of the liquid droplet, as it disperses rapidly in the cold liquid. Displacement of the droplet in another fluid also requires energy in the form of work done against the hydrodynamic drag force. For the droplet to move a distance L, the work required is given by

$$E_D = L(\pi r^2) C_D \rho u^2/2$$

where C_D denotes the hydrodynamic drag coefficient. The surface energy and the kinetic energy of the droplet are usually small compared to the work done against drag. Thus, if a volume of the hot liquid V_F fragments into N

droplets of radius r (i.e., $N4\pi r^3/3 = V_F$) and disperses in the cold liquid in a time t_D to form a volume of fragments-liquid mixture of dimension L (where $L \sim V_F^{1/3}$), the total energy requirement can be written as

$$\frac{3}{8}\rho C_D V_F^2/t_D^2 r$$

where we have taken $u \approx L/t_D$ and $V_F \approx L^3$. Since C_D is on the order of unity, we may write the energy requirement per unit mass as

$$e_T \approx V_F/t_D^2 r$$

As an example, consider the energy requirement for fragmentation and dispersion of 10^3 kg of molten UO_2 ($\rho \approx 10^4$ kg/m^3) in water. If we consider the molten metal to break into droplets of size $r = 100$ μm, in a time $t_D = 10^{-3}$ s, the energy requirement is of the order of 10^9 J/kg! This is far in excess of the thermal energy of the molten CO_2 itself ($\sim 2 \times 10^6$ J/kg at 3000 K). Thus, it is very unlikely that the vapor explosion occurs in a single step. On the other hand, if the process occurs in two stages, with the first stage being one of coarse fragmentation ($r \approx 10^{-2}$ m) and slow dispersion ($t_D \approx 1$ s), followed by the second stage in which the individual coarse fragments break up into fine particles 100 μm in size in a time scale $t_D \approx 10^{-3}$ s, the energy requirement is reduced drastically. For example, for 10^3 kg of UO_2, the coarse fragmentation stage now requires only 10 J/kg instead of 10^9 J/kg, a reduction of eight orders of magnitude. The second stage, in which coarse fragments of the order 10^{-2} m break up into 100-μm particles in a time scale of 10^{-3} s now requires 10^4 J/kg, two orders of magnitude less than the available internal energy of the molten UO_2 (i.e., $\sim 2 \times 10^6$ J/kg). If more than two stages are involved, the energy requirements are further reduced.

The Leidenfrost Effect

If the explosion involves more than one stage, with the first stage being the formation of a coarse fragment mixture in a long time scale of the order of seconds, there must exist a mechanism to thermally insulate the coarse fragments and reduce the heat-transfer rate during the formation of the coarse fragment mixture. Without this mechanism, a stable coarse fragment mixture cannot be formed because of the violent boiling heat transfer and turbulence generated by rapid local vaporization around the fragments. If a stable vapor film exists around the fragments as they are formed and dispersed, no direct contact between the two liquids occurs. The vapor film has a much lower thermal conductivity than the liquids, which limits the heat transfer from the hot fragments, thus allowing the fragments to be dispersed slowly in the cold liquid to form the mixture. The existence of a stable vapor film around the hot fragments is referred to as the Leidenfrost effect. The existence of a stable vapor film requires the temperature at the interface between the hot and cold liquid to exceed a certain critical value T_C, referred to as the Leidenfrost temperature. From thermodynamics (e.g., see Spiegler et al. 1963), an approximate estimate of the Leidenfrost temperature is given by

$$T_L \approx \frac{27}{32} T_C$$

where T_C is the critical temperature of the cold liquid. More accurate correlation formulas are given by Baumeister and Simon (1973). For water, where $T_C \approx 647.3$ K, the Leidenfrost temperature is $T_L \approx 546.2$ K. From the one dimensional heat conduction equation, the interface temperature T_I between a hot and a cold media can be obtained as

$$T_I = \frac{T_H \beta_H + T_C \beta_C}{\beta_H + \beta_C}$$

where T_H and T_C refer to the temperatures of the hot and cold mediums, respectively, and $\beta = k/(a)^{1/2}$, where k is the thermal conductivity and a the thermal diffusivity. For molten UO_2 at 3000 K and water at 300 K, the interface temperature can be calculated, $T_I \approx 1500$ K. Thus, we see that the criterion for the existence of a stable vapor film is satisfied and the coarse fragment mixture can be formed.

The Phases of a Vapor Explosion

From the considerations given in the previous sections, it is clear that a violent vapor explosion must follow four essential phases; coarse mixing, triggering, propagation, and expansion.

In the coarse mixing stage the hot liquid breaks up into relatively large fragments, which intermingle with the cold liquid to form a quasistable coarse mixture. Typical coarse fragment size is of the order of 10^{-2} m. The formation of the coarse mixture may occur on the order of seconds and, in this phase, the heat-transfer rate is relatively low because of the Leidenfrost effect, i.e., the existence of a stable vapor film around the hot fragments. Although a certain amount of violence and turbulence may be generated during this coarse mixing phase, the pressure generation is insignificant. The vapor production and turbulence may assist the fragmentation and dispersion of the coarse fragments. Thus the internal energy of the hot liquid can be converted to mechanical work required for the coarse fragmentation and dispersion process.

After the coarse mixture has been formed, further breakup of coarse fragments into micron-size particles must involve a destabilization of the stable vapor film that surrounds the coarse fragments. Thus, the triggering process involves the breakdown of the vapor film, leading to direct liquid-to-liquid contact for rapid heat transfer to the cold surrounding liquid and fine fragmentation. The localized perturbation that initiates the vapor film collapse may be a local increase in the pressure or hydrodynamic shear due to turbulence leading to Kelvin-Helmholtz instability of the liquid-vapor interface.

When triggering occurs, the coarse fragment breaks up into particles, and the rapid heat transfer gives rise to very rapid local vapor formation. The shock wave and the hydrodynamic flow associated with the vapor expansion destabilizes the vapor film around neighboring coarse fragments. This initiates fine fragmentation of these neighboring fragments when the vapor film collapses, and thus the fine fragmentation process propagates through the mixture. If the fine fragmentation process occurs rapidly enough to follow the triggering shock wave, a coherent vaporization front is coupled to the shock, and the phenomenon is similar to that of a detonation wave, where the reaction front is coupled to the triggering leading shock and propagates with it. Such a detonation wave analogy has been proposed by Board et al. (1975) for the vapor explosion wave in a coarse mixture. If the delay time is long, fine fragmentation cannot follow the triggering shock wave. Although propagation, or sequential triggering of the coarse fragments, still occurs, the shock waves generated by fine fragmentation and rapid vapor formation of the triggered coarse fragments are radiated away at a much higher velocity than that of the fragmentation or vaporization front.

The final stage of a vapor explosion is the subsequent expansion of the vapor formed after the vaporization process has propagated throughout the entire volume of the coarse mixture. This final phase is of interest in the evaluation of the mechanical damage to structural elements where the overpressure time history and the impulse generated by the expansion are required. Laboratory and large-scale experiments seem to confirm the existence of the four stages of a vapor explosion described here. However, depending on the initial and boundary conditions, not all of the four stages are of equal importance. For example, if the hot fluid fragments are explosively dispersed, it may not be necessary to have the thermal isolation requirement because the heat losses during the rapid dispersion may be negligibly small if the dispersion time is very short. Also, an external triggering mechanism may not be necessary if the vapor film is highly unstable.

Fragmentation Mechanisms

Perhaps the most important process in the vapor explosion phenomenon is the fragmentation process, i.e., the initial breakup of the volume of that liquid into coarse fragments and the subsequent fine fragmentation of the coarse fragments into micron-size particles during the propagation phase. The breakup of a volume of liquid (e.g., a droplet) is due to the inertia force generated when the liquid droplet is suddenly introduced into a flow with a relative velocity u. The stabilizing force that tends to prevent the disintegration of the liquid droplet is due to surface tension. Thus the fragmentation can best be described by some critical value of a dimensionless parameter involving the ratio of the inertia and surface tension forces, i.e., the Weber number,

$$We = \rho u^2 d/\sigma$$

A similar dimensionless parameter called the Bond number, giving the ratio of the acceleration force to the surface tension force, may also be used; i.e.,

$$Bo = \rho a d^2/\sigma$$

where a is the induced acceleration resulting from the sudden exposure of the droplet to the flowfield. The Bond and Weber numbers are related if the acceleration a is induced by the hydrodynamic drag, i.e.,

$$\frac{\pi d^2}{4} C_D \frac{\rho u^2}{2} = \frac{4}{3} \pi \left(\frac{d}{2}\right)^3 \rho a$$

Dividing both sides by the surface tension gives

$$We = \frac{4}{3} Bo/C_D$$

The Weber number is more convenient because it is difficult, in general, to define the acceleration.

Based on the Weber number, Pilch et al. (1981) gave five distinct mechanisms for the fragmentation of a liquid droplet in a gas flow. They also concluded that drop breakup in liquid flowfields differs little from drop breakup in gas flowfields. For low Weber numbers, We ≤ 12, the breakup is due to the forced vibration of the droplet at its natural frequencies driven by the flowfield. When breakup occurs, only a few large fragments are generally observed, and the breakup time is relatively long from the buildup of the oscillation amplitude.

For a range of Weber numbers, 12 ≤ We ≤ 50, the fragmentation is described as a *bag breakup*. The droplet deforms in the flow like a soap bubble blown from a ring. A thin, hollow bag is formed downstream until the bag bursts into a large number of small fragments. The remaining toroidal rim of liquid then disintegrates into a few larger fragments.

For higher Weber numbers, 50 ≤ We ≤ 100, the breakup geometry changes slightly. Apart from a thin bag anchored around a toroidal rim, a central column of liquid (stamen) is also formed along the drop axis parallel to the flow. Again, the bag disintegrates, first into small fragments, with subsequent breakup of the rim and the stamen. This type of breakup is referred to as the *bag-and-stamen*, or *umbrella breakup*.

At still larger Weber numbers, i.e., 100 ≤ We ≤ 350, the breakup is now observed to be distinctly different from the previous configurations. Instead of the formation of a bag, a thin sheet of liquid is drawn continuously from the deformed droplet. The jet sheet then disintegrates a short distance downstream from the deformed drop. A coherent residual drop exists during the entire duration of the stripping and sheet breakup. This mode of fragmentation is referred to as *sheet stripping*.

At very large Weber numbers, i.e., 350 ≤ We, surface waves develop on the windward surface of the drop. As the amplitude grows, the wave crests are continuously eroded by the action of the flow over the surface of the drop. This mode of breakup is referred to as *wave-crest stripping*. Longer waves also develop, however, and as their amplitude grows, the droplet is then divided into several large fragments, increasing the surface area on which the small-wavelength surface waves can grow, hence augmenting the wave crest stripping rate. This mode is referred to as *catastrophic breakup*. The different modes are illustrated schematically in Fig. 1.

Based on the Weber number, the different types of breakup modes of a liquid globule can be classified. The vibrational breakup for We ≤ 12 is not of interest in general in vapor explosions because of the very long time scale involved. For the other modes corresponding to higher Weber numbers, the basic mechanism responsible for the breakup itself can be credited to the growth of Rayleigh-Taylor instability waves at the interface. It is well known that a density interface is unstable under acceleration if the direction of the acceleration is from the lighter to the heavier fluid. In other words, if a heavy fluid rests on top of a lighter fluid in a downward gravity field (i.e., equivalent to the interface being accelerated upward toward the heavier fluid), the interface is unstable. For a liquid globule suddenly exposed to a flow, the drag force accelerates the windward interface in the direction of the flow. If the density of the liquid globule is greater than the flow medium, Rayleigh-Taylor instability sets in. Thus for a liquid in an airstream or molten UO_2 in water, the interface is unstable. The bag breakup corresponds to one unstable Rayleigh-Taylor wave on the windward surface, and the bag and stamen mode is due to the formation of two unstable Rayleigh-Taylor waves. The wave crest stripping and the catastrophic modes can similarly be ascribed to the growth of Rayleigh-Taylor waves of higher frequencies (or a combination of high and low frequencies).

The sheet stripping mode is distinctly different from the other modes that appear to be described by the growth of Rayleigh-Taylor instability. The sheet stripping mode may be due to the droplet deformation leading to the formation of an annular jet sheet. Kelvin-Helmholtz instability of the jet sheet may result in the growth of surface perturbations, leading to the subsequent fragmentation of the sheet itself. No model has been proposed currently to explain the sheet stripping mode; however, Pilch et al. (1981) pointed out that the observed breakup times and

STEAM EXPLOSIONS

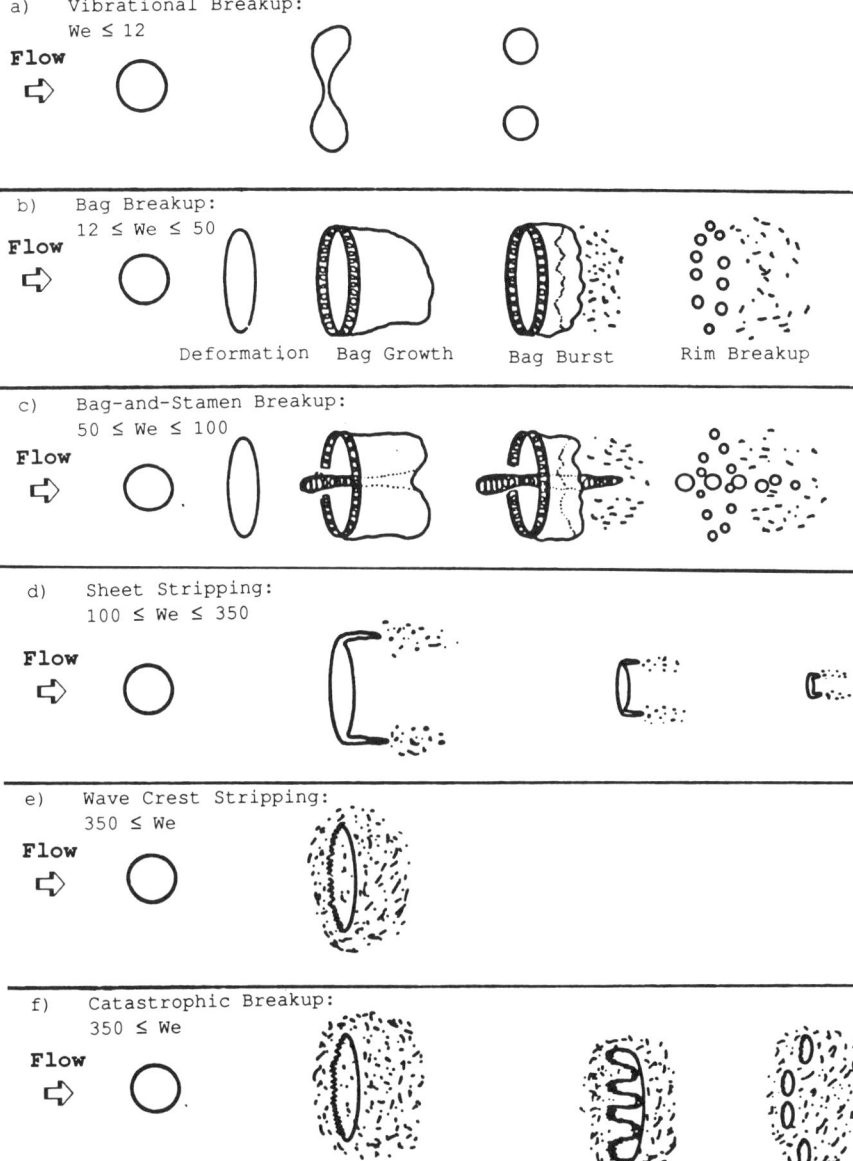

Fig. 1 Fragmentation of a drop as a function of Weber number (from Pilch et al. 1981).

fragment sizes correlate with parameters suggested by the Rayleigh-Taylor instability theory.

For a molten globule of UO_2 falling a distance of a few meters, the Weber number upon entry into water is generally large enough for catastrophic breakup. For example 100 kg of UO_2 melt have a diameter of about 0.27 m ($\rho \approx 10^4$ kg/m^3). If the entry velocity of the melt is 2 m/s, the Weber number is about 2×10^4 ($\sigma \approx 0.5$ N/m), much in excess of the limiting value for catastrophic breakup of about 350. Thus, a large globule will break up into smaller fragments until the Weber number corresponding to the fragment size reaches the critical value We_c (i.e., $We_c \approx 12$). The final stable fragment size d can be estimated as follows:

$$We_c = \frac{\rho(u - u_d)^2 d}{\sigma} = \frac{\rho u^2 d_o}{\sigma} \left(1 - \frac{u_d}{u}\right)^2 \frac{d}{d_o}$$

or

$$\frac{d}{d_o} = \frac{We_c}{We} \left(1 - \frac{u_d}{u}\right)^{-2}$$

where u is the velocity of the flow around the fragment, u_d is the final droplet velocity, and d_o the initial size of the globule. Although such a maximum stable fragment size concept makes qualitative sense, it is difficult to predict d quantitatively because of the difficulty in evaluating the complex turbulent flow around the droplet during the breakup process itself. Experiments indicate that the final fragment size (i.e., the coarse fragment size prior to triggering) is of the order of 10^{-2} m.

Triggering and propagation involve the further breakup of the coarse fragments ($\approx 10^{-2}$ m) to finer fragments ($\approx 10^{-4}$ m). The time scale involved for the finer fragmentation process is in general much more rapid (< 1 ms). The coarse fragments in a stable mixture are surrounded by a vapor film. Triggering, in general, involves the destabilization of the film, bringing about direct liquid-liquid contact and rapid boiling. Destabilization of the vapor film is essentially an interface instability problem brought about by Rayleigh-Taylor or Kelvin-Helmholtz instability mechanisms. The growth of the R-T waves leads eventually to liquid-liquid contact at localized points on the hot fragment surface. The rapid heat transfer associated with liquid-liquid contact immediately causes vapor bubbles to form at these liquid-liquid contact points, which grow and collapse. The collapse of cavitation bubbles near a boundary results in the formation of cavitation jets. It appears that the very fine rapid fragmentation of the coarse fragments is most probably due to this powerful mechanism of cavitation jet penetration. The impact of two-phase jets generated during the rapid vaporization (observed by Frost, 1987, during evaporation of highly superheated liquids) on the coarse fragments may also contribute to the fragmentation process. The destabilization of the vapor film from R-T and K-H instability could be caused by shock waves or shear flow turbulence.

In a propagating vaporization front, the fine fragmentation is due to the shock wave or the flow associated with the shock waves generated by the expansion of the vapor from the explosion of the coarse fragments. The time scale requirement for a strong coherent wave is that fine fragmentation be sufficiently rapid to follow the propagating shock. This is equivalent to a normal chemical detonation, an analogy advanced by Board et al. (1975). If the time scale for the fine fragmentation is longer, the fine fragmentation cannot be coupled to the triggering shock and although propagation, i.e., sequential triggering is still possible, the velocity of the vaporization front is small with the leading triggering shock decoupled from the explosion process. This is equivalent to the deflagration mode of propagation in a chemical medium.

Outstanding Problems

Although the vapor explosion phenomenon is fairly well understood on a qualitative basis, a number of difficult problems remain that must be resolved before any quantitative description is possible. The first question involves the formation of the coarse mixture itself for a given accident scenario. For example, if a spherical globule of hot liquid is dropped from a certain height or a gravity-driven jet of the hot liquid penetrates the cold liquid, a quantitative prediction of the subsequent breakup and mixing processes is extremely difficult. Complex hydrodynamic codes and large-scale experiments for code validation are probably required. The initial conditions of a coarse mixture that can support a propagating vaporization front (i.e., a violent vapor explosion) are not known. For example, consider two coarse fragments of dimension d spaced a distance λ apart. It is clear that if $d/\lambda \ll 1$, i.e., the spacing between the fragments is far apart, the effects (i.e., shock wave and flow perturbations) generated by one will not be strong enough to trigger the other and so there must exist a critical minimum spacing for sequential triggering to sustain the propagation process. This critical value of the d/λ parameter has not yet been established. Also, it is not clear how the d/λ ratio varies with the size of the fragments since the fragmentation time and effects generated do not scale linearly with the size of the fragments.

Although fragmentation of liquid droplets has been studied extensively, the triggering criterion of a coarse fragment has yet to be determined. In other words, the complex mechanism of cavitation jet breakup, described earlier, must be characterized quantitatively.

The different modes of propagation have not been determined as yet. As suggested by Board et al. (1975), it seems feasible for a detonation mode to exist in which the fine fragmentation can follow and be coupled to the shock front. However, the detonation velocity, detonation states, and detonation limits (i.e., conditions required for the detonation mode to be possible) have not been analyzed. The slower mode of propagation (where the shock propagates ahead of the fine fragmentation and vaporization process) has not been studied extensively. Frost and Ciccarelli (1987) have conclusively demonstrated the existence of this propagation mode. But the theory for the propagation speed and so on has yet to be developed.

Perhaps the most important problem to be studied is the influence of confinement. From work in chemical explosives, it is well known that confinement assists the propagation of a detonation wave. In other words, the detonability limits are narrower for an unconfined detonation. The influence of confinement is important in the scaling problem since a spherical wave can be thought of as a self-confined wave (i.e., the explosion of a particle is confined by the pressure developed by the explosion of the neighboring particles). Recent work by Frost and Ciccarelli (1988) suggests that the propagation speed of the vaporization wave in a linear array of coarse fragments increases when the fragments are laterally confined by sidewalls. However, the precise mechanism due to the influence of confinement on the propagation has yet to be established. In addition, the effect of the amplitude of the trigger on the propagation mechanism and velocity has not been explored in detail.

STEAM EXPLOSIONS

Concluding Remarks

Vapor explosions is a fairly new area to the chemical explosion community. Although the energy source is thermal rather than chemical, the basic mechanisms and the phenomenon itself appear to be very similar to that of deflagration and detonation propagation in a chemical reacting medium. In view of its importance to the metallurgical and nuclear industries, the growth of activity in this field is expected to be rapid in the future.

Acknowledgments

The authors would like to express their gratitude to Dr. A. Omar of Atomic Energy Control Board of Canada, and Dr. M. Berman of Sandia National Laboratories for their interest in and encouragement of the steam explosion work at McGill University.

References

Baumeister, K. J. and Simon, F. F. (1973) Leidenfrost temperature—its correlation for liquid metals, cryogens, hydrocarbons and water. *J. Heat Transfer* 95(2), 166-173.

Board, S. J., Hall, R. W., and Hall, R. S. (1975) Detonation of fuel coolant explosions, *Nature* 254, 319-321.

Buxton, L. D. and Nelson, L. S. (1975) Steam Explosions: Core-Meltdown Experimental Review, Sandia National Laboratories, Albuquerque, NM, SAND74-0382, chap. 6.

Cho, D. H., Fauske, H. K., and Grolmes, M. A. (1976) Some aspects of mixing in large-mass, energetic fuel-coolant interactions. *Proceeding of the International Meeting on Fast Reactor Safety*, American Nuclear Society, Chicago, IL, 1852-1861.

Corradini, M. L., Kim, B. J., and Oh, M. K. Vapor Explosions: A Review of Theory and Modelling, to be published.

Cronenberg, A. W. and Benz, R. (1980) Vapor Explosion Phenomena with Respect to Nuclear Reactor Safety Assessment. *Advances Nucl. Sci. Tech.* 12, 247-334.

Frost, D. L. (1987) Dynamics of explosive boiling of a droplet. *Proceedings of the 2nd ASME/JSME Thermal Engineering Joint Conference*, Vol. 2 (edited by P.J. Marto and I. Tanasawa), ASME Press, New York, 447-454.

Frost, D. L., and Ciccarelli, G. (1987) Dynamics of explosive interactions between multiple drops of tin and water. *Proceedings of the 11th International Colloquium on Dynamics of*

Explosions and Reactive Systems, Warsaw, Poland, Aug. 2-7, 1987.

Frost, D. L. and Ciccarelli, G. (1988) Propagation of explosive boiling in molten tin—water mixtures. *Proceedings of the ASME/AIChE Nat. Heat Transfer Conference*, Houston, TX, July, 1988.

Pilch, M., Erdman, C. A. and Reynolds, A. B. (1981) Acceleration Induced Fragmentation of Liquid Droplets. Prepared for the Nuclear Regulatory Commission, NUREG/CR-2247.

Reid, R. C. (1983) Rapid phase transitions from liquid to vapor. *Adv. Chem. Eng.* 12, 105-208.

Spiegler, P., Hoperfeld, J., Silberberg, M., Bumpus, C. V. and Norman, A. (1963) Onset of stable film boiling and the foam limit. *Int. J. Heat Mass Transfer* 6, 987-994.

Dynamics of Explosive Interactions Between Multiple Drops of Tin and Water

David L. Frost* and Gaby Ciccarelli†
McGill University, Montreal, Quebec, Canada

Abstract

The dynamics of the explosive interaction between multiple drops of molten tin and water has been investigated experimentally. The objective of the present study is to investigate the initial conditions required for the propagation of a vapor explosion in a linear array of drops immersed in water and to identify and characterize the propagation mechanism involved. In situations in which stable film boiling would otherwise occur, explosive boiling and fragmentation of the drop was externally triggered by contact with a wire mesh as well as by the impact of an underwater shock wave. In experiments with multiple drops in a linear array, a triggered explosive interaction was observed to propagate at an average velocity of 5-10 m/s. The explosion propagates as a result of the interaction between adjacent drops. It was found that the pressure wave and convective disturbance produced by the collapse of a steam explosion bubble from one drop serves to trigger the explosion of the neighboring drop. In this way, the explosion propagates from one drop to the next with a propagation velocity that depends only on the characteristic bubble growth/collapse time and drop spacing. Coupling between the explosive interaction zone and the initial shock was not observed.

Copyright © 1985 by the American Institute of Aeronautics and Astronautics, Inc. All rights reserved.
*Assistant Professor, Mechanical Engineering Department.
†Graduate Student, Mechanical Engineering Department.

Introduction

It is well known that a violent destructive vapor explosion can result when a hot liquid is suddenly brought into contact with a cold volatile liquid. This fact has been substantiated by the occurrence of numerous accidental physical explosions of a nonchemical nature (as far as energy release is concerned) in the steel and aluminum, pulp and paper, and cryogenic industries. The serious consequences of a thermal explosion in a nuclear reactor from a molten fuel/coolant interaction following a loss of coolant scenario has brought the subject into the forefront in recent years. An extensive summary of the occurrence of accidental vapor explosions in a variety of industries has been given by Reid (1983). The hazards associated with industrial vapor explosions (also called thermal explosions, rapid phase transitions, or fuel/coolant interactions) provide the motivation for conducting exploratory experiments at small scale to investigate the basic phenomena. The physical processes involved in the escalation of a small-scale vapor explosion into a large-scale coherent explosion are poorly understood at present. A knowledge of the propagation mechanisms is necessary for understanding the dynamics of large-scale thermal explosions. In this respect, key experiments are required to define the initial and boundary conditions and processes required to generate and support a propagating vapor explosion.

The thermodynamic requirements for a physical vapor explosion to occur are fairly well established (i.e., contact between hot and cold fluids with the temperature of the interface greater than the homogeneous nucleation temperature of the volatile cold liquid). Direct liquid-liquid contact and a rapid increase in interfacial area are necessary to provide the efficient heat transfer for the rapid conversion of liquid to vapor that characterizes a vapor explosion. One common scenario for a *large-scale* explosive interaction involves the pouring or spilling of a hot fluid into a coolant. The hot fluid breaks up into fragments that distribute themselves in the cold fluid to form a mixture of hot fragments and coolant. When an explosion is triggered locally (due to the breakdown of the vapor film surrounding the hot fragments), the shock wave and convective disturbance associated with the expansion and collapse of the vapor bubble from the local explosion induce the collapse of the vapor layer of the neighboring fragments and cause them to explode. Thus, a vaporization front propagates throughout the volume of the mixture of fragments and coolant. The propagation of such a vaporization front has not been studied previously in great detail. The objective of the present study is to investigate the initial conditions required for the propagation of a thermal interaction between molten metal and water at small scale and to identify and characterize the propagation mechanism involved.

Board et al. (1975) first proposed that vapor explosions may propagate in a manner analogous to a detonating chemical explosion. Initially, a coarse mixture of fuel and coolant undergoing quiescent film boiling is assumed to be present. A triggering event such as a shock wave then leads to local liquid-liquid contact, producing explosive expansion and further fragmentation that supports the propagation of the explosion front through the medium. If the subsequent explosions are coupled to the triggering shock, then the situation is analogous to a chemical detonation wave with a classical steady-state structure of a shock followed by an energy release zone. The limiting conditions required to sustain a steady-state thermal detonation are presently not known. It is likely that a variety of mechanisms and propagation speeds for vapor explosions exists, depending on the initial and boundary conditions. This is analogous with the propagation of a flame in a combustible mixture, in that a continuous spectrum of flame speeds is possible, depending on the initial and boundary conditions of the mixture and the strength of the ignition source.

In earlier experimental work, many investigators have studied the spontaneous thermal explosions that may result when single drops of molten metal are dropped into water. Corradini et al. (1978) surveyed a variety of molten metal/water experiments and accumulated experimental data from nine different reports to determine the effects of independent variables on the occurrence of explosions. For tin/water interactions, most investigators found that explosive reactions are observed over a fairly wide range of initial tin and water temperatures. Furthermore, at a given drop temperature within this explosive range, if the water temperature is increased, a threshold temperature is reached above which no explosions occur (e.g., see Dullforce et al. 1976). There is considerable scatter in the experimental data due to variations in the drop size and drop height. The latter two parameters determine the Weber number, $N_W = \rho D V^2/\sigma$, where ρ is the density of water, D the drop diameter, V the drop velocity at impact with the water, and σ the surface tension of tin. The Weber number provides a measure of the tendency for hydrodynamic fragmentation to occur as the drop falls through the water. The amount of gas that is entrained after the drop contacts the water surface is also a function of the drop height. Permanent gases surrounding the drop tend to stabilize the vapor film and prevent spontaneous explosions (Nelson and Duda 1982). Spontaneous explosions may also be suppressed by increasing the coolant viscosity (Nelson and Guay 1986). The use of an external disturbance for triggering a vapor explosion in circumstances that would otherwise involve stable film boiling has been investigated by a number of investigators, e.g., Board et al. (1974), Arakeri et al. (1978), Fröhlich and Anderle (1980), and Nelson and Duda (1982). The latter investigators used a shock wave to initiate an explosion and found, for the case of an iron oxide drop in water, that a threshold overpressure is required for triggering an explosion. With large peak triggering pressures, they observed immediate explosions whereas with lower pressures, the drops exploded only after a delay of up to 100 ms.

Several earlier investigators have observed the propagation of a vapor explosion. At small scale, Fröhlich and Alisch (1982) injected a jet of molten tin into water. In one photographic series, they observed a pulsating interaction that propagated up the column of tin at an average speed of several m/sec. Fröhlich (1987) also found that the lateral propagation of an explosion through a series of vertical tin jets traveled at speeds on the order of 5 m/s. Baines (1984) recorded the propagation of a pressure pulse during the explosion of a mixture of tin and water in a constrained one-dimensional geometry. He measured propagation speeds in the range of 50-250 m/s. In large-scale experiments at Sandia National Laboratories, explosive interactions were generated by pouring molten material into an open water tank (Mitchell et al. 1981). A propagating explosion wave was observed in some cases that traveled at speeds between 250 and 560 m/s. In large-scale experiments, it is difficult to characterize the initial conditions that are required for a self-sustained propagating thermal explosion. During a vapor explosion, the rapid production of vapor also hinders the direct observation of the details of the explosion process.

In the present paper, the results of several experiments in which single drops of tin are dropped into water will be first described briefly. Both spontaneous and triggered explosions have been investigated. This will be followed by a discussion of results concerning the dynamics of multiple-drop/water interactions. A selection of photographs and pressure traces exhibits the propagation mechanism of the explosive interaction.

Experimental

A schematic diagram of the experimental apparatus used in the present study is shown in Fig. 1. The water is contained in a tank with glass sides (5 mm thick, 61.0 cm long x 20.3 cm wide x 20.3 cm high) and aluminum base. Fast-response (1 µs risetime) piezoelectric pressure transducers (PCB model 113A24, 5 mV/psi, nominal) are used to record the pressure history during an explosive interaction. The transducers are mounted in delrin plugs, which in turn are flush-mounted in the base of the tank at intervals of 6.4 cm. Pressure signals are recorded using an HP Vectra computer in conjunction with a multichannel A/D board capable of digitizing at 1 MHz with 12 bit resolution. To generate the shock wave used for triggering an explosion, electrodes are inserted into the water at one end of the tank. A spherical shock wave is generated when a switching spark gap is triggered, discharging a HV capacitor (0.2 µF, charged to 20 kV) through the electrodes.

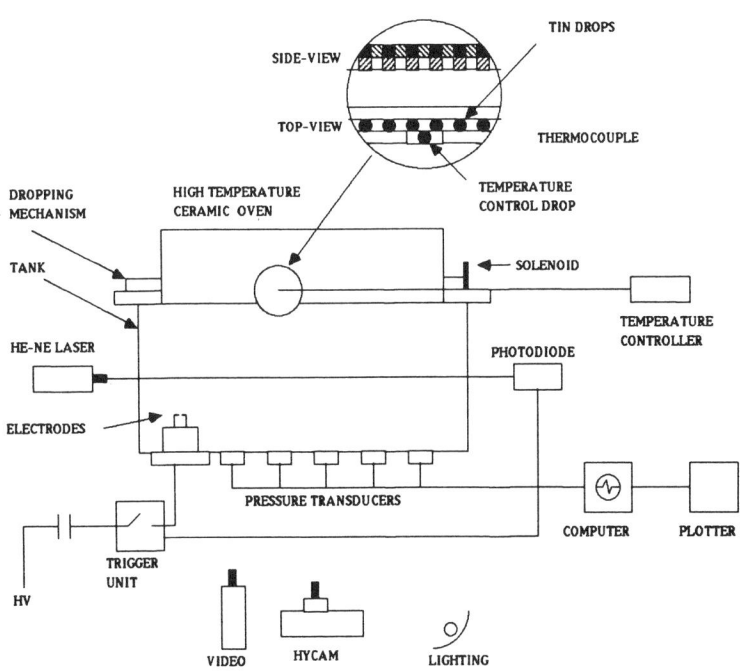

Fig. 1 Schematic diagram of the apparatus

Triggering of the spark discharge is facilitated with the use of a light beam (provided by an 0.8 mW He-Ne laser) that passes horizontally through the tank at a preset vertical location. When a drop intercepts the beam, a photodiode detector circuit produces a pulse that triggers the underwater spark.

The mechanism for simultaneously dropping multiple drops consists of upper and lower stainless steel plates through which multiple holes were drilled (0.64 cm diam., 1.9 cm spacing). With the plates displaced relative to each other, equal-sized drops (typically 2-3 g each) are placed in the holes in the upper plate. The entire array is then placed within a 1250 W semicylindrical ceramic oven. A chromel-alumel thermocouple monitors the temperature of a control drop in the center of the oven and a temperature controller maintains the temperature in the oven at a preset level. When the oven reaches the operating temperature, a solenoid is used to slide the upper plate until the holes in both plates are aligned, allowing the drops to fall through into the water. Before each set of trials, the tank is filled with fresh tap water and heated to the desired temperature with a 1200 W incoloy resistive heater. No special precautions were taken in purifying the water. Prior to each test, the water is mixed with a mechanical stirrer to ensure a uniform temperature distribution. The oven and dropper mechanism are then placed on top of the tank and an experimental test is carried out.

The explosive interactions were visualized using video photography and high-speed cinephotography. All trials were recorded with a standard video camcorder, with a framing rate of 64 frames/s, for immediate feedback. To investigate the detailed dynamics of an explosive interaction as well as the propagation of a thermal explosion, a Hycam 16 mm camera was used at framing rates up to 5000 frames/s during selected trials. Kodak high-speed (300 ASA) 7292 color movie film was used. Lighting for the photography included three 1000 W quartz-halogen lamps equipped with reflectors placed in front of the water tank and one 500 W tungsten lamp with reflector placed behind the tank. Graph paper (with 6.35 mm squares) was mounted on the outside of the back face of the tank for dispersing the backlighting. The grid pattern is visible in each photograph and provides a convenient scale for analysis.

Results

Single Drop Experiments

A series of single drop experiments was performed over a range of water and drop temperatures. Spontaneous explosions were observed over a temperature range consistent with earlier investigators. For example, with water temperatures of 15-40°C, spontaneous explosions resulting in fine fragmentation were observed consistently for 2.0 g drops heated to temperatures between 600 and 800°C. Explosive interactions were observed for drop temperatures as low as 435°C, although for temperatures less than 600°C often only partial fragmentation of the drop occurred. In some cases, the drops exploded immediately upon impact with the water surface. In each trial in the present experiments, the drops fell from a height of 3.2 cm above the water surface. This corresponds to a Weber number of 11 for a 2.5 g drop, which is just within the critical Weber number range of 10-20 that is often quoted for drop breakup. However, in the present experiments, all of the drops remain intact as they enter the water. The drop height has a strong effect on the amount of noncondensible air that is entrained along with the drop as it enters the water surface. The presence of a gas "bag" trailing the drop as it falls through the water tends to suppress the occurrence of spontaneous steam explosions. To provide a more reliable measure of the stability of the vapor/gas film surrounding the drop, a wire mesh (6.35 mm square, 0.5 mm thick) was placed in the path of a falling drop near the center of the tank. When the drop contacts the grid, the gas bag attached to the drop is dislodged and the drop often splits into several smaller drops. If an explosive interaction is triggered when the drop contacts the mesh, it is assumed that the drop is only marginally stable for that specific combination of drop and water temperatures. However, if the vapor film reforms and stable film boiling persists, the drop is said to be stable to a finite perturbation.

To determine the water temperature range for which stable film boiling persists, a series of tests was carried out with 2.5 g drops heated to a temperature of 800°C. Spontaneous explosions were observed to occur above the mesh typically with water temperatures less than about 50°C. For water temperatures above 53°C, explosive interactions were not observed, even after the drop contacted the mesh. At the critical water temperature (about 52°C), explosions sometimes occurred after a short dwell time following the contact of the drop with the mesh. Apparently, the impact of the drop with the grid perturbs the vapor film surrounding the drop sufficiently that the film becomes unstable and collapses after a short time interval. Having established a temperature range over which explosive interactions consistently do *not* spontaneously occur, the remaining trials were run within this range (drop temperature at 800°C, water temperature at 54°C or above). It should be noted that the use of a wire mesh for triggering would be ideal in the study of propagation mechanisms except for the fact that the resulting explosion is not derived from the entire drop, since it breaks up into several pieces upon contact with the mesh.

To study the propagation of a vapor explosion, explosive interactions were triggered by a shock wave generated by underwater spark discharge. Figure 2 shows the pressure recorded during spark discharge by a transducer mounted on the base of the tank at a distance of 14.5 cm from the electrodes. The first pressure spike is noise that is generated during the spark discharge. The amplitude of the initial shock wave that arrives at the transducer after a transit time varies somewhat from one trial to the next and is a function of the high voltage level as well as the electrode gap distance. The sharp pressure pulses that are visible at about 3.5 and 5.2 ms are caused by the initial and secondary collapses of the vapor bubble generated by the spark, respectively. The small-scale pressure oscillations are due to internal reflections within the tank. Systematic studies of the threshold shock strength required

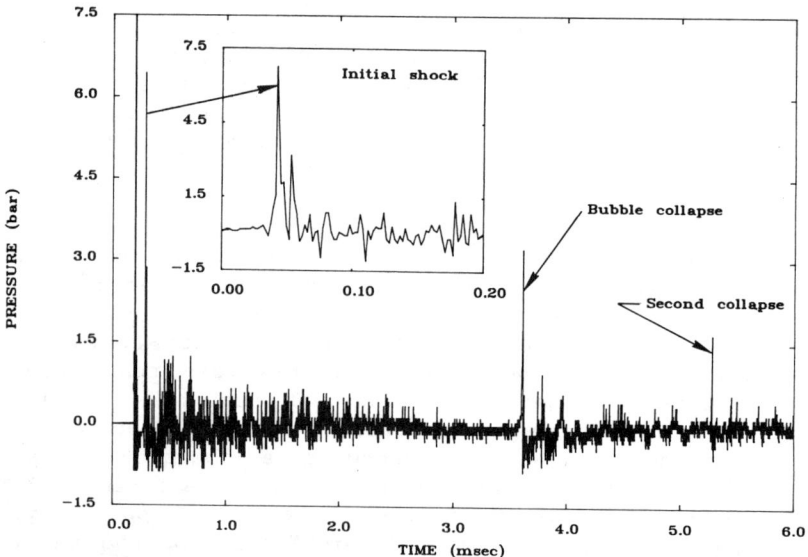

Fig. 2 Pressure trace recorded during spark discharge showing shock waves generated during initial expansion and first and second collapses of electrode vapor bubble (electrode/transducer distance is 14.5 cm).

to trigger a vapor explosion have been carried out by earlier investigators (e.g., Nelson and Duda 1982). In the present experiments, the high-voltage discharge capacitor was charged to a level (20 kV) such that the resulting shock wave triggered immediate explosions only in drops very close (within 4 cm) to the electrodes. Beyond about 9 cm from the electrodes, the shock overpressure is attenuated to a level below the threshold required for triggering an explosion and there is no visible effect of the shock on the drop. At intermediate distances from the electrodes, a transitional regime occurs, similar to that observed by Nelson and Duda (1982) for molten iron oxide drops exploding in water. In this case, the impact of the shock with the drop perturbs the vapor film, causing it to undulate and locally contact the drop surface. Fine metal fragments are ejected from the drop but the drop remains largely intact. In some cases, after a delay time often as long as 100 ms, an explosive interaction occurs generating fine fragmentation of the drop.

Figure 3 shows a series of photographs that illustrates the shock triggering of a single 2.5 g drop of tin. The initial conditions for this drop, and for the drops in each subsequent trial reported in this paper, were found to be stable to a finite perturbation and correspond to an initial drop temperature of 800°C and water temperature of 54°C. Discharge of the electrodes produces a vapor bubble that grows to a maximum diameter of about 4.0 cm at 1.6 ms. The initial shock wave generated at discharge causes the profile of the drop to expand, but an explosive interaction does not occur immediately. After collapsing, the electrode bubble rebounds to a second maximum diameter at 4.4 ms and at this time small metal fragments are visible around the drop. The fully developed explosion begins (at about 5.2 ms) only following the second collapse of the electrode bubble. The steam bubble generated during the explosion expands to a maximum diameter at 7.8 ms, although it is obscured by the distributed tin debris that by this time is finely fragmented. The steam bubble undergoes several more mild bubble oscillations that reach maxima at 10.6 and 13.0 ms, respectively.

An example of the explosion of a drop in the transitional stability regime described above is shown in Fig. 4. The photographs illustrate a common feature, i.e., the explosion of a drop consists of a number of bubble growth/collapse cycles that escalate until the drop has been totally fragmented. The drops shown were located about 8 cm from the electrodes when the shock wave was

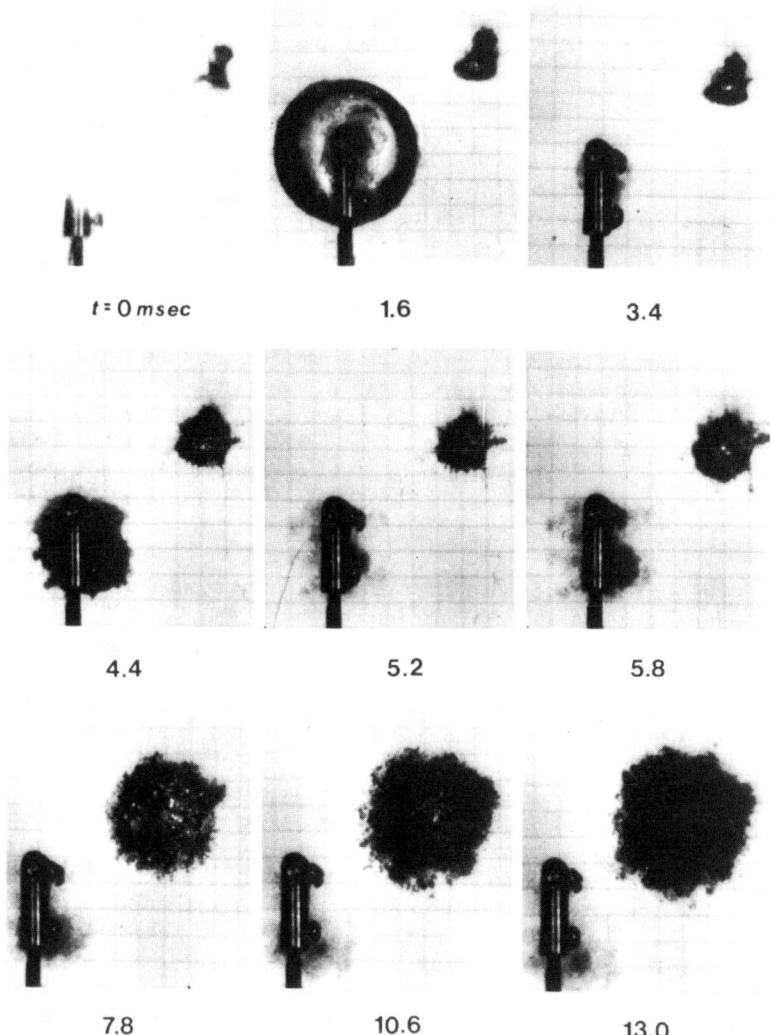

Fig. 3 Shock triggering of a single 2.5 g drop of tin at an initial temperature of 800°C and water temperature of 54°C (scale of background grid = 0.64 cm).

generated. The explosion of the lowest drop in the photographs began at a time (labeled t=0) corresponding to a delay time of 44 ms after the spark discharge. At this time, a local collapse of the vapor film on the lower portion of the drop caused a small steam bubble (indicated with an arrow in the photograph) to grow to a diameter of 3 mm after 0.3 ms. The collapse of this small bubble at 0.7

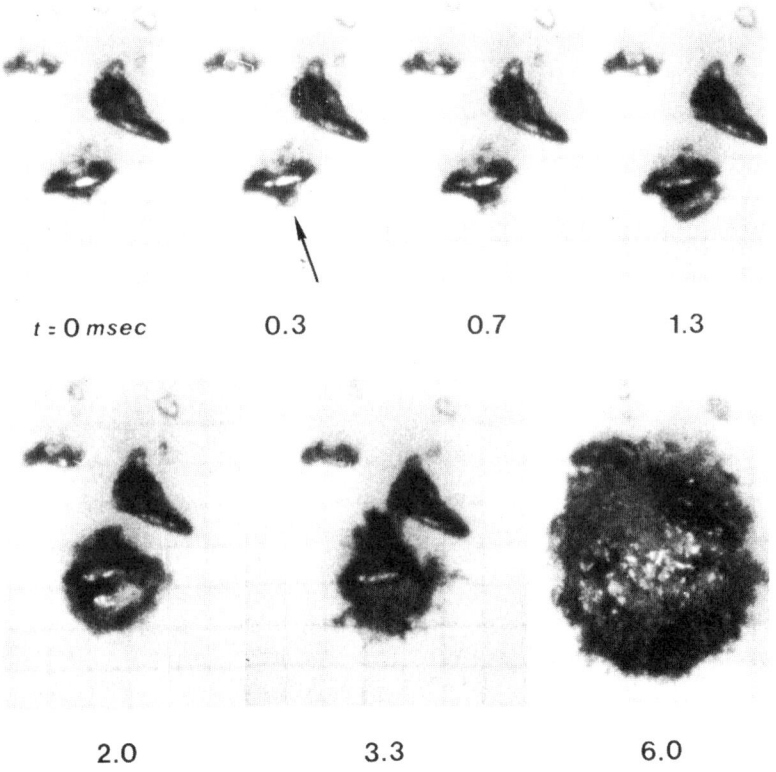

Fig. 4 Explosion of a tin drop in the transitional stability regime. Note the localized initiation of the explosion (indicated by the arrow).

ms drives the growth of a second steam bubble that reaches a maximum size at 2.0 ms. This bubble collapses at 3.3 ms; at this time, the original shape of the drop is still visible, indicating that the drop is still largely intact. The explosion following the collapse of this bubble shatters the remainder of the drop and by 6.0 ms the fragments are scattered over a large volume. The two drops above the exploding drop are triggered, exploding violently shortly after the initial explosion.

Multiple Drop Experiments

Tests were carried out in which a linear array of molten drops was dropped into water. High-speed films taken during each trial clearly show that the explosion propagates as a series of pulsating interactions. The ex-

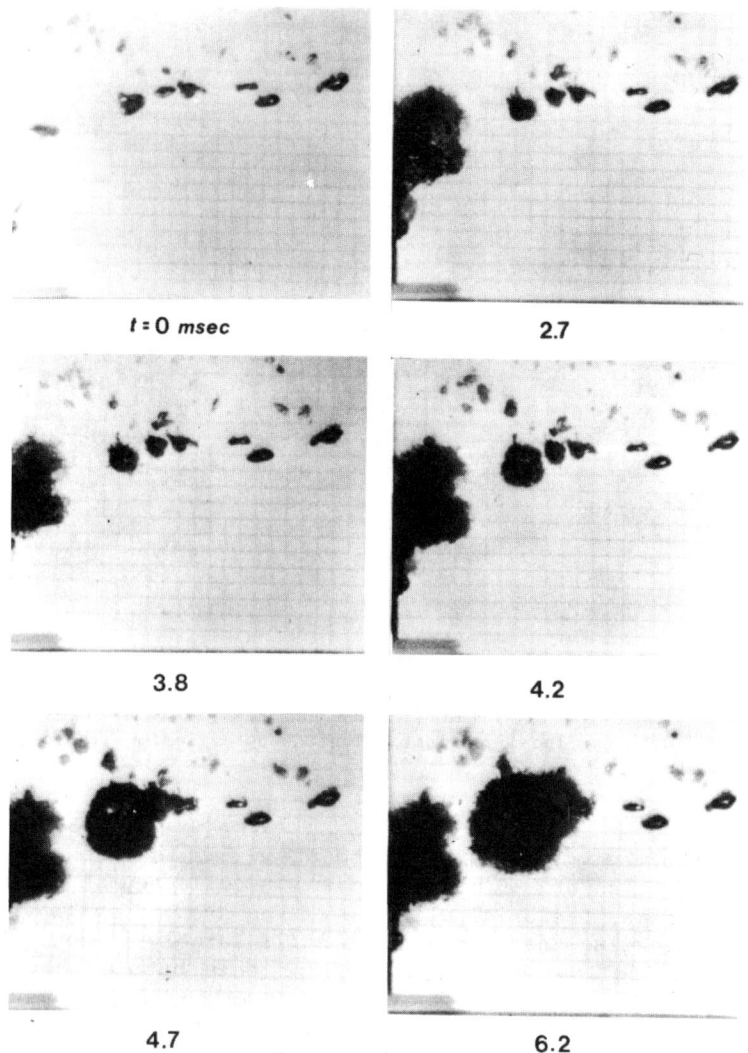

Fig. 5a Propagation of a vapor explosion in which bubble collapse causes triggering of adjacent drop (initial drop temperature 800°C, water temperature 54°C, and initial drop spacing 1.91 cm).

EXPLOSIVE MOLTEN-TIN/WATER INTERACTIONS

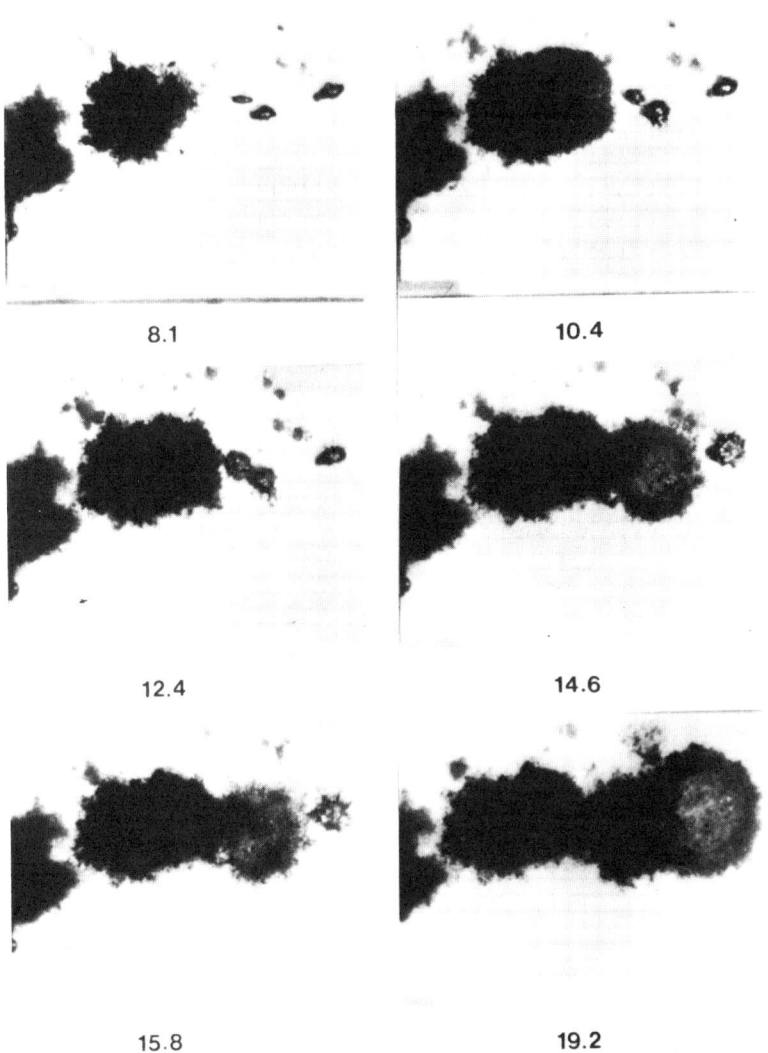

Fig. 5b See fig. 5a for caption.

plosion of a drop generates a steam bubble that upon collapse triggers the explosion of the adjacent drop. In this way, the explosion propagates through all the drops on a time scale of the order of tens of milliseconds. This propagation mechanism is illustrated by the sequence of photographs shown in Figs. 5a,b. Seven drops (denoted 1-7, from left to right) with an initial spacing of 1.9 cm are visible in the first photograph, which was taken shortly after the spark discharge. The small objects trailing above the drops are bubbles that have detached from the drops. The initial shock wave triggers the explosion of the nearest drop and the resulting vapor bubble reaches a maximum diameter at 2.7 ms. At this time, the profile of drops 2-4 have expanded slightly due to the disturbance created by the initial shock and the explosion of the first drop. The collapse of the vapor bubble from drop 1 at 3.8 ms coincides with the explosion of drop 2. Shortly afterward (at 4.2 ms), drop 3 explodes nearby and the vapor bubble from drops 2 and 3 reaches a maximum diameter at 6.2 ms. The collapse of this bubble at 8.1 ms is followed immediately by the explosion of drop 4. This cycle of bubble growth and collapse continues several more times until the final drop explodes at 15.8 ms.

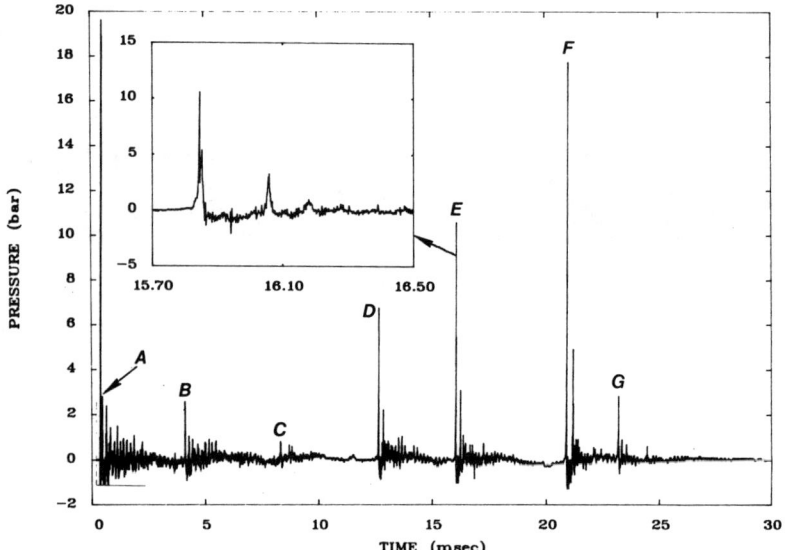

Fig. 6 Pressure trace recorded for the trial shown in Fig. 5. Note the discrete pressure spikes corresponding to each bubble collapse.

EXPLOSIVE MOLTEN-TIN/WATER INTERACTIONS 465

The far-field pressure history during the above trial as recorded by a transducer located 14.5 cm from the electrodes on the tank base, is shown in Fig. 6. The pressure peaks (labeled A-G in Fig. 6) are generated by the collapse of the various vapor bubbles. The drop explosion coincident with each bubble collapse also produces a compressive disturbance. By comparing the pressure data with the photographic record, the pressure peaks can be associated with the following events: peak A, shock wave from the spark discharge; peak B, collapse of the electrode bubble; peak C, collapse of the steam explosion bubble from drops 2 and 3; peak D, collapse of the bubble from the explosion of drop 4; peak E, collapse of the bubble from the simultaneous explosion of drops 5 and 6; peak F, collapse of the bubble from the explosion of drop 7; peak G, second collapse of the bubble from drop 7. In Fig. 7, the time at which each drop starts to explode is plotted as a function of the electrode/drop distance. From the slope of the resulting curve, the average propagation velocity of the explosive interaction is 7.8 m/s.

Essential to the propagation mechanism is the requirement that the drop spacing be small enough so that the pressure wave and convective disturbance generated by

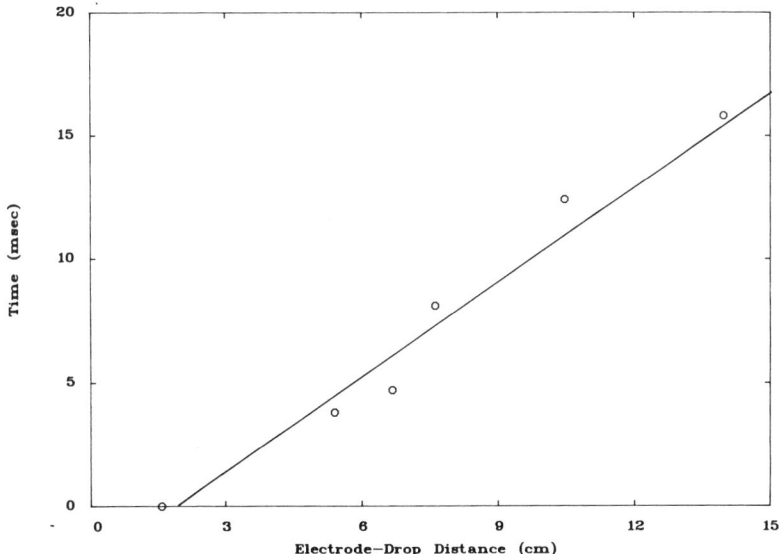

Fig. 7 Explosion time vs distance from transducer showing the average propagation velocity corresponding to trial shown in Fig. 5.

the explosion of one drop is sufficient to trigger the explosion of an adjacent drop. The series of photographs shown in Figs. 8a,b illustrate a case in which the propagation of the explosion is interrupted at some point. In this experiment, 10 drops again with an initial spacing of 1.9 cm were dropped. Following the spark discharge the first five drops explode in a sequential fashion and then drops 6 and 7 explode simultaneously at 16.0 ms. The disturbance from the subsequent vapor bubble collapse at 22.4 ms perturbs drops 8 and 9 and small metal fragments are ejected from the drops. However, the drops remain intact for 100 ms until the explosion of drop 9 triggers the explosion of the remaining drops. The delayed explosion of the final three drops is similar to the behavior observed in the transitional regime described earlier for the single drop trials. The average propagation velocity for the interactions up to drop 7 was determined to be 7.9 m/s.

Discussion

The present experiments have demonstrated that a linear array of molten drops in water can support the propagation of an explosive interaction. The propagation mechanism is due to the interaction between adjacent drops. The explosion of a drop generates a vapor bubble that upon collapse, triggers the explosion of a neighboring drop. The large pressure pulse produced during the bubble collapse destabilizes the vapor film that surrounds nearby drops. The violent collapse of the vapor film may lead to the formation of jets of water that impact and penetrate the molten drop. When the water contacts the molten metal, it rapidly superheats and boils violently. In this respect, the evaporative instability and jetting observed by Shepherd and Sturtevant (1982) and Frost (1987) during the explosive boiling of liquids heated to the superheat limit may play a role in the fragmentation of the drop.

The propagation velocity is a function of the drop spacing and the average time for a bubble growth/collapse cycle. In the present experiments, the appropriate values lead to a propagation velocity on the order of 5-10 m/s. If the drop spacing is too large, the cooperative effects between the drops are not sufficient to sustain the propagation. In the present experiments, it was found for 2.5 g drops that, if the ratio of the average drop spacing to the effective drop diameter was greater than about 4, the explosion failed to propagate.

EXPLOSIVE MOLTEN-TIN/WATER INTERACTIONS

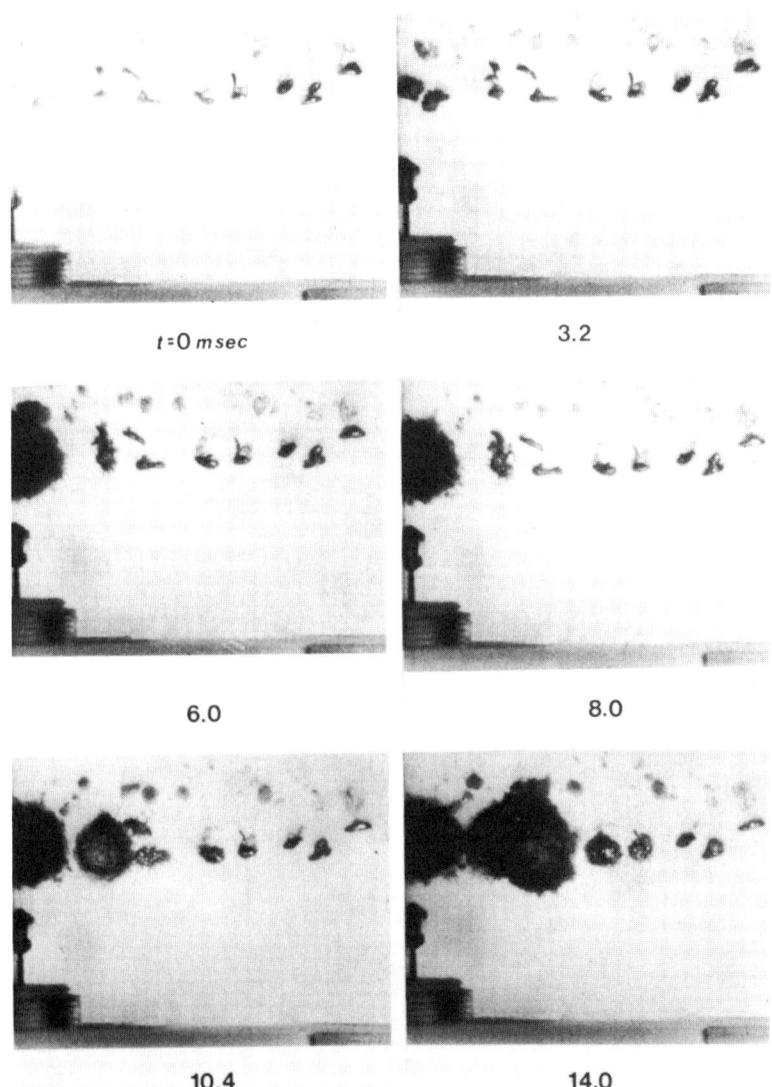

Fig. 8a Sudden halt in the propagation of the explosion due to the incomplete explosion of drops 6 and 7.

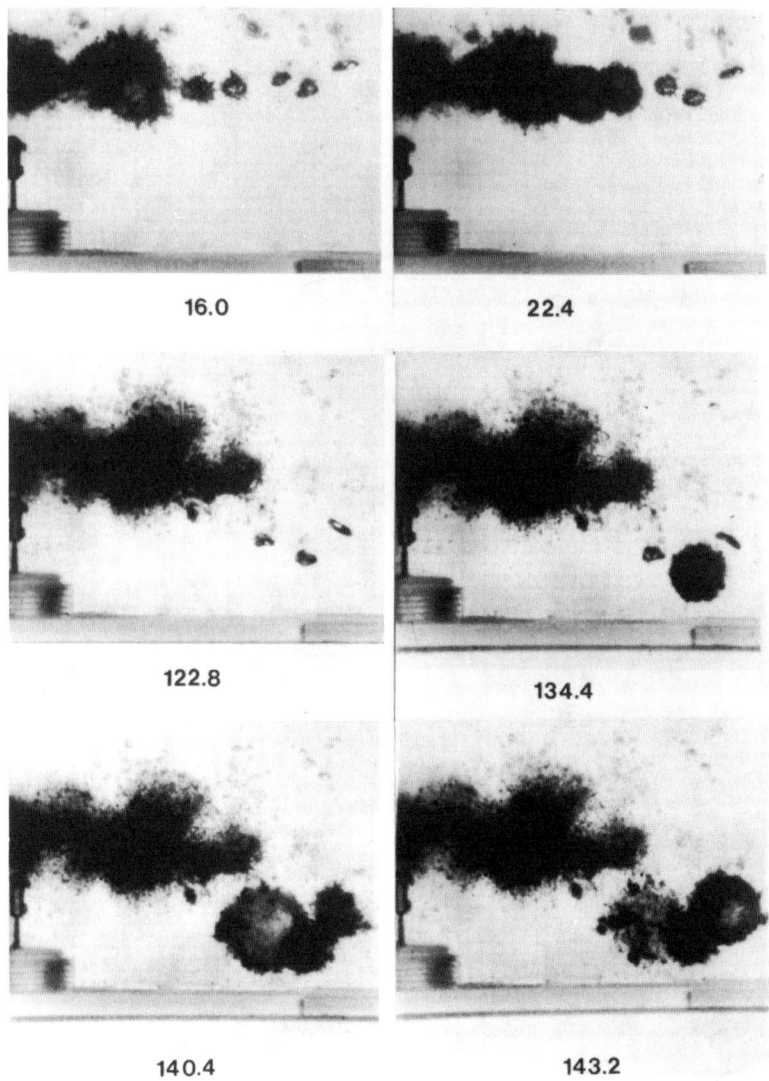

Fig. 8b See fig. 8a for caption.

EXPLOSIVE MOLTEN-TIN/WATER INTERACTIONS

High-speed movies taken of many different multiple-drop experiments reveal that a second triggering mechanism may play a role in the explosion propagation for a linear drop array in which the drops are closely spaced. The films show that when the vapor bubble of a triggered drop collapses, the neighboring drop will be pulled toward the collapsing bubble, if it is sufficiently close. This motion is a result of the drag forces imparted on the adjacent drop due to the flowfield induced by the collapsing vapor bubble. If the induced flow is substantial, such that the Weber number is above a critical value, vapor film stripping and hydrodynamic fragmentation of the neighboring drop can be initiated.

To investigate this mechanism, the bubble dynamics of the single drop trial shown in Fig. 3 were analyzed. A radius time history for the vapor bubble generated by the exploding drop was determined using the photographic data. Using the camera framing rate (5000 frames/s), the radial bubble velocity was determined and plotted as a function of time for one cycle together with the radius in Fig. 9. As expected, the bubble grows rapidly at first, as the thermal energy from the molten hot drop is deposited into the cold liquid creating a high-pressure vapor bubble. As the bubble expands, the pressure inside the bubble decreases and the radial velocity of the bubble decelerates. Due to the

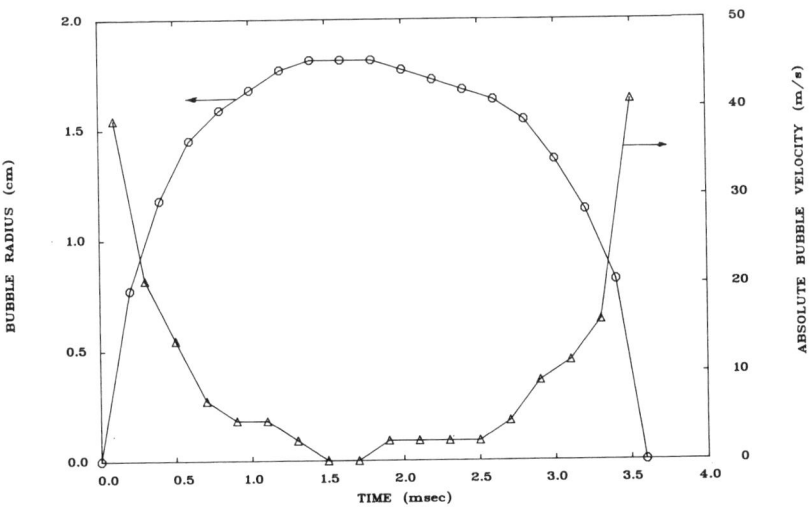

Fig. 9 Vapor bubble radius and radial bubble velocity vs time obtained from the high-speed film record corresponding to the exploding single drop shown in Fig. 3.

inertia of the displaced water, the bubble overshoots the equilibrium radius and subsequently collapses, completing the cycle.

Using the results from Fig. 9 as characteristic for an exploding 2.5 g drop, far field velocities can be calculated using the conservation of mass, i.e.,

$$U(R) \ R^2 = U(r) \ r^2$$

where $U(R)$ and R are the radial bubble velocity and bubble radius, respectively, and $U(r)$ the fluid velocity at a distance r from the center of the bubble. Using this expression and solving for $U(r)$, an instantaneous Weber number time history corresponding to the neighboring drop at a distance r can be derived. Such a profile is shown in Fig. 10 for a drop spacing of 1.905 cm. The profile is independent of drop spacing, although the magnitude is inversely proportional to the fourth power of the distance r. The profile consists of two peaks, one in each of the expansion and collapse phases. The minimums correspond to when the bubble radius reaches a minimum and maximum. The magnitude of the Weber number is relatively high, reaching a maximum of 700 at 0.5 ms, which is well above the critical value of 10-20 required for hydrodynamic breakup. This critical

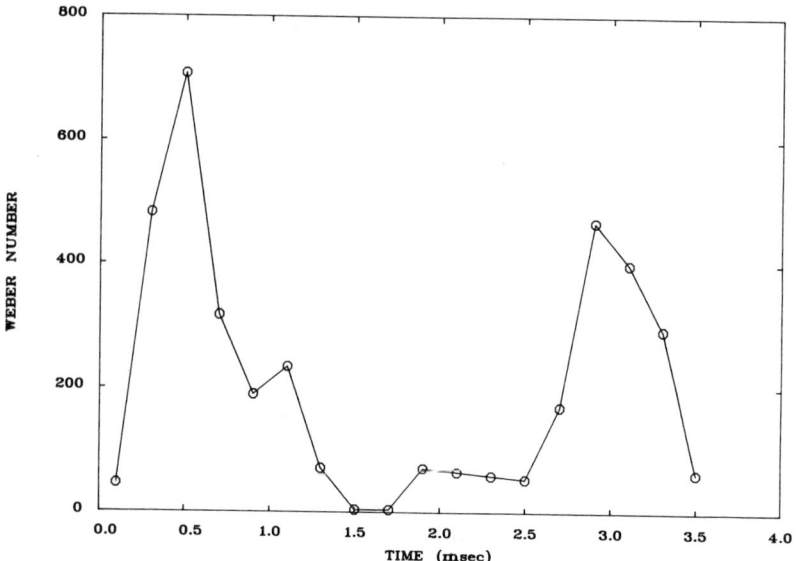

Fig.10 Instantaneous Weber number time history felt by an adjacent drop located 1.91 cm from the center of the exploding drop shown in Fig. 3.

value is based on a steady flow velocity, whereas, in the present case, the flow velocity is of course highly unsteady.

It appears from Fig. 10 that, if there is any influence on the neighboring drop, it would occur during the time at which the Weber number is at a maximum. Furthermore, the effect would be accentuated for closely spaced drops. The interaction of the second and third drops in Fig. 5 is a good illustration of this mechanism. This particular interaction departs from the characteristic trigger mechanism due to bubble collapse that prevails in all the other drop interactions in this trial. Instead, the drop is triggered prematurely. The time difference between the explosion of the second drop and the third drop is approximately 0.5 ms which corresponds roughly with the first maximum in the Weber number time profile.

Conclusions

In a series of experiments, the interaction between multiple drops of tin and water has been investigated. From experiments with single drops, a combination of drop and water temperatures was determined at which a drop undergoing film boiling was stable to a finite perturbation. For drops heated to 800°C, it was found that, for water temperatures above 53°C, contact between the drop and a wire mesh failed to initiate a vapor explosion. In shock triggering tests with single 2.5 g drops heated to 800°C and dropped into water at 54°C, it was found that beyond a critical distance (about 9 cm) the shock had no effect on the drop. In many cases, the explosion of a drop consists of an escalating series of local bubble growth and collapse cycles until the drop is completely fragmented.

In experiments with multiple drops in a linear array, an explosive interaction was observed to propagate at an average velocity of 5-10 m/s. The explosion propagates as a result of the interaction between adjacent drops. It was found that the pressure wave produced by the collapse of a steam explosion bubble from one drop served to trigger the explosion of the neighboring drop. In this way, the explosion propagates from one drop to the next with a propagation velocity that depends only on the characteristic bubble growth/collapse time and the drop spacing. For some cases in which the drop spacing is small, the convective disturbance created by the explosive vapor bubble growth is sufficient for triggering an explosion. Calculations estimating the effective Weber number felt by a drop adjacent to an expanding or collapsing bubble support the validity of this hydrodynamic triggering mechanism. Failure of a self-sustained propagating explosion may occur as the result of the incomplete explosion of a drop within the linear array or if the local drop spacing exceeds a critical amount.

The slow propagation mode observed in the present experiments differs from the thermal detonation model in that the reaction zone is not coupled to the initial shock wave. It is likely that the mode of propagation is highly dependent on the initial and boundary conditions. It is possible that, if the degree of confinement and the strength of the trigger were increased, a linear drop array may support a faster mode of propagation. More work is required to understand the effect of scale on the mechanisms of triggering and propagation of a vapor explosion.

Acknowledgments

This work was supported by the Natural Sciences and Engineering Research Council of Canada and by Sandia National Laboratories. The authors would also like to express their gratitude to Prof. J. H. S. Lee for many useful discussions.

References

Arakeri, V. H., Catton, I., and Kastenberg, W. E. (1978) An experimental study of the molten glass/water thermal interaction under free and forced conditions. Nucl.Sci. Eng. 66, 153-166.

Baines, M. (1984) Preliminary measurements of steam explosion work yields in a constrained system. First U.K. National Conference on Heat Transfer, Institution of Chemical Engineers Symposium Series, No. 86, pp. 97-108.

Board, S. J., Farmer, C. L., and Poole, D. H. (1974) Fragmentation in thermal explosions. Int. J. Heat Mass Transfer 17, 331-339.

Board, S. J., Hall, R. W., and Hall, R. S. (1975) Detonation of fuel coolant explosions. Nature 254, 319-321.

Corradini, M. L., Rohsenow, W. M., and Todreas, N. E. (1978) A proposed model for tin-water interactions. Topics in Two-Phase Heat Transfer and Flow, ASME Winter Annual Meeting, ASME, New York, 17-28.

Dullforce, T. A., Buchanan, D. J., and Peckover, R. S. (1976) Self-triggering of small-scale fuel-coolant interactions: I: experiments J. Phys. D: Appl. Phys. 9, 1295-1303.

Fröhlich, G. and Alisch, S. (1982) Optische Erfassung der Wechselwirkung von heissen Zinnschmelzestrahlen mit Wasser bei selbst und Stosswellentriggerung. Rep. IKE2-58, *Institut für Kernenergetik und Energie systeme*, Universität Stuttgart, Stuttgart, FRG.

Fröhlich, G. and Anderle, M. (1980) Experiments for studying the initiation mechanisms for steam explosions. Rep. IKE2-51 *Institute für Kernenergetik und Energiesysteme*, Universität Stuttgart, Stuttgart, FRG.

Fröhlich, G. (1987) Private communication. *Institute für Kernenergetic und Energiesysteme*, Universität Stuttgart, Stuttgart, FRG.

Frost, D. L. (1987) Dynamics of explosive boiling of a droplet. *Proc. ASME-JSME Thermal Engineering Joint Conference*, Vol. 2, pp. 447-454. ASME, New York.

Mitchell, D. E., Corradini, M. L., and Tarbell, W. W. (1981) Intermediate scale steam explosion phenomena: experiments and analysis. SAND81-0124, Sandia National Laboratories, Albuquerque, NM.

Nelson, L. S. and Duda, P. M. (1982) Steam explosion experiments with single drops of iron oxide melted with a CO_2 laser. *High Temp.-High Pressures* 14, 259-281.

Nelson, L. S. and Guay, K. P. (1986) Suppression of steam explosions in tin and $Fe-Al_2O_3$ melts by increasing the viscosity of the coolant. *High Temp.-High Pressures* 18, 107-111.

Reid, R. C. (1983) Rapid phase transitions from liquid to vapor. *Adv. Chem. Eng.* 12, 105-208.

Shepherd, J. E. and Sturtevant, B. (1982) Rapid evaporation at the superheat limit. *J. Fluid Mech.* 121, 379-402.

Chapter VI. Vapor-Cloud Explosions and Safety Applications

Dispersion of Dense Gaseous Fuels Released into the Atmosphere

O. M. F. Elbahar* and M. M. Kamel†
Cairo University, Giza, Egypt

ABSTRACT

The prediction of the behavior of dense gases released into the atmosphere is of great practical importance in many branches of engineering. Recently, the potential hazards expected during the transportation of liquefied gaseous fuels by giant tankers makes the prediction of the behavior of a spilled amount of the flammable cryogenic cargo even more important. The assessment of the risks associated with engineering systems for transporting and storing such a cargo is impossible without having efficient models for predicting the behavior of the gas on its spillage. In the present contribution, a mathematical model is presented for the calculation of the evaporation, spread, and dispersion of a quantity of a liquefied gaseous fuel released into the atmosphere on an accident. The model facilities the calculation of the time required for the processes of evaporation, spread, and dispersion of the spilled cryogenic fuel. Furthermore, the model facilitated of calculation of the distance required by the fuel-air cloud formed by the dispersion of the evaporated cryogenic fuel into the surrounding atmosphere until it is diluted below the lower flammability limit.

Copyright © 1988 by the American Institute of Aeronautics and Astronautics, Inc. All rights reserved.
  * Assoc. Professor, Mechanical Power Department, Faculty of Engineering.
  † Professor, Mechanical Power Department, Faculty of Engineering.

I. Introduction

The assessment of the potential hazards associated with large-scale releases of hazardous materials into the atmosphere requires efficient mathematical models for the prediction of the behavior of the released material. An example of great practical relevance in this context is the spill of a large quantity of the liquefied gaseous fuel (LGF) during its transportation through inland waterways. With spills of this magnitude, risk analysis becomes imperative.

In the present contribution, a mathematical model is presented for the prediction of the behavior of a large quantity of LGF spilled onto water. The model is intended to help quantify the potential risk associated with the passage of giant LGF tankers through inland waterways. The parameters considered in the model are the spill size, the spill rate, and the meteorological conditions.

II. Mathematical Modeling of Evaporation and Dispersion of Liquefied Gaseous Fuels

The mathematical models available in the literature for the prediction of the behavior of LGF spills range from simple empirical relations gained from small-scale experiments to complex models comprising the solution of the three-dimensional, time-dependent partial differential equations expressing the conservations of mass, momentum, energy, and species under turbulent-flow conditions using fnite-difference or finite-element techniques [Khalil and Kamel (1985); Havens (1977); Science Applications (1975); Chan (1981)].

The application of complex models has been justified by the inaccuracy of simpler models in predicting the dispersion of gaseous species into the atmosphere. The Gaussian plume model is a good example in this context.

The model developed in the present work for LGF spills onto water is based on subdividing this process into three separate stages, as shown in Fig. 1. In the first stage, the spilled liquid boils, thus forming a vapor cloud at the boiling point without entraining air from the surrounding atmosphere. This is followed by the expanding-vapor phase in which the vapor cloud expands by two mechanisms: heat transfer and entrainment of air from the surrounding atmosphere. The third phase contains

DISPERSION OF DENSE GASEOUS FUELS

the turbulent dispersion of the fuel vapor into the atmosphere, forming the fuel-air cloud (FAC).

In the present contribution, a new version of the Gaussian plume model is proposed for the calculation of the turbulent dispersion of the fuel vapor into the atmosphere. This version combines both accuracy and economy because it can be easily programmed on a personal computer. The accuracy of the model is verified by comparing its predictions with experimental results of finite-rate spills available in the literature.

III. Mathematical Modeling of Evaporation and Dispersion of Liquefied Gaseous Fuels

A. Evaporation and Gravity-Spread Phases

On the instantaneous spill of a volume V_0 of liquefied natural gas (LNG), a liquid pool is formed which immediately begins to evaporate, thus increasing in diameter and height. If an average constant value of heat flux $\dot{q}$ is assumed (Briscoe and Shaw 1980), the maximum radius r of the pure LNG pool is given by

$$r = [(h_{fg}/v_f\dot{q})^2 \cdot V_0^3 \cdot (gG)]^{0.125} \quad (1)$$

and the evaporation time t is given by

$$t = 0,675 \, [(h_{fg}/v_f\dot{q})^2 \, (V_0/gG)]^{0.25} \quad (2)$$

where h_{fg} is the latent heat of vaporization of LGF, v_f is the specific volume, g is the earth gravitational acceleration and $G = (v_a/v_f - 1)$, and v_a is the specific volume of ambient air.

Because of the negative buoyancy, this cloud starts to gravity-spread entraining air. The entrainment coefficient is generally a function of the Richardson number [Cox (1980); National Materials Advisory Board (1980)] and has a value of about 0.1 [National Materials Advisory Board (1980)]. Therefore,

$$u = dr/dt = [kgh \, (v_a/v_c - 1)]^{0.5} \quad (3)$$

where k is a constant ($1 \leq k \leq 2$), h is the cloud height, and the subscripts c and a refer to cloud and ambient air, respectively, and u is the radial spread velocity.

Assuming that the radial spread velocity varies linearly with radius, being zero at the cloud center, than the mass entrained, through an element of area of radius r and thickness dr, is given by [National Materials Advisory, Board (1980)]

$$d\dot{m}_a = 2\pi \cdot r' \, dr' \cdot a_t \cdot u \cdot r' \, (v_a) \qquad (4)$$

where m_a is the mass of air entrained per unit time, and a_t is the entrainment coefficient. Integration of the above equation gives

$$d\dot{m}_a/dt = 2 \cdot \pi \cdot a_t \cdot r^2 \cdot u/(3v_a) \qquad (5)$$

The gravity spreading ceases when $v_a \simeq v_c$.

Noting that the rate of mass increase of the whole cloud is equal to the rate of air entrained; since no vapor escapes the cloud because of gravity, then the application of the first law of thermodynamics gives

$$(\dot{m}c_p \cdot dT/dt)_c = (\dot{m}c_p)_a (T_a - T_c) + \dot{Q} \qquad (6)$$

where the subscripts c and a refer to the cloud and the air, respectively, T is the temperature, and $\dot{Q}$ is the rate of sensible energy addition to the cloud (heat addition plus the energy resulting from the condensation or even freezing of the water vapor in air).

The system of Eqs. (1-5) is solved numerically using the Runge-Kutta-IV method to give the instantaneous values of the cloud radius, height, temperature, and concentration, as well as the amount of entrained air.

B. Dispersion Phase

Dispersion of the cloud is assumed to start at the end of the gravity-spread phase. The distribution in space and time of the fuel vapor is governed by the species conser-

DISPERSION OF DENSE GASEOUS FUELS

vation equation:

$$DC/Dt = \text{div}(K^\circ C) + S_f \qquad (7)$$

where C is the concentration of the fuel vapor, K is the turbulent diffusivity, and S_f is a source term.
Analytical solutions to Eq. (6) exist for some simple cases of release from point and line sources assuming a Gaussian distribution of the concentration in the lateral (y) and vertical (z) directions at each downwind location x (Fig. 1). These solutions may be written as

$$C(x,y,z,t) = \{2N/[(2 \cdot \pi)^{1.5} s_x s_y s_z]\} \qquad (8)$$
$$\cdot \exp(0,5 \{[(x - ut)/s_x]^2 + (y/s_y)^2 + (z/s_z)^2\})$$

for instantaneous releases from point sources (puffs) and

$$C(y,z) = (dN/dt)/[(2 \cdot \pi) s_y s_z u)] \qquad (9)$$
$$\exp\{-0.5 \ [(y/s_y)^2 + (z/s_z)^2]$$

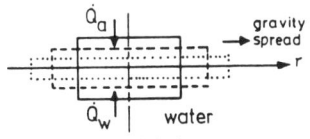
1. Boiling liquid stage

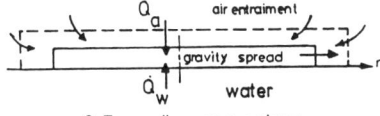
2. Expanding vapour stage

Fig. 1 Stages of fuel-air cloud formation

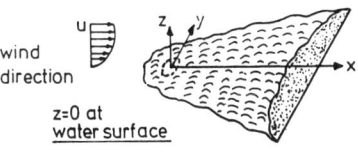
3. Dispersion stage

for continous releases from point sources (plumes), where s_x, s_y, s_z are the plume standard deviations and u is the wind speed.

Thus, the cloud dimensions and the distance to the lower flammability limit can be calculated.

IV. Results and Discussions

For an instantaneous spill onto water with $V_0 = 25,000$ m^3, Eqs. (1) and (2) predict the evaporation time to be about 264 s and the radius of pure fuel vapor cloud at the end of evaporation phase to be 385 m. At the end of this phase, the gravity-spread phase starts. The cloud, which is assumed to be cylindrical, expands, thus radially entraining air and receiving heat from the surroundings. Figure 2 shows the development of the cloud size with time during the gravity-spread phase. Here, the time reference is taken to be zero at the start of the gravity spread. From Fig. 2 it will be clear that, within a short period of time in the order of 100 s, the cloud radius drastically increases from 385 to about 950 m, whereas its height increases from 13 to 23 m. The average concentration within the cloud at the end of gravity spread is about 23%, which is still above the upper flammability limit of 15%.

The size of the cloud at the end of gravity-spread period following an instantaneous spill is heavily depen-

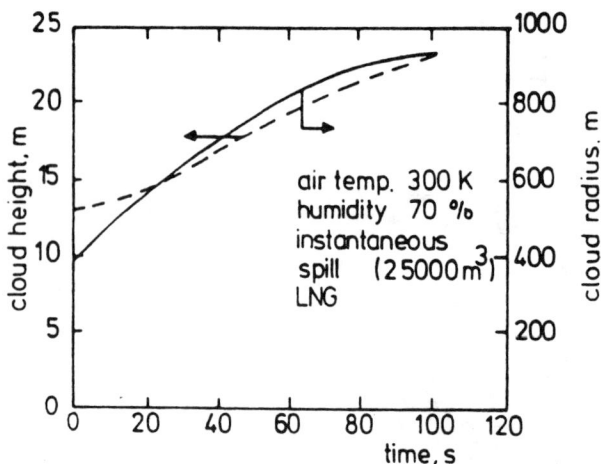

Fig. 2 Development of the cloud size with time for a 25.000 m^3 spill

dent on the initial spill size. This is depicted in Fig. 3. The predictions show the dependence of the cloud radius and hence its heigt on the initial spill size. Furthermore, the computations show that the average concentration of the cloud at the end of gravity spread is almost independent of the spill size and assumes values of around 20%.

The downwind distance to the lower flammability limit (LFL) is one of the important parameters in risk analysis, for it separates the safe area from the endangered one.

The computed distance to the LFL as the function of the initial spill size is presented in Fig. 4a, where the results are compared with other predictions available in the literature [Anon. (1978)]. The calculations were performed for Pasquill class E weather condition with wind speeds of 2.24 m/s, which is believed to be the most adverse weather condition [Anon. (1978)]. The expected effect of the spill size on the cloud propagation is obvious.

Although instantaneous spills represent the worst hazard situation, there is a greater likelihood of encountering spills with finite rates. This is also the prevailing situation in experiments. The application of the dispersion model, as given by Eqs. (7) and (8) leads to huge discrepancies between predictions and measurements. This was observed by Ermak et al. (1982), who applied Eq. (8) to predict the maximum distance to the LFL of the Burros series of LNG spill experiments [Koop-

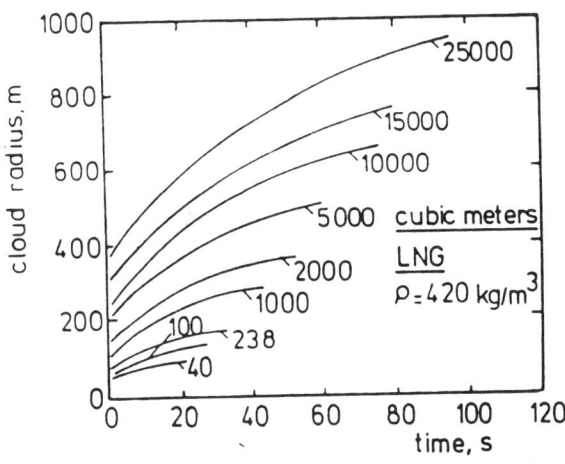

Fig. 3 Effect of spill size on cloud radius at the end of gravity-spread stage

mann et al. (1981)]. Not only do the discrepancies between predictions and measurements present a crippling drawback to the analysis, but one also finds that the steady-state solution gives a single value of the distance to the LFL that is independent of time, which contradicts experimental observations.

For this reason, the analytical solution to Eq. (5), the so-called Gaussian plume model, is considered inadequate for predicting this class of problems. However, extensive analysis of the performance of this model throughout the present investigation showed that, although the model itself is quite adequate, the discrepancies arise from applying it to the improper problem.

Common to most of the known investigations in the literature where the Gaussian plume model has been used to predict finite-rate, finite-duration spills, is the use of the steady-state formulation [Eq. (8)]; thus treating a transient problem as if it were steady. In the following, a new version of the Gaussian plume model is presented for finite-rate, finite-duration spills:

1) The spill period is subdivided into a number of intervals n, each lasting dt seconds. Each time interval is treated separately as an instantaneous release (i.e. a puff).

2) Thus, the local concentration $C(x,y,z,t)$ is given by the summation over all the time intervals of the analytical solution of Eq. (5), [i.e. Ep. (8)] so that

$$C(z,y,z,t) = \sum_{i=1}^{n} [2dN (2 \cdot \pi)^{-1.5} \tag{10}$$

$$/(s_x\ s_y\ s_z)]$$

$$\cdot \exp\{-0.5\ [(x - u\ [t - (i - 1)\ dt]$$

$$/(s_x)^2 + (y/s_y)^2$$

$$+ (z/s_z)^2]\}$$

where dN is the number of moles released within time interval dt.

DISPERSION OF DENSE GASEOUS FUELS 485

The proposed version is tested by applying it to predict the distance to the LFL as a function of time for the Burro series of LNG spill tests [(Koopman et al. (1981)]. Figure 4b compares the predictions with the measurements for the three spill tests: Burro 3, 8, and 9. Excellent agreement is obtained for the three tests that cover a wide range of spill rates and weather conditions. For comparison, the original Gaussian plume model would predict maximum distances to the lower flammability limit of 126, 661, and 255 m for Burro 3, 8, and 9, respectively [Ermak et al. (1981)].

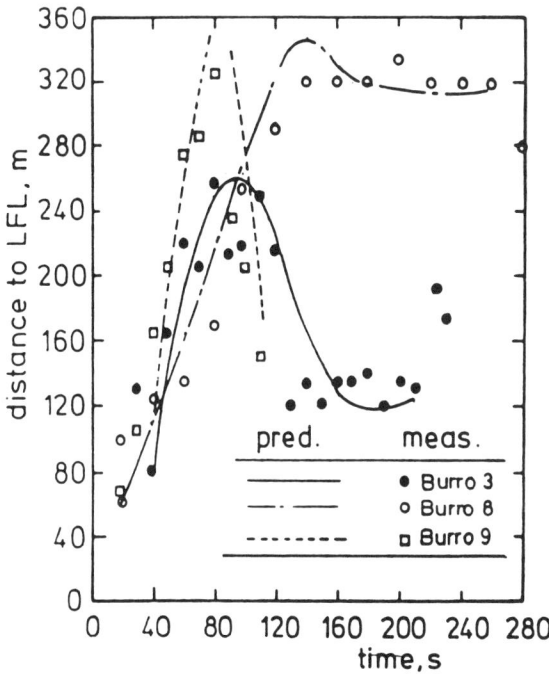

Fig. 4a) Prediction of the Burro Series of LNG spill experiments with the proposed extension of the Gaussian plume model

Conditions of experiments:

Burro 3: $V = 34$ m^3, $\dot{V} = 12.2$ m^3/min, $u = 5.7$ m/s
Burro 8: $V = 28.4$ m^3, $\dot{V} = 16.0$ m^3/min, $u = 1.8$ m/s
Burro 9: $V = 24.4$ m^3, $\dot{V} = 18.4$ m^3/min, $u = 5.7$ m/s

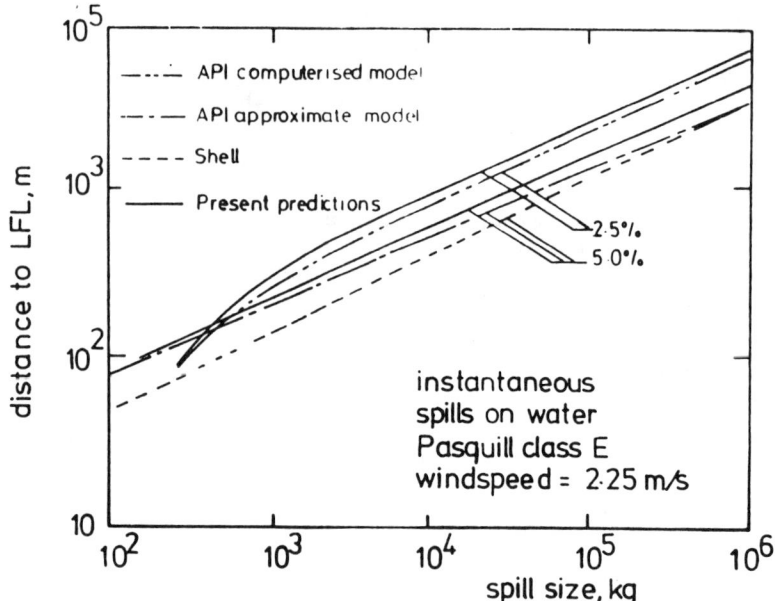

Fig. 4b) Comparison of present predictions of instantaneous spill onto water with predictions of API and Shell [Anonymous (1978)].

VI. Conclusions

A mathematical model has been presented for the analysis of the potential hazards of LNG tankers during the loaded voyage through the Suez Canal. The model predictions were successfully compared with available predictions in the literature for instantaneous spills onto water, and good agreement was found.

Furthermore, a new version of the Gaussian plume model was developed that ensures good accuracy of predicting the finite-rate, finite-duration spills of LGF. Comparisons of the predictions of the developed model with the well-documented Burro series of LNG spill experiments [Koopmann et al. (1981)] showed excellent agreement.

Further, analysis for the calculation of radiation effects from the FAC, and the quantification of the expected loss of life and property is currently under way.

Acknowledgment

The authors gratefully acknowledge the financial support offered by the Foreign Relations Coordination Unit of the Supreme Council of Egyptian Universities through Grant No. 81015.

References

Abou-Arab, T. W., Enayat, M. M., and Kamel, M. M. (1985) FRCU Project MS/842087, Rept. 2, Cairo University, Egypt.

Anonymous (1978) 'Canvey: an investigation of potential hazards from operation in the Canvay Island/Thurrock area'. Her Majesty's Stationery Office, London, UK.

Anonymous (1976) Federal Power Commission, USA.

Briscoe, F., and Shaw, P. (1980) Progress in Energy and Combustion Science, Vol. 6, pp. 127-140.

Chan, S. T., Rodean, H. C., and Ermak, D. L. (1981) Report UCRL-87256, Lawrence Livermore National Laboratory, CA., USA.

Cox, R. A. (1980) Progress in Energy and Combustion Science, Vol. 6, pp. 141-149.

Ermak, D. L., Chan, S. T., Morgan, D. L., and Morris, L. K. (1982) J. Hazardous Materials, 6, 129-160.

Havens, J. A. (1977) Federal Power Commission (1976) Report AD-525, University of Arkansas.

Khalil, A. K. and Kamel, M. M. (1985). Project FRCU 81015, 11th Progress Rept., Cairo University, Egypt.

Koopman, R. P., et al. (1981). Report UCRL-86704, Lawrence Livermore National Laboratory, CA., USA.

Lees, F. P. (1980) Loss Prevention in Process Industries, Vol. 1, Butterworth.
National Bureau of Standards.
Report No. NMAB-354, USA.

Raj, P. P. C. and Kalelkar, A. S. (1974). U.S. Coast Guard Report CG-446-3.

Science Application (1975), Rept. SAI-75-614-LJ., USA (1980).

Experimental Investigations into the Deflagration of Flat, Premixed Hydrocarbon/Air Gas Clouds

H. Pförtner* and H. Schneider†

Fraunhofer-Institut für Chemische Technologie (ICT), Pfinztal-Berghausen, Federal Republic of Germany

Abstract

In a test program consisting of a series of 17 tests, experimental studies of the combustion of "pancake-shaped" free clouds of near-stoichiometric methane/ethane/air and propane/air mixtures of a volume of up to 13,000 m³ under different test conditions were carried out. The parameters were the geometry and size of the cloud, geometry of the source of ignition (are, line, and point ignition), geometry and arrangement of obstacles (cubic and/or cylindrical), and generation of different turbulence fields. The flame propagation and the pressure/time course of the pressure waves produced by the flame were measured. In undisturbed clouds the maximum flame speeds did not exceed 6-8 m/s and the maximum overpressures were 1-2 mbar. The results indicate a slight increase in flame speed as a function of the distance traveled by the flame front due to flame-induced turbulence. Additional turbulence and obstacles increase the flame speed significantly. For a rms turbulent velocity of u´=0,56 m/s (20% turbulence level), typical flame speeds of up to 20 m/s and corresponding overpressures of about 5 mbar were measured. At obstacles, gaps, and entries and exits of lanes, higher turbulence levels were generated, resulting in local increases in flame speeds and overpressures of up to 80 m/s and 80 mbar, respectively. The peak overpressures were found to be approximately proportional to the square of the flame speed.

Copyright © 1988 by the American Institute of Aeronautics and Astronautics, Inc. All rights reserved.
* Project Manager, Safety Technology Section.
† Senior Project Engineer, Thermochemistry Section.

DEFLAGRATION OF HYDROCARBON/AIR GAS CLOUDS

Introduction

A lot of experimental work has already been carried out to simulate combustion processes in unconfined clouds. A large amount of laboratory scale experiments has also been conducted to study in more detail certain aspects of the combustion process. Furthermore, theoretical studies have been made to interpret the experimental work and to try to scale it to the size of a real incident (Lind 1974; Pförtner 1975; Kuhl et al. 1973; Guirao et al. 1975; Strehlow 1975). The results of these studies have led to the belief that it is very improbable that the combustion of an unconfined cloud of liquid natural gas (LNG) vapor, for example, will produce damaging pressure waves. But all of the studies carried out so far have failed to explain the serious accidents known to have occurred in the past with fuels other than natural gas. It has been suggested that the flame speed increases continuously as the distance traveled by the flame increases and that the height of the combustible mixture has an influence on the deflagration behavior and on its overpressure. From these suggestions, it can be concluded that high flame speeds and high overpressures cannot be excluded in the case of large, flat clouds.

Therefore, further experiments have been carried out to obtain a better knowledge of the complex combustion process of large, flat clouds. The objectives of this work were to investigate the relationship between overpressure and flame speed and the effect of turbulence in the unburned mixture in front of the flame on flame speed and overpressure.

Experimental Program

The program consisted of a series of 17 tests involving different cloud geometries: 1) Flat clouds of rectantgular shape (length 40 m, width 40 and 20 m, height 8 m) and 2) Flat clouds of cylindrical shape (diameter 40 m, height 8 m).

The gas/air mixtures were produced inside enclosures made of thin polyethylene sheets. The tests were conducted with homogeneous near-stoichiometric methane/air mixtures, methane/ethane/air mixtures (90% methane, 10% ethane), oxygen-enriched methane/air mixtures, and propane/air mixtures.

The mixtures were ignited by weak ignition sources of different geometries to produce a plane, cylindrical, or hemispherical flame: 1) Area ignition (100 pyrotechnical fuses, 75 J each). 2) Vertical line ignition along the centerline of the enclosure. 3) Point ignition on the ground in the center or at a side of the enclosure (exploding wire with an ignition energy of 640 J).

To produce turbulence, obstacles were used with a rectangular shape measuring 5 x 2 x 2.5 m and/or cylindrical shape of a diameter of 1.25 m and a height of 2.5 m. Additional turbulence was generated by one to four fans with a capacity of up to 72 000 m³/h, each producing a vertical or horizontal flow the velocity and turbulence of which was measured by the hot-wire technique.

Flame propagation was recorded by four top- or side-view high-speed cameras. The pressure wave was measured by 12 transducers connected with a microcomputer-controlled multichannel transient recorder. The pressure gages were

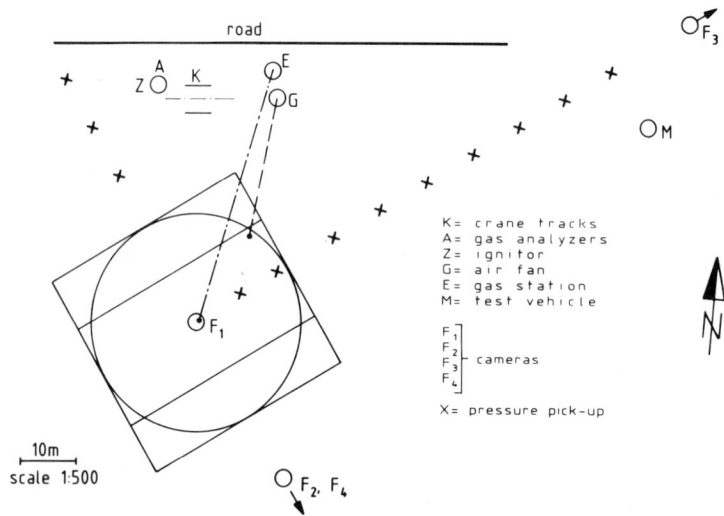

Fig. 1 Test arrangement.

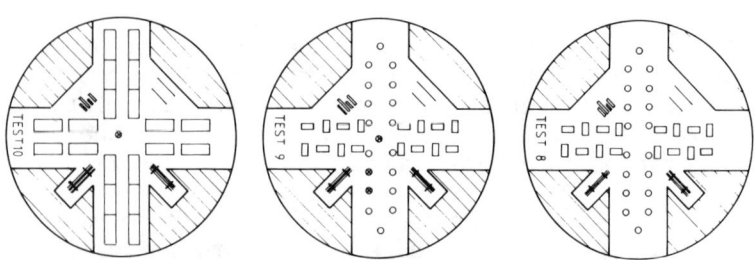

Fig. 2 Arrangement of obstacles.

DEFLAGRATION OF HYDROCARBON/AIR GAS CLOUDS

flushmounted in the ground inside and outside the enclosure.

Figure 1 shows the test arrangement and Fig. 2 the arrangement of the cylindrical and the cube-shaped obstacles and the fan.

Results

As already mentioned, the test program consisted of 17 tests carried out under different conditions. In Table 1 the significant results are summarized for point ignition in the center of the enclosure, in Table 2 for point ignition at the side of the enclosure, in Table 3 for line ignition, and in Table 4 for area ignition at the side of the enclosure without and with obstacles and/or additional turbulence produced by one or several fans at different fan speeds and flow directions.

Flame propagation in a flat cloud following point ignition on the ground was initially hemispherical with slight flame acceleration. A constant increase in overpressure during the hemispherical propagation of the flame was observed. The overpressure peaked when the flame reached the upper cloud boundary. Hemispherical propagation ended at this time.

In cases where the flame speed was low, the combustion of the remaining uncombusted mixture resulted in a second overpressure peak comparable to the maximum overpressure measured during hemispherical propagation because of the confining effect of the enclosure used in the experimental setup. In spite of the fact that it is not possible to quantify the confining effect of the enclosure, in an unconfined cloud, this second peak is likely to be lower due to the cylindrical geometry of the residual combustion. The overpressure in the region of hemispherical flame propagation was nearly the same at all measurement points. Outside this region, the overpressure decayed at a rate inversely proportional to the distance from the point of ignition as predicted by existing acoustic models.

In the case of line ignition on the centerline of the enclosure, flame propagation was initially cylindrical. The flame speed was lower than in the case of spherical flame propagation. Because ignition occurred up to the top of the enclosure, the combustion products were forced away from the top of the cloud and, as a result, peak overpressures were lower than in the point ignition experiments.

If more than one combustion center is present, as in the case of area ignition simulated by 100 equally distributed ignition points at one side of the enclosure, the pressure waves generated by these centers interact. Since

Table 1   Test results for point ignition in the center of the enclosure

Test No	1	2	4	8
Shape (m)	40×20×8	Φ40×8	Φ40×8	Φ40×8
Volume (m^3)	6,400	10,053	10,053	10,053
Gas Composition (Vol%)	11.3 CH_4 88.7 air	8.86 CH_4 0.98 C_2H_6 90.16 air	8.18 CH_4 0.93 C_2H_6 90.89 air	9.2 CH_4 1.0 C_2H_6 89.8 air
Ignition Energy (J)	640	640	640	715
Obstacles	no	no	no	cylindrical
Turbulence	no	no	no	no
Flame Speed (near the ground) $S_{F,max}$ (m/s)	7.8	11.8	10.5	10.5 10 11
Overpressure Δp_{max} (mbar)	1.25	2.7	2.4	1.4 2.6 2.1
Cause of pressure increase	hemispherical flame	hemispherical flame	hemispherical flame	cylindrical flame during burnout / hemispherical flame / interaction with obstacles

Table 1 (cont.)   Test results for point ignition in the center of the enclosure

Test No	10				17	
Shape (m)	Φ40×8				Φ40×8	
Volume (m^3)	10,053				10,053	
Gas Composition (Vol%)	8.11 CH_4 0.85 C_2H_6 91.04 air				8.29 CH_4 0.91 C_2H_6 90.80 air	
Ignition Energy (J)	715				715	
Obstacles	cube-shaped				cube-shaped + 1 cylindrical	
Turbulence	1 fan, speed 2 vertical downward flow				4 fans, speed 3 horizontal flow	
Flame Speed (near the ground) $S_{F,max}$ (m/s)	53	20	20	50	56	100
Overpressure Δp_{max} (mbar)	25.6	8.7	3.6	20.9	53.6	78.5
Cause of pressure increase	flame due to fans	hemispherical flame	cylindrical flame during burnout	flame through the lane between obstacles	flame due to fans	flame through the gaps between obstacles

DEFLAGRATION OF HYDROCARBON/AIR GAS CLOUDS

Table 1 (cont.) Test results for point ignition in the center of the enclosure

Test No.	7		9	
Shape (m)	$\phi 40 \times 8$		$\phi 40 \times 8$	
Volume (m³)	10,053		10,053	
Gas Composition (Vol%)	8.41 CH_4 0.93 C_2H_6 90.66 air		8.57 CH_4 0.96 C_2H_6 90.47 air	
Ignition Energy (J)	715		715	
Obstacles	no		cylindrical	
Turbulence	1 fan, speed 2 vertical downward flow		1 fan, speed 2 vertical downward flow	
Flame Speed (near the ground) $S_{F,max}$ (m/s)	20	45	22	60
Overpressure Δp_{max} (mbar)	6.9	18.5	7.7	(30)
Cause of pressure increase	hemispherical flame	flame due to fan	hemispherical flame	flame due to fan

Table 1 (cont.) Test results for point ignition in the center of the enclosure

Test No.	11		15	
Shape (m)	$\phi 40 \times 8$		$\phi 40 \times 8$	
Volume (m³)	10,053		10,053	
Gas Composition (Vol%)	16.80 CH_4 30.33 O_2 52.87 N_2		3.11 C_3H_8 1.26 C_3H_6 0.15 C_xH_y 95.48 air	
Ignition Energy (J)	715		715	
Obstacles	no		cube-shaped	
Turbulence	no		4 fans, speed 2 horizontal flow	
Flame Speed (near the ground) $S_{F,max}$ (m/s)	86	68	100	20
Overpressure Δp_{max} (mbar)	~100	76.7	72	8.0
Cause of pressure increase	hemispherical flame	hemispherical flame	flame due to fans and lanes	hemispherical flame until burnout at the upper edge

Table 2 Test results for point
ignition at the side of the enclosure

Test No.	3	13
Shape (m)	40×20×8	40×20×8
Volume (m^3)	6,400	6,400
Gas Composition (Vol%)	7.22 CH$_4$ 0.81 C$_2$H$_6$ 91.97 air	8.50 CH$_4$ 0.93 C$_2$H$_6$ 90.57 air
Ignition Energy (J)	640	715
Obstacles	no	cube-shaped+ 2 cylindrical
Turbulence	no	4 fans, speed 2 horizontal flow
Flame Speed (near the ground) $S_{F,max}$ (m/s)		55
Overpressure Δp_{max} (mbar)	0.45	29.7
Cause of pressure increase	hemispherical flame	flame due to fans and lanes

Table 3 Test results for line ignition

Test No.	5		16	
Shape (m)	Φ40×8		Φ40×8	
Volume (m^3)	10,053		10,053	
Gas Composition (Vol%)	7.48 CH$_4$ 0.90 C$_2$H$_6$ 91.62 air		8.07 CH$_4$ 0.86 C$_2$H$_6$ 91.07 air	
Ignition Energy (J)	2290 Vertical center line		1500 4×near the periphery	
Obstacles	no		cube-shaped	
Turbulence	no		4 fans, speed 2 horizontal flow	
Flame Speed (near the ground) $S_{F,max}$ (m/s)	6.7	7.0	49	39
Overpressure Δp_{max} (mbar)	0.95	0.63	22.7	24.1
Cause of pressure increase	hemispherical flame(?)	cylindrical flame	flame due to fans	hemispherical flame

the combustion products can expand upward, to the sides, and backward, pressure relief occurs. The overpressures measured were lowest under these conditions, which were experimentally realized by a plane flame front.

In the case of undisturbed flame propagation, the maximum flame speeds were 6-8 m/s, corresponding to maximum

DEFLAGRATION OF HYDROCARBON/AIR GAS CLOUDS   495

Table 4   Test results for area ignition
at the side of the enclosure

Test No	6	12	14
Shape (m)	40×40×8	40×40×8	40×40×8
Volume (m³)	12,800	12,800	12,800
Gas Composition (Vol%)	8.51 CH_4 0.83 C_2H_6 90.66 air	7.00 CH_4 0.82 C_2H_6 92.18 air	8.37 CH_4 0.87 C_2H_6 90.76 air
Ignition Energy (J)	7500	7500	7500
Obstacles	no	2 cylindrical	cube-shaped+ 2 cylindrical
Turbulence	no	4 fans, speed1 horizontal flow	4 fans, speed2 horizontal flow
Flame Speed (near the ground) $S_{F,max}$ (m/s)	27	22 14	55
Overpressure Δp_{max} (mbar)	7.9	5.5 2.7	30
Cause of pressure increase	almost plane flame front	flame due to fans	almost plane flame front	flame through the lane between obstacles

overpressures of about 1-2 mbar. Oxygen enrichment increased the flame speed considerably. In a methane/oxygen/nitrogen mixture with 16.8% methane, 30.3% oxygen, and 52.9% nitrogen, the maximum flame speed was about 86 m/s with an maximum overpressure of about 100 mbar.

The results of the whole experimental program indicate a slight increase in flame speed as a function of the distance traveled by the flame front because of the flame-induced turbulence. It should be added that theoretical work and experimental data from other combustible mixtures suggest an upper limit (Wagner 1981; Strehlow 1980). In the case of a spherical flame, the increase of flame speed due to flame-induced turbulence will not exceed the value of the expansion factor, i.e., the ratio of densities of the unburned and burned mixture.

Obstacles and additional turbulence increased the flame speed. The influence of obstacles on the flame speed and on overpressure was relatively small, if no additional turbulence was present. The lowest increase in flame speed was produced by a few discrete cylindrical obstacles. Typical flame speeds of about 10 m/s were measured and the corresponding overpressures were 2-3 mbar.

The action of a fan generated a flowfield in which turbulence was highest near the ground. For this reason, a cylindrical deformation on the hemispherical geometry of the undisturbed flame occurred.

The resulting overpressures were nearly constant and in qualitative agreement with the data from undisturbed cylindrical or plane flame propagation. For a rms turbulent velocity of $u'=0.56$ m/s (20% turbulence level) without obstacles, flame speeds of up to 20 m/s and overpressures of 5 ± 2 mbar were observed.

An obstacle configuration involving lanes and gaps between obstacles and additional turbulence produced a flame speed that was two to three times higher than the speed of the flame during similar tests without obstacles. At obstacles, gaps, and entries or exits of lanes, higher turbulences are generated, resulting in local increases in flame speed.

These regions of higher energy release rates are the origins of short, sharp overpressures, the decrease of which was inversely proportional to the distance from the

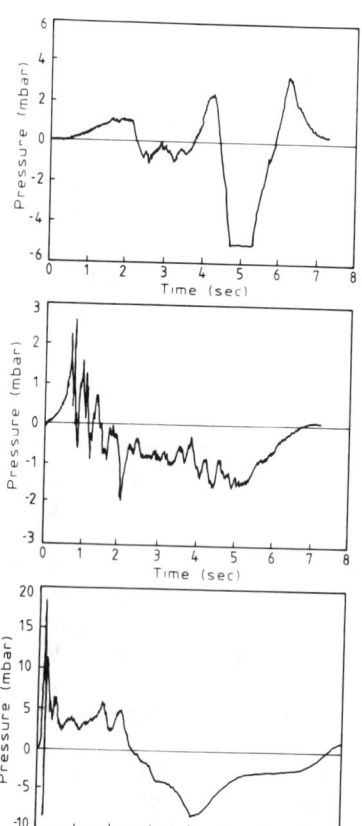

Fig. 3a Pressure signal without obstacles and additional turbulence.

Fig. 3b Pressure signal with obstacles only.

Fig. 3c Pressure signal with additional turbulence only.

DEFLAGRATION OF HYDROCARBON/AIR GAS CLOUDS

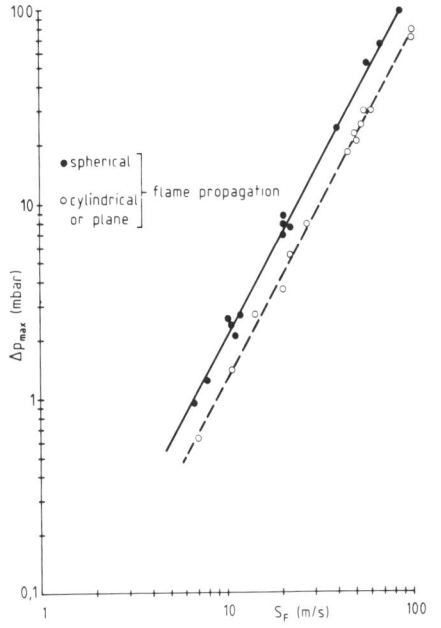

Fig. 4 Peak overpressure Δp_{max} as a function of flame speed S_F.

pressure source. A sequence of obstacles produced a corresponding sequence of sharp peaks, which are locally higher than the overpressures produced by the deflagration of the bulk of the cloud. Figure 3 compares the pressure signals for the unobstructed test (3a), for a test with obstacles (3b), and another test with additional turbulence (3c).

In Fig. 4, the peak overpressures are plotted as a function of flame speed. For hemispherical flame propagation, the overpressure is approximately proportional to the square of the flame speed, whereas, for cylindrical and plane flame propagation, the slope is the same at a lower pressure level. The straight lines were obtained by least squares. It is important not to extrapolate these data and the curves of the diagram to obtain flame speeds and overpressures not covered by the experiments.

This result is in agreement with theoretical predictions that for cylindrical and plane flames the overpressure is about half the pressure of a hemispherical flame when the flame speeds are the same (Strehlow 1980; Leyer 1982).

Conclusions

The work has shown that the flame speed reached during the deflagration of near-stoichiometric hydrocarbon/air mixtures consisting of 90% methane and 10% ethane are considerably below 20 m/s. Flame speeds are also lower than 20 m/s, even if major turbulence occurs in the uncombusted gas cloud. The maximum overpressures measured during the tests were between 1 and 25 mbar and are thus very low.

The presence of obstacles in the more or less unconfined gas cloud does not interact with combustion to produce significantly higher overpressures at any point, except for some points in the immediate vicinity of the obstacles.

The flat shape of the gas cloud terminating the phase of hemispherical flame propagation is a further reason for the low flame speeds and overpressures.

The experiments have shown that the ignition of stoichiometric mixtures of hydrocarbons such as LNG and air forming flat unconfined clouds due to their higher density relative to air initiates a deflagration process with overpressures that do not imply any risks to the neighborhood.

References

Guirao, C.M., Bach, G.G., and Lee, J.H. (1976) Pressure waves generated by spherical flames. Combust. Flame 27, 341.

Kuhl, A.L., Kamel, M.M., and Oppenheim, A.K. (1973) Pressure waves generated by steady flames. 14th Symposium (International) on Combustion, The Combustion Institute, Pittsburgh, p. 1201.

Leyer, J.C. (1982) An experimental study of pressure fields by exploding cylindrical clouds. Combust. Flame 48, 251.

Lind, C.D. (1974) Explosion Hazards associated with spills of large quantities of hazardous materials. U.S. Coast Guard Rept. CG-D-30-75.

Pförtner, H. (1975) Gas cloud explosions and resulting blast effects. Paper presented at the International Seminar ELCALAP in connection with the 3rd International Conference on Structural Mechanics in Reactor Technology, Sept. 8 - 11, 1975, Berlin, Germany.

Strehlow, R.A. (1975) Pressure waves generated by constant velocity flames - a simplified approach. Combust. Flame 24, 257.

Strehlow, R.A. (1980) The blast wave from deflagrative explosions - an acoustic approach. Paper presented at the 13th Loss Prevention Symposium, American Institute of Chemical Engineers, Philadelphia, June 8 - 12, 1980.

Wagner, H.Gg. (1981) Flammenbeschleunigung - Ein zentrales Problem bei der Entstehung von Explosionen. PTB - Mitt. 91, 248.

Analysis of a Damage Scenario and Potential Hazards of Liquefied Gaseous Fuel Carriers in Inland Waterways

O. M. F. Elbahar* and M. M. Kamel†
Cairo University, Giza, Egypt

Abstract

The passage of liquefied gaseous fuel carriers in inland waterways presents unprecedented hazards. Because of the uncertainty of events leading to and resulting from possible large-scale cargo spills, risk analysis becomes imperative. The present contribution deals with the analysis of potential hazards of giant liquefied gaseous fuel tankers during their voyage through inland waterways, with special emphasis on the Suez Canal. Special attention is focused here on hazards caused by component(s) failure of the liquefied gaseous fuel storage- and- handling system, rather than those resulting from external causes. The investigation presented here is carried out for a 125,000 m^3, fully refrigerated liquefied natural gas (LNG) tanker, typical of the modern fleet. A damage scenario in which the different events that could lead to a large-scale spill with possible consequences of that spill is proposed and discussed. Furthermore, a systematic analysis of the various elements of the fuel storage- and- handling system is made using the fault-tree analysis technique, which facilitates the quantitative determination of failure probabilities.

Introduction

Large-scale spills from giant tankers belong to a class of undesired events characterized by a low

---

Copyright © 1988 by the American Institute of Aeronautics and Astronautics, Inc. all rights reserved.
*Associate Professor, Mechanical Power Department, Faculty of Engineering.
†Professor, Mechanical Power Department, Faculty of Engineering.

probability of occurence and disastrous consequences. About 270 tankers carrying liquefied gaseous fuels pass through the Suez Canal annually (Kamel et al. 1984). Typical of the modern fleet are tankers that carry about 125,000 m^3 of cryogenic liquefied gaseous fuels.

Risk assessment is a multidisciplinary task that involves mainly tow parts. The first is the hazard probability analysis, and the second is the analysis of the consequences of such a hazard. This paper is concerned with only the former.

It is always desirable for each engineering system to operate safely. During operation, each system is subjected to a probable failure of one or more of its components or subsystems. Any dynamic change in the condition of the system or one of its components is termed an "event". If this change does not affect the ability of the system to perform the required task, it is assumed to be a normal event. On the other hand, if the system's ability is impaired, the event is considered to be a "fault". If, for a specific system, n events could contribute to its failure, then the "risk" is defined as

$$\text{risk} = \sum_{i=1}^{n} P_i \cdot \text{consequence of event i} \quad (1)$$

where P_i is the probability of occurrence of event i.

Among several methods for system safety analysis [Lambert 1973; van der Horst and van der Schaaf 1983; Vesely et al. 1981], the fault-tree analysis technique was chosen because of its versatility and capability of handling large, complicated systems with an acceptable degree of accuracy. It is a formalized, deductive procedure that facilitates the quantification of the probability of occurrence of an undesired event. This requires the analysis of all possible combinations of event sequences that could lead to the undesired event.

Damage Scenario

In attempting to identify the events that could lead to the spill of a considerable quantity of liquefied gaseous fuel, it is found that there is a multitude of interacting events with a large number of varying parameters. Although a large, catastrophic spill in an inland waterway is yet to occur, it is possible

to identify the potential sources of a hazard resulting
in a large-scale spill. This is accomplished by thoroughly
investigating the endangered system (here the liquefied
gaseous fuel carrier) together with the operating,
topographical and meteorological conditions pertinent
to the hazard site. These investigations and analyses
are then cast in a so-called 'damage scenario', which
includes the possible combinations of events that could
lead to the top undesired event and the circumstances
under which these conditions may take place. Thus,
a finite probability of occurrence for the top undesired
event can be estimated.

It can be assumed that the first condition for the
occurence of a large-scale hazard is that the liquefied
gaseous fuel carrier gets involved in a damaging incident
of some kind. In its very general definition, the "incident"
here means a change in the normal operating mode of
the liquefied gaseous fuel carrier. This could occur
either because of an external source, such as the collision
with another ship, or because of an internal event,
such as the failure of one or more of the components
of the subsystems of the ship. Figure 1a depicts the
part of the damage scenario concerned with the causes
of a liquefied gaseous fuel carrier accident in an
inland waterway and the information required for the risk
analysis and damage quantification. Detailed information
is required concerning the size of the liquefied gaseous
fuel carrier and the ship or any other obstacle that
was involved in the collision and the speeds and directions
of both vessels before the damage resulting from the
collision can be quantified. Furthermore, the navigational,
meteorological, and topographical conditions at the
accident site are of great value, for it is expected
that these parameters would have a great influence
on the hazard probability and on the consequences thereof.
Another parameter is the position of the ship in the
convoy, because this has an influence on further propagation of the hazard, since the damage caused by the
spill could extend to other ships in its neighborhood.

On the other hand, events that could lead to a spill
of considerable size involve failures of the fuel handling-
and- storage systems. Numerous events lead to such
a consequence, including the failure of the insulation
of the tanks and the rupture of the fuel storage tanks
due to fatigue stressing, etc (Fig. 1a).

If the accident results in a spill of considerable
size, many factors would affect the severity of the
consequences of the spill. Among these factors are

the quantity of the fuel spilled and the rate at which the release into the atmosphere takes place and the location of the spill relative to the water line. Another important factor is the ignition of the spilled cryogenic fuel. Therefore, there are three possibilities: the cryogen does not ignite at all, the cryogen iginites as a pool fire in the vicinity of the spill site, or the cryogen forms a pool of evaporating liquid topped by a cloud of expanding vapor (Figs. 1b-1e), which ends with the formation of a cloud of fuel vapor and

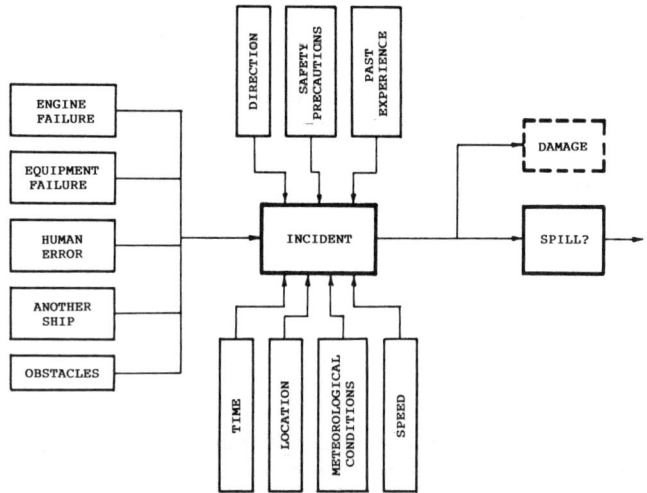

Fig. 1a Causes and consequences of a spill. Spill damage scenario.

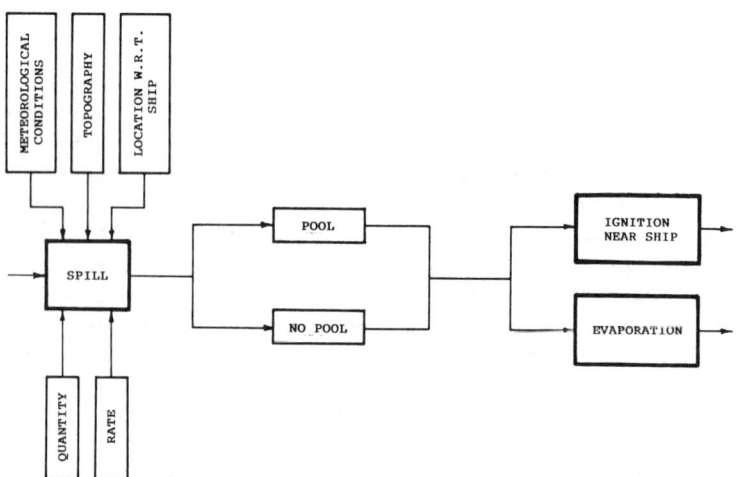

Fig. 1b Causes and consequences of a spill. Spill damage scenario.

ANALYSIS OF A DAMAGE SCENARIO

atmospheric air [called the fuel-air cloud (FAC) or the vapor cloud].
The ignition of the cryogenic fuel in the vicinity of the spill site (Fig. 1c) could be in the form of a pool fire. Intense radiation effects are to be expected, with possible effects on the population besides its effects on other ships in the convoy.

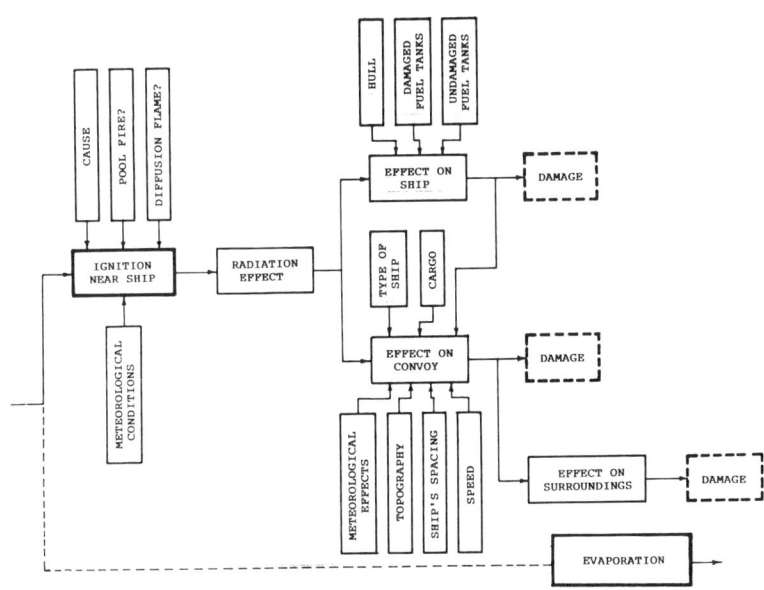

Fig. 1c Causes and consequences of a spill. Spill damage scenario.

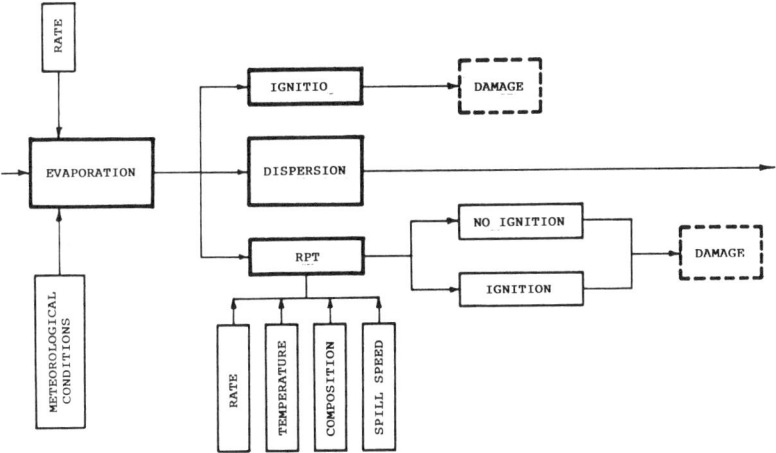

Fig. 1d Causes and consequences of a spill. Spill damage scenario.

On the other hand, the evaporation of the spilled cryogen in a boiling liquid/expanding vapor regime can result in the formation of a cloud of fuel vapor and atmospheric air [the FAC, which moves in wind direction until it ignites or gets diluted below its lower flammability limit (Figs. 1d and 1e)]. During the phase of violent evaporation of the cryogen, rapid-phase transition (RPT) may take place.

The mode with which the spilled cryogenic fuel ignites depends on many parameters, such as the presence of active ignition sources with sufficient power, the meteorological and topographical conditions determining the rate of spread and evaporation of the cryogen,

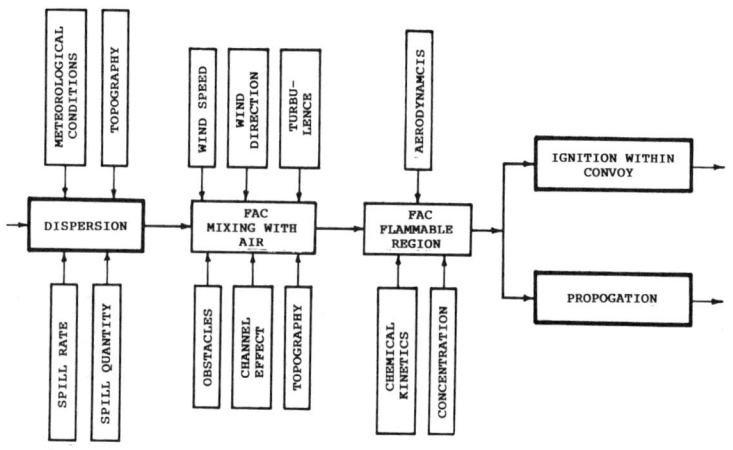

Fig. 1e Causes and consequences of a spill. Spill damage scenario.

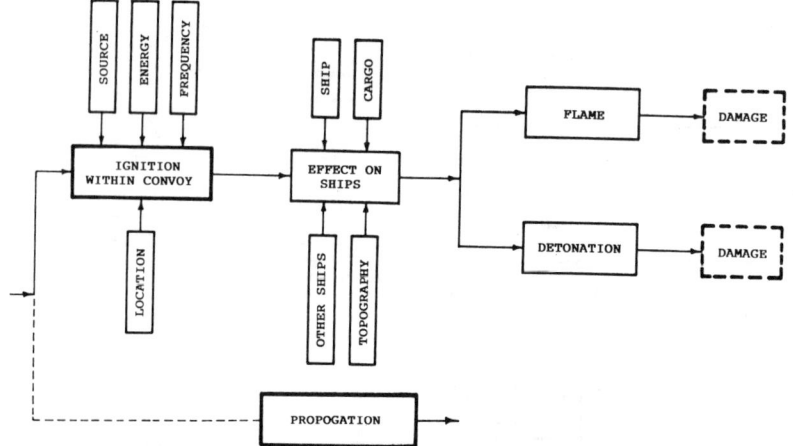

Fig. 1f Causes and consequences of a spill. Spill damage scenario.

ANALYSIS OF A DAMAGE SCENARIO

Fig. 1g Causes and consequences of a spill. Spill damage scenario.

Fig. 1h Causes and consequences of a spill. Spill damage scenario.

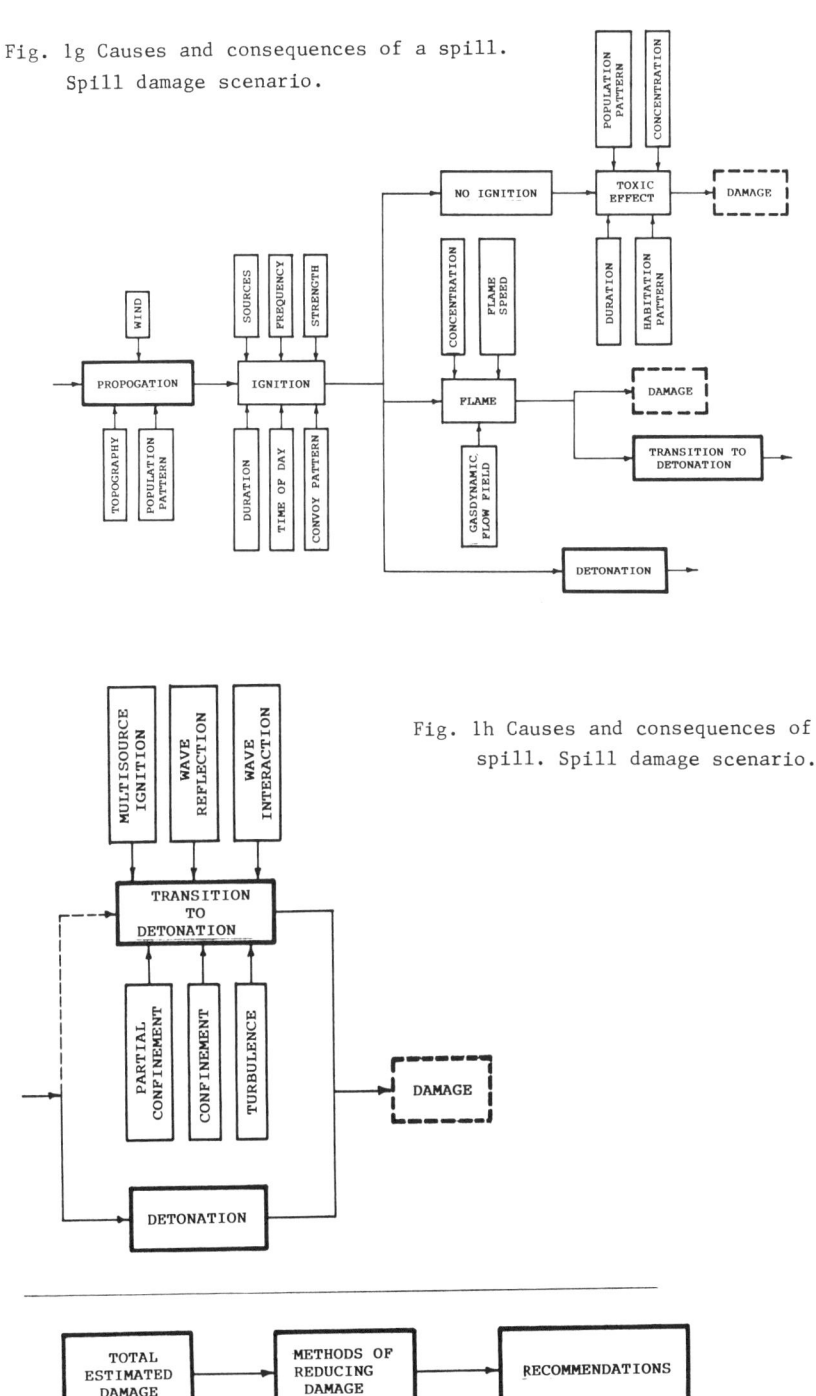

and the demographic circumstances (Figs. 1f-1h). Unfortunately, the Suez Canal zone is a densely inhabited area, where a large-scale fire could cause numerous fatalities. A separate analysis conducted by the authors showed that the ignition of the fuel-air cloud is almost certain if the wind drives it over an inhabited area.

If the spilled cryogenic fuel ignites in the form of a pool fire, intense radiation effects as well as suffocation effects for the crew of the ship are likely to occur. However, detonation is not expected to occur. On the other hand, if an FAC is formed, it would certainly ignite upon drifting over an inhabited area. The ignition here could be as a flash fire or a fireball. Detonation of the cloud requires the existence of an active ignition source with sufficient energy and power. On the other hand, the flame could be accelerated to high speeds as a result of confinement, interaction with other flame fronts, or existence of objects in the direction of its spread that could increase the turbulence level in the flowfield.

Another important effect is the suffocation of population as the cloud drifts over the inhabited area before its ignition.

Analysis of the navigational procedures in the Suez Canal reveals that the probility of collision between a liquefied gaseous fuel carrier and another ship may be classified as extremely remote, since liquefied gaseous fuel carriers do not stop during their voyage through the Suez Canal and are not allowed to enter the Canal under bad weather conditions. Furthermore, the probability of collision at the two approaches of the Canal (Port Said and Suez) is severely reduced by the stringent manuver regulations and by the definition of special paths for the ships moving in north and south directions. For this reason, this paper focuses on accidents caused by system component failures rather than by ship collision. However, the damage scenario holds for any other inland waterway.

The accident history of the past 15 years (van der Horst and van der Schaaf, 1984) shows that the confidence band for the accidents of LNG carriers is extremely wide ($0.00024 <$ incident rate < 0.023). Fortunately, however, no accident involving a liquefied gaseous fuel carrier has yet been recorded by the Suez Canal Authorities.

Thus, based on the previously mentioned damage scenario, a hazardous spillage of considerable amount of liquefied

ANALYSIS OF A DAMAGE SCENARIO

gaseous fuel (e.g., in the order of 10,000 m^3 or more) can be summarized in the following steps: 1) a cause of spill, either internal (e.g., component failure) or external (e.g., ship collision); 2) a release of a quantity of the cryogen into the atmosphere, either on land or into water, and 3) the evaporation, dispersion, and probable ignition of the flammable cloud.

In the following discussion a fault-tree analysis of the spills from liquefied gaseous fuel carriers is presented in which attention is focused on spills onto water from fully refrigerated LNG carriers.

Spill Probability Analysis of Liquefied Natural Gas (LNG) Carriers

The loaded-voyage mode is the relevant mode during the passage of tankers through inland waterways, since no loading or unloading take place during the voyage. The fuel storage- and handling system is depicted in Fig. 2. A detailed examination of this system by Elbahar and Kamel (1985) resulted in the identification of the following causes of a spill of considerable size: 1) failure of the tank structure; 2) pressure build-up inside the cargo tank, which is not designed to withstand high pressures, resulting from several events such as insulation failure, venting blockage, or the failure of a relief valve; and 3) failure of or leakage from the vapor-handling system that would lead to a spill of very limited size, since the density of the vapor phase of natural gas is two orders of magnitude less than that of LNG.

The failure-rate database was taken from the US Coast Guard (1976). To compute the failure probability p_f of a component from its failure rate, it is assumed that failures have a Poisson distribution in time and that the intervals between events have an exponential distribution. The relation between the p_f and the failure rate L is thus given by:

$$p_f = 1 - \exp(-Ldt) \tag{2a}$$

or, if L is small,

$$p_f = Ldt \tag{2b}$$

Elbahar and Kamel (1985) conducted a detailed analysis of the various subsystems that might affect the rupture

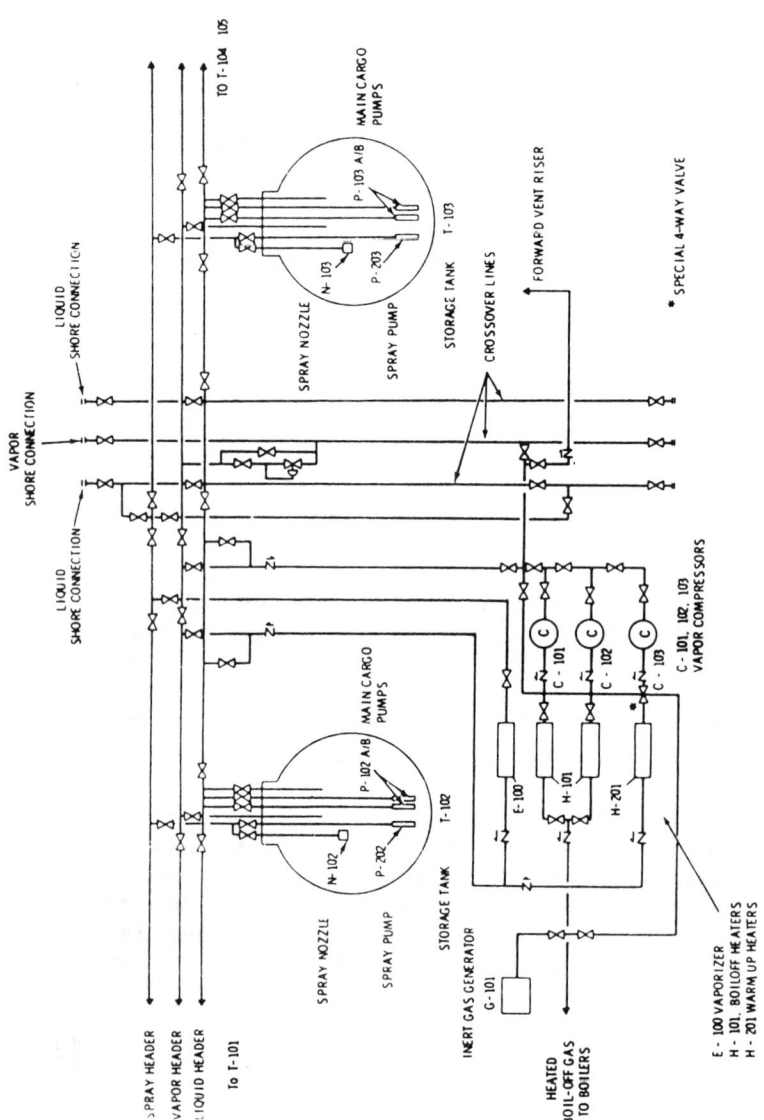

Fig. 2 Fuel storage-and-handling system of the LNG carrier.

of one of the five spherical cargo tanks, thus resulting in a top undesired event that is equivalent to the spill of 25,000 m³ of liquefied natural gas. They deduced that the probability that the boiloff would not be relieved is 0.0000024/h (the voyage of the liquefied gaseous fuel tanker through the Suez Canal takes about 24 h). Furthermore, the suppression of the relief of the boiloff does not represent a hazard source in the case of navigation through the Suez Canal, since the build-up of pressure inside the liquefied natural gas tank becomes dangerous after about 12 days of supressing

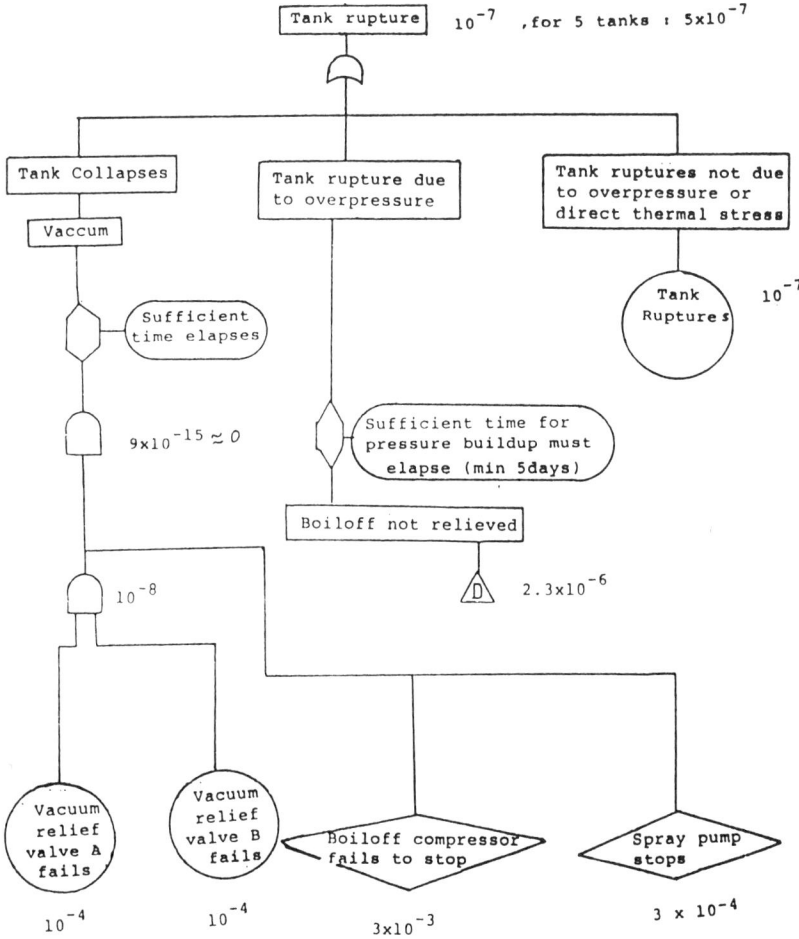

Fig. 3 Last stage of the fault tree with the top event. (rupture of LNG tank).

the relief of the boiloff (Elbahar and Kamel 1985), whereas the voyage through the Suez Canal takes only about 1 day. Thus, tank rupture due to fatigue and material defects remains the major cause of a large-scale spill hazard. Fig. 3 summarizes the last stage of the fault-tree analysis for the tank rupture. The results show that the probability of tank rupture is $5 \cdot 10^{-7}$/h for each of the five storage tanks of the ship.

Results and Conclusions

In this paper the relevance of application of the system safety analysis techniques to the analysis of the potential risks of explosion hazards of liquefied gaseous fuel transportation in inland waterways and their approaches has been emphasized. Among the available methods, the fault-tree analysis technique was chosen on the basis of it versatility.

The damage scenario proposed in this paper, together with the combinations of events presented that lead to an accident having serious consequences, throws light on the mechanism of accident occurrence and the information needed for the hazard probability analysis.

A fault-tree analysis with the top undesired event being the rupture of the LNG storage tank was conducted. Fig. 3 summarizes the final stage of the analysis. The probability of the tank rupture a computed from the analysis is 10^{-7}/h; i.e., $5 \cdot 10^{-7}$ for any of the five cargo tanks of the ship. A tank rupture due to the build-up of pressure inside the cargo tank as a result of supresison of venting the boiloff is shown to be highly improbable because of the short duration of the voyage through the Suez Canal, which generally takes about 24 h. These results show the importance of the systematic inspection of the cargo tanks.

Acknowledgment

The authors gratefully acknowlege the financial support offered by the Foreign Relations Coordination Unit of the Supreme Council of Egyptian Universities through Grant FRCU 81015.

References

Blything, K. W. and Edmonson, J. N. (1976)<u>I. Chem. E.</u>, Symposium Series CG-94-76, US Coast Guard.

Elbahar, O. M. F. and Kamel, M. M. (1985) Second Anual Report, edited by M. M. Kamel et al., Project FRCU 81015, Cairo University, Giza, Egypt.

Kamel, M. M. et al. (1984) First Annual Report, Project 81015, Cairo University, Giza, Egypt.

Lambert, H. E. (1973) Proc. SIAM, 77-100.

Pelto, P. J., Baker, E. G., Holter, G. M., and Powers, T. B. (1982) Battelle Memorial Institute, Rep. PNL-4014.

US Coast Guard (1976) Fire Safety Aboard LNG Vesels, Final Rep. CG-D-94-76.

Van der Horst, J. and van der Schaaf, J. (1983) I. Chem. E., Symposium Series 81, pp. B49-B110.

Vesely, V. E., Goldberg, F., Roberts, I. H., and Haasl, I. F. (1981) Fault Tree Handbook, US NUREG.

Influence of Obstacles on the Rate of Pressure Rise in Closed Vessel Explosions

G. E. Andrews* and P. Herath†
University of Leeds, Leeds, England, United Kingdom

Abstract

The influence of a four hole grid plate baffle on the rate of pressure rise in a 1.5 m long by 0.76 m diameter closed-tube explosion was determined for methane/air gas mixtures. The variation of the maximum rate of pressure rise with the baffle blockage and position relative to the spark at the base of the tube was investigated. It was shown that with no baffle the tube explosion was turbulent with a maximum flame speed much greater than that of a spherical laminar flame. The maximum rate of pressure rise increased as the blockage increased due to the higher turbulence levels generated in the explosion. Mixtures near the flammability limits had significant rates of pressure rise, which when normalized to the rate of pressure rise without the baffle were similar to the stoichiometric mixtures. This was considered to be due to the high turbulence levels generated in the explosions which were much higher than the laminar burning velocites.

Introduction

The rate of pressure rise in a closed vessel explosion is a function of the vessel shape and volume, the mixture composition, and any turbulence promoting blockages in the path of the flame travel [Harris (1983)]. It is this latter factor that is the objective of the present investigations and for which there has been relatively little previous

---

Copyright © 1988 by the American Institute of Aeronautics and Astronautics, Inc. All rights reserved.
*Senior Lecturer, Department of Fuel and Energy.
†Research Student, Department of Fuel and Energy (currently with British Gas, Midlands Research Station, Solihull).

work. In the design of vents for explosions, the creation of turbulence, due to a blockage to the flame, results in a larger vent area requirement for a specified maximum overpressure [Harris (1983); Lunn (1985)]. This is due to the influence of the turbulence, created by the blockage, on the flame acceleration and hence on the mass burn rate and rate of pressure rise. The increase in vent area is in direct proportion to the increase in flame velocity due to the blockage. In dust explosions, vent design is based on a measurement of the maximum rate of pressure rise in a closed vessel explosion [Lunn (1985)] and this was the procedure adopted in the present work for gaseous explosions. The influence of a blockage on the maximum rate of pressure rise was investigated and used to assess the flame velocity turbulence factor used in vent design procedures.

Most previous work on flame acceleration due to obstacles has been for flame acceleration in long tubes, generally open at one end, and the flame speed was the main parameter measured. Mason and Wheeler (1919) showed that a 90 m long tube of 0.3 m diameter open at both ends with two single hole restrictions produced a flame acceleration to near detonation. Chapman and Wheeler (1926) showed that methane/air flames rapidly accelerated to speeds of about 420 m/s when a sequence of restricting rings (orifice plates), spaced about one diameter apart along a 2.4 m tube of 5 cm diameter, were placed in the path of the flame. The corresponding flame speed in an obstacle-free, smooth walled tube was about 0.7m/s. Robinson and Wheeler (1933) repeated the experiments in a larger tube (30.5 cm diam. and 32.3 m long) and obtained similar flame speeds. Flame acceleration in very-rough-walled tubes was studied by Shelkin (1940), who showed that the transition distance from flame to detonation was drastically reduced when a spiral coiled helix was inserted into the tube.

More recently, Moen et al (1980, 1982), Hjertager (1983) and Knystautas et al (1984) have demonstrated that if the turbulence intensity is maintained by placing a series of obstacles in the path of a freely propagating flame, then the rate of combustion and the degree of turbulence become highly coupled so as to promote a strong feedback mechanism that in partially confined geometries, can lead to violent explosions. Meon et al. (1980) obtained values as high as 800m/s for methane/air in a 5cm diameter tube with a 44% blockage. In many situations, the multibaffle tube explosions had peak overpressures as high as the adiabatic closed vessel explosion pressure, even though the tubes were open at one end and in some situations the detonation over pressure was reached. There has also recently been interest in the influence of obstacles on partially confined duct explosions [Urtiew et al (1983), Taylor (1985,1987)] and fast flames were obtained for multibaffle situations. Harrison and Eyre (1987) have recently shown that multibaffles can strongly accelerate unconfined explosions.

Less work has been undertaken on the influence of single baffles on flame acceleration in tubes. Chapman and Wheeler (1926) investigated the influence of single orifice plates and showed that flame speeds up to 10 times those without a baffle could be achieved. They also showed that the influence of the baffle was a function of its distance from the spark. Evans et al. (1949) also observed significant flame acceleration when the flame propagated past single turbulence-inducing obstructions in the form of grid plates, the increase in flame speed being a factor of ten. Evans et al. also found considerable variablility of the influence of a grid plate on the flame speed. Schmidt et al. (1951) investigated the influence of a single baffle in a square duct and used high speed schlieren photography to show the very strong acceleration of the flame downstream of the baffle. They also demonstrated the sudden transition from a laminar to turbulent flame caused by the baffle.

For geometries other than tubes the influence of baffles has had less study. Dorge et al. (1976) demonstrated the strong influence of hemispherical wire mesh grid plates on the turbulent acceleration of flames in large volume explosions. A maximum flame speed enhancement of 6 was found, although the screens were of relatively low blockage and small turbulent scale and positioned quite close to the spark relative to the explosion vessel diameter. Kumar et al (1983) studied the effects of baffles for spherical explosion vessels and found a relatively small influence for lean hydrogen-air explosions using a 50% blockage and two flat grid plates across the sphere above and below the spark. They attributed the relatively small effect to the quenching effect of the grid plate. Stel´chuk et al. (1984) investigated the influence of a high blockage single hole baffle in a large volume vented explosion. They found a very strong influence of the baffle on the maximum explosion pressure, due to the turbulence created by the unburned gas pushed through the dividing wall hole when the ignition was on one side and well away from the opening in the baffle. Blockages of up to 90% will be studied in the present work and are relevant to this divided chamber or interconnecting vessel application which is an item of practical concern.

INFLUENCE OF OBSTACLES ON PRESSURE RISE

Few investigators have used closed-vessel explosions with a long tube configuration, which is the objective of the present work. Also, very few investigators have reported the influence of obstacles on the rate of pressure rise, which was a further objective of the present work. Kirkby and Wheeler (1931) used 10 orifice plate restrictions in the central portion of a 1.7 m long 10 cm diam. closed tube. Strong flame acceleration was found with an almost instantaneous pressure rise once the flame reached the baffles. The work of Schmidt et al. (1951) referred to above was carried out for both open and closed tube explosions with and without single baffles, however the baffle part of the investigations was a small part of a much larger investigation of accelerating flames in long tubes.

The influence of turbulence on closed vessel explosions has also been studied using fans to create the turbulence and one of the authors [Andrews (1972,1975)] has developed a four-fan cylindrical explosion vessel for such work. Similar work has also been carried out by Harris (1967), Kumar et al. (1983), and Al-Khishali et al. (1983). However, the application of this type of data to the influence of baffles on explosions is very difficult as the turbulence created by the baffle has first to be calculated, which requires a calculation of the gas velocity ahead of the moving flame in the presence of the baffle. Also, the turbulence created by a baffle has to be known if the fan-stirred turbulent explosion data is to be applied. It is thus much simpler and more relevant to the eventual applicaton of the data to determine directly the influence of blockage effects on the maximum rate of pressure rise. This is the objective of the present work.

Experimental Equipment

The long cylindrical tube, closed-vessel geometry is used, as it can be easily arranged to have different blockages at various distances from the spark. Also, most previous work on accelerated flames has used the long-tube geometry as discussed above. However, this previous work has generally involved the extreme situation of multiple baffles of fixed blockage with, in many cases, a full transition to detonation. In the present work, the much simpler geometry of a single blockage is investigated, because this is considered to be much closer to the practical situation where baffles accelerate flames, but rarely to detonation levels. A 76 mm diam. stainless steel closed tube, 1.5 m long, was used. This had a conventional automotive spark plug ignitor, with a 16 J ignition energy. This high ignition energy was used to ensure good ignition in near flammable limit mixtures. This ignitor was mounted in the center of the bottom flange and there were two further flanges at 0.5 and 1.0 m from the bottom. These flanges were used to mount the baffles so that the influence of the baffle position relative to the spark could be studied.

In the present work, a simple four hole flat-grid plate was used and the blockage was varied over the range 50-90%. The details of each grid plate is given in Table 1. The grid plates were identical to those used previously for stabilized premixed turbulent flames [Al-Dabbagh and Andrews (1984)] and the discharge coefficients Cd for the grid plates were known from this work. This work also showed that, in the 76 mm diameter tube, the four-hole grid plate was the minimum number of holes that gave an adequate combustion efficiency. Therefore, it was considered that the more uniform turbulence distribution and large length scales generated by the four-hole grid plate could have a stronger influence on flame acceleration than the single hole baffles that have been used in most previous work. A program investigating different grid plate geometries, from single to large numbers of holes, is under investigation by the authors.

The pressure transducer was a Druick PDCR 10/T and was mounted in the top flange of the rig, 1.5 m from the spark. The flame speed was determined using the time of travel between two exposed junction mineral insulated type K thermocouples. The validification of this technique will be discussed elsewhere, but it has been compared with high-speed photographic measurements on spherical flame explosions and shown to give excellent agreement. Unfortunately, insufficient data recording equipment was available to take flame speed measurements at different positions along the tube. Global flame speeds were also deduced from the pressure record. The explosion pressure

Table 1 Four-hole grid plate baffles

Blockage, (%)	Diam. (D,mm)	Thickness, (t,mm)	Cd
49.7	27.0	3.2	0.90
56.8	25.0	3.2	0.85
78.3	17.7	3.2	0.80
83.1	15.7	3.2	0.69
93.5	9.7	3.2	0.64

and thermocouple output records were recorded on storage oscilloscopes.

Explosions in Long Tubes without Baffles

Ellis (1928) was the first to investigate closed-vessel tube explosions with modest length-to-diameter ratios. He showed that, for a tube length to diameter ratio of the order of 10 (20 in the present work), there was an initial high velocity flame, occupying over half of the tube length, followed by a lower velocity distorted tulip flame in the final period of travel. This tulip type of flame propagation has been recently investigated using LDA measurements by Dunn-Rankine and Sawyer (1988) and Starke and Roth (1986). A typical flame shape history is reproduced in Fig.1 [Dunn-Rankine and Sawyer (1988)] which is identical to the multi-exposure photographs of Ellis (1928). Figure 1 has been analysed to give an initial flame speed of 6.7 m/s and a flame speed of 2.4 m/s in the end third of the tube. In the present work, the first baffle position 0.5 m from the spark would have a high approach flame speed corresponding to initial zone. The second baffle position 1 m from the spark would have a lower approach flame speed as it would be in the end zone. The gas velocity across the baffle creates the turbulence, which is a function of the flame speed. Thus, lower turbulence and hence less flame acceleration would be expected at the 1 m baffle position. The experimental results confirm this, as discussed later. The direct LDA gas velocity measurements of Starke and Roth (1986) show much higher velocities in the initial period, with a factor of four velocity difference between the

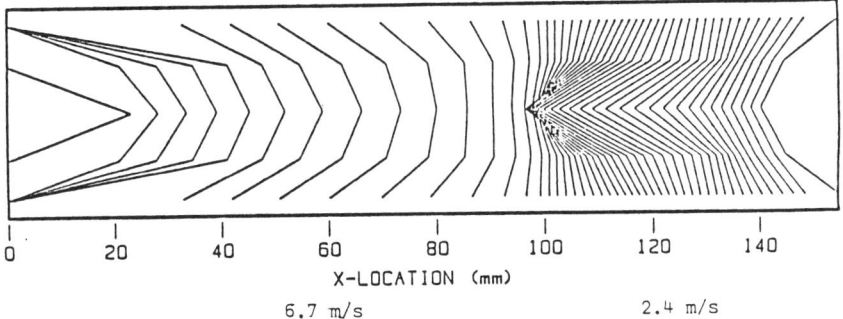

Fig.1 Flame shape history [from Dunn-Rankine and Sawyer (1986)]

initial and final periods for a stoichiometric ethylene air mixture.

In the present work, the rate of pressure rise with a baffle was compared to that without a baffle. If the flame propagation without a baffle was laminar, then this ratio of rates of pressure rise may be shown to be equivalent to a ratio of turbulent to laminar burning velocities. To access the flame situation in the absence of a baffle, flame speed measurements were made in the initial period of the tube propagation using the two thermocouple technique and compared with the equivalent flame speeds for a laminar spherical flame explosion. These were measured on a 0.5 m diam. by 0.5 m long cylindrical explosion vessel using the same thermocouple technique. The flame speed ratios for methane/air are shown in Fig.2 and clearly show that in the absence of a baffle the flame must be turbulent in the tube. Similar ratios have been measured for hydrogen/air and have a peak value of 6. Reynolds number calculations for the unburned gases ahead of the flame show that this will be turbulent and a function of the mixture composition, due to the variation in flame speed with the

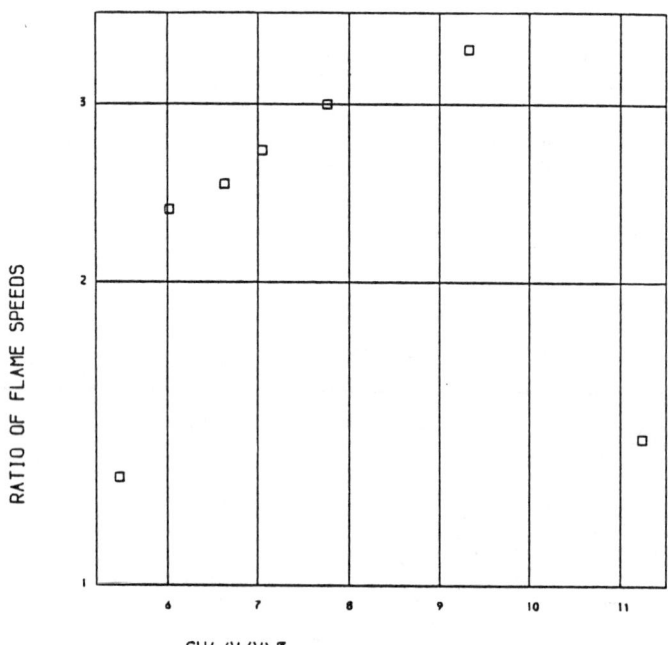

Fig.2 Ratio of the tube flame speed with no baffle to that of a spherical laminar flame speed, as a function of mixture composition.

mixture composition. Thus, the ratio for hydrogen is much greater than the lower speed methane results, which are in turn a maximum at stoichiometric and much closer to unity near the flammability limits. Thus, the present ratios of the rates of pressure rise for configurations with and without a baffle may be applicable only to such tube situations as large volume explosions, where laminar flames prior to the baffle may have a greater influence. Work is in progress to investigate spherical baffle effects on large vessel explosions.

Explosions in Long Tubes with a Four Hole Baffle

Typical Pressure Traces

A typical pressure record is shown in Fig.3. The strong increase in the rate of pressure rise due to the baffle is clearly demonstrated. The pressure traces were analyzed, as shown in Fig.4, to determine the peak pressure, the maximum rate of pressure rise, the proportion of the total pressure rise that occurred at this maximum rate, and the global flame speeds before and after the baffle. The sharp increase in the rate of pressure rise was taken as the flame arrival at the baffle and the subsequent time to peak pressure was used to compute the global flame speed after the baffle.

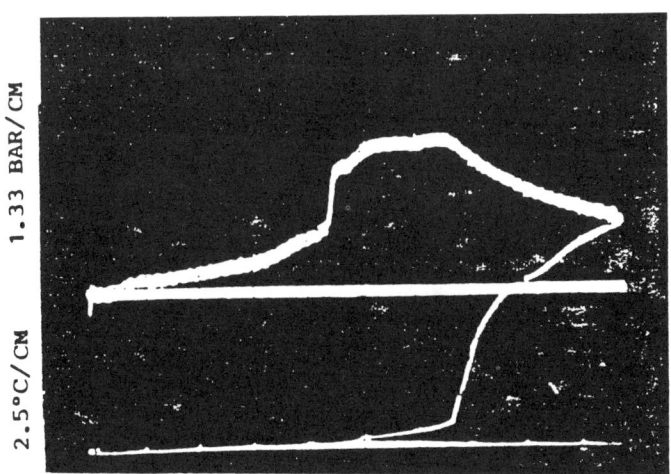

0.0625 sec/cm

Fig.3 Typical pressure trace and top thermocouple trace for a 10% methane/air explosion with a four hole baffle 0.5 m from the bottom of the tube and with a 78% blockage.

A thermocouple was mounted at the top of the tube to detect the flame arrival as shown in Fig.3. Comparison of the flame arrival time with the pressure trace shows that it was preceded by a period of slow pressure rise extending over an appreciable fraction of the total pressure rise time. It is considered that this was due to the decay of the turbulence created by the baffle, which results in flame deceleration some distance from the baffle. This feature was influenced by the baffle position. For a baffle close to the spark only a small fraction of

$$\frac{(P_i - P_1)}{(t_i - t_1)} = \text{Maximum rate of pressure rise, bar/sec.}$$

$$\frac{(P_i - P_1)(t_i - t_1)_{\text{WITH BAFFLES}}}{(P_i - P_1)(t_i - t_1)_{\text{NO BAFFLES}}} = \text{Normalised pressure rise rate, dimensionless.}$$

$(P_i - P_1)/(P_m - P_o)$ = Fractional pressure rise at maximum rate, dimensionless.

P_i/P_m = Fractional pressure rise, dimensionless.

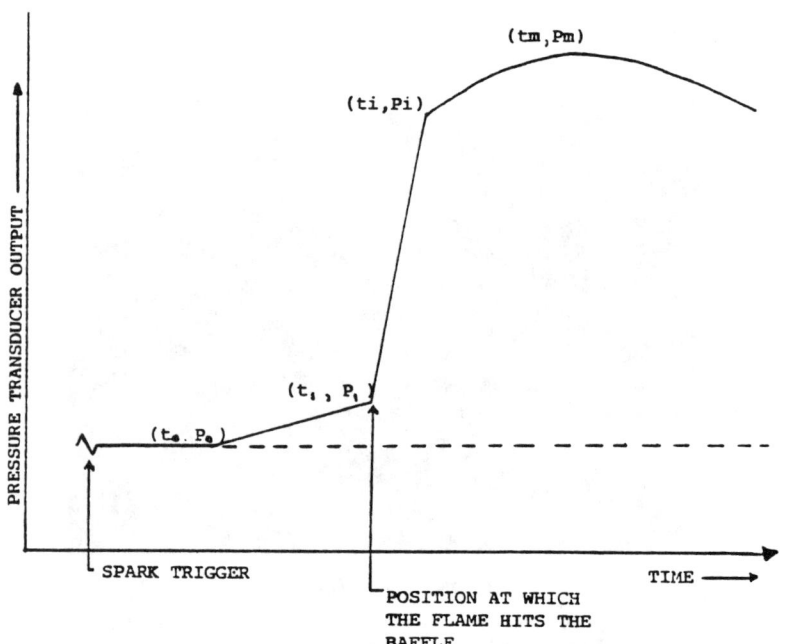

Fig.4 Measurements taken from the pressure rise traces.

the total mixture can be pushed through the baffle ahead of the flame. Thus, the turbulence generated may not affect the whole tube. Hence, the flame may initially accelerate and then decelerate, as shown in Fig.3 for the 0.5 m baffle position. However, for the 1 m baffle position, although the maximum rates of pressure rise was lower, it generally ended in the peak pressure with no flame deceleration period.

Peak Explosion Pressure

The influence of the four hole baffle on the peak pressure is shown in Figs. 5 and 6 for the two baffle positions and a range of blockages B, together with the no baffle situation. There was no major influence of the baffles on the peak pressure compared with the no-baffle situation; however, there were two small but significant influences. First, peak pressures were somewhat lower with baffles for near-stoichiometric mixtures and somewhat higher for mixtures near the flammability limits. For

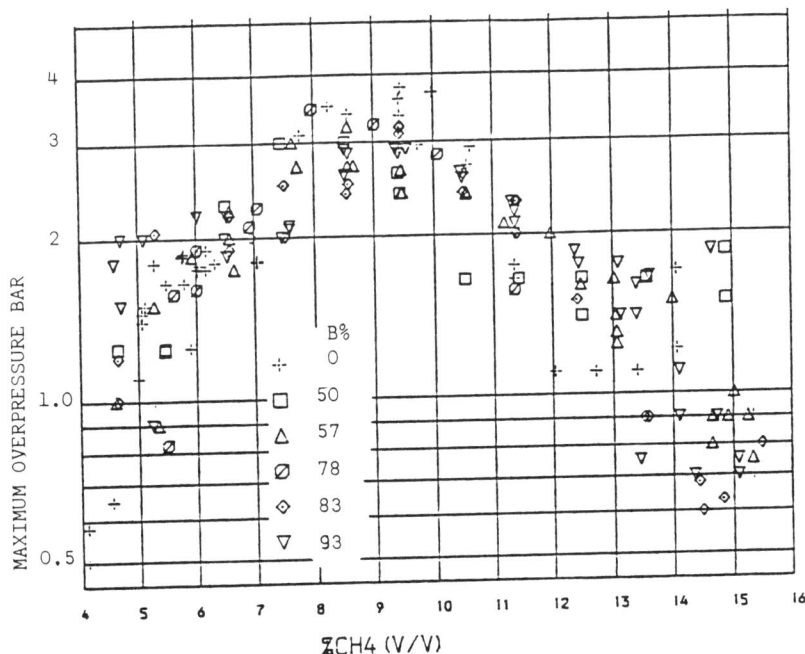

Fig.5 Maximum overpressure (bar) as a function of mixture composition for a range of baffle blockage 0.5 m from the spark.

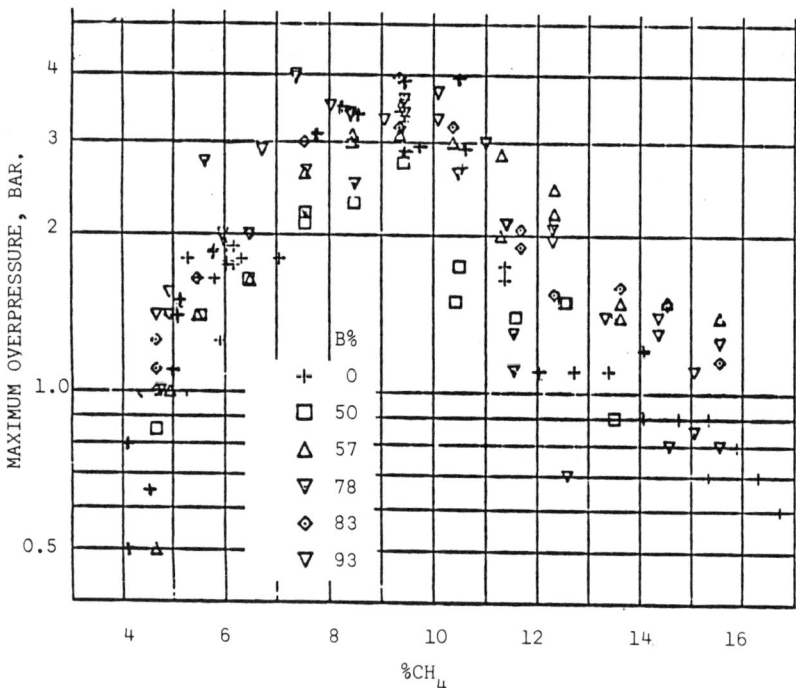

Fig.6 Maximum overpressure (bar) as a function of mixture composition for a range of baffle blockage 1.0 m from the spark.

near-stoichiometric mixtures, the flame acceleration will be high due to the higher levels of turbulence. This turbulence will increase the wall heat losses, which affects the peak pressure. For near-limit mixtures, the accelerated flames will have a greater combustion efficiency, giving a higher peak pressure. Also, the faster flames will have a shorter contact time with the walls and a lower heat loss. Second, the peak pressures with the baffle at the 1 m position are slightly higher than at the 0.5 m position. This was probably due to the lower time spent in the highly turbulent flame situation and hence lower heat losses.

Rates of Pressure Rise

The measured maximum rates of pressure rise are shown in Fig. 7 for the 1 m baffle position and as a ratio to the no-baffle situation in Figs. 8 and 9 for the 0.5 and 1m baffle positions, respectively. The accuracy of the rates

INFLUENCE OF OBSTACLES ON PRESSURE RISE

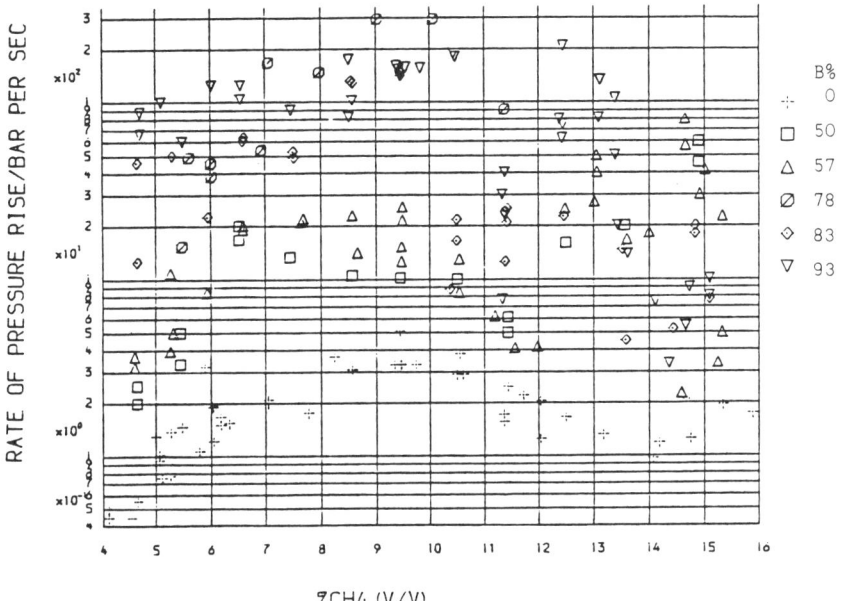

Fig.7 Rate of pressure rise as a function of mixture composition for a range of baffle blockage 0.5 m from the spark.

of pressure rise measurements was relatively poor at approximately 20% and decreased as the rate of pressure rise increased. This was due to the use of a simple storage oscilloscope for recording the pressure rise data. More recently digital data aquisition systems have been used to improve this accuracy. However, the variation from explosion to explosion was much greater than these measurement errors could account for and may be assumed to be a real effect.

There are four major features of these results:
1) For a given geometry there was considerable variation from one explosion to the next, which was greater with the baffles than without, as shown in Figs. 7 and 8. This is an inherent feature of highly turbulent explosions and is also a cause of cycle to cycle variations in SI engines. In an explosion the flame sees the instantaneous velocity fluctuations and not any time averaged turbulence. Al-Khishali et al. (1983) have investigated the variation in the turbulent burning velocity over 100 explosions in a fan-stirred turbulent explosion vessel. They showed that there was considerable variablity, with a ratio of maximum-to-minimum burning velocity of over 3, similar to

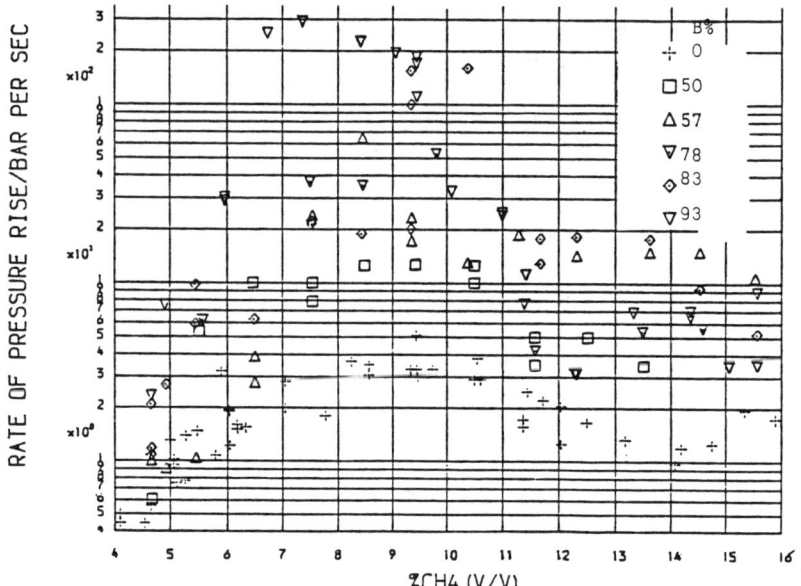

Fig.8 Rate of pressure rise as a function of mixture composition for a range of baffle blockage 1.0 m from the spark.

that of the present work for the rate of pressure rise which is related to the turbulent burning velocity. From a safety viewpoint, it is the maximum values that are of interest, not the mean of several explosions. For the lowest blockage of 50%, the maximum normalised rates of pressure rise in Fig.9 are in the range of 8-10, with the lowest values in the range of 2-4. This covers the whole spectrum of turbulent enhancement factors used in vent designs [Harris (1983)] and indicates that for safe design a tubulence factor of close to 10 may be required for quite modest flow blockages.

2) The rate of pressure rise shows a clear influence of the blockage B, with values as high as 100 times the rate with no baffle for very high blockage. This is due to the dependence of the turbulence generated by the baffle on the blockage. There would be serious problems in adequately venting such high blockage explosions. The 93% blockage corresponds to the interconnecting vessel situation with a small pipe, door, or any opening connecting two larger explosion vessels. The extemely high rates of pressure rise shown in Figs. 9 and 10 for 93% blockage clearly demonstrate the devestating consequences of this type of

INFLUENCE OF OBSTACLES ON PRESSURE RISE

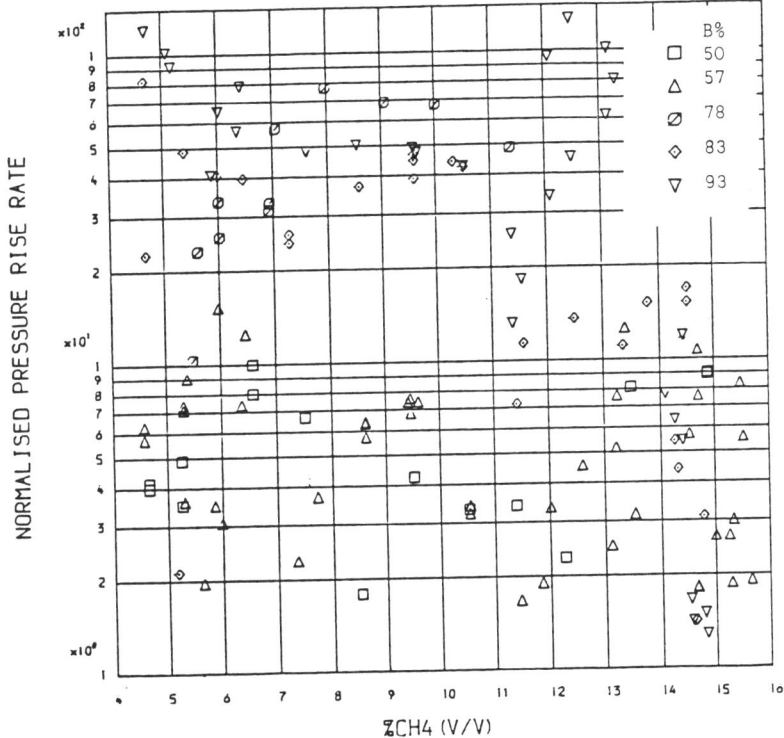

Fig.9 Ratio of the maximum rate of pressure rise with a baffle to that without, as a function of mixture composition for a range of baffle blockages 0.5 m from the spark.

explosion situation. Rates of pressure rise of 100 times the pressure rise with no baffle are impossible to vent.

The results of Andrews (1972) for the normalized rates of pressure rise in a fan-stirred explosion vessel compare reasonably well with the present results for a 10% methane/air mixture. For fan speeds of 500-5000rpm, the normalized rates of pressure rise increased from 2.8 to 19 with a linear relationship with fan speed. This is of the same order as the present results, although much higher peak normalised rates of pressure rise have been found at high blockage. The turbulent velocity over this fan speed range varied linearly 1-9 m/s [Abdel-Gayed et al. (1984)]. Using the flame speed to estimate the unburned gas velocity, the authors have estimated, using the techniques described by Al-Dabbagh and Andrews (1984), the turbulence created by the baffles to vary over the range 2-9 m/s as a function of the blockage. Consequently, it can be seen that

the influence of the baffles is to create intense turbulence, which increases with the baffle blockage.
3) Mixtures near the flammability limits have rates of pressure rises that are quite significant with the baffle. Figure 7 shows the near-limit rates of pressure rise with a baffle that are much higher than for a stoichiometric mixture with no blockage. Figures 9 and 10 show that the normalized rates of pressure rise were independent of the mixture composition. This indicates that the turbulent combustion was very intense and was in the region where the turbulent velocity u´, was much greater than the laminar burning velocity Su. In this turbulent combustion region where Su<<u´, the turbulent burning velocity is dominated by u´ and hence becomes relatively insensitive to Su and thus to the mixture composition. A similar situation has been found using these same baffles as flame stabilizers in

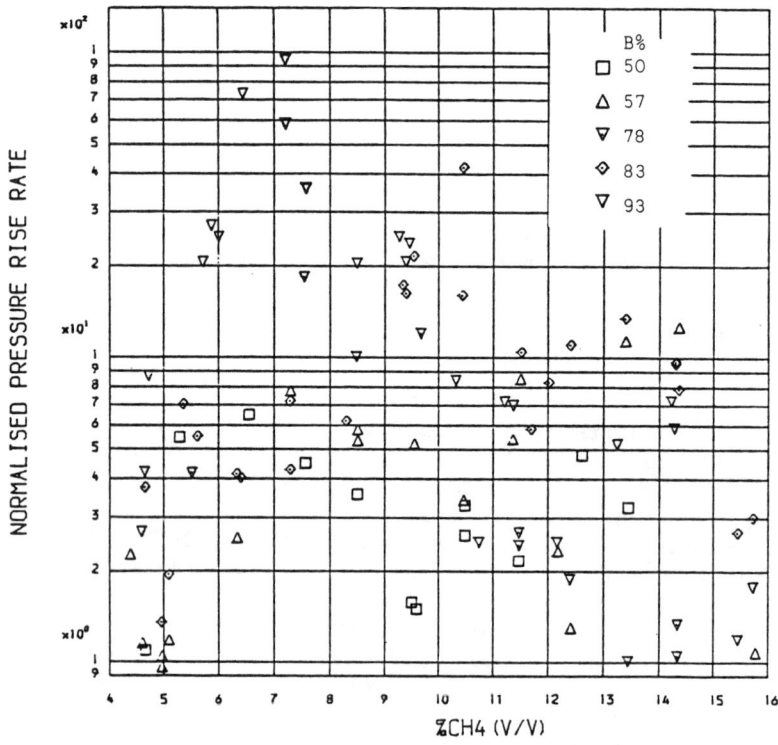

Fig.10 Ratio of the maximum rate of pressure rise with a baffle to that without, as a function of mixture composition for a range of baffle blockages 1.0 m from the spark.

a premixed flow situation [Al-Dabbagh and Andrews (1984)]. For near-limit mixtures, flames could be stabilized in very high velocity flows (25m/s) even though the Su value was of the order of 0.1 m/s.

4) Comparison between Figs. 9 and 10 shows that the initial 0.5 m baffle position had a greater influence on the rate of pressure rise than for the 1 m position. This was due to the slower flame speed prior to the more distant baffle, as discussed above.

Fraction of Maximum Pressure Rise.

The results for the proportion of the pressure rise that occured at the maximum rate are shown in Figs. 11 and 12 for the two baffle positions. There was a very wide data scatter, for the reasons discussed above in relation to the data scatter for the rates of pressure rise. However, it

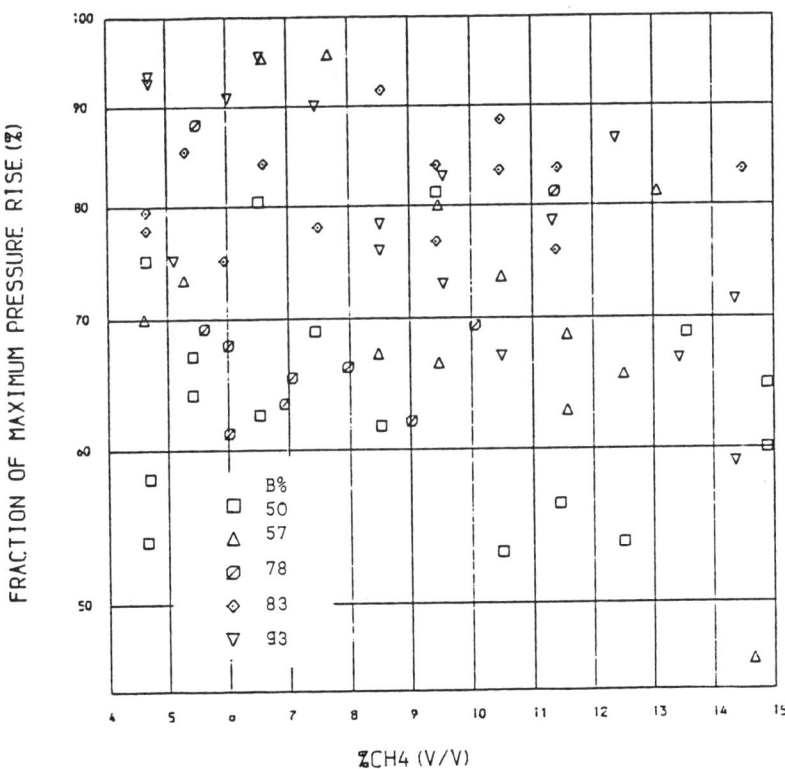

Fig.11 Fraction of the total pressure rise that occurred at the maximum rate as a function of the mixture composition for a range of baffle blockages 0.5 m from the spark.

can be seen that there was a trend for the lower blockages to have a lower proportion of the total pressure rise at the maximum rate. The higher blockages almost always gave a very sharp pressure rise to the peak pressure, with none of the peak pressure plateau shown in Fig.3. The reason for this was twofold: 1) due to turbulence decay at low blockages, as discussed above and 2) due to the higher gas velocities and hence turbulent Reynolds numbers ahead of the accelerating flame. This latter effect increased as the blockage increased until the whole of the mixture after the baffle was burned in a fast flame mode. Comparison of Figs.11 and 12 shows that there was a trend for the 1.0 m baffle position to have a greater fraction of the total pressure rise at the maximum rate, especially at low blockages. The reason for this was the greater proportion of the mixture that was made turbulent at the second baffle position.

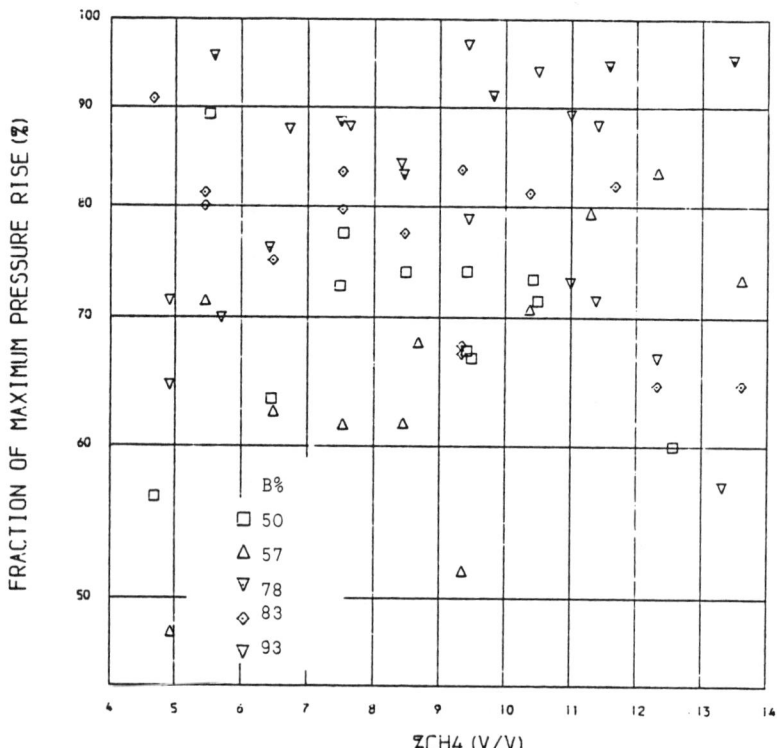

Fig.12 Fraction of the total pressure rise that occurred at the maximum rate as a function of the mixture composition for a range of baffle blockages 1.0 m from the spark.

Flame Speeds

The flame speeds were deduced from the pressure records as a simple global speed, given by the time to peak pressure divided by the 1.5m tube length. This was found to give reasonable agreement with the direct thermocouple measurements. Direct flame speed data were not obtained in most of the explosions, due to inadequate transient data recording facilities at the time. The results are shown in Figs. 13 and 14 for the two baffle positions. In general, the speeds are higher for a given blockage for the initial 0.5 m baffle position, due to the higher initial flame speed. At high blockage flame speeds in the 10-20 m/s range were found. These were well short of the 500 m/s typical of multibaffle tube exolosions, but still very significant as an explosion hazard. Limited direct thermocouple measurements showed that locally higher flame speeds were recorded inside the tube.

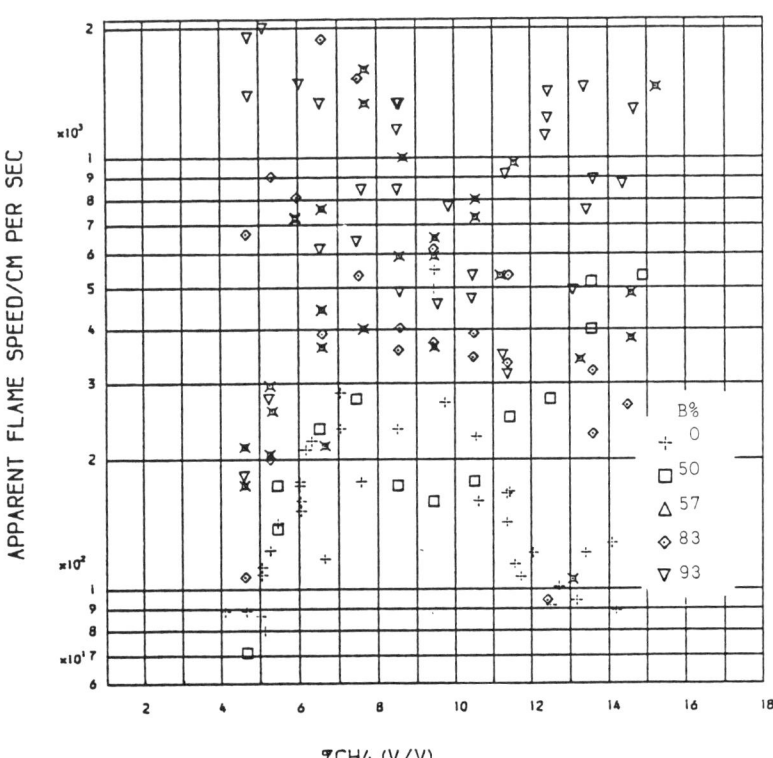

Fig.13 Apparent flame speed as a function of the mixture composition for a range of baffle blockages 0.5 m from the spark.

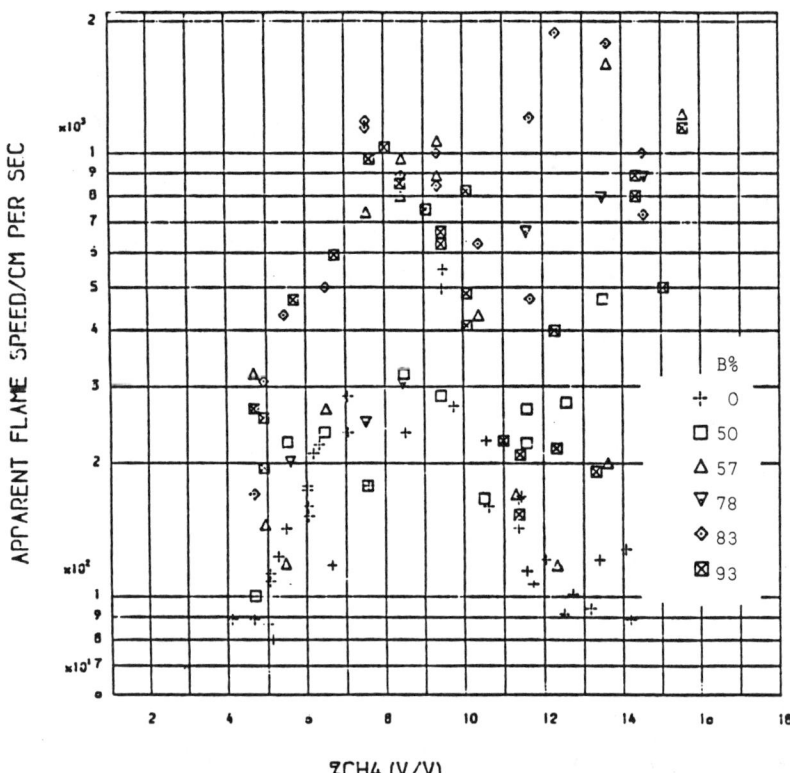

Fig.14 Apparent flame speed as a function of the mixture composition for a range of baffle blockages 1.0 m from the spark.

Acknowledgments

We would like to thank the UK Science and Engineering Research Council for a research grant (GR/D/81855) in support of this work and for a studentship to P. Herath. I. Langstaff installed and commissioned the explosions test facility.

References

Abdel-Gayed,R.G., Al-Khishali,K.J. and Bradley,D. (1984) Turbulent burning velocities and flame straining in explosions. Proc.R.Soc. London A391, 393-414.

Al-Dabbagh,N.A. and Andrews, G.E. (1981) The influence of premixed combustor flame stabilizer geometry on flame stability and emissions. Trans. ASME, J.Eng.Power 103, 749-758.

Al-Dabbagh, N.A. and Andrews, G.E. (1984) Weak extinction and turbulent burning velocity for grid plate stabilized premixed flames. Combust. Flame 55, 31-52.

Al-Khishali, K.J., Boston, P.M., Bradley, D., Lawes,M., and Pegg,M.J. (1983) The influence of fluctuations in turbulence upon fluctuations in turbulent burning velocity. Combustion in Engineering Conference, pp.175-180. Institution of Mechanical Engineers, London.

Andrews, G.E. (1972) Laminar and turbulent combustion. Ph.D. Thesis, University of Leeds, Leeds, U.K.

Andrews,G.E. Bradley,D., and Lwakabamba,S. (1975) Measurement of turbulent burning velocity for large turbulent Reynolds numbers. Fifteenth Symposium (International) on Combustion, pp.655-664, The Combustion Institute, Pittsburgh, PA.

Chapman, W.R. and Wheeler, R.V. (1926) The propagation of flame in mixtures of methane and air, Part 1V: The effect of restrictions in the path of the flame. J.Chem.Soc. 2139-2147.

Dorge,K.J., Pangritz,D., and Wagner,H.Gg. (1976) Experiments on velocity augmentation of spherical flames by grids. Acta Astron. 3, 1067-1076.

Dunn-Rankine, D. and Sawyer, R.F. (1986) The interaction of laminar flame with its self generated flow during constant volume combustion. Dynamics of Reactive Systems, Part 1: Flames and Configurations (edited by J.R.Bowen, J.C.Leyer and R.I.Soloukhin), pp.115, AIAA, New York.

Ellis, O.C. de (1928) Flame movement in gaseous explosive mixtures. Fuel 7, 195,336,502.

Evans, M.W., Scheer, M.D., Schoen, L.J. and Miller, E.L. (1949) A study of high velocity flames developed by grids in tubes. Third Symposium (International) on Combustion, pp.168-176. The Combustion Institute, Pittsburgh, PA.

Harris, G.F.P. (1967) The effect of vessel size and turbulence on maximum explosion pressure and rates of pressure rise. Combust. Flame 11, 17-25.

Harris, R.J. (1983) The investigation and control of gas explosions in buildings and heating plant, E&FN Spon, London, New York.

Harrison,A.J. and Eyre,J.A. (1987) The effect of obstacle arrays on the combustion of large premixed gas/air clouds. Combust.Sci.Technol. 52, 121-137.

Hjertager,B.H. (1983) Influence of turbulence on gas explosions. Proc. Conf. on Control and Prevention of Gas Explosions, London, pp. 39-79.

Kirby and Wheeler R.V. (1931) Explosions in closed cylinders. Part V: the effect of restrictions. J.Chem.Soc.pp.2303-2306.

Knystautas,R., Lee, J.H., and Chan,C.K. (1984) Turbulent flame propagation in obstacle filled tubes. Twentieth Symposium (International) on Combustion, pp.1663-1672. The Combustion Institute, Pittsburg, PA.

Kumar, R.K., Tamm,H. and Harrison, W.C.,(1983) Combustion of hydrogen at high concentrations including the effect of obstacles. Combust. Sci. Technol 35, 175-186.

Lunn, G.A. (1985) Venting Gas and Dust Explosions - A Review, (1985). Institution of Chemical Engineers, Rugby, England.

Mason,S. and Wheeler,R.V. (1919) The propagation of flame in mixtures of methane and air. J.Chem.Soc. 117, 47

Moen, I.O., Donato,M., Knystautas,R. and Lee, J.H. (1980) Flame acceleration due to turbulence produced by obstacles. Combust. and Flame 39, 21-32.

Moen, I.O., Lee,J.H.S., Hjertager,B.H., Fuhre,K. and Eckhoff,R.K. (1982) Pressure developement due to turbulent flame propagation in large scale methane air explosions. Combust. and Flame 47, 31-52.

Robinson, H. and Wheeler, R.V. (1933) Explosions of methane and air: propagation through a restricted tube. J.Chem.Soc. 758-760.

Schelkin, K.I. (1940), J.Exp.Theor.Phys. (USSR) 10, 823-827.

Schmidt,E., Steine,H. and Neubert,U. (1951) Flame and schlieren photographs of the combustion of gas air mixtures in tubes. VDI-Forschungsh 431.

Starke,R. and Roth,P. (1986) An experimental investigation of flame behaviour during cylindrical vessel explosions. Combust. and Flame 66, 249-259.

Strel´chuk,N.A., Mishuev A.V., Nikitin,A.G., and Ovakhelashvili,N.V. (1984) Gas dynamics of combustion in a gas-air mixture in a semiclosed volume with pressure release into an adjacent volume. Combust. Explos. Shock Waves (USSR) 20, 59-62.

Taylor, P.H. (1985) On the role of partial confinement in the generation of fast flames. Combust. Flame 44, 161-178.

Taylor, P.H. (1988) Fast flames in a vented duct. Twentyfirst Symposium (International) on Combustion, pp. 1601-1608. The Combustion Institute, Pittsburgh, PA.

Urtiew,P.A., Brandeis,J. and Hogan,W.J. (1983) Experimental study of flame propagation in semiconfined geometries with obstacles. Comb.Sci.Technol. 30, 105-119.

Author Index

Andrews, G. E. 512
Attetkov, A. V. 303
Bailly, P. 389
Bauer, P. A. 64
Bennett, C. A. 45
Benz, F. J. 45
Bishop, C. V. 45
Borisov, A. A. 124, 211, 303
Brossard, J. 389
Chick, K. M. 372
Ciccarelli, G. 451
Desbordes, D. 170, 419
Desrosier, C. 389
Dupré, G. 45, 248
Ebert, F. 3
Elbahar, O. M. F. 477, 499
Ermolaev, B. S. 303, 322
Fenton, D. L. 45
Fouad, M. A. 401
Frolov, S. M. 99, 211
Frost, D. L. 436, 451
Gelfand, B. E. 99, 211
Gu, L. S. 232
Gubin, S. A. 331
Guirguis, R. H. 155
Herath, P. 512
Herrmann, K.-P. 362
Heuzé, O. 64
Kailasanath, K. 155
Kamel, M. M. 401, 477, 499
Kauffman, C. W. 264
Khasainov, B. A. 284, 303
Khrapovski, V. E. 322
Knystautas, R. 32, 45, 232, 248
Krier, H. 341
Kuhl, A. L. 419
Lee, J. H. 32, 45, 232, 248, 436
Lisyanskii, V. V. 124

Liu, J. C. 264
Mailkov, A. E. 211
Manson, N. 64
McClenagan, R. D. 45
Odintsov, V. V. 331
Ogawa, Y. 140
Oran, E. S. 155
Pedley, M. D. 45
Peng, J. 201
Pepekin, V. I. 331
Peraldi, O. 45, 248
Pförtner, H. 488
Phillips, H. 77
Plewinsky, B. 362
Polenov, A. N. 99
Powers, J. M. 341
Renard, J. 389
Salem, H. 401
Schneider, H. 488
Schöffel, S. U. 3
Schott, G. L. 372
Sergeev, S. S. 331
Sichel, M. 264
Shepherd, J. E. 45
Skachkov, G. I. 124
Soloviev, V. S. 303
Stewart, D. S. 341
Sulimov, A. A. 322
Taki, S. 140
Tang, M. 201
Troshin, K. Y. 124
Tsyganov, S. M. 99
Vandermeiren, M. 186
Van Tiggelen, P. J. 186
Veyssière, B. 284
Vidal, P. 64
Wegener, W. 362
Zamanskii, V. M. 124
Zel'dovich, Y. B. 99, 211

PROGRESS IN ASTRONAUTICS AND AERONAUTICS SERIES VOLUMES
VOLUME TITLE/EDITORS

*1. Solid Propellant Rocket Research (1960)
Martin Summerfield
Princeton University

*2. Liquid Rockets and Propellants (1960)
Loren E. Bollinger
The Ohio State University
Martin Goldsmith
The Rand Corporation
Alexis W. Lemmon Jr.
Battelle Memorial Institute

*3. Energy Conversion for Space Power (1961)
Nathan W. Snyder
Institute for Defense Analyses

*4. Space Power Systems (1961)
Nathan W. Snyder
Institute for Defense Analyses

*5. Electrostatic Propulsion (1961)
David B. Langmuir
Space Technology Laboratories, Inc.
Ernst Stuhlinger
NASA George C. Marshall Space Flight Center
J.M. Sellen Jr.
Space Technology Laboratories, Inc.

*6. Detonation and Two-Phase Flow (1962)
S.S. Penner
California Institute of Technology
F.A. Williams
Harvard University

*7. Hypersonic Flow Research (1962)
Frederick R. Riddell
AVCO Corporation

*8. Guidance and Control (1962)
Robert E. Roberson
Consultant
James S. Farrior
Lockheed Missiles and Space Company

*9. Electric Propulsion Development (1963)
Ernst Stuhlinger
NASA George C. Marshall Space Flight Center

*10. Technology of Lunar Exploration (1963)
Clifford I. Cummings
Harold R. Lawrence
Jet Propulsion Laboratory

*11. Power Systems for Space Flight (1963)
Morris A. Zipkin
Russell N. Edwards
General Electric Company

12. Ionization in High-Temperature Gases (1963)
Kurt E. Shuler, Editor
National Bureau of Standards
John B. Fenn, Associate Editor
Princeton University

*13. Guidance and Control—II (1964)
Robert C. Langford
General Precision Inc.
Charles J. Mundo
Institute of Naval Studies

*14. Celestial Mechanics and Astrodynamics (1964)
Victor G. Szebehely
Yale University Observatory

*15. Heterogeneous Combustion (1964)
Hans G. Wolfhard
Institute for Defense Analyses
Irvin Glassman
Princeton University
Leon Green Jr.
Air Force Systems Command

16. Space Power Systems Engineering (1966)
George C. Szego
Institute for Defense Analyses
J. Edward Taylor
TRW Inc.

17. Methods in Astrodynamics and Celestial Mechanics (1966)
Raynor L. Duncombe
U.S. Naval Observatory
Victor G. Szebehely
Yale University Observatory

18. Thermophysics and Temperature Control of Spacecraft and Entry Vehicles (1966)
Gerhard B. Heller
NASA George C. Marshall Space Flight Center

*Out of print.

*19. Communication Satellite Systems Technology (1966)
Richard B. Marsten
Radio Corporation of America

20. Thermophysics of Spacecraft and Planetary Bodies: Radiation Properties of Solids and the Electromagnetic Radiation Environment in Space (1967)
Gerhard B. Heller
NASA George C. Marshall Space Flight Center

21. Thermal Design Principles of Spacecraft and Entry Bodies (1969)
Jerry T. Bevans
TRW Systems

22. Stratospheric Circulation (1969)
Willis L. Webb
Atmospheric Sciences Laboratory, White Sands, and University of Texas at El Paso

23. Thermophysics: Applications to Thermal Design of Spacecraft (1970)
Jerry T. Bevans
TRW Systems

24. Heat Transfer and Spacecraft Thermal Control (1971)
John W. Lucas
Jet Propulsion Laboratory

25. Communication Satellites for the 70's: Technology (1971)
Nathaniel E. Feldman
The Rand Corporation
Charles M. Kelly
The Aerospace Corporation

26. Communication Satellites for the 70's: Systems (1971)
Nathaniel E. Feldman
The Rand Corporation
Charles M. Kelly
The Aerospace Corporation

27. Thermospheric Circulation (1972)
Willis L. Webb
Atmospheric Sciences Laboratory, White Sands, and University of Texas at El Paso

28. Thermal Characteristics of the Moon (1972)
John W. Lucas
Jet Propulsion Laboratory

29. Fundamentals of Spacecraft Thermal Design (1972)
John W. Lucas
Jet Propulsion Laboratory

30. Solar Activity Observations and Predictions (1972)
Patrick S. McIntosh
Murray Dryer
Environmental Research Laboratories, National Oceanic and Atmospheric Administration

31. Thermal Control and Radiation (1973)
Chang-Lin Tien
University of California at Berkeley

32. Communications Satellite Systems (1974)
P.L. Bargellini
COMSAT Laboratories

33. Communications Satellite Technology (1974)
P.L. Bargellini
COMSAT Laboratories

34. Instrumentation for Airbreathing Propulsion (1974)
Allen E. Fuhs
Naval Postgraduate School
Marshall Kingery
Arnold Engineering Development Center

35. Thermophysics and Spacecraft Thermal Control (1974)
Robert G. Hering
University of Iowa

36. Thermal Pollution Analysis (1975)
Joseph A. Schetz
Virginia Polytechnic Institute

37. Aeroacoustics: Jet and Combustion Noise; Duct Acoustics (1975)
Henry T. Nagamatsu, Editor
General Electric Research and Development Center
Jack V. O'Keefe, Associate Editor
The Boeing Company
Ira R. Schwartz, Associate Editor
NASA Ames Research Center

38. Aeroacoustics: Fan, STOL, and Boundary Layer Noise; Sonic Boom; Aeroacoustic Instrumentation (1975)
Henry T. Nagamatsu, Editor
General Electric Research and Development Center
Jack V. O'Keefe, Associate Editor
The Boeing Company
Ira R. Schwartz, Associate Editor
NASA Ames Research Center

39. Heat Transfer with Thermal Control Applications (1975)
M. Michael Yovanovich
University of Waterloo

40. **Aerodynamics of Base Combustion** (1976)
S.N.B. Murthy, Editor
Purdue University
J.R. Osborn, Associate Editor
Purdue University
A.W. Barrows
J.R. Ward, Associate Editors
Ballistics Research Laboratories

41. **Communications Satellite Developments: Systems** (1976)
Gilbert E. LaVcan
Defense Communications Agency
William G. Schmidt
CML Satellite Corporation

42. **Communications Satellite Developments: Technology** (1976)
William G. Schmidt
CML Satellite Corporation
Gilbert E. LaVean
Defense Communications Agency

43. **Aeroacoustics: Jet Noise, Combustion and Core Engine Noise** (1976)
Ira R. Schwartz, Editor
NASA Ames Research Center
Henry T. Nagamatsu, Associate Editor
General Electric Research and Development Center
Warren C. Strahle, Associate Editor
Georgia Institute of Technology

44. **Aeroacoustics: Fan Noise and Control; Duct Acoustics; Rotor Noise** (1976)
Ira R. Schwartz, Editor
NASA Ames Research Center
Henry T. Nagamatsu, Associate Editor
General Electric Research and Development Center
Warren C. Strahle, Associate Editor
Georgia Institute of Technology

45. **Aeroacoustics: STOL Noise; Airframe and Airfoil Noise** (1976)
Ira R. Schwartz, Editor
NASA Ames Research Center
Henry T. Nagamatsu, Associate Editor
General Electric Research and Development Center
Warren C. Strahle, Associate Editor
Georgia Institute of Technology

46. **Aeroacoustics: Acoustic Wave Propagation; Aircraft Noise Prediction; Aeroacoustic Instrumentation** (1976)
Ira R. Schwartz, Editor
NASA Ames Research Center
Henry T. Nagamatsu, Associate Editor
General Electric Research and Development Center
Warren C. Strahle, Associate Editor
Georgia Institute of Technology

47. **Spacecraft Charging by Magnetospheric Plasmas** (1976)
Alan Rosen
TRW Inc.

48. **Scientific Investigations on the Skylab Satellite** (1976)
Marion I. Kent
Ernst Stuhlinger
NASA George C. Marshall Space Flight Center
Shi-Tsan Wu
The University of Alabama

49. **Radiative Transfer and Thermal Control** (1976)
Allie M. Smith
ARO Inc.

50. **Exploration of the Outer Solar System** (1976)
Eugene W. Greenstadt
TRW Inc.
Murray Dryer
National Oceanic and Atmospheric Administration
Devrie S. Intriligator
University of Southern California

51. **Rarefied Gas Dynamics, Parts I and II (two volumes)** (1977)
J. Leith Potter
ARO Inc.

52. **Materials Sciences in Space with Application to Space Processing** (1977)
Leo Steg
General Electric Company

53. **Experimental Diagnostics in Gas Phase Combustion Systems** (1977)
Ben T. Zinn, Editor
Georgia Institute of Technology
Craig T. Bowman, Associate Editor
Stanford University
Daniel L. Hartley, Associate Editor
Sandia Laboratories
Edward W. Price, Associate Editor
Georgia Institute of Technology
James G. Skifstad, Associate Editor
Purdue University

54. **Satellite Communications: Future Systems** (1977)
David Jarett
TRW Inc.

55. **Satellite Communications: Advanced Technologies** (1977)
David Jarett
TRW Inc.

56. **Thermophysics of Spacecraft and Outer Planet Entry Probes** (1977)
Allie M. Smith
ARO Inc.

57. **Space-Based Manufacturing from Nonterrestrial Materials** (1977)
Gerard K. O'Neill, Editor
Princeton University
Brian O'Leary, Assistant Editor
Princeton University

58. **Turbulent Combustion** (1978)
Lawrence A. Kennedy
State University of New York at Buffalo

59. **Aerodynamic Heating and Thermal Protection Systems** (1978)
Leroy S. Fletcher
University of Virginia

60. **Heat Transfer and Thermal Control Systems** (1978)
Leroy S. Fletcher
University of Virginia

61. **Radiation Energy Conversion in Space** (1978)
Kenneth W. Billman
NASA Ames Research Center

62. **Alternative Hydrocarbon Fuels: Combustion and Chemical Kinetics** (1978)
Craig T. Bowman
Stanford University
Jorgen Birkeland
Department of Energy

63. **Experimental Diagnostics in Combustion of Solids** (1978)
Thomas L. Boggs
Naval Weapons Center
Ben T. Zinn
Georgia Institute of Technology

64. **Outer Planet Entry Heating and Thermal Protection** (1979)
Raymond Viskanta
Purdue University

65. **Thermophysics and Thermal Control** (1979)
Raymond Viskanta
Purdue University

66. **Interior Ballistics of Guns** (1979)
Herman Krier
University of Illinois at Urbana-Champaign
Martin Summerfield
New York University

*67. **Remote Sensing of Earth from Space: Role of "Smart Sensors"** (1979)
Roger A. Breckenridge
NASA Langley Research Center

68. **Injection and Mixing in Turbulent Flow** (1980)
Joseph A. Schetz
Virginia Polytechnic Institute and State University

69. **Entry Heating and Thermal Protection** (1980)
Walter B. Olstad
NASA Headquarters

70. **Heat Transfer, Thermal Control, and Heat Pipes** (1980)
Walter B. Olstad
NASA Headquarters

71. **Space Systems and Their Interactions with Earth's Space Environment** (1980)
Henry B. Garrett
Charles P. Pike
Hanscom Air Force Base

72. **Viscous Flow Drag Reduction** (1980)
Gary R. Hough
Vought Advanced Technology Center

73. **Combustion Experiments in a Zero-Gravity Laboratory** (1981)
Thomas H. Cochran
NASA Lewis Research Center

74. **Rarefied Gas Dynamics, Parts I and II (two volumes)** (1981)
Sam S. Fisher
University of Virginia at Charlottesville

75. **Gasdynamics of Detonations and Explosions** (1981)
J.R. Bowen
University of Wisconsin at Madison
N. Manson
Université de Poitiers
A.K. Oppenheim
University of California at Berkeley
R.I. Soloukhin
Institute of Heat and Mass Transfer, BSSR Academy of Sciences

76. **Combustion in Reactive Systems** (1981)
J.R. Bowen
University of Wisconsin at Madison
N. Manson
Université de Poitiers
A.K. Oppenheim
University of California at Berkeley
R.I. Soloukhin
Institute of Heat and Mass Transfer, BSSR Academy of Sciences

77. **Aerothermodynamics and Planetary Entry** (1981)
A.L. Crosbie
University of Missouri-Rolla

78. **Heat Transfer and Thermal Control** (1981)
A.L. Crosbie
University of Missouri-Rolla

79. **Electric Propulsion and Its Applications to Space Missions** (1981)
Robert C. Finke
NASA Lewis Research Center

80. **Aero-Optical Phenomena** (1982)
Keith G. Gilbert
Leonard J. Otten
Air Force Weapons Laboratory

81. **Transonic Aerodynamics** (1982)
David Nixon
Nielsen Engineering & Research, Inc.

82. **Thermophysics of Atmospheric Entry** (1982)
T.E. Horton
The University of Mississippi

83. **Spacecraft Radiative Transfer and Temperature Control** (1982)
T.E. Horton
The University of Mississippi

84. **Liquid-Metal Flows and Magnetohydrodynamics** (1983)
H. Branover
Ben-Gurion University of the Negev
P.S. Lykoudis
Purdue University
A. Yakhot
Ben-Gurion University of the Negev

85. **Entry Vehicle Heating and Thermal Protection Systems: Space Shuttle, Solar Starprobe, Jupiter Galileo Probe** (1983)
Paul E. Bauer
McDonnell Douglas Astronautics Company
Howard E. Collicott
The Boeing Company

86. **Spacecraft Thermal Control, Design, and Operation** (1983)
Howard E. Collicott
The Boeing Company
Paul E. Bauer
McDonnell Douglas Astronautics Company

87. **Shock Waves, Explosions, and Detonations** (1983)
J.R. Bowen
University of Washington
N. Manson
Université de Poitiers
A.K. Oppenheim
University of California at Berkeley
R.I. Soloukhin
Institute of Heat and Mass Transfer, BSSR Academy of Sciences

88. **Flames, Lasers, and Reactive Systems** (1983)
J.R. Bowen
University of Washington
N. Manson
Université de Poitiers
A.K. Oppenheim
University of California at Berkeley
R.I. Soloukhin
Institute of Heat and Mass Transfer, BSSR Academy of Sciences

89. **Orbit-Raising and Maneuvering Propulsion: Research Status and Needs** (1984)
Leonard H. Caveny
Air Force Office of Scientific Research

90. **Fundamentals of Solid-Propellant Combustion** (1984)
Kenneth K. Kuo
The Pennsylvania State University
Martin Summerfield
Princeton Combustion Research Laboratories, Inc.

91. **Spacecraft Contamination: Sources and Prevention** (1984)
J.A. Roux
The University of Mississippi
T.D. McCay
NASA Marshall Space Flight Center

92. **Combustion Diagnostics by Nonintrusive Methods** (1984)
T.D. McCay
NASA Marshall Space Flight Center
J.A. Roux
The University of Mississippi

93. **The INTELSAT Global Satellite System** (1984)
Joel Alper
COMSAT Corporation
Joseph Pelton
INTELSAT

94. **Dynamics of Shock Waves, Explosions, and Detonations** (1984)
J.R. Bowen
University of Washington
N. Manson
Université de Poitiers
A.K. Oppenheim
University of California
R.I. Soloukhin
Institute of Heat and Mass Transfer, BSSR Academy of Sciences

95. **Dynamics of Flames and Reactive Systems** (1984)
J.R. Bowen
University of Washington
N. Manson
Université de Poitiers
A.K. Oppenheim
University of California
R.I. Soloukhin
Institute of Heat and Mass Transfer, BSSR Academy of Sciences

96. **Thermal Design of Aeroassisted Orbital Transfer Vehicles** (1985)
H.F. Nelson
University of Missouri-Rolla

97. **Monitoring Earth's Ocean, Land, and Atmosphere from Space— Sensors, Systems, and Applications** (1985)
Abraham Schnapf
Aerospace Systems Engineering

98. **Thrust and Drag: Its Prediction and Verification** (1985)
Eugene E. Covert
Massachusetts Institute of Technology
C.R. James
Vought Corporation
William F. Kimzey
Sverdrup Technology AEDC Group
George K. Richey
U.S. Air Force
Eugene C. Rooney
U.S. Navy Department of Defense

99. **Space Stations and Space Platforms — Concepts, Design, Infrastructure, and Uses** (1985)
Ivan Bekey
Daniel Herman
NASA Headquarters

100. **Single- and Multi-Phase Flows in an Electromagnetic Field Energy, Metallurgical, and Solar Applications** (1985)
Herman Branover
Ben-Gurion University of the Negev
Paul S. Lykoudis
Purdue University
Michael Mond
Ben-Gurion University of the Negev

101. **MHD Energy Conversion: Physiotechnical Problems** (1986)
V.A. Kirillin
A.E. Sheyndlin
Soviet Academy of Sciences

102. **Numerical Methods for Engine-Airframe Integration** (1986)
S.N.B. Murthy
Purdue University
Gerald C. Paynter
Boeing Airplane Company

103. **Thermophysical Aspects of Re-Entry Flows** (1986)
James N. Moss
NASA Langley Research Center
Carl D. Scott
NASA Johnson Space Center

104. **Tactical Missile Aerodynamics** (1986)
M.J. Hemsch
PRC Kentron, Inc.
J.N. Nielsen
NASA Ames Research Center

105. **Dynamics of Reactive Systems Part I: Flames and Configurations; Part II: Modeling and Heterogeneous Combustion** (1988)
J.R. Bowen
University of Washington
J.-C. Leyer
Université de Poitiers
R.I. Soloukhin
Institute of Heat and Mass Transfer, BSSR Academy of Sciences

106. **Dynamics of Explosions** (1986)
J.R. Bowen
University of Washington
J.-C. Leyer
Université de Poitiers
R.I. Soloukhin
Institute of Heat and Mass Transfer, BSSR Academy of Sciences

107. **Spacecraft Dielectric Material Properties and Spacecraft Charging** (1986)
A.R. Frederickson
U.S. Air Force Rome Air Development Center
D.B. Cotts
SRI International
J.A. Wall
U.S. Air Force Rome Air Development Center
F.L. Bouquet
Jet Propulsion Laboratory, California Institute of Technology

108. **Opportunities for Academic Research in a Low-Gravity Environment** (1986)
George A. Hazelrigg
National Science Foundation
Joseph M. Reynolds
Louisiana State University

109. **Gun Propulsion Technology** (1988)
Ludwig Stiefel
U.S. Army Armament Research, Development and Engineering Center

110. **Commercial Opportunities in Space** (1988)
F. Shahrokhi
K.E. Harwell
University of Tennessee Space Institute
C.C. Chao
National Cheng Kung University

111. **Liquid-Metal Flows: Magnetohydrodynamics and Applications** (1988)
Herman Branover, Michael Mond, and Yeshajahu Unger
Ben-Gurion University of the Negev

112. **Current Trends in Turbulence Research** (1988)
Herman Branover, Michael Mond, and Yeshajahu Unger
Ben-Gurion University of the Negev

113. **Dynamics of Reactive Systems
Part I: Flames;
Part II: Heterogeneous Combustion and Applications** (1988)
A.L. Kuhl
R & D Associates
J.R. Bowen
University of Washington
J.-C. Leyer
Université de Poitiers
A. Borisov
USSR Academy of Sciences

114. **Dynamics of Explosions** (1988)
A.L. Kuhl
R & D Associates
J.R. Bowen
University of Washington
J.-C. Leyer
Université de Poitiers
A. Borisov
USSR Academy of Sciences

(Other Volumes are planned.)